Absorption and Scattering of Light by Small Particles

Absorption and Scattering of Light by Small Particles

CRAIG F. BOHREN
Associate Professor of Meteorology
The Pennsylvania State University

DONALD R. HUFFMAN
Professor of Physics
The University of Arizona

A Wiley-Interscience Publication

JOHN WILEY & SONS

New York • Chichester • Brisbane • Toronto • Singapore

Library of Congress Cataloging in Publication Data:

Bohren, Craig, F., 1940–
   Absorption and scattering of light by small particles.

   "A Wiley-Interscience publication."
   Includes bibliographical references and index.
   1. Aerosols—Optical properties. 2. Absorption of
light. 3. Light—Scattering. I. Huffman, Donald R.,
1935–   II. Title.

QC882.B63 1983   535'.3   82-20312
ISBN 0-471-05772-X

Printed in the United States of America

10 9 8 7 6 5 4

To our families

Preface

When asked during the writing of this book what topic could divert us for so long from the pleasures of a normal life we would answer: "It is about how small particles absorb and scatter light." "My goodness," would be the response, "who could possibly be interested in that?" As it happens, scientists and engineers from a surprising variety of disciplines—solid-state physics, electrical engineering, meteorology, chemistry, biophysics, astronomy—make forays into this field, some never to escape. To completely satisfy such diverse groups, each with its peculiar conventions, notation, terminology, and canons, is an impossible task: physicists prefer the language of elementary excitations—phonons, plasmons, and all that; electrical engineers are more comfortable with the language of antennas and waveguides; chemists and biophysicists might not like either. We have therefore striven for the middle ground with the hope of, if not pleasing everyone, at least not antagonizing anyone. Ultimately, however, our point of view is that of physicists. Quantum-mechanical concepts are introduced where they serve to elucidate physical phenomena, but otherwise our approach is primarily classical.

Like so many other books, this one began its existence as lecture notes. Separately and jointly we have given lectures to graduate students and researchers with the kinds of diverse backgrounds and interests we expect our readers will have. Although more of an advanced monograph than a textbook, this book has a pedagogical flavor because of its origins. Indeed, many of the topics covered are in direct response to questions asked either in classrooms or by our colleagues.

There is one important idea, the *raison d'être* of this book, that we should like to implant firmly in the minds of our readers: scattering theory divorced from the optical properties of bulk matter is incomplete. Solving boundary-value problems in electromagnetic theory may be great fun and often requires considerable skill; but the full physical ramifications of mathematical solutions are hidden to those with little knowledge of how refractive indices of various solids and liquids depend on frequency, the values they take, and the constraints imposed on them. Accordingly, this book is divided into three parts.

Part 1, Chapters 1 through 8, is primarily scattering theory. After an introduction there is a chapter on those topics from electromagnetic theory essential to an understanding of the succeeding six chapters on exact and approximate solutions to various scattering problems. Because uninterrupted strings of mathematical formulas tend to pall, computational and experimental results are interspersed throughout these chapters.

Bulk matter, rather than particles, is the subject of Part 2. In Chapter 9 we discuss classical theories of optical properties based on idealized models. Such models rarely conform strictly to reality, however, so Chapter 10 presents measurements for three representative materials over a wide range of frequencies, from radio to ultraviolet: aluminum, a metal; magnesium oxide, an insulator; and water, a liquid.

Part 3 is a marriage of Parts 1 and 2, the offspring of which are chapters on extinction (Chapter 11), surface modes (Chapter 12), and angular scattering (Chapter 13). Applications are not totally absent from the first thirteen chapters, but there is a greater concentration of them in Chapter 14.

We did not attempt an exhaustive list of references, even assuming that were possible. Instead, we concentrated on the years since publication of Kerker's book (1969), which cites nearly a thousand references. Even with this restriction we were selective, guided by our tastes rather than some ideal notion of completeness.

We avoided irritating statements such as "it can be shown"; while implying calm, they usually signal rough sailing ahead. Of course, we do not give all the details of lengthy derivations, but we do provide enough guideposts so that a reader can, with a bit of effort, duplicate our results. We always chose the simplest derivations, preferring physical plausibility over mathematical rigor. Those who demand the latter are reminded that one man's rigor is another man's mortis.

This book was not written with scissors: all derivations are our own, as are most of the figures, many of them generated with the computer programs in the appendixes. Even much of the experimental data was taken with an eye toward examples for the book. Any errors, therefore, are solely ours.

CRAIG F. BOHREN
DONALD R. HUFFMAN

University Park, Pennsylvania
Tucson, Arizona
January 1983

Acknowledgments

During much of the writing of this book I was a wandering scholar. At each institution I visited I widened the circle of those to whom I am indebted for suggestions, comments, and encouragement. Although fading memory prevents me from adequately expressing my gratitude to all of them, there are many whose contributions remain fixed in my mind.

Daya Gilra, my office-mate in the Department of Applied Mathematics and Astronomy at University College, Cardiff, Wales, suggested that I give a course of lectures on light scattering, the notes for which subsequently formed some of the raw material for this book. For this suggestion and for much more, I am grateful. My thanks also go to those who faithfully attended these lectures, particularly Harry Abadi and Indra Dayawansa, my collaborators, and Joachim Köppen. I would be remiss if I did not acknowledge the assistance of two members of the Pure Mathematics Department at Cardiff, W. D. (Des) Evans and George Greaves.

Louis Battan provided me with a haven for over a year in the Institute of Atmospheric Physics at the University of Arizona, and his support of my endeavors has never flagged, although I have not always followed his sage advice. Sean Twomey was an incisive critic, a fertile source of ideas, and an arbiter of disputes. Without his constant goading—"how many pages did you write today, Craig?"—the writing of this book would have continued into the hereafter. Margaret Sanderson Rae patiently answered hundreds of questions about matters of style and scrutinized some of the first chapters to keep me from straying too far from good usage. E. Philip Krider and Michael Box also read parts of the manuscript, and I thank them for their suggestions.

At Arizona I was partly supported by the Department of Physics through the generosity of Robert Parmenter. I am also indebted to other members of this department, particularly John Kessler, Michael Scadron, Bernard Bell, Rein Kilkson, and William Bickel.

A grant from the Institute of Occupational and Environmental Health in Montreal, obtained through the kind assistance of George Wright, enabled me to return to Wales, where I worked with Vernon Timbrell in the Medical Research Council Pneumoconiosis Unit at Llandough Hospital. It was there that most of Chapter 8 was written as well as the first version of Appendix C.

Further work on this book was undertaken at Los Alamos Scientific Laboratory, for which I must thank Paul Mullaney. I am also grateful to Gary Salzman, who helped in many ways, and to Sally Wilkins, who inspected the programs in the appendixes and made several suggestions for improving them; she is not, however, responsible for the numbers they produce.

My colleagues at Pennsylvania State University, Alistair Fraser, John Olivero, and Timothy Nevitt, are to be thanked for helping me to keep this book as free from errors as possible.

Others who deserve a word of thanks are Alfred Holland and Arlon Hunt.

At Wiley, Beatrice Shube, in the words of the Beatles, "was like a mum to us."

My gratitude is deepest, however, to Nanette Malott Bohren, who followed me without complaint on my wanderings, was neglected for over three years, but who nevertheless read every page of the manuscript—several times—thereby improving its readability.

C.F.B.

For helping me to learn about the interaction of light with small particles I sincerely thank the other students and co-workers who have been my colleagues over the years: James L. Stapp, Terry Steyer, Roger Perry, Janice Rathmann, Otto Edoh, Lin Oliver, Wolfgang Krätschmer, and Kenrick Day. Special thanks are due to Arlon Hunt for the work we shared in the days when everything about small particles was new to us, exciting, and occasionally explosive.

D.R.H.

Contents

Chapter 4.   Absorption and Scattering by a Sphere, 82

Chapter 5.   Particles Small Compared with the Wavelength, 130

Chapter 6.   Rayleigh–Gans Theory, 158

Chapter 7.   Geometrical Optics, 166

Chapter 8.   A Potpourri of Particles, 181

PART 2—OPTICAL PROPERTIES OF BULK MATTER

Chapter 9. Classical Theories of Optical Constants, 227

Chapter 10. Measured Optical Properties, 268

PART 3—OPTICAL PROPERTIES OF PARTICLES

Chapter 11. Extinction, 287

Chapter 12. Surface Modes in Small Particles, 325

Chapter 13. Angular Dependence of Scattering, 381

Chapter 14. A Miscellany of Applications, 429

APPENDIXES   COMPUTER PROGRAMS

Part 1

Basic Theory

Chapter 1

Introduction

Cumulus clouds in the summer afternoon sky present a striking contrast of white against a bright blue sky. During a sudden thundershower the primary and secondary rainbows display their multicolored arches. Other colors in nature are the dark green of forest foliage and the red and orange hues of the Grand Canyon in early morning. High in the mountains or on the desert when the air is clean one can clearly see dark patches in the bright band of the Milky Way. Chimney smut turns all it touches to dirty blackness, and iridescent opal shimmers with a variety of colors. All these visual phenomena and many more are manifestations of scattering and absorption of light by small particles, which is the subject of this book. We do not, however, restrict ourselves to visible light.

The study of light scattering and its applications is an enormous field, much too large to be treated successfully in one book. We limit our treatment, therefore, to scattering by single particles within the framework of classical electromagnetic theory and linear optics. Not all the examples given above fall within these restrictions. The colors of rocks and foliage, for example, involve complicated interactions of light with many densely packed centers of scattering and absorption. Nevertheless, all detailed treatments of the more complicated phenomena begin with those that we shall study.

1.1 PHYSICAL BASIS FOR SCATTERING AND ABSORPTION

Scattering of electromagnetic waves by any system is related to the heterogeneity of that system: heterogeneity on the molecular scale or on the scale of aggregations of many molecules. Regardless of the type of heterogeneity, the underlying physics of scattering is the same for all systems. Matter is composed of discrete electric charges: electrons and protons. If an obstacle, which could be a single electron, an atom or molecule, a solid or liquid particle, is illuminated by an electromagnetic wave (Fig. 1.1), electric charges in the obstacle are set into oscillatory motion by the electric field of the incident wave. Accelerated electric charges radiate electromagnetic energy in all directions; it is this *secondary radiation* that is called the radiation *scattered* by the

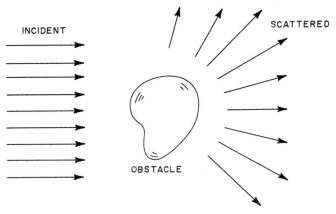

Figure 1.1   Scattering by an obstacle.

obstacle:

$$\text{Scattering} = \text{excitation} + \text{reradiation}$$

In addition to reradiating electromagnetic energy, the excited elementary charges may transform part of the incident electromagnetic energy into other forms (thermal energy, for example), a process called *absorption*. Scattering and absorption are not mutually independent processes, and although, for brevity, we often refer only to scattering, we shall always mean absorption as well.

1.2   SCATTERING BY FLUCTUATIONS AND BY PARTICLES

Everything except a vacuum is heterogeneous in some sense. Even in media that we usually consider to be homogeneous (e.g., pure gases, liquids, or solids) it is possible to distinguish the individual heterogeneities (atoms and molecules) with a sufficiently fine probe. Therefore, all media scatter light. In fact, many phenomena that are not usually referred to as scattering phenomena are ultimately the result of scattering. Among these are: (1) diffuse reflection by rough surfaces; (2) diffraction by slits, gratings, and edges; and (3) specular reflection and refraction at optically smooth interfaces. Let us examine the third example in more detail. The directions of the reflected and refracted rays in Fig. 1.2 are specified by the law of specular reflection and Snell's law. These laws have been known empirically for a long time. They can, however, be derived by using the Maxwell equations, and in so doing one also obtains the amplitudes and phases of the reflected and refracted rays (i.e., the Fresnel equations). It is also possible, although far from easy, to derive these laws by explicitly considering the molecular nature of matter. The transparent medium

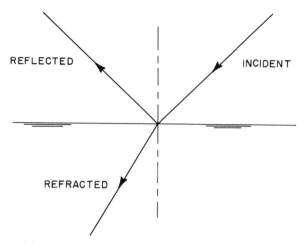

Figure 1.2 Reflection and refraction at an optically smooth interface.

on which light is incident (Fig. 1.2) is an aggregation of very many molecules. The field in the neighborhood of a given molecule induces an oscillating dipole moment in the molecule, which in turn gives rise to secondary dipole radiation. Solids, liquids, and many gases are *optically dense*: the molecular separation is much less than the wavelength of the incident light. In solids and liquids the molecular separation is about 2–3 Å, whereas for gases at standard temperature and pressure (STP) the average separation is about 30 Å. Thus, each molecule is acted on not only by the incident field but also by the resultant of the secondary fields of all the other molecules. But the secondary field of a molecule depends on the field to which it is exposed. Therefore, we have an electromagnetic many-body problem: the molecules are coupled. The net result of the solution to this problem, subject to suitable approximations, is that inside the medium the secondary waves superpose on each other and on the incident wave to give just a refracted wave with propagation velocity c/n, where c is the speed of light *in vacuo* and n is the refractive index. The incident wave is completely extinguished inside the medium; this is called the Ewald–Oseen extinction theorem. Outside the medium the secondary waves superpose to give a specularly reflected wave. As we shall see in Chapter 9, n depends on the number of molecules per unit volume and the polarizability of a single molecule; this underscores the assertion that refraction is a scattering phenomenon in its essentials: the refractive index is a manifestation of scattering by the many molecules that comprise the medium.

Careful observation of the configuration shown in Fig. 1.2 reveals a phenomenon that contradicts the familiar law of refraction. Suppose that we completely darken the surroundings and illuminate a transparent medium, which could be pure water, with an intense laser beam. Even if the medium is free from all particulate contamination, the path of the beam in the medium can be dimly perceived (this might require a suitable detector) if we look in directions other than in the plane of incidence. This is not compatible with Snell's law, which asserts that the refracted ray lies in the plane of incidence and which takes no account of light other than in this plane. Therefore, Snell's law is only a first approximation, and we must probe deeper to discover the origins of this light weakly scattered in all directions, which is superposed on the more intense unidirectional refracted beam.

It is assumed in the usual analysis of the interaction of a beam of light with an optically smooth interface that the refracting medium is perfectly homogeneous, whereas, in fact, it is only *statistically homogeneous*. That is, the average number of molecules in a given volume element is constant, but at any instant the number of molecules in this element will be different compared with any other instant. It is these *density fluctuations* which give rise to scattering in optically dense media. Although we, and others, refer to scattering *by* fluctuations for brevity, we must emphasize that molecules are the scattering agents. But one can ignore them and imagine that their scattering is the result of local density fluctuations in an otherwise homogeneous medium. It would be more precise, therefore, to refer to the fluctuation theory of scattering by molecules rather than to scattering by fluctuations.

There are other types of fluctuations. For example, if sugar is dissolved in water, after thorough stirring the sugar concentration will be statistically homogeneous, but *concentration fluctuations* will give rise to scattering. If the molecules are nonspherical, there will be *orientation fluctuations*.

All such scattering by fluctuations is excluded here; we confine ourselves to scattering by particles, and a fluctuation is not a particle in our sense. It is important to distinguish between scattering by fluctuations and scattering by particles because there is a large body of literature on "light scattering" which is clearly distinct from the subject of this book. Although the mathematical expressions are often similar, the underlying physics is somewhat different: scattering by fluctuations, for example, involves thermodynamic arguments, whereas scattering by particles does not. Moreover, there is common terminology, which is a possible source of confusion. For example, scattering by density fluctuations in ideal gases has the same functional form as scattering by dilute suspensions of particles small compared with the wavelength. We shall call the latter type of scattering Rayleigh scattering, but in the theory of scattering by fluctuations the term may have a somewhat different meaning. We refer the reader to an article by Young (1982) in which he attempts to sort out all the different ways in which "Rayleigh scattering" is used and misused.

The basic problem to which we confine our attention is the interaction of light of arbitrary wavelength with a single particle (i.e., a well-defined aggregate of very many atoms or molecules), which is embedded in an otherwise homogeneous medium (Fig. 1.3). By homogeneous is meant that the atomic or molecular heterogeneity is small compared with the wavelength of the incident light; we also ignore scattering by fluctuations, which is usually much less than

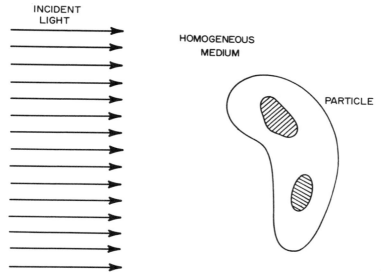

INCIDENT LIGHT

HOMOGENEOUS MEDIUM

PARTICLE

Figure 1.3 Interaction of light with a single particle.

scattering by particles. Although the particle may be complicated in shape and may have several homogeneous components, we assume that it is composed of matter that is at every point describable in macroscopic terms. That is, the optical properties of a particle or regions thereof are completely specified by frequency-dependent optical constants; the interaction of photons with elementary quantum excitations need not be considered explicitly.

We restrict our treatment to *elastic* scattering: the frequency of the scattered light is the same as that of the incident light. This excludes from study *inelastic* scattering phenomena such as Mandel'stam–Brillouin and Raman scattering. Elastic scattering is sometimes denoted as *coherent* scattering, but elastic is more physically descriptive and the notion of coherence as a definite phase relation between different sources of radiation is firmly established in optics. To add to the confusion, Rayleigh scattering is sometimes used to denote scattering in which there is no change in frequency. Again, see the article by Young (1982) for clarification.

1.3 PHYSICS OF SCATTERING BY A SINGLE PARTICLE

We can acquire a qualitative understanding of the physics of scattering by a single particle without invoking a specific particle or doing any computations. Consider an arbitrary particle, which we conceptually subdivide into small regions (Fig. 1.4). An applied oscillating field (e.g., an incident electromagnetic wave) induces a dipole moment in each region. These dipoles oscillate at the frequency of the applied field and therefore scatter secondary radiation in all

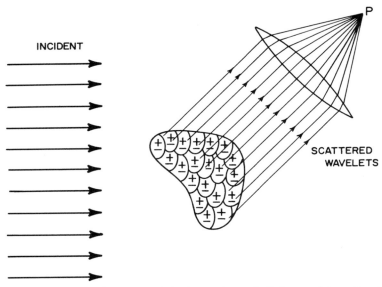

Figure 1.4 The total scattered field at P is the resultant of all the wavelets scattered by the regions into which the particle is subdivided.

directions. In a particular direction (i.e., at a distant point P), the total scattered field is obtained by superposing the scattered wavelets, where due account is taken of their phase differences: scattering by the dipoles is coherent. In general, these phase relations change for a different scattering direction; we therefore expect the scattered field to vary with scattering direction. If the particle is small compared with the wavelength, all the secondary wavelets are approximately in phase; for such a particle we do not expect much variation of scattering with direction. That this is indeed so will be shown in Chapter 5. As the particle size is increased, however, the number of possibilities for mutual enhancement and cancellation of the scattered wavelets increases. Thus, the larger the particle, the more peaks and valleys in the scattering pattern. Shape is also important: if the particle in Fig. 1.4 is distorted, all the phase relations, hence the scattering pattern, are different.

The phase relations among the scattered wavelets depend on geometrical factors: scattering direction, size, and shape. But the amplitude and phase of the induced dipole moment for a given frequency depend on the material of which the particle is composed. Thus, for a full understanding of scattering and absorption by small particles, we need to know how bulk matter responds to oscillatory electromagnetic fields; this is the subject of Chapters 9 and 10.

Methods for calculating scattering by particles are physically equivalent to the procedure outlined above, although their mathematical form may obscure the underlying physics. For certain classes of particles, however, the scattered field may be approximated by subdividing the particle into dipole scatterers and superposing the scattered wavelets; this is what is done in the Rayleigh–Gans approximation (Chapter 6), where interactions among the dipoles are ignored. A more general computational technique in which interactions among the dipoles are accounted for is that of Purcell and Pennypacker (1973) (see Section 8.6).

1.4   COLLECTIONS OF PARTICLES

The fundamental problem under consideration is scattering and absorption by single particles; in natural environments, however, we are usually confronted with collections of very many particles. Even in the laboratory, where it is possible to do experiments with single particles, it is more usual to make measurements on many particles. A rigorous theoretical treatment of scattering by many particles is indeed formidable (see, e.g., Borghese et al., 1979). But if certain conditions are satisfied, a collection poses no more analytical problems than does a single isolated particle.

Particles in a collection are electromagnetically coupled: each particle is excited by the external field and the resultant field scattered by all the other particles; but the field scattered by a particle depends on the total field to which it is exposed. Considerable simplification results if we assume *single scattering*: the number of particles is sufficiently small and their separation sufficiently large that, in the neighborhood of any particle, the total field scattered by all the particles is small compared with the external field. With this assumption the total scattered field is just the sum of the fields scattered by the individual particles, each of which is acted on by the external field in isolation from the other particles. It is difficult to state precise general conditions under which the single scattering criterion is satisfied; it is not satisfied, for example, by clouds, where *multiple scattering* can be appreciable. In laboratory experiments, however, it is usually possible to prepare dilute suspensions of sufficiently small size to ensure single scattering.

We shall assume, in addition to single scattering, that the particles are many and their separations random, which implies *incoherent scattering*. That is, there is no systematic relation among the phases of the waves scattered by the individual particles; thus, the total irradiance scattered by the collection is just the sum of the irradiances scattered by the individual particles. Even, however, in a collection of randomly separated particles, the scattering is coherent in the forward direction, a subject to which we shall return in Chapter 3.

1.5   THE DIRECT AND INVERSE PROBLEM

There are two general classes of problems in the theory of the interaction of an electromagnetic wave with a small particle.

The Direct Problem.   Given a particle of specified shape, size, and composition, which is illuminated by a beam of specified irradiance, polarization, and

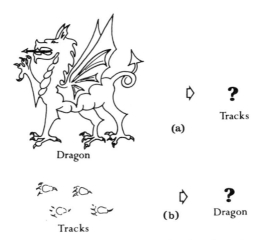

Figure 1.5 (*a*) The direct problem: Describe the tracks of a given dragon. (*b*) The inverse problem: Describe a dragon from its tracks.

frequency, determine the field everywhere. This is the "easy" problem; it consists of describing the tracks of a given dragon (Fig. 1.5*a*).

The Inverse Problem.   By a suitable analysis of the scattered field, describe the particle or particles that are responsible for the scattering. This is the "hard" problem; it consists of describing a dragon from an examination of its tracks (Fig. 1.5*b*).

Unfortunately, the problem that is most frequently of interest is the inverse problem, the type of problem to which Sherlock Holmes might have directed his powers had he been a physical scientist rather than a detective. For example, the composition of interstellar dust is one of the major unsolved mysteries in astronomy. Although there is little doubt about the existence of this dust, it has defied complete identification despite considerable effort. The only means for its investigation is analysis of light of various wavelengths that traverses the dust without undergoing scattering or absorption and, less commonly, the light scattered in various directions by the dust. In laboratory investigations, light scattering techniques are often used to determine the size of particles of known shape and composition. Radar backscattering can be used to discriminate between rain and hail. In all these applications the basic problem is to describe a dragon (or flock of dragons) by detailed examination of its tracks. To understand why we have labeled this the "hard" problem, consider that the information necessary to specify a particle uniquely is (1) the vector amplitude and phase of the field scattered in all directions, and (2) the field inside the particle (Hart and Gray, 1964). The field inside a particle is not usually accessible to direct measurement, although under certain conditions, which are not likely to be met except in the laboratory, this field can be approximated by the incident field (see Chapter 6). Even in this special case, however, the amplitude and phase of the scattered field are required; although this is not impossible in principle, it is rarely achieved in practice. The measurements usually available for analysis are the irradiance of the scattered light for a set of directions. We are therefore almost always faced with the task of trying to describe a particle (or worse yet, a collection of particles) with a less than theoretically ideal set of data in hand. But this is not necessarily cause for despair. Often, supplementary information about the particles, some of which is obtained by means other than light scattering techniques, is sufficient to enable them to be described. Thus, we should not be too ready to discard what little information is available or might become available. One source of information sometimes neglected is the polarization properties of the scattered light, a subject we treat in detail in succeeding chapters.

NOTES AND COMMENTS

An excellent concise treatment of scattering—by molecules and particles, single and multiple—at an intermediate level is Chapter 14 of Stone (1963). Among the books devoted entirely to scattering by particles, that by Shifrin (1951) most closely resembles ours in that it discusses optical properties of bulk matter as well. But the two books that have influenced us most are those of van de Hulst (1957) and Kerker (1969); we are indebted to both authors. Another book on scattering, which emphasizes polydispersions, is by Deirmendjian (1969).

There are also more limited treatments of scattering. McCartney (1976, Chaps. 4–6) confines his attention to scattering by atmospheric particles. This is also discussed by Twomey (1977, Chaps. 9–10) in his treatise on atmospheric aerosols. In Goody (1964, Chap. 7) there are discussions of absorption by gases and, in less detail, extinction by molecules and by droplets. Parts of books on electromagnetic theory or optics include the theory of scattering by a sphere, most notably Stratton (1941, pp. 563–573) and Born and Wolf (1965, pp. 633–664). The latter also derive the Ewald–Oseen extinction theorem and apply it to reflection and refraction at a plane interface (pp. 98–104).

The fluctuation theory of scattering by molecules is treated in books by Bhagavantam (1942), Fabelinskii (1968), and Chu (1974, Chap. 3).

Two of the more venerable works on multiple scattering are the review article by Milne (1930) and the book by Chandrasekhar (1950). A long-awaited treatise on this subject by van de Hulst (1980) recently made its appearance. Two reports, with many references, edited by Lenoble (1977) and by Fouquart et al. (1980), have been published by the International Association of Meteorology and Atmospheric Physics.

The inverse problem in scattering theory is discussed in the collections of papers edited by Baltes (1978, 1980).

Chapter 2

Electromagnetic Theory

The treatment of absorption and scattering of light by small particles is a problem in electromagnetic theory, a subdivision of which is optics. Various aspects of this theory are found scattered throughout books on electricity and magnetism, optics, and polarization of light. To rely on such existing books as the basis for our development might, however, result in a confusing tangle of conflicting assumptions, notation, and conventions. Also, it is convenient to gather together in one place the mathematical and physical apparatus that underlies succeeding chapters. Thus, we present in this chapter our version of those concepts and equations from electromagnetic theory that are germane to the subject at hand; we shall have need of them throughout the remainder of the book.

2.1 FIELD VECTORS AND THE MAXWELL EQUATIONS

As stated in the introductory chapter, we adopt a macroscopic approach to the problem of determining absorption and scattering of electromagnetic waves by particles. Therefore, the logical point of departure is the Maxwell equations for the macroscopic electromagnetic field at interior points in matter, which in SI units may be written

$$\nabla \cdot \mathbf{D} = \rho_F, \tag{2.1}$$

$$\nabla \times \mathbf{E} + \frac{\partial \mathbf{B}}{\partial t} = 0, \tag{2.2}$$

$$\nabla \cdot \mathbf{B} = 0, \tag{2.3}$$

$$\nabla \times \mathbf{H} = \mathbf{J}_F + \frac{\partial \mathbf{D}}{\partial t}, \tag{2.4}$$

where $\mathbf{E}$ is the electric field and $\mathbf{B}$ the magnetic induction. The electric displacement $\mathbf{D}$ and magnetic field $\mathbf{H}$ are defined by

$$\mathbf{D} = \varepsilon_0 \mathbf{E} + \mathbf{P}, \tag{2.5}$$

$$\mathbf{H} = \frac{\mathbf{B}}{\mu_0} - \mathbf{M}, \tag{2.6}$$

where **P** is the electric polarization (average electric dipole moment per unit volume), **M** the magnetization (average magnetic dipole moment per unit volume), ε_0 the permittivity, and μ_0 the permeability of free space. Implicit in (2.5) is the assumption that quadrupole and higher moments are negligible compared with the dipole moment. In free space the polarization and magnetization vanish identically. The charge density ρ_F and current density $\mathbf{J}_F$ are associated with so-called "free" charges. The terms "free" and "bound" are sometimes set in quotation marks, which indicates that they are slightly suspect. Indeed, Purcell (1963, pp. 342–347) has simply but convincingly shown that it is not always possible to unambiguously distinguish between free and bound charges in matter. Nevertheless, we shall assume in time-honored fashion that the ambiguity in the meanings of free and bound leads to no observable consequences in the problems with which we shall be concerned.

Although there appears to be nearly universal agreement about the *microscopic* Maxwell equations, it is not trivial to provide a derivation of the macroscopic equations that will satisfy everyone. The process of so doing was begun by Lorentz some 100 years ago, and new derivations appear regularly in the literature, each of which claims to be more general, freer from ambiguity, and more logically consistent than its predecessors. Without wishing to enter the lists of combatants, we direct the interested reader to Russakoff (1970) and Robinson (1973) for further discussion of the transition from the microscopic to the macroscopic field equations. We accept, therefore, (2.1)–(2.6) as *the* macroscopic field equations without further comment; any attempt to justify this choice rigorously and to define all terms precisely would lead to a volume that would dwarf the present work.

Equations (2.1)–(2.6) are not sufficient in themselves; they must be supplemented with *constitutive relations*, which are assumed to have the form

$$\mathbf{J}_F = \sigma\mathbf{E}, \tag{2.7}$$

$$\mathbf{B} = \mu\mathbf{H}, \tag{2.8}$$

$$\mathbf{P} = \varepsilon_0\chi\mathbf{E}, \tag{2.9}$$

where σ is the *conductivity*, μ the *permeability*, and χ the electric *susceptibility*. The *phenomenological coefficients* σ, μ, and χ depend on the medium under consideration, but will be assumed to be independent of the fields (the medium is *linear*), independent of position (the medium is *homogeneous*), and independent of direction (the medium is *isotropic*). There are many classes of materials for which these assumptions are *not* valid. Equations (2.7)–(2.9) are not universal laws of nature; they merely describe a particular class of materials which, fortunately, has a large number of members.

We shall not, however, assume that the phenomenological coefficients are independent of *frequency*. To provide some insight into this assertion, we must digress briefly. Consider, for example, the susceptibility. The polarization **P** is the average dipole moment per unit volume of the medium, that is, the vector sum of the dipole moments in a unit volume. An isolated sample of matter (excluding electrets) is unpolarized ($P = 0$). But when placed in an external field, which we may assume to be time harmonic, it becomes polarized: the electric field induces a net dipole moment. For a linear homogeneous isotropic medium, the relation (2.9) between P and E indicates that χ is a measure of how easily the material can be polarized; it represents the response of the material to the field E. In particular, χ may be interpreted as the amplitude of the response to a unit field. Now it is well known (to those who know it well) that the response of a mechanical system to a periodic driving force is a sensitive function of the frequency. Therefore, by analogy, it is plausible that χ is a function of the frequency. The frequency dependence of χ is discussed in much greater detail in Chapters 9 and 10.

2.2   TIME-HARMONIC FIELDS

The general time-harmonic field F has the form

$$F = A \cos \omega t + B \sin \omega t, \tag{2.10}$$

where ω is the *angular frequency*. The *real* vector fields A and B are independent of time but may depend on position. We note that F may be written as the real part of a complex vector: $F = \text{Re}\{F_c\}$, where

$$F_c = C \exp(-i\omega t), \qquad C = A + iB. \tag{2.11}$$

The vector F_c is a *complex representation* of the real field F. If all our operations on time-harmonic fields are *linear* (e.g., addition, differentiation, integration), it is more convenient to work with the complex representation. The reason this may be done is as follows. Let $\mathcal{L}$ be any linear operator; we can operate on the field (2.10) by operating on the complex representation (2.11) and then take the real part of the result:

$$\mathcal{L}F = \mathcal{L}\,\text{Re}\{F_c\} = \text{Re}\{\mathcal{L}F_c\}.$$

Note that there is a degree of arbitrariness associated with the complex representation of a real field: F could just as easily have been written $F = \text{Re}\{F_c^*\}$, where $F_c^* = C^* \exp(i\omega t)$ and the asterisk denotes the complex conjugate. Thus, there are two possible choices for the time-dependent factor in a complex representation of a time-harmonic field: $\exp(i\omega t)$ and $\exp(-i\omega t)$. It makes no difference which choice is made: the quantities of physical interest are always real. But once a sign convention has been chosen it must be used consistently in all analysis. We shall take the time-dependent factor to be $\exp(-i\omega t)$; this is the convention found in standard books on optics (Born and Wolf, 1965) and electromagnetic theory (Stratton, 1941; Jackson, 1975) as well as being nearly universal in solid-state physics. If comparison is made between our expressions and similar expressions in the scientific literature, the sign convention must be carefully noted.

If $\exp(-i\omega t)$ time dependence is assumed for all fields, and the constitutive relations (2.7)–(2.9) are substituted into (2.1)–(2.4), we obtain

$$\nabla \cdot (\varepsilon \mathbf{E}_c) = 0, \tag{2.12}$$

$$\nabla \times \mathbf{E}_c = i\omega\mu\mathbf{H}_c, \tag{2.13}$$

$$\nabla \cdot \mathbf{H}_c = 0, \tag{2.14}$$

$$\nabla \times \mathbf{H}_c = -i\omega\varepsilon\mathbf{E}_c, \tag{2.15}$$

where the complex *permittivity* is

$$\varepsilon = \varepsilon_0(1 + \chi) + i\frac{\sigma}{\omega}. \tag{2.16}$$

If $\varepsilon \neq 0$, the electric field is divergence free; this is the general condition for a *transverse* field. Except possibly at frequencies where $\varepsilon = 0$, therefore, the medium cannot support *longitudinal* fields.

Equations (2.12)–(2.16) will usually be our point of departure in scattering problems. However, to avoid a cluttered notation, we shall often omit the subscript c from the complex fields. In those instances where confusion might result, the subscript will be retained, although it should usually be clear from the context if we are dealing with real fields or their complex representations.

2.3 FREQUENCY-DEPENDENT PHENOMENOLOGICAL COEFFICIENTS

Although we gave some physical justification, the seemingly *ad hoc* introduction of frequency-dependent phenomenological coefficients implies that they have a well-defined meaning only if the fields are harmonic in time. But this is very unsatisfactory: it is certainly not a law of nature that all electromagnetic fields must be harmonic in time. Thus, we must answer the questions: How are **P** and **E** (or **B** and **H**) related in general, and what is the meaning of frequency-dependent phenomenological coefficients? To answer these questions we must invoke the mathematical apparatus of Fourier transforms.

2.3.1 Fourier Transforms

Consider a *real* function of time $F(t)$. The Fourier transform $\mathscr{F}(\omega)$ of $F(t)$ is defined as

$$\mathscr{F}(\omega) = \int_{-\infty}^{\infty} F(t)e^{i\omega t} \, dt, \tag{2.17}$$

where ω is real. We assume that the integral in (2.17) converges in some sense for all functions of interest. The *inverse Fourier transform* is given by

$$F(t) = \frac{1}{2\pi} \int_{-\infty}^{\infty} \mathscr{F}(\omega) e^{-i\omega t} \, d\omega. \qquad (2.18)$$

According to (2.18), an arbitrary time-dependent function can be expressed as a superposition of time-harmonic functions $\exp(-i\omega t)$, where the complex amplitude $\mathscr{F}(\omega)$ depends on the frequency ω. The condition that $F(t)$ be real is that $\mathscr{F}^*(\omega) = \mathscr{F}(-\omega)$; therefore, $F(t)$ can be expressed as a superposition of time-harmonic functions with positive frequency:

$$F(t) = \text{Re}\{F_c(t)\},$$

$$F_c(t) = \frac{1}{\pi} \int_0^{\infty} \mathscr{F}(\omega) e^{-i\omega t} \, d\omega.$$

Note that (2.17) and (2.18) are not unique: all complex functions could be replaced by their complex conjugates, and the factor $1/2\pi$ could appear either in (2.17) or (2.18). If we want our expressions to appear more symmetrical, both integrals can have the common multiplicative factor $1/\sqrt{2\pi}$. There is no universally accepted convention for Fourier transforms. However, once the form of the Fourier transform has been specified, the corresponding expression for the inverse Fourier transform is uniquely determined.

If we Fourier analyze the Maxwell equations (2.1)–(2.4), with $\rho_F = 0$, and assume that the operations of integration and differentiation may be interchanged, we obtain

$$\nabla \cdot \mathscr{D}(\omega) = 0, \qquad (2.19)$$

$$\nabla \cdot \mathscr{B}(\omega) = 0, \qquad (2.20)$$

$$\nabla \times \mathscr{E}(\omega) = i\omega \mathscr{B}(\omega), \qquad (2.21)$$

$$\nabla \times \mathscr{H}(\omega) = \mathscr{J}_F(\omega) - i\omega \mathscr{D}(\omega), \qquad (2.22)$$

where

$$\mathscr{E}(\omega) = \int_{-\infty}^{\infty} \mathbf{E}(t) e^{i\omega t} \, dt; \quad \text{etc.}$$

The Fourier transforms of the fields are, in general, position dependent, although this is not explicitly indicated. Because the fields are real, we need consider only positive frequencies. Let us now assume that the Fourier transforms of the fields, rather than the fields themselves, are connected by linear constitutive relations:

$$\mathscr{P}(\omega) = \varepsilon_0 \chi(\omega)\mathscr{E}(\omega), \tag{2.23}$$

$$\mathscr{J}_F(\omega) = \sigma(\omega)\mathscr{E}(\omega), \tag{2.24}$$

$$\mathscr{B}(\omega) = \mu(\omega)\mathscr{H}(\omega), \tag{2.25}$$

where $\chi^*(\omega) = \chi(-\omega)$, and so on. We also assume that $\chi(\omega)$, $\sigma(\omega)$, and $\mu(\omega)$ are scalar quantities independent of position. As we shall see, the constitutive relations (2.23)–(2.25) reduce to familiar expressions when the fields are time harmonic but, in addition, answer the question posed above: How are arbitrarily time-dependent fields related?

Before we proceed we need one more result from Fourier transform theory. If the Fourier transform $\mathscr{X}(\omega)$ of a function $X(t)$ is expressed as a product of Fourier transforms

$$\mathscr{X}(\omega) = \mathscr{Y}(\omega)\mathscr{Z}(\omega),$$

the *convolution theorem* states that

$$X(t) = \int_{-\infty}^{\infty} Y(t - t')Z(t')\,dt',$$

where

$$Y(t) = \frac{1}{2\pi}\int_{-\infty}^{\infty} \mathscr{Y}(\omega)e^{-i\omega t}\,d\omega; \quad \text{etc.}$$

If we apply the convolution theorem to the constitutive relation (2.23), for example, we obtain

$$\mathbf{P}(t) = \int_{-\infty}^{\infty} G(t - t')\mathbf{E}(t')\,dt', \tag{2.26}$$

where

$$G(t) = \frac{1}{2\pi}\int_{-\infty}^{\infty} \varepsilon_0 \chi(\omega)e^{-i\omega t}\,d\omega. \tag{2.27}$$

Thus, a frequency-dependent susceptibility $\chi(\omega)$ implies that the polarization $\mathbf{P}$ at time t depends on the electric field $\mathbf{E}$ at all other times t'. This conclusion is consistent with simple physical reasoning. If, for example, a steady electric field is applied to a sample of matter for a sufficient period of time, a steady polarization will be induced in the sample. However, if the electric field were to be suddenly removed, the polarization would not immediately drop to zero but would decay according to characteristic times associated with microscopic processes. In this example it is clear that the polarization is not proportional to the instantaneous field.

Let us now assume that the fields are time harmonic with angular frequency ω_0:

$$\mathbf{E}(t) = \mathbf{A}\cos\omega_0 t + \mathbf{B}\sin\omega_0 t = \text{Re}\{\mathbf{E}_c\}.$$

The Fourier transform of $\mathbf{E}(t)$ is

$$\mathcal{E}(\omega) = \pi\{(\mathbf{A} + i\mathbf{B})\delta(\omega - \omega_0) + (\mathbf{A} - i\mathbf{B})\delta(\omega + \omega_0)\},$$

where δ is the Dirac delta function. From the Fourier inversion formula (2.18) and the constitutive relation (2.23) it follows that

$$\mathbf{P}(t) = \frac{1}{2\pi}\int_{-\infty}^{\infty} \varepsilon_0\chi(\omega)\mathcal{E}(\omega)e^{-i\omega t}\,d\omega$$

$$= \frac{\varepsilon_0}{2}\{(\mathbf{A} + i\mathbf{B})\chi(\omega_0)e^{-i\omega_0 t} + (\mathbf{A} - i\mathbf{B})\chi(-\omega_0)e^{i\omega_0 t}\}.$$

If we use the relation $\chi(-\omega_0) = \chi^*(\omega_0)$, then

$$\mathbf{P}(t) = \text{Re}\{\mathbf{P}_c(t)\} = \text{Re}\{\varepsilon_0\chi(\omega_0)\mathbf{E}_c(t)\}.$$

Thus, the perhaps unfamiliar constitutive relations (2.23)–(2.25) yield familiar results when the fields are time harmonic; moreover, because of (2.26) and (2.27), physical meaning can now be attached to the phenomenological coefficients even for arbitrarily time-dependent fields.

Although we concentrated attention on the electric susceptibility, similar remarks hold for the conductivity and permeability. In particular, we note that μ can be frequency dependent and is not restricted to be real. Complex phenomenological coefficients imply differences in *phase* between the various time-harmonic fields. For example, if the permeability $\mu = \mu' + i\mu''$ is complex, then for a real magnetic field $\mathbf{H} = \mathbf{H}_0\cos\omega t$, the corresponding magnetic induction $\mathbf{B}$ is $\bar{\mu}\mathbf{H}_0\cos(\omega t - \phi)$, where the phase angle ϕ is given by $\tan\phi = \mu''/\mu'$ and $\bar{\mu} = \{(\mu')^2 + (\mu'')^2\}^{1/2}$. In Section 2.6 we shall show that if the imaginary part of any of the phenomenological coefficients of a medium is nonzero, the amplitude of a plane wave will decrease as it propagates through such a medium because of *absorption* of electromagnetic energy. Thus, complex phenomenological coefficients or, alternatively, a phase difference between $\mathbf{P}$ and $\mathbf{E}$ (or $\mathbf{B}$ and $\mathbf{H}$) are physically manifested by absorption, the detailed nature of which is hidden from view in a macroscopic approach.

If the Fourier-transformed Maxwell equations (2.19)–(2.22) and the constitutive relations (2.23)–(2.25) are combined, we obtain

$$\nabla \cdot \mathcal{E}(\omega) = 0, \qquad (2.28)$$

$$\nabla \cdot \mathcal{H}(\omega) = 0, \qquad (2.29)$$

$$\nabla \times \mathcal{E}(\omega) = i\omega\mu(\omega)\mathcal{H}(\omega), \qquad (2.30)$$

$$\nabla \times \mathcal{H}(\omega) = -i\omega\varepsilon(\omega)\mathcal{E}(\omega), \qquad (2.31)$$

where $\varepsilon(\omega) = \varepsilon_0\{1 + \chi(\omega)\} + i\sigma(\omega)/\omega$. Note that the two sets of equations (2.12)–(2.15) and (2.28)–(2.31) are formally identical although interpreted differently. In the former set, $(\mathbf{E}_c, \mathbf{H}_c)$ is the complex representation of a time-harmonic field; whereas in the latter set, $(\mathcal{E}, \mathcal{H})$ is the Fourier transform of an arbitrarily time-dependent electromagnetic field. But the same equations have the same solutions, so we need consider only time-harmonic fields; the general time-dependent field can be constructed by Fourier synthesis.

Both the conductivity and the susceptibility contribute to the imaginary part of the permittivity: $\operatorname{Im}\{\varepsilon\} = \operatorname{Im}\{\chi\} + \operatorname{Re}\{\sigma/\omega\}$. A nonzero value for $\operatorname{Im}\{\varepsilon\}$ manifests itself physically by absorption of electromagnetic energy in the medium. We may associate $\operatorname{Im}\{\chi\}$ with the "bound" charge current density and $\operatorname{Re}\{\sigma/\omega\}$ with the "free" charge current density. Absorption is determined by the *sum* of these two quantities, however, and it is not possible to determine by absorption measurements their relative contributions. This underscores our assertion that there is no clearly defined distinction between "free" and "bound" charges.

2.3.2 Kramers–Kronig Relations

As an alternative we could have begun our discussion of constitutive relations with the *assumption* that $\mathbf{P}(t)$ and $\mathbf{E}(t)$ are related through the linear functional equation

$$\mathbf{P}(t) = \int_{-\infty}^{\infty} G(t, t')\mathbf{E}(t')\, dt'. \tag{2.32}$$

Suppose that the electric field is a delta function applied at time t_0: $\mathbf{E}(t) = \delta(t - t_0)\mathbf{E}_0$; the corresponding polarization is therefore

$$\mathbf{P}(t) = G(t, t_0)\mathbf{E}_0.$$

Thus, G is the polarization resulting from a unit amplitude delta function. If the properties of the medium do not change with time, the polarization must depend only on the time elapsed between t_0 and t:

$$G(t, t_0) = G(t - t_0).$$

Therefore, we obtain (2.26), which when inverted yields the constitutive relation (2.23). We note that *causality*—the system cannot squeal before it is hurt—requires that $G(\tau) = 0$ for $\tau < 0$.

The susceptibility is the Fourier transform of $G(t)$:

$$\varepsilon_0\chi(\omega) = \int_{-\infty}^{\infty} G(t)e^{i\omega t}\, dt = \int_{0}^{\infty} G(t)e^{i\omega t}\, dt, \tag{2.33}$$

and is a complex-valued function of the real variable ω. Let us *define* a complex-valued function of the *complex* variable $\tilde{\omega}$ by

$$\varepsilon_0 \chi(\tilde{\omega}) = \int_0^\infty G(t) e^{i\tilde{\omega}t} \, dt,$$

where $\tilde{\omega} = \omega_r + i\omega_i$. The function $\chi(\tilde{\omega})$ coincides with $\chi(\omega)$ when $\tilde{\omega}$ is a point on the real axis. For any $t \geqslant 0$, $G(t)e^{i\tilde{\omega}t}$ is an analytic function of $\tilde{\omega}$, and $|G(t)e^{i\tilde{\omega}t}| \leqslant |G(t)|$ if $\omega_i > 0$. Therefore, if the integral

$$\int_0^\infty |G(t)| \, dt \tag{2.34}$$

converges, then

$$\int_0^\infty G(t) e^{i\tilde{\omega}t} \, dt$$

converges to an analytic function in the upper half of the complex $\tilde{\omega}$ plane. Convergence of (2.34) is assured if $\chi(0)$ is finite; this follows from (2.33). If $\chi(\tilde{\omega})$ is analytic, so is $\chi(\tilde{\omega})/(\tilde{\omega} - \omega)$ except at the pole $\tilde{\omega} = \omega$. Cauchy's theorem states that

$$\int_C f(\tilde{\omega}) \, d\tilde{\omega} = 0,$$

provided that the closed contour C encloses no poles of the analytic function $f(\tilde{\omega})$. Let us apply Cauchy's theorem to the function $\chi(\tilde{\omega})/(\tilde{\omega} - \omega)$, where ω is a point on the real axis, and the contour C, shown in Fig. 2.1, is the union of four curves with parametric representations

$$C_1: \quad \tilde{\omega} = \Omega \qquad\qquad (-A \leqslant \Omega \leqslant \omega - a),$$

$$C_2: \quad \tilde{\omega} = \omega - ae^{-i\Omega} \qquad (0 \leqslant \Omega \leqslant \pi),$$

$$C_3: \quad \tilde{\omega} = \Omega \qquad\qquad (\omega + a \leqslant \Omega \leqslant A),$$

$$C_4: \quad \tilde{\omega} = Ae^{i\Omega} \qquad\qquad (0 \leqslant \Omega \leqslant \pi).$$

Therefore, from Cauchy's theorem

$$\int_{-A}^{\omega-a} \frac{\chi(\Omega)}{\Omega - \omega} \, d\Omega + \int_{\omega+a}^A \frac{\chi(\Omega)}{\Omega - \omega} \, d\Omega + \int_0^\pi \frac{iAe^{i\Omega}\chi(Ae^{i\Omega})}{Ae^{i\Omega} - \omega} \, d\Omega$$

$$= \int_0^\pi i\chi(\omega - ae^{i\Omega}) \, d\Omega.$$

The integral over the curve C_4 vanishes as A tends to infinity if $\lim_{|\tilde{\omega}| \to \infty} \chi(\tilde{\omega})$

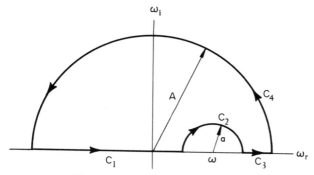

Figure 2.1   Contour of integration.

$= 0$; this implies that

$$i\pi\chi(\omega) = P\int_{-\infty}^{\infty} \frac{\chi(\Omega)}{\Omega - \omega} d\Omega, \qquad (2.35)$$

where the symbol P denotes the *Cauchy principal value* of the integral, defined by

$$\lim_{a \to 0} \int_{-\infty}^{\omega - a} \frac{\chi(\Omega)}{\Omega - \omega} d\Omega + \int_{\omega + a}^{\infty} \frac{\chi(\Omega)}{\Omega - \omega} d\Omega.$$

We were slightly careless in our derivation of (2.35): $\chi(\tilde{\omega})$ was required to be analytic in the upper half of the complex $\tilde{\omega}$ plane, whereas part of the contour C included the real axis. But this can easily be remedied, and a more general form of (2.35) is

$$i\pi \lim_{\eta \to 0} \chi(\omega + i\eta) = \lim_{\eta \to 0} P\int_{-\infty}^{\infty} \frac{\chi(\Omega + i\eta)}{\Omega - \omega} d\Omega,$$

from which (2.35) follows if $\chi(\tilde{\omega})$ is continuous.

The fundamental relation (2.35) can be written as two real integral relations:

$$\chi'(\omega) = \frac{2}{\pi} P\int_{0}^{\infty} \frac{\Omega \chi''(\Omega)}{\Omega^2 - \omega^2} d\Omega, \qquad (2.36)$$

$$\chi''(\omega) = \frac{-2\omega}{\pi} P\int_{0}^{\infty} \frac{\chi'(\Omega)}{\Omega^2 - \omega^2} d\Omega, \qquad (2.37)$$

where $\chi = \chi' + i\chi''$ and we have invoked the *crossing condition* $\chi^*(\omega) = \chi(-\omega)$. Equations (2.36) and (2.37) are an important example of a rather remarkable class of mathematical relations called *Kramers–Kronig* or *dispersion relations*. Their implication, which is unexpected on physical grounds and therefore has an aura of black magic, is that the real and imaginary parts of χ are not independent but are connected by integral relations; this imposes a constraint on physically realizable susceptibilities. Moreover, if χ' is known over a sufficiently large range of frequencies around ω, $\chi''(\omega)$ can be obtained by integration, and vice versa. An interesting corollary of (2.36) is the *sum rule*:

$$\chi'(0) = \frac{2}{\pi} \int_0^\infty \frac{\chi''(\Omega)}{\Omega} d\Omega.$$

Although the Kramers–Kronig relations do not follow directly from physical reasoning, they are not devoid of physical content: underlying their derivation are the assumptions of *linearity* and *causality* and restrictions on the asymptotic behavior of χ. As we shall see in Chapter 9, the required asymptotic behavior of χ is a physical consequence of the interaction of a frequency-dependent electric field with matter.

The derivation of Kramers–Kronig relations for the susceptibility was relatively easy, perhaps misleadingly so. With a bit of extra effort, however, we can often derive similar relations for other frequency-dependent quantities that arise in physical problems. Suppose that we have two time-dependent quantities of unspecified origin, which we may call the *input* $X_i(t)$ and the *output* $X_o(t)$; the corresponding Fourier transforms are denoted by $\mathfrak{X}_i(\omega)$ and $\mathfrak{X}_o(\omega)$. If the relation between these transforms is linear,

$$\mathfrak{X}_o(\omega) = \mathfrak{R}(\omega)\mathfrak{X}_i(\omega),$$

and causal (i.e., the output cannot precede the input in time), then $\mathfrak{R}(\tilde{\omega})$ is an analytic function in the top half of the complex $\tilde{\omega}$ plane. It is also necessary that $\mathfrak{R}(\tilde{\omega})$ vanish on the circular arc C_4 (Fig. 2.1) as A approaches infinity; if it does not, we are permitted to fiddle with it until it does. That is, we can change the asymptotic behavior of $\mathfrak{R}(\tilde{\omega})$ by multiplying it by some analytic function $g(\tilde{\omega})$, or adding $g(\tilde{\omega})$ to it, without changing its analyticity. Of course, in so doing, we may also change the crossing condition, and the resultant dispersion relations may be different from (2.36) and (2.37). Techniques of fiddling with $\mathfrak{R}$ until it behaves properly are best illustrated with specific examples, which we shall encounter later in this chapter and elsewhere in the book.

2.4   SPATIAL DISPERSION

We have shown that a frequency-dependent susceptibility implies *temporal dispersion*: the polarization at time t depends on the electric field at all times previous to t. It is also possible under some circumstances to have *spatial dispersion*: the polarization at point x depends on the values of the electric field at points in some neighborhood of x. This nonlocal relation between **P** and **E** leads to the concept of a susceptibility that depends on frequency *and* wave vector. To understand this it is helpful to introduce the three-dimensional Fourier transform of an arbitrary real function $F(\mathbf{x})$:

$$\mathscr{F}(\mathbf{k}) = \int F(\mathbf{x})e^{-i\mathbf{k}\cdot\mathbf{x}}\,d\mathbf{x},$$

where the integration is taken over all space. The inverse Fourier transform is

$$F(\mathbf{x}) = \frac{1}{(2\pi)^3}\int \mathscr{F}(\mathbf{k})e^{i\mathbf{k}\cdot\mathbf{x}}\,d\mathbf{k}.$$

Let us Fourier analyze the polarization and the electric field:

$$\mathscr{P}(\mathbf{k},\omega) = \int\int \mathbf{P}(\mathbf{x},t)e^{-i(\mathbf{k}\cdot\mathbf{x}-\omega t)}\,d\mathbf{x}\,dt,$$

$$\mathscr{E}(\mathbf{k},\omega) = \int\int \mathbf{E}(\mathbf{x},t)e^{-i(\mathbf{k}\cdot\mathbf{x}-\omega t)}\,d\mathbf{x}\,dt.$$

If we assume that there is a linear relation $\mathscr{P}(\mathbf{k},\omega) = \varepsilon_0\chi(\mathbf{k},\omega)\mathscr{E}(\mathbf{k},\omega)$, the convolution theorem yields

$$\mathbf{P}(\mathbf{x},t) = \int\int G(\mathbf{x}-\mathbf{x}',t-t')\mathbf{E}(\mathbf{x}',t')\,d\mathbf{x}'\,dt',$$

where

$$G(\mathbf{x},t) = \frac{1}{(2\pi)^4}\int\int \varepsilon_0\chi(\mathbf{k},\omega)e^{i(\mathbf{k}\cdot\mathbf{x}-\omega t)}\,d\mathbf{k}\,d\omega.$$

Thus, a susceptibility that depends on frequency *and* wave vector implies that the relation between $\mathbf{P}(\mathbf{x},t)$ and $\mathbf{E}(\mathbf{x},t)$ is nonlocal in time *and* space. Such spatially dispersive media lie outside our considerations. However, spatial dispersion can be important when the wavelength is comparable to some characteristic length in the medium (e.g., mean free path), and it is well at least to be aware of its existence: it can have an effect on absorption and scattering by small particles (Yildiz, 1963; Foley and Pattanayak, 1974; Ruppin, 1975, 1981).

2.5 POYNTING VECTOR

Consider an electromagnetic field $(\mathbf{E},\mathbf{H})$, which is not necessarily time harmonic. The *Poynting vector* $\mathbf{S} = \mathbf{E}\times\mathbf{H}$ specifies the magnitude and direction of the rate of transfer of electromagnetic energy at all points of space; it is of fundamental importance in problems of propagation, absorption, and scattering of electromagnetic waves. If the orientation of a plane surface with area A is specified by a unit normal vector $\hat{n}$, the rate at which electromagnetic energy is transferred across this surface is $\mathbf{S} \cdot \hat{n} A$, provided that $\mathbf{S}$ is constant over the surface. When $\mathbf{S}$ is a function of position and the surface has arbitrary shape, this can be generalized to

$$\int \mathbf{S} \cdot \hat{n} \, dA.$$

The net rate W at which electromagnetic energy crosses the boundary of a *closed* surface A which encloses a volume V is

$$W = -\int_A \mathbf{S} \cdot \hat{n} \, dA.$$

Why the minus sign? There are two choices for the unit normal to a closed surface: inward and outward. We have chosen the outward normal. If $\mathbf{S}$ and $\hat{n}$ are oppositely directed at a given point ($\mathbf{S} \cdot \hat{n} < 0$), the minus sign ensures that the contribution to W is positive. Therefore, W is positive if there is a net transfer of electromagnetic energy *into* the volume. A positive W implies that electromagnetic energy is *absorbed* in V; that is, electromagnetic energy is converted into other forms of energy (e.g., thermal energy) within V.

The formation of the vector product of two vectors is not a linear operation. Therefore, if the electromagnetic field is time harmonic, it is *not* true that $\mathbf{S} = \text{Re}\{\mathbf{E}_c \times \mathbf{H}_c\}$, although it is true that

$$\mathbf{S} = \text{Re}\{\mathbf{E}_c\} \times \text{Re}\{\mathbf{H}_c\}. \tag{2.38}$$

The instantaneous Poynting vector (2.38) is a rapidly varying function of time for frequencies that are usually of interest. Most instruments are not capable of following the rapid oscillations of the instantaneous Poynting vector, but respond to some time average $\langle \mathbf{S} \rangle$:

$$\langle \mathbf{S} \rangle = \frac{1}{\tau} \int_t^{t+\tau} \mathbf{S}(t') \, dt',$$

where τ is a time interval long compared with $1/\omega$. The time-averaged Poynting vector for time-harmonic fields is given by

$$\langle \mathbf{S} \rangle = \frac{1}{2} \text{Re}\{\mathbf{E}_c \times \mathbf{H}_c^*\}.$$

When it is clear from the context that it is the time-averaged Poynting vector with which we are dealing, the brackets enclosing $\mathbf{S}$ will be omitted.

2.6   PLANE-WAVE PROPAGATION IN UNBOUNDED MEDIA

Let us "look for" plane-wave solutions to the Maxwell equations (2.12)–(2.15). What does this statement mean? We know that the electromagnetic field $(\mathbf{E}, \mathbf{H})$ cannot be arbitrarily specified. Only certain electromagnetic fields, those that satisfy the Maxwell equations, are physically realizable. Therefore, because of their simple form, we should like to know under what conditions plane electromagnetic waves

$$\mathbf{E}_c = \mathbf{E}_0 \exp(i\mathbf{k} \cdot \mathbf{x} - i\omega t); \qquad \mathbf{H}_c = \mathbf{H}_0 \exp(i\mathbf{k} \cdot \mathbf{x} - i\omega t), \qquad (2.39)$$

where $\mathbf{E}_0$ and $\mathbf{H}_0$ are constant vectors, are compatible with the Maxwell equations. The wave vector $\mathbf{k}$ may be complex

$$\mathbf{k} = \mathbf{k}' + i\mathbf{k}'', \qquad (2.40)$$

where $\mathbf{k}'$ and $\mathbf{k}''$ are real vectors. If (2.40) is substituted in (2.39), we obtain

$$\mathbf{E}_c = \mathbf{E}_0 \exp(-\mathbf{k}'' \cdot \mathbf{x}) \exp(i\mathbf{k}' \cdot \mathbf{x} - i\omega t),$$

$$\mathbf{H}_c = \mathbf{H}_0 \exp(-\mathbf{k}'' \cdot \mathbf{x}) \exp(i\mathbf{k}' \cdot \mathbf{x} - i\omega t).$$

$\mathbf{E}_0 \exp(-\mathbf{k}'' \cdot \mathbf{x})$ and $\mathbf{H}_0 \exp(-\mathbf{k}'' \cdot \mathbf{x})$ are the *amplitudes* of the electric and magnetic waves, and $\phi = \mathbf{k}' \cdot \mathbf{x} - \omega t$ is the *phase* of the waves. An equation of the form $\mathbf{K} \cdot \mathbf{x} = $ constant, where $\mathbf{K}$ is any real vector, defines a plane surface the normal to which is $\mathbf{K}$. Therefore, $\mathbf{k}'$ is perpendicular to the *surfaces of constant phase*, and $\mathbf{k}''$ is perpendicular to the *surfaces of constant amplitude*. If $\mathbf{k}'$ and $\mathbf{k}''$ are parallel, which includes the case $\mathbf{k}'' = 0$, these surfaces coincide and the waves are said to be *homogeneous*; if $\mathbf{k}'$ and $\mathbf{k}''$ are not parallel, the waves are said to be *inhomogeneous*. For example, waves propagating in a vacuum are homogeneous.

Let us briefly consider propagation of surfaces of constant phase. Choose an arbitrary origin O and a plane surface over which the phase ϕ is constant (Fig. 2.2). At time t the distance from the origin O to the plane is z, where $\mathbf{k}' \cdot \mathbf{x} = k'z$ and $k'z - \omega t = \phi$. In a time interval Δt the surface of constant phase will have moved a distance Δz, where

$$k'z - \omega t = k'(z + \Delta z) - \omega(t + \Delta t) = \phi.$$

Thus, the velocity of propagation of surfaces of constant phase, the *phase velocity* v, is

$$v = \frac{\Delta z}{\Delta t} = \frac{\omega}{k'},$$

and the vector $\mathbf{k}'$ specifies the direction of propagation.

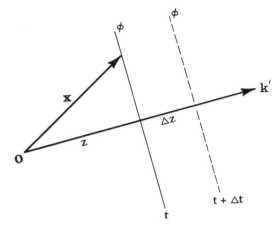

Figure 2.2 Propagation of constant phase surfaces.

The Maxwell equations for plane waves are

$$\mathbf{k} \cdot \mathbf{E}_0 = 0, \tag{2.41}$$

$$\mathbf{k} \cdot \mathbf{H}_0 = 0, \tag{2.42}$$

$$\mathbf{k} \times \mathbf{E}_0 = \omega\mu\mathbf{H}_0, \tag{2.43}$$

$$\mathbf{k} \times \mathbf{H}_0 = -\omega\varepsilon\mathbf{E}_0. \tag{2.44}$$

Equations (2.41) and (2.42) are the conditions for *transversality*: $\mathbf{k}$ is perpendicular to $\mathbf{E}_0$ and $\mathbf{H}_0$. It is also evident from (2.43) or (2.44) that $\mathbf{E}_0$ and $\mathbf{H}_0$ are perpendicular. However, $\mathbf{k}$, $\mathbf{E}_0$, and $\mathbf{H}_0$ are, in general, complex vectors, and the interpretation of the term "perpendicular" is not simple unless the waves are homogeneous; for such waves, the real fields $\mathbf{E}$ and $\mathbf{H}$ lie in a plane the normal to which is parallel to the direction of propagation.

If we take the vector product of both sides of (2.43) with $\mathbf{k}$,

$$\mathbf{k} \times (\mathbf{k} \times \mathbf{E}_0) = \omega\mu\mathbf{k} \times \mathbf{H}_0 = -\omega^2\varepsilon\mu\mathbf{E}_0,$$

and use the **BAC–CAB** vector identity:

$$\mathbf{A} \times (\mathbf{B} \times \mathbf{C}) = \mathbf{B}(\mathbf{A} \cdot \mathbf{C}) - \mathbf{C}(\mathbf{A} \cdot \mathbf{B})$$

together with (2.41), we obtain

$$\mathbf{k} \cdot \mathbf{k} = \omega^2\varepsilon\mu. \tag{2.45}$$

Our analysis shows that plane waves (2.39) are compatible with the Maxwell equations provided that $\mathbf{k}$, $\mathbf{E}_0$, and $\mathbf{H}_0$ are perpendicular:

$$\mathbf{k} \cdot \mathbf{E}_0 = \mathbf{k} \cdot \mathbf{H}_0 = \mathbf{E}_0 \cdot \mathbf{H}_0 = 0.$$

In addition, the wave vector must satisfy (2.45); this equation may be written

$$k'^2 - k''^2 + 2i\mathbf{k}' \cdot \mathbf{k}'' = \omega^2 \varepsilon \mu. \tag{2.46}$$

Note that $\varepsilon \mu$ is a property of the *medium* in which the wave propagates; the vectors $\mathbf{k}'$ and $\mathbf{k}''$, however, are properties of the *wave*. If $\mathbf{k}'$ and $\mathbf{k}''$ satisfy (2.46), they may otherwise be arbitrary: ε and μ do not uniquely specify the details of wave propagation.

The wave vector of a homogeneous wave may be written $\mathbf{k} = (k' + ik'')\hat{\mathbf{e}}$, where k' and k'' are nonnegative and $\hat{\mathbf{e}}$ is a real unit vector in the direction of propagation. Equation (2.45) requires that

$$k = k' + ik'' = \frac{\omega N}{c},$$

where c is the speed of light *in vacuo* and the *complex refractive index* N is

$$N = c\sqrt{\varepsilon \mu} = \sqrt{\frac{\varepsilon \mu}{\varepsilon_0 \mu_0}}. \tag{2.47}$$

We shall write

$$N = n + ik, \tag{2.48}$$

where n and k are nonnegative. Other notation commonly encountered for the complex refractive index is $n' + in''$ and $n(1 + i\kappa)$; if we had chosen $\exp(i\omega t)$ time dependence, these would be written $n' - in''$ and $n(1 - i\kappa)$. The symbol k, where $\omega\sqrt{\varepsilon\mu} = $ k, is widely used to denote the wave number. On the other hand, the use of k to denote the imaginary part of the complex refractive index is equally sacrosanct, particularly among those who actually undertake the task of its measurement. We shall adhere to both of these conventions; however, to differentiate between the two quantities, we shall use italic type for the imaginary part of the refractive index and Roman type for the wave number. The use of similar symbols to denote two physically distinct but not unrelated quantities may possibly lead to confusion; but even more befuddlement would undoubtedly result to many readers if we were to meddle with well-established conventions for the sake of symbolic precision.

The free-space wave number is $\omega/c = 2\pi/\lambda$, where λ is the wavelength *in vacuo*. Therefore, a plane homogeneous wave has the form

$$\mathbf{E}_c = \mathbf{E}_0 \exp\left(-\frac{2\pi k z}{\lambda}\right)\exp\left(\frac{i2\pi n z}{\lambda} - i\omega t\right),$$

where $z = \hat{\mathbf{e}} \cdot \mathbf{x}$. Thus, the imaginary part of the complex refractive index determines the attenuation of the wave as it propagates through the medium; the real part determines the phase velocity $v = c/n$. The pair of quantities n and k are often referred to as the *optical constants*, terminology that is as widespread as it is misleading: the optical "constants" are not constant, they often strongly depend on frequency. It is not unusual for k of many common solids to range over six orders of magnitude within a relatively narrow range of frequencies.

Although refractive indices of many transparent materials at visible wavelengths have been accurately known for a long time, experimental determination of optical constants is by no means trivial in wavelength regions where a solid or liquid is appreciably absorbing. Special techniques have had to be developed to measure optical constants for various materials in different portions of the electromagnetic spectrum, and lack of reliable optical constants is still a serious impediment to small-particle scattering and absorption calculations in several areas of applied physics.

The real and imaginary parts of the complex refractive index satisfy Kramers–Kronig relations; sometimes this can be used to assess the reliability of measured optical constants. $N(\omega)$ satisfies the same crossing condition as $\chi(\omega)$: $N^*(\omega) = N(-\omega)$. However, it does not vanish in the limit of indefinitely large frequency: $\lim_{\omega \to \infty} N(\omega) = 1$. But this is a small hurdle, which can be surmounted readily enough by minor fiddling with $N(\omega)$: the quantity $N(\omega) - 1$ has the desired asymptotic behavior. If we now assume that $N(\tilde{\omega})$ is analytic in the top half of the complex $\tilde{\omega}$ plane, it follows that

$$n(\omega) - 1 = \frac{2}{\pi} P \int_0^\infty \frac{\Omega k(\Omega)}{\Omega^2 - \omega^2} d\Omega, \tag{2.49}$$

$$k(\omega) = \frac{-2\omega}{\pi} P \int_0^\infty \frac{n(\Omega)}{\Omega^2 - \omega^2} d\Omega, \tag{2.50}$$

where we used

$$P \int_0^\infty \frac{d\Omega}{\Omega^2 - \omega^2} = 0 \tag{2.51}$$

in the derivation of (2.50).

2.6.1 Absorption of Electromagnetic Energy

The Poynting vector of a plane wave is

$$\mathbf{S} = \tfrac{1}{2}\mathrm{Re}\{\mathbf{E} \times \mathbf{H}^*\} = \mathrm{Re}\left\{ \frac{\mathbf{E} \times (\mathbf{k}^* \times \mathbf{E}^*)}{2\omega\mu^*} \right\},$$

where $\mathbf{E} \times (\mathbf{k}^* \times \mathbf{E}^*) = \mathbf{k}^*(\mathbf{E} \cdot \mathbf{E}^*) - \mathbf{E}^*(\mathbf{k}^* \cdot \mathbf{E})$. If the wave is homogeneous, $\mathbf{k} \cdot \mathbf{E} = 0$ implies that $\mathbf{k}^* \cdot \mathbf{E} = 0$; for such a wave propagating in the $\hat{\mathbf{e}}$

direction, we have

$$S = \tfrac{1}{2}\text{Re}\left\{\sqrt{\frac{\varepsilon}{\mu}}\right\}|E_0|^2\exp\left(-\frac{4\pi kz}{\lambda}\right)\hat{e}.$$

Not surprisingly, S is in the direction of propagation. The magnitude of S, which we shall denote by the symbol I, is called the *irradiance* and its dimensions are energy per unit area and time. (The term *intensity* is often used to denote irradiance; however, intensity is also used for other radiometric quantities, and we shall therefore tend to avoid this term because of possible confusion. E is now the recommended symbol for irradiance, but this hardly seems appropriate in a book where the electric field and irradiance often appear side by side.) As the wave traverses the medium, the irradiance is exponentially attenuated:

$$I = I_0 e^{-\alpha z},$$

where the *absorption coefficient* α is

$$\alpha = \frac{4\pi k}{\lambda} \tag{2.52}$$

and I_0 is the irradiance at $z = 0$. Throughout this chapter we adopt the viewpoint that bulk matter is homogeneous. However, this is only approximately true; even in media that are usually considered to be homogeneous, such as bulk samples of a pure liquid or solid, a beam of light is attenuated both by absorption and by scattering. Although absorption is usually the dominant mode of attenuation in such media, scattering is not entirely absent, and unless special techniques are used a measurement of attenuation unavoidably yields the combined effect of absorption and scattering. In later chapters we shall discuss attenuation by collections of material particles; such attenuation may or may not be dominated by absorption depending on the size and optical properties of the particles.

The rate at which electromagnetic energy is removed from the wave as it propagates through the medium is determined by the imaginary part of the complex refractive index. If the irradiances I_0 and I_t (or rather their ratio) are measured at two different positions $z = 0$ and $z = h$, then α, and hence k, can be obtained in principle from the relation

$$\alpha h = \ln\frac{I_0}{I_t}. \tag{2.53}$$

This equation is strictly valid only if the detector is optically identical with the medium for which the absorption coefficient is to be measured (Fig. 2.3a), a condition that is difficult to satisfy. The usual experimental configuration is

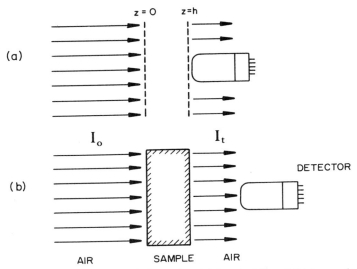

Figure 2.3 Measurement of absorption: (a) in principle and (b) in practice.

shown schematically in Fig. 2.3b. The transmission I_t/I_0 may be determined from the detector response with and without the sample interposed between source and detector. But α cannot be obtained from such a measurement unless reflections at the two interfaces are negligible.

2.7 REFLECTION AND TRANSMISSION AT A PLANE BOUNDARY

The considerations in the preceding section make it worthwhile to discuss reflection and transmission at plane boundaries: first, one plane boundary separating infinite media, then in the next section two successive plane boundaries forming a slab. In addition to providing useful results for bulk materials, these relatively simple boundary-value problems illustrate methods used in more complicated small-particle problems. Also, the optical properties of slabs often will be compared to those of small particles—both similarities and differences—to develop intuitive thinking about particles by way of the more familiar properties of bulk matter.

2.7.1 Normal Incidence

Consider a plane wave propagating in a nonabsorbing medium with refractive index $N_2 = n_2$, which is incident on a medium with refractive index $N_1 = n_1 + ik_1$ (Fig. 2.4). The amplitude of the incident electric field is $\mathbf{E}_i$, and we assume that there are transmitted and reflected waves with amplitudes $\mathbf{E}_t$ and $\mathbf{E}_r$, respectively. Therefore, plane-wave solutions to the Maxwell equations at

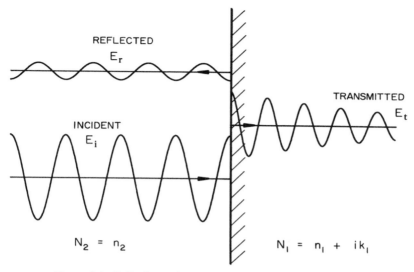

Figure 2.4   Reflection and transmission of normally incident light.

interior points on both sides of the boundary between the media are

$$\mathbf{E}_t \exp\left[i\omega\left(\frac{N_1 z}{c} - t\right)\right] \qquad (z > 0),$$

$$\mathbf{E}_i \exp\left[i\omega\left(\frac{N_2 z}{c} - t\right)\right] + \mathbf{E}_r \exp\left[-i\omega\left(\frac{N_2 z}{c} + t\right)\right] \qquad (z < 0).$$

The tangential components of the electric field are required to be continuous across the boundary $z = 0$:

$$\mathbf{E}_i + \mathbf{E}_r = \mathbf{E}_t. \tag{2.54}$$

Continuity of the tangential magnetic field yields the condition

$$\mathbf{E}_i - \mathbf{E}_r = \frac{N_1}{N_2}\mathbf{E}_t, \tag{2.55}$$

where we have used (2.43) and also assumed that $\mu_1 = \mu_2$. Equations (2.54) and (2.55) are readily solved for the amplitudes $\mathbf{E}_r$ and $\mathbf{E}_t$:

$$\mathbf{E}_r = \tilde{r}\mathbf{E}_i, \qquad \mathbf{E}_t = \tilde{t}\mathbf{E}_i, \tag{2.56}$$

where the *reflection* and *transmission coefficients* are

$$\tilde{r} = \frac{1 - m}{1 + m}, \qquad \tilde{t} = \frac{2}{1 + m}, \tag{2.57}$$

and $N_1/N_2 = m = n + ik$ is the refractive index of medium 1 relative to medium 2. The *reflectance* R for normally incident light, defined as the ratio of reflected to incident irradiance, is

$$R = |\tilde{r}|^2 = \left| \frac{1 - m}{1 + m} \right|^2 = \frac{(n - 1)^2 + k^2}{(n + 1)^2 + k^2}. \tag{2.58}$$

Note that $R \times 100\%$ is close to 100% if either $n \gg 1$ or $n \ll 1$ or $k \gg 1$. One might think that a material with $k \gg 1$ would be highly absorbing. But such a material is highly reflecting, and an incident wave cannot "get into" the material to be absorbed.

The amplitudes $\mathbf{E}_i$ and $\mathbf{E}_r$ in (2.56) may be interpreted as the Fourier transforms of the incident and reflected fields; therefore, from the general considerations at the end of Section 2.3, it follows that the real and imaginary parts of $\tilde{r}$ satisfy Kramers–Kronig relations of the form (2.36) and (2.37). Although this is perhaps interesting, it is not terribly useful; but with a bit of effort we can derive Kramers–Kronig relations of great practical utility. To do so we must express $\tilde{r}$ in terms of its amplitude (or modulus) r and phase (or argument) Θ:

$$\tilde{r} = re^{i\Theta}, \tag{2.59}$$

where $r = \sqrt{R}$. The reflection coefficient may also be written as an explicit function of n and k:

$$\tilde{r} = \frac{1 - n^2 - k^2}{(1 + n)^2 + k^2} - i\frac{2k}{(1 + n)^2 + k^2}, \tag{2.60}$$

from which it follows that

$$n = \frac{1 - R}{1 + R + 2\sqrt{R}\cos\Theta}, \qquad k = \frac{-2\sqrt{R}\sin\Theta}{1 + R + 2\sqrt{R}\cos\Theta}. \tag{2.61}$$

For k to be positive, $\sin\Theta$ must be negative. The function $\log\tilde{r}$, defined by

$$\log\tilde{r} = \text{Log}\, r + i\Theta,$$

where $\text{Log}\, r = \ln r$, is analytic in the domain $r > 0$, $\pi \leqslant \Theta < 2\pi$. Because of causality, the function $\tilde{r}(\tilde{\omega})$ is analytic in the top half of the complex $\tilde{\omega}$ plane. If the values of $\tilde{r}(\tilde{\omega})$ lie in the domain of definition of $\log\tilde{r}$, then the function $\log\tilde{r}(\tilde{\omega})$ is also analytic in the top half of the complex $\tilde{\omega}$ plane. However, the asymptotic behavior of $\log\tilde{r}$ is unacceptable: we shall show in Chapter 9 that $\lim_{\omega \to \infty}\tilde{r}(\omega) = 1$. Once again, we are permitted to do some fiddling: the function

$$F(\omega) = \frac{\log\tilde{r}(\omega)}{\omega}$$

vanishes as ω increases without limit. Moreover, $F(\tilde{\omega})$ is analytic in the top half of the complex $\tilde{\omega}$ plane. Therefore, $F(\omega)$ satisfies the relation (2.35). The *phase shift dispersion relation* then follows from the crossing condition $F^*(\omega) = -F(-\omega)$:

$$\Theta(\omega) = \frac{-2\omega}{\pi} P \int_0^\infty \frac{\text{Log } r(\Omega)}{\Omega^2 - \omega^2} d\Omega. \tag{2.62}$$

Equation (2.62) is not merely a mathematical curiosity: from (2.61) it follows that at a given frequency ω the optical constants are determined by the reflectance and phase; however, if the reflectance is measured over a sufficiently large range of frequencies about ω, the phase can be obtained from (2.62).

2.7.2   Oblique Incidence

All plane waves normally incident on a plane boundary are reflected and transmitted according to (2.56) and (2.57) independently of their state of polarization (i.e., the direction of $\mathbf{E}_i$). This is analogous to scattering in the forward or backward directions by an isotropic sphere or a collection of randomly oriented particles, where the polarization is of no importance. However, when a plane wave is obliquely incident on a plane boundary, or when scattering angles other than forward or backward are considered, the polarization of the incident wave is indeed important. Incident unpolarized light may, upon reflection from a plane boundary or scattering by a particle, become highly polarized at certain angles. In treating reflection of a wave incident at an arbitrary angle on a plane boundary we therefore consider two polarizations: electric field vectors parallel and perpendicular to the *plane of incidence* (Fig. 2.5), defined by the direction of propagation of the incident wave and the normal to the boundary. An arbitrary wave may be written as the superposition of waves of these two kinds. Moreover, the two polarizations are independent of each other: if the incident wave is polarized parallel to the plane of incidence, for example, the reflected and transmitted waves are so polarized.

We first consider the electric vectors to be parallel to the plane of incidence. The tangential components of the electric and magnetic fields are required to be continuous across the boundary:

$$E_{\parallel i}\cos\Theta_i + E_{\parallel r}\cos\Theta_r = E_{\parallel t}\cos\Theta_t, \tag{2.63}$$

$$H_{\perp i} + H_{\perp r} = H_{\perp t}. \tag{2.64}$$

If we use (2.43) and also take the permeabilities to be equal in both media, (2.64) becomes

$$E_{\parallel i} - E_{\parallel r} = \frac{N_1}{N_2} E_{\parallel t}. \tag{2.65}$$

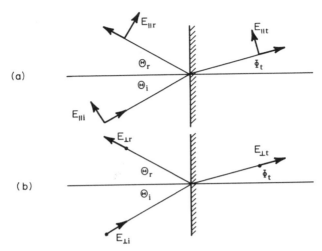

Figure 2.5 Reflection and transmission of obliquely incident light for electric vector parallel (*a*) and perpendicular (*b*) to the plane of incidence.

We also have $\Theta_r = \Theta_i$ from the law of specular reflection, and Θ_t is determined from the *generalized Snell's law*:

$$\sin \Theta_t = \frac{\sin \Theta_i}{m}. \tag{2.66}$$

Note that $\sin \Theta_t$, and hence Θ_t, is *complex* when medium 1 is absorbing. For such media the transmitted wave is inhomogeneous, although the incident and reflected waves are homogeneous (N_2 is real). In Fig. 2.5, Φ_t is the angle between the normal to the boundary and the direction of propagation of the surfaces of constant phase. The surfaces of constant amplitude are planes parallel to the boundary. If medium 1 is nonabsorbing, $\Theta_t = \Phi_t$. In general, however, the relation between Θ_t and geometrical properties of the transmitted wave is complicated. Rather than sink into the morass of trying to interpret a complex angle of refraction physically, it is much less frustrating to look upon Θ_t in (2.63) and (2.66) as merely a mathematical quantity.

Equations (2.63) and (2.65) are readily solved for the reflection and transmission coefficients

$$\tilde{r}_\| = \frac{E_{\|r}}{E_{\|i}} = \frac{\cos \Theta_t - m \cos \Theta_i}{\cos \Theta_t + m \cos \Theta_i}, \tag{2.67}$$

$$\tilde{t}_\| = \frac{E_{\|t}}{E_{\|i}} = \frac{2 \cos \Theta_i}{\cos \Theta_t + m \cos \Theta_i}. \tag{2.68}$$

When the electric vectors are perpendicular to the plane of incidence a similar

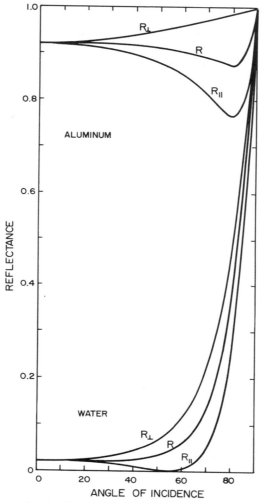

Figure 2.6 Reflectances for electric vector parallel ($R_\parallel$) and perpendicular ($R_\perp$) to the plane of incidence. R is the reflectance for unpolarized incident light.

analysis yields

$$\tilde{r}_\perp = \frac{E_{\perp r}}{E_{\perp i}} = \frac{\cos \Theta_i - m \cos \Theta_t}{\cos \Theta_i + m \cos \Theta_t}, \tag{2.69}$$

$$\tilde{t}_\perp = \frac{E_{\perp t}}{E_{\perp i}} = \frac{2 \cos \Theta_i}{\cos \Theta_i + m \cos \Theta_t}. \tag{2.70}$$

Equations (2.67)–(2.70) are the *Fresnel formulas* for reflection and transmission of light obliquely incident on a plane boundary.

The ratio of reflected to incident irradiance is more readily measured than the ratio of amplitudes; the reflectances for the two polarization states of the incident light are

$$R_{\parallel} = |\tilde{r}_{\parallel}|^2, \qquad R_{\perp} = |\tilde{r}_{\perp}|^2. \qquad (2.71)$$

One sometimes encounters the notation R_p and R_s for $R_{\parallel}$ and $R_{\perp}$, where the p and s stand for the German *parallel* and *senkrecht* (perpendicular); R_{π} and R_{σ} are also used. The two reflectances (2.71) and the reflectance $R = \frac{1}{2}(R_{\parallel} + R_{\perp})$ for incident unpolarized light are shown as a function of the angle of incidence Θ_i in Fig. 2.6. The refractive indices are those for liquid water at visible wavelengths ($N_1 \simeq 1.33$) and aluminum at a wavelength of 4958 Å($N_1 = 0.771 + i5.91$); medium 2 is taken to be a vacuum ($N_2 = 1$) in both cases. Note that for water, an insulator, $R_{\perp}$ is always nonzero, but there is an angle of incidence $\Theta_i = \Theta_p$ for which $R_{\parallel} = 0$. This angle is called the *polarizing* or *Brewster angle*, and it follows from (2.66) and (2.67) that

$$\tan \Theta_p = n.$$

For aluminum, a conductor, $R_{\parallel}$ is always nonzero, although it does have a minimum at a particular angle. If the angle of incidence is other than 0 or 90°, $R_{\perp}$ is greater than $R_{\parallel}$ for an insulator or a conductor; therefore, if unpolarized light is incident on a plane boundary, the reflected light is partially polarized perpendicularly to the plane of incidence. At the Brewster angle (for an insulator) the reflected light is completely polarized.

2.8   REFLECTION AND TRANSMISSION BY A SLAB

We now consider reflection and transmission of a wave $E_i\exp[i\omega(N_2 z/c - t)]$ normally incident on a plane-parallel slab of arbitrary material embedded in a nonabsorbing medium (Fig. 2.7). The reflected and transmitted waves are

$$E_r\exp\left[-i\omega\left(\frac{N_2 z}{c} + t\right)\right], \qquad E_t\exp\left[i\omega\left(\frac{N_2 z}{c} - t\right)\right],$$

and to satisfy all the boundary conditions we must postulate waves inside the slab that propagate in the $+z$ and $-z$ directions:

$$E_1^+\exp\left[i\omega\left(\frac{N_1 z}{c} - t\right)\right], \qquad E_1^-\exp\left[-i\omega\left(\frac{N_1 z}{c} + t\right)\right].$$

The field amplitudes are written as scalars because reflection and transmission at normal incidence are independent of polarization. At the first boundary ($z = 0$), the amplitudes satisfy the usual boundary conditions:

$$E_i + E_r = E_1^+ + E_1^-, \qquad E_i - E_r = \frac{N_1}{N_2}(E_1^+ - E_1^-),$$

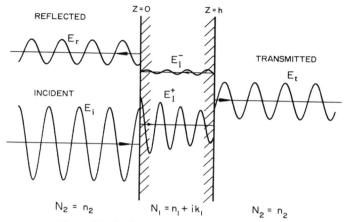

Figure 2.7   Reflection and transmission by a slab.

and at the second boundary $(z = h)$ we have

$$E_1^+ \exp(ikN_1 h) + E_1^- \exp(-ikN_1 h) = E_t \exp(ikN_2 h),$$

$$E_1^+ \exp(ikN_1 h) - E_1^- \exp(-ikN_1 h) = \frac{N_2}{N_1} E_t \exp(ikN_2 h),$$

where $k = \omega/c = 2\pi/\lambda$. The reflection and transmission coefficients are obtained by solving these four simultaneous equations:

$$\tilde{r}_{\text{slab}} = \frac{E_r}{E_i} = \frac{\tilde{r}[1 - \exp(i2kN_1 h)]}{1 - \tilde{r}^2 \exp(i2kN_1 h)}, \tag{2.72}$$

$$\tilde{t}_{\text{slab}} = \frac{E_t}{E_i} = \frac{4m}{(m+1)^2} \frac{\exp(-ikN_2 h)}{[\exp(-ikN_1 h) - \tilde{r}^2 \exp(ikN_1 h)]}, \tag{2.73}$$

where $\tilde{r}$ is the reflection coefficient (2.57). More algebra yields the slab transmittance:

$$T_{\text{slab}} = |\tilde{t}_{\text{slab}}|^2 = \frac{(1 - R)^2 + 4R \sin^2 \psi}{R^2 e^{-\alpha h} + e^{\alpha h} - 2R \cos(\zeta + 2\psi)}, \tag{2.74}$$

where

$$\psi = \tan^{-1}\left(\frac{2n_2 k_1}{n_1^2 + k_1^2 - n_2^2}\right) \qquad 0 \leqslant \psi \leqslant \pi,$$

$$\zeta = \frac{4\pi n_1 h}{\lambda}, \qquad \alpha = \frac{4\pi k_1}{\lambda}, \qquad R = |\tilde{r}|^2.$$

Transmission experiments are usually carried out with the slab in air ($n_2 \simeq 1$) and are feasible only if a measurable amount of light is transmitted. This in turn requires that for all but very thin samples (say, $h \ll \lambda$) k_1 must be small compared with unity. With this restriction ψ is small compared with ζ, and the slab transmittance is to good approximation given by

$$T_{\text{slab}} = \frac{(1 - R)^2}{R^2 e^{-\alpha h} + e^{\alpha h} - 2R \cos \zeta}. \qquad (2.75)$$

Interference bands in slab transmission *may* be observed because of the oscillatory term $\cos \zeta$ in (2.75). Transmission maxima occur when $\zeta = 2\pi p$ ($p\lambda = 2n_1 h$), where $p = 1, 2, 3, \ldots$. If a maximum occurs at a wavelength λ, neighboring maxima occur at $\lambda + \Delta\lambda$, where

$$\frac{\Delta\lambda}{\lambda} = \pm \frac{\lambda}{2n_1 h}\left(1 \mp \frac{\lambda}{2n_1 h}\right)^{-1},$$

provided that n_1 does not vary greatly over this interval.

We have taken the incident beam to be *perfectly monochromatic* up to this point. Such beams are produced easily enough on paper—a few strokes of the pen are sufficient! However, real beams with which experimenters must contend are composed of a finite spread, albeit narrow, of wavelengths. Thus, for practical purposes, T_{slab} in (2.74) should be looked upon as an average over a range of wavelengths. If the wavelength spread of the beam is $\delta\lambda$, interference bands are observable only if $\delta\lambda \ll |\Delta\lambda|$. Another tacit assumption is that k_1 does not vary greatly over the wavelength region of interest; if not, the absorption spectrum may obscure transmission maxima and minima that are the result of interference. This is obvious from inspection of (2.75): for sufficiently large absorption (αh) the oscillatory term is negligible compared with $\exp(\alpha h)$. Interference bands in transmission by a thin MnS crystal are shown in Fig. 2.8, where *optical density*, defined as $\log_{10}(1/T_{\text{slab}})$, is plotted as a function of wavelength. Note that in the vicinity of the absorption band at about 6000 Å the interference bands are damped somewhat.

If $\delta\lambda \gg |\Delta\lambda|$, distinct interference bands in the transmission spectrum will not be observed; in this case T_{slab} averaged over the wavelength interval $\delta\lambda$ is

$$T_{\text{slab}} = \frac{(1 - R)^2 e^{-\alpha h}}{1 - R^2 e^{-2\alpha h}}. \qquad (2.76)$$

The strict validity of (2.76) also requires that $\delta\lambda/\lambda \ll 1$, where λ is the average wavelength in the interval $\delta\lambda$; this ensures that the maxima of $\cos \zeta$ are evenly spaced.

Even if $\delta\lambda \ll |\Delta\lambda|$, this is not sufficient to ensure that the interference bands predicted by (2.75) are experimentally observable. This equation also

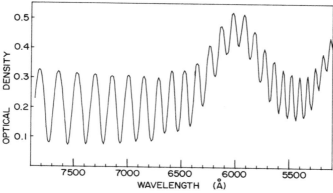

Figure 2.8 Transmission by an MnS crystal ($h \simeq 6.3$ μm, $n_1 \simeq 2.68$) measured with a Cary 14 spectrophotometer ($\delta\lambda \simeq 5$ Å).

contains the implicit requirement that $\delta h \ll \lambda/4n_1$, where δh is the average departure from parallelism of the faces of the slab over the area illuminated by the incident beam. We shall encounter a similar restriction when we deal with scattering by spheres: maxima and minima in the scattering cross section for a suspension of spheres are observable only if the dispersion of sphere radii is sufficiently narrow.

It is sometimes desirable to eliminate interference bands from transmission spectra; this is particularly true in the measurement of absorption spectra (α as a function of wavelength), where interference bands can be a nuisance. If $\delta h \gg \lambda/4n_1$, a condition that can be achieved by using a wedge-shaped sample, for example, then the observed transmittance is given by (2.76).

Yet another implicit assumption underlying both (2.75) and (2.76), and indeed all expressions in this section, is that the surfaces are *optically smooth*; that is, the surface roughness is sufficiently small compared with the wavelength of the incident light. It is difficult to state a precise criterion for smoothness, but according to the Rayleigh criterion a surface is reckoned to be smooth if $d < \lambda/(8\cos\Theta)$, where d is the height of surface irregularities (Beckmann and Spizzichino, 1963, p. 9). If the surfaces are not smooth, incident light may be *diffusely reflected* over a distribution of angles rather than *specularly reflected* at a single angle.

Equations (2.75) and (2.76) represent two extremes. The former applies to a "perfectly" monochromatic beam incident on a "perfectly" parallel, smooth slab (although perfection in this sense is not absolute but lies within certain tolerance limits). On the other hand, the latter equation is appropriate to what might be called the perfectly imperfect case: a slab–beam combination that has been carefully prepared to eliminate *all* interference effects. Theory has little to say about cases intermediate between these two extremes. Thus, if quantitative data are to be extracted from transmission measurements, some care must be taken. It takes little experimental ability to insert a sample into a spectropho-tometer and press the scan button. Some kind of spectrum will dutifully emerge from the instrument. But is is an entirely different matter to extract from such spectra accurate numerical values for n and k.

Although (2.76) can be derived from (2.75) by integration, a more physically satisfying approach is to consider the multiple reflections and transmissions of an incident beam. The assumption of nonparallel slab faces or a sufficient spread of wavelengths is equivalent to assuming that we may deal with the beam without regard to its phase. Consider a beam with irradiance I_i that is incident on a slab (Fig. 2.9). A fraction R of the incident light is reflected at the first interface and the unreflected fraction traverses the slab undergoing attenuation by a factor $\exp(-\alpha h)$. At the second interface part of the light is reflected and an amount $I_i(1 - R)^2\exp(-\alpha h)$ is transmitted. The reflected light traverses the slab, is reflected at the first interface, again traverses the slab, and an amount $I_i(1 - R)^2 R^2\exp(-3\alpha h)$ emerges—and so on into the long hours of night. In this manner we obtain the total transmitted irradiance

$$I_t = I_i(1 - R)^2 e^{-\alpha h}\left(1 + R^2 e^{-2\alpha h} + R^4 e^{-4\alpha h} + \cdots\right),$$

and the infinite series is readily summed to yield (2.76). One further approxi-mation is useful: if $R^2 e^{-2\alpha h}$ is small compared with unity, which is true for the

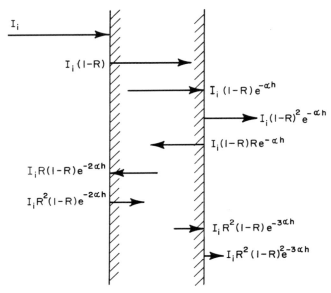

Figure 2.9 Transmission by a slab. Only the first two components that contribute to the total are shown. There is no systematic relationship among the phases of the various components.

large class of nonconductors at visible wavelengths, then (2.76) becomes

$$T_{\text{slab}} = (1 - R)^2 e^{-\alpha h}.$$

2.9   EXPERIMENTAL DETERMINATION OF OPTICAL CONSTANTS

The optical constants n and k are not directly measurable but must be derived from measurable quantities (e.g., reflection and transmission coefficients, reflectances and transmittances, and angles of refraction) by interposing a suitable theory. Most of the necessary theory has been discussed in the preceding two sections. Therefore, before considering the analogy between a particle and a slab, it is worthwhile to consider briefly how the solutions to electromagnetic boundary-value problems for plane surfaces can be used to solve the inversion problem for homogeneous media, that is, how to determine the optical constants. At a given wavelength there are, in general, two optical constants to be determined. This is turn implies that two or more measurements are required. The following methods may be used.

1. Measurement of refraction angles, such as the angle of minimum deviation of a prism; n is obtained from Snell's law. This requires samples of high transparency ($k \simeq 0$).

2. Measurement of the transmittance and reflectance of a slab for light at near-normal incidence. The samples must be sufficiently transparent for measurable transmission in thin slabs, but not as transparent as required in method 1.

3. Measurement of reflectance at near-normal incidence over a wide range of frequencies. The phase shift of the reflected light is obtained from a Kramers–Kronig analysis. This technique is of great value in spectral regions where the sample is highly opaque, but requires measurements over an extended region and extrapolations into unmeasured regions.

4. Ellipsometric techniques in which amplitude ratios *and* phase shifts for reflected light are directly measured as opposed to the previous technique in which the phase shift is indirectly obtained. This is difficult to do over large wavelength regions because of requirements on optical elements such as polarizers and retarders.

5. Measurement of reflectances for incident light of various polarization states and two oblique angles of incidence; the results are analyzed with the Fresnel formulas. Large angles are required for high accuracy, and this requires large sample surfaces.

These methods and their variations are the principal means of measuring optical constants. As noted, none of them is clearly superior in all instances.

Further details may be found in the references listed at the ends of this chapter and Chapter 10.

2.10   THE ANALOGY BETWEEN A SLAB AND A PARTICLE

We have seen that the response of a plane surface to an incident beam can be rather complicated. Yet we are familiar with many of the resulting phenomena because of everyday experiences with them: reflection of sunlight from a smooth pond or a glass window; a shiny piece of metal; the colors of a soap film. Many readers probably have performed simple laboratory experiments such as measuring the dispersion of natural light by a prism or determining the Brewster angle using polarizers. All these contacts help to build intuition about bulk optical effects. By way of contrast, we have fewer opportunities to observe directly scattering and absorption by small particles the size of which may be comparable with the wavelength. Because of useful analogies between the more often experienced and inherently simpler phenomena and those less commonly experienced, we encourage the use of such analogies while recommending a good bit of caution. Many have come to grief by adhering too strongly to the analogy between small particles and plane surfaces. A classical example is the unnecessary confusion that occurred during the period in which the blue color and polarization of skylight were topics of keen scientific interest. Because of Tyndall's experiments, small particles, which we now know to be the air molecules themselves, were a strong candidate for the source of the blue sky. But one of the obstacles to the acceptance of this explanation was the polarizing angle. It was observed that the light scattered by small particles at 90° to the direction of the incident beam was highly polarized regardless of the composition of the particles. In the language of planar surfaces we would say that the polarizing angle for such particles is always 45°. This was considered by some to be a problem: the polarizing angle for planar surfaces is a function of refractive index and, hence, composition. Rayleigh's response (1871) to this objection is worth quoting:

> I venture to think that the difficulty is imaginary, and is caused mainly by misuse of the word reflection. Of course there is nothing in the etymology of reflection and refraction to forbid their application in this sense; but the words have acquired technical meanings, and become associated with certain well-known laws called after them. Now a moment's consideration of the principles according to which reflection and refraction are explained in the wave theory is sufficient to show that they have no application unless the surface of the disturbing body is larger than many square wave-lengths; whereas the particles to which the sky is supposed to owe its illumination must be *smaller* than the wave-length or else the explanation of the colour breaks down. The idea of polarization by reflection is therefore out of place, and that "the law of Brewster does not apply to matter in this condition" (of extreme fineness) is only what might have been inferred from the principles of the wave theory.

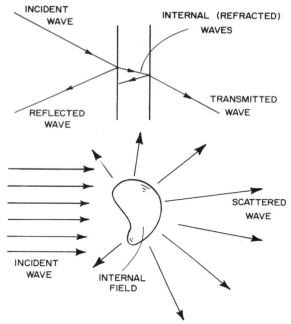

Figure 2.10 Analogy between scattering by a particle and reflection–transmission by a slab.

Rayleigh adds in a footnote that "in many departments of science a tendency may be observed to extend the field of familiar laws beyond their proper limits." It is well to keep Rayleigh's words in mind when comparison is made between slabs and particles.

The formal analogy between scattering by a particle and reflection–transmission by a slab, in its most general aspect, is shown schematically in Fig. 2.10 and in Table 2.1. We mentioned in Section 2.7 that unpolarized incident light may become polarized upon reflection from a plane interface *or* by scattering from a particle; throughout the rest of this book we shall be alert to further analogies, both as an aid to understanding and as a guide into the investigation of new and unfamiliar phenomena.

Table 2.1 Analogy between Slab and Particle

Slab	Particle
Incident wave	Incident wave
Reflected wave + transmitted wave	Scattered wave
Internal (refracted) waves	Internal field

2.11   POLARIZATION

In addition to irradiance and frequency, a monochromatic (i.e., time-harmonic) electromagnetic wave has a property called its *state of polarization*, a property that was briefly touched on in Section 2.7, where it was shown that the reflectance of obliquely incident light depends on the polarization of the electric field. In fact, polarization would be an uninteresting property were it not for the fact that two waves with identical frequency and irradiance, but different polarization, can behave quite differently. Before we leave the subject of plane waves it is desirable to present polarization in a systematic way, which will prove to be useful when we discuss the polarization of scattered light.

Consider a plane monochromatic wave with angular frequency ω and wave number k which is propagating in the z direction in a nonabsorbing medium. In discussions of polarization it is customary to focus attention on the electric field **E**:

$$\mathbf{E} = \mathrm{Re}\{\mathbf{E}_c\} = \mathrm{Re}\{(\mathbf{A} + i\mathbf{B})\exp(ikz - i\omega t)\}$$

$$= \mathbf{A}\cos(kz - \omega t) - \mathbf{B}\sin(kz - \omega t), \tag{2.77}$$

where the real vectors **A** and **B** are independent of position. The electric field vector at any point lies in a plane the normal to which is parallel to the direction of propagation. In a particular plane, say $z = 0$ for convenience, the tip of the electric vector traces out a curve:

$$\mathbf{E}(z = 0) = \mathbf{A}\cos \omega t + \mathbf{B}\sin \omega t. \tag{2.78}$$

Equation (2.78) describes an ellipse, the *vibration ellipse* (Fig. 2.11). If $\mathbf{A} = 0$ (or $\mathbf{B} = 0$), the vibration ellipse is just a straight line, and the wave is said to be *linearly polarized*; the vector **B** then specifies the *direction of vibration*. (The term *plane polarized* is also used, but it has become less fashionable in recent years.) If $|\mathbf{A}| = |\mathbf{B}|$ and $\mathbf{A} \cdot \mathbf{B} = 0$, the vibration ellipse is a circle, and the wave is said to be *circularly polarized*. In general, a monochromatic wave of the form (2.77) is *elliptically polarized*.

A given vibration ellipse can be traced out in two opposite senses: clockwise and anticlockwise. The vibration ellipse in Fig. 2.11 is rotating clockwise as viewed from above the page. However, as viewed from the opposite direction, it is rotating in the anticlockwise sense. Thus, these terms do not have absolute meaning but depend on the direction from which the ellipse is observed. The two opposite senses of rotation lead to a classification of vibration ellipses according to their *handedness*, and herein lies a problem: there are two conventions for assigning handedness to vibration ellipses. On the one hand, the vibration ellipse may be designated as right-handed if the rotation is clockwise as viewed by an observer who is looking toward the source of light. That is, if the direction of the vector $\mathbf{A} \times \mathbf{B}$ is opposite to the direction of

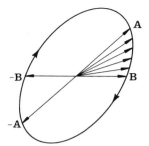

Figure 2.11 Vibration ellipse.

propagation, the vibration ellipse is said to be right-handed. Among the adherents to this convention are chemists (Djerassi, 1960) and optical physicists (Jenkins and White, 1957; Shurcliff, 1962; Stone, 1963). This convention might be called "traditional" and, as such, has been adopted by Born and Wolf (1965, p. 28), who nonetheless are slightly uncomfortable with the seeming unnaturalness of their choice—it is not compatible with the rotation behavior of a right-handed screw. Clarke (1974), however, has argued that the traditional convention need not tread on notions of what is or is not natural if we lay aside our screwdrivers and concentrate instead on the helix traced out in space by the electric field. At any instant of time (say, $t = 0$) the locus of all points described by the tip of the electric field vector is

$$\mathbf{E}(t = 0) = \mathbf{A} \cos kz - \mathbf{B} \sin kz, \qquad (2.79)$$

which is the equation of a helix; (2.79) is a "snapshot" of the electric field at a particular time. With increasing time the helix moves in the direction of propagation and, in so doing, its intersection with any plane $z = $ constant describes a vibration ellipse (Fig. 2.12). If the helix is right-handed, the corresponding vibration ellipse is also right-handed according to the traditional convention. Because the handedness of a helix is independent of the direction from which it is observed, the helix associated with a given wave unambiguously assigns a handedness to that wave. We therefore adopt the traditional convention according to which an elliptically polarized wave is reckoned right-handed if the vibration ellipse is rotating in the clockwise sense as viewed by an observer looking toward the source. The opposite convention seems to be favored by astronomers (van de Hulst, 1957; Hansen and Travis, 1974; Gehrels, 1974a).

We shall be concerned primarily with media through which plane waves of arbitrary polarization propagate without change of polarization state. However, there are many materials that do not possess this property. For example, there are materials which, for a given direction of propagation, have different refractive indices depending on the state of linear polarization of the wave. If the real parts of the refractive indices are different, the material is said to be *linearly birefringent*; if the imaginary parts of the refractive indices are differ-

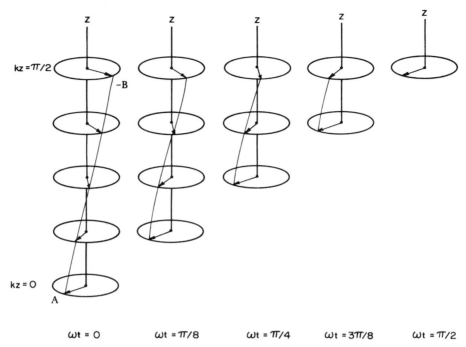

$kz = \pi/2$

$-B$

$kz = 0$

A

$\omega t = 0$ $\omega t = \pi/8$ $\omega t = \pi/4$ $\omega t = 3\pi/8$ $\omega t = \pi/2$

Figure 2.12 A series of snapshots of the electric field.

ent, it is said to be *linearly dichroic*. Similarly, there are *circularly birefringent* and *circularly dichroic* media, those for which the complex refractive index depends on handedness. The terms "birefringent" and "dichroic" are often used without qualification, particularly if their meaning is clear from the context. To describe such birefringent and dichroic media at the phenomenological level, the constitutive relations (2.7)–(2.9) must be modified somewhat. We shall encounter specific examples in later chapters.

In addition to its handedness, a vibration ellipse is characterized by its *ellipticity*, the ratio of the length of its semiminor axis to that of its semimajor axis, and its *azimuth*, the angle between the semimajor axis and an arbitrary reference direction (Fig. 2.13). Handedness, ellipticity, and azimuth, together with irradiance, are the *ellipsometric parameters* of a plane wave.

2.11.1 Stokes Parameters

Although the ellipsometric parameters completely specify a monochromatic wave of given frequency and are readily visualized, they are not particularly conducive to understanding the transformations of polarized light. Moreover, they are difficult to measure directly (with the exception of irradiance, which can easily be measured with a suitable detector) and are not adaptable to a

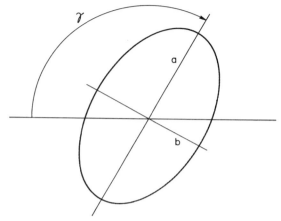

Figure 2.13   Vibration ellipse with ellipticity b/a and azimuth γ.

discussion of partially polarized light. The irradiances of two incoherently superposed beams are additive, but no such additivity exists for the other three ellipsometric parameters. As we shall see, the *Stokes parameters* are an equivalent description of polarized light, but one of greater usefulness, particularly in scattering problems.

In previous sections we stated that the polarization state of a wave may be changed by interaction with a suitable optical system (e.g., reflection at the polarizing angle or transmission through a dichroic medium). An arbitrary monochromatic wave may be expressed as a superposition of two *orthogonal* components: horizontal and vertical; right-circular and left-circular; and so on. This decomposition is more than just a mathematical device: we can construct polarizers that have the property of transmitting only one of these components. For the moment we regard such a polarizer P as a black box into which light is fed.

Let us consider a series of hypothetical experiments which can be performed with an arbitrary monochromatic beam, a detector, and various polarizers (Fig. 2.14). The detector responds to irradiance independently of the polarization state, and the polarizers are assumed to be ideal: they do not change the amplitude of the transmitted component. The electric field $\mathbf{E}$ referred to orthogonal axes $\hat{\mathbf{e}}_{\parallel}$ and $\hat{\mathbf{e}}_{\perp}$, which we shall call "horizontal" and "vertical," respectively, is

$$\mathbf{E} = \mathbf{E}_0 \exp(ikz - i\omega t); \qquad \mathbf{E}_0 = E_{\parallel} \hat{\mathbf{e}}_{\parallel} + E_{\perp} \hat{\mathbf{e}}_{\perp},$$

$$E_{\parallel} = a_{\parallel} e^{-i\delta_{\parallel}}; \qquad E_{\perp} = a_{\perp} e^{-i\delta_{\perp}}.$$

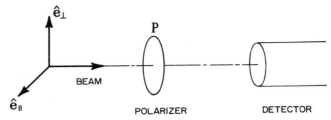

Figure 2.14   The detector measures the irradiance of the beam transmitted by the polarizer P.

Experiment I. No Polarizer.   If there is no polarizer in the beam the irradiance I recorded by the detector is $E_\parallel E_\parallel^* + E_\perp E_\perp^*$, where for convenience we omit the factor $k/2\omega\mu_0$.

Experiment II. Horizontal and Vertical Polarizers.   (1) Let P be a horizontal polarizer; the amplitude of the transmitted wave is $E_\parallel$ and the irradiance $I_\parallel$ recorded by the detector is $E_\parallel E_\parallel^*$. (2) Let P be a vertical polarizer; the amplitude of the transmitted wave is $E_\perp$ and the irradiance $I_\perp$ recorded by the detector is $E_\perp E_\perp^*$. The difference between these two measured irradiances is

$$I_\parallel - I_\perp = E_\parallel E_\parallel^* - E_\perp E_\perp^* .$$

Experiment III. +45° and −45° Polarizers.   To analyze this experiment it is convenient to introduce another orthonormal set of basis vectors $\hat{e}_+$ and $\hat{e}_-$, which are obtained by rotating $\hat{e}_\parallel$ by $+45°$ and $-45°$ (Fig. 2.15):

$$\hat{e}_+ = \frac{1}{\sqrt{2}}(\hat{e}_\parallel + \hat{e}_\perp), \qquad \hat{e}_- = \frac{1}{\sqrt{2}}(\hat{e}_\parallel - \hat{e}_\perp).$$

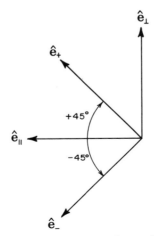

Figure 2.15   Basis vectors $\hat{e}_+$ and $\hat{e}_-$.

The electric field $\mathbf{E}_0$ may be written $\mathbf{E}_0 = E_+\hat{\mathbf{e}}_+ + E_-\hat{\mathbf{e}}_-$, where

$$E_+ = \frac{1}{\sqrt{2}}(E_\parallel + E_\perp), \qquad E_- = \frac{1}{\sqrt{2}}(E_\parallel - E_\perp).$$

(1) Let P be a $+45°$ polarizer; the amplitude of the transmitted wave is $(E_\parallel + E_\perp)/\sqrt{2}$ and its irradiance I_+ is $(E_\parallel E_\parallel^* + E_\parallel E_\perp^* + E_\perp E_\parallel^* + E_\perp E_\perp^*)/2$. (2) Let P be a $-45°$ polarizer; the irradiance of the transmitted wave is $I_- = (E_\parallel E_\parallel^* - E_\parallel E_\perp^* - E_\perp E_\parallel^* + E_\perp E_\perp^*)/2$. The difference between these two irradiances is

$$I_+ - I_- = E_\parallel E_\perp^* + E_\perp E_\parallel^*.$$

Experiment IV. Circular Polarizers. We need to introduce one more set of basis vectors $\hat{\mathbf{e}}_R$ and $\hat{\mathbf{e}}_L$:

$$\hat{\mathbf{e}}_R = \frac{1}{\sqrt{2}}(\hat{\mathbf{e}}_\parallel + i\hat{\mathbf{e}}_\perp), \qquad \hat{\mathbf{e}}_L = \frac{1}{\sqrt{2}}(\hat{\mathbf{e}}_\parallel - i\hat{\mathbf{e}}_\perp).$$

These basis vectors represent right-circularly and left-circularly polarized waves and are orthonormal in the sense that

$$\hat{\mathbf{e}}_R \cdot \hat{\mathbf{e}}_R^* = 1, \qquad \hat{\mathbf{e}}_L \cdot \hat{\mathbf{e}}_L^* = 1, \qquad \hat{\mathbf{e}}_R \cdot \hat{\mathbf{e}}_L^* = 0.$$

The incident field may be written $\mathbf{E}_0 = E_R\hat{\mathbf{e}}_R + E_L\hat{\mathbf{e}}_L$, where

$$E_R = \frac{1}{\sqrt{2}}(E_\parallel - iE_\perp), \qquad E_L = \frac{1}{\sqrt{2}}(E_\parallel + iE_\perp).$$

(1) Let P be a right-handed polarizer; the transmitted irradiance I_R is $(E_\parallel E_\parallel^* - iE_\parallel^* E_\perp + iE_\perp^* E_\parallel + E_\perp E_\perp^*)/2$. (2) Let P be a left-handed polarizer; the transmitted irradiance I_L is $(E_\parallel E_\parallel^* + iE_\perp E_\parallel^* - iE_\parallel E_\perp^* + E_\perp E_\perp^*)/2$. The difference between these two irradiances is

$$I_R - I_L = i(E_\perp^* E_\parallel - E_\parallel^* E_\perp).$$

We have now done enough thought experiments to determine the *Stokes parameters* I, Q, U, V:

$$I = E_\parallel E_\parallel^* + E_\perp E_\perp^* = a_\parallel^2 + a_\perp^2,$$

$$Q = E_\parallel E_\parallel^* - E_\perp E_\perp^* = a_\parallel^2 - a_\perp^2,$$

$$U = E_\parallel E_\perp^* + E_\perp E_\parallel^* = 2a_\parallel a_\perp \cos \delta, \qquad (2.80)$$

$$V = i(E_\parallel E_\perp^* - E_\perp E_\parallel^*) = 2a_\parallel a_\perp \sin \delta,$$

where the phase difference δ is $\delta_{\parallel} - \delta_{\perp}$. Note that we have omitted the factor $k/2\omega\mu_0$ from (2.80); it is unimportant because relative, rather than absolute, irradiances are what are usually measured. Our notation for the Stokes parameters is by no means universal: there are many other symbols in use. Stokes (1852) himself used A, B, C, D; I, Q, U, V are favored by Walker (1954), Chandrasekhar (1950), and van de Hulst (1957); Perrin (1942) and Shurcliff (1962) find I, M, C, S more to their liking; Collett (1968) prefers s_0, s_1, s_2, s_3. Rozenberg (1960) makes no attempt to conceal his contempt for the "irrationality" of the symbols usually employed in the "foreign literature," and he offers us S_1, S_2, S_3, S_4. To add to the confusion, the definition of the Stokes parameters can be changed without serious damage: various linear combinations of the Stokes parameters (2.80), particularly I and Q, can be, and are, used as suitable Stokes parameters. Caution is therefore in order when one leaves the pages of this book.

The Stokes parameters are related to the ellipsometric parameters as follows:

$$I = c^2,$$

$$Q = c^2 \cos 2\eta \cos 2\gamma,$$

$$U = c^2 \cos 2\eta \sin 2\gamma, \tag{2.81}$$

$$V = c^2 \sin 2\eta,$$

where

$$c^2 = a^2 + b^2 = (\text{semimajor axis})^2 + (\text{semiminor axis})^2,$$

$$\gamma = \text{clockwise angle between } \hat{\mathbf{e}}_{\parallel} \text{ and major axis } (azimuth)\ (0 \leqslant \gamma \leqslant \pi),$$

$$|\tan \eta| = \frac{b}{a}\ (ellipticity)\left(-\frac{\pi}{4} \leqslant \eta \leqslant \frac{\pi}{4}\right).$$

The sign of V specifies the handedness of the vibration ellipse: positive denotes right-handed and negative denotes left-handed. We also have the relations

$$\tan 2\gamma = \frac{U}{Q}, \qquad \tan 2\eta = \frac{V}{\sqrt{Q^2 + U^2}}. \tag{2.82}$$

Thus, the Stokes parameters are equivalent to the ellipsometric parameters; although less easily visualized, they are operationally defined in terms of measurable quantities (irradiances). Additional advantages of the Stokes parameters will become evident as we proceed. Note that Q and U depend on the choice of horizontal and vertical directions. If the basis vectors $\hat{\mathbf{e}}_{\parallel}$ and $\hat{\mathbf{e}}_{\perp}$

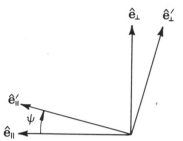

Figure 2.16   Rotation of basis vectors.

are rotated through an angle ψ (Fig. 2.16), the transformation from (I, Q, U, V) to Stokes parameters (I', Q', U', V') relative to the rotated axes $\hat{\mathbf{e}}'_\parallel$ and $\hat{\mathbf{e}}'_\perp$ is

$$\begin{pmatrix} I' \\ Q' \\ U' \\ V' \end{pmatrix} = \begin{pmatrix} 1 & 0 & 0 & 0 \\ 0 & \cos 2\psi & \sin 2\psi & 0 \\ 0 & -\sin 2\psi & \cos 2\psi & 0 \\ 0 & 0 & 0 & 1 \end{pmatrix} \begin{pmatrix} I \\ Q \\ U \\ V \end{pmatrix}. \tag{2.83}$$

It is fairly obvious from either (2.81) or (2.83) that there are three quantities associated with the Stokes parameters that are invariant under rotation of the reference directions: I, $Q^2 + U^2$, and V. In addition, the Stokes parameters are not all independent:

$$I^2 = Q^2 + U^2 + V^2.$$

A few representatives sets of Stokes parameters, written as column vectors, are shown in Table 2.2; the irradiance I is normalized to unity.

Although a strictly monochromatic wave, one for which the time dependence is $\exp(-i\omega t)$, has a well-defined vibration ellipse, not all waves do. Let us consider a nearly monochromatic, or *quasi-monochromatic* beam:

$$\mathbf{E} = \mathbf{E}_0(t)\exp(ikz - i\omega t), \qquad \mathbf{E}_0(t) = E_\parallel(t)\hat{\mathbf{e}}_\parallel + E_\perp(t)\hat{\mathbf{e}}_\perp,$$

where the complex amplitudes $E_\parallel$ and $E_\perp$ are now functions of time but vary slowly over time intervals of the order of the period $2\pi/\omega$. However, for time intervals long compared with the period, the amplitudes fluctuate in some manner, perhaps independently of each other, or perhaps with some correlation. If $E_\parallel(t)$ and $E_\perp(t)$ are completely *uncorrelated*, the beam is said to be *unpolarized*; so-called natural light (e.g., light from the sun, incandescent and fluorescent lamps) is unpolarized. In such a beam of light the electric vector traces out a vibration ellipse the parameters of which—handedness, ellipticity, and azimuth—vary slowly in time. Moreover, there is no preferred vibration ellipse: over a sufficiently long period of time vibration ellipses of all shapes, orientations, and handedness will have been traced out. Hurwitz (1945) has

Table 2.2   Stokes Parameters for Polarized Light

Linearly Polarized

0°	90°	+45°	−45°	γ
↔	↕	↘	↗	
$\begin{pmatrix} 1 \\ 1 \\ 0 \\ 0 \end{pmatrix}$	$\begin{pmatrix} 1 \\ -1 \\ 0 \\ 0 \end{pmatrix}$	$\begin{pmatrix} 1 \\ 0 \\ 1 \\ 0 \end{pmatrix}$	$\begin{pmatrix} 1 \\ 0 \\ -1 \\ 0 \end{pmatrix}$	$\begin{pmatrix} 1 \\ \cos 2\gamma \\ \sin 2\gamma \\ 0 \end{pmatrix}$

Circularly Polarized

Right	Left
↻	↺
$\begin{pmatrix} 1 \\ 0 \\ 0 \\ 1 \end{pmatrix}$	$\begin{pmatrix} 1 \\ 0 \\ 0 \\ -1 \end{pmatrix}$

discussed the statistical properties of unpolarized light in an interesting and instructive paper. If $E_\parallel$ and $E_\perp$ are *completely correlated*, the light is said to be *polarized*. This definition of polarization includes strictly monochromatic light but is somewhat more general: $a_\parallel$, $a_\perp$, $\delta_\parallel$, $\delta_\perp$ may separately fluctuate provided that the ratio $a_\parallel/a_\perp$ of the real amplitudes and the phase difference $\delta_\parallel - \delta_\perp$ are independent of time. If $E_\parallel$ and $E_\perp$ are *partially correlated*, the light is said to be *partially polarized*. A partially polarized beam exhibits a preference for handedness, or ellipticity, or azimuth. But this preference is not perfect: there is some statistical fluctuation.

The Stokes parameters of a quasi-monochromatic beam are given by

$$I = \langle E_\parallel E_\parallel^* + E_\perp E_\perp^* \rangle,$$

$$Q = \langle E_\parallel E_\parallel^* - E_\perp E_\perp^* \rangle,$$

$$U = \langle E_\parallel E_\perp^* + E_\perp E_\parallel^* \rangle, \qquad (2.84)$$

$$V = i \langle E_\parallel E_\perp^* - E_\perp E_\parallel^* \rangle,$$

where the angular brackets indicate time averages over an interval long compared with the period. From (2.84) it follows that

$$Q^2 + U^2 + V^2 = I^2 - 4\left(\langle a_\parallel^2 \rangle \langle a_\perp^2 \rangle - \langle a_\parallel a_\perp e^{i\delta} \rangle \langle a_\parallel a_\perp e^{-i\delta} \rangle \right),$$

which implies that

$$I^2 \geqslant Q^2 + U^2 + V^2. \qquad (2.85)$$

Equality holds if the light is polarized; if the light is unpolarized, $Q = U = V = 0$. The inequality (2.85) leads naturally to the notion of *degree of polarization* $\sqrt{Q^2 + U^2 + V^2}/I$, as well as *degree of linear polarization* $\sqrt{Q^2 + U^2}/I$, and *degree of circular polarization* V/I. For a partially polarized beam the sign of V indicates the preferential handedness of the vibration ellipses traced out by the electric vector: positive indicates right-handed and negative indicates left-handed. We may interpret the quantities U/Q and $V/\sqrt{Q^2 + U^2}$ [see (2.82)] as specifying the preferential azimuth and ellipticity of the vibration ellipses.

If two or more quasi-monochromatic beams propagating in the same direction are superposed *incoherently*, that is to say, there is no fixed relationship among the phases of the separate beams, the total irradiance is merely the sum of the individual beam irradiances. Because the definition of the Stokes parameters involves only irradiances, it follows that the Stokes parameters of a collection of incoherent sources are additive.

In the derivations above of the Stokes parameters we began with monochromatic light and then extended our results to the more general case of quasi-monochromatic light. However, the operational definition of the Stokes parameters in terms of a set of elementary experiments involving a detector and various polarizers, as opposed to the formal mathematical definitions (2.80) and (2.84), is independent of any assumed properties of the beam. Unless otherwise stated, we shall assume that all beams of interest are quasi-monochromatic, which includes as a special case monochromatic light.

2.11.2 Mueller Matrices

We may represent a beam of arbitrary polarization, including partially polarized light, by a column vector, the *Stokes vector*, the four elements of which are the Stokes parameters. In general, the state of polarization of a beam is changed on interaction with an optical element (e.g., polarizer, retarder, reflector, scatterer). Thus, it is possible to represent such optical elements by a 4 × 4 matrix (Mueller, 1948). The *Mueller matrix* describes the relation between "incident" and "transmitted" Stokes vectors; by "incident" is meant before interaction with the optical element, and by "transmitted" is meant after interaction. As an example, consider the Mueller matrix for an *ideal linear polarizer*. Such a polarizer transmits, without change of amplitude, only electric field components parallel to a particular axis called the *transmission axis*. Electric field components in other directions are completely removed from the transmitted beam by some means which we need not explicitly consider. The relation between incident field components ($E_{\parallel i}$, $E_{\perp i}$) and field components ($E_{\parallel t}$, $E_{\perp t}$) transmitted by the polarizer is

$$\begin{pmatrix} E_{\parallel t} \\ E_{\perp t} \end{pmatrix} = \begin{pmatrix} \cos^2 \xi & \sin \xi \cos \xi \\ \sin \xi \cos \xi & \sin^2 \xi \end{pmatrix} \begin{pmatrix} E_{\parallel i} \\ E_{\perp i} \end{pmatrix}, \qquad (2.86)$$

where ξ is the (smallest) angle between $\hat{e}_\parallel$ and the transmission axis. After a bit of algebra, we obtain from (2.86) the Mueller matrix for an ideal linear polarizer:

$$\frac{1}{2}\begin{pmatrix} 1 & \cos 2\xi & \sin 2\xi & 0 \\ \cos 2\xi & \cos^2 2\xi & \cos 2\xi \sin 2\xi & 0 \\ \sin 2\xi & \sin 2\xi \cos 2\xi & \sin^2 2\xi & 0 \\ 0 & 0 & 0 & 0 \end{pmatrix}. \tag{2.87}$$

The irradiance transmitted by the linear polarizer (2.87) is

$$I_t = \tfrac{1}{2}(I_i + Q_i \cos 2\xi + U_i \sin 2\xi).$$

Thus, as the polarizer is rotated so that ξ varies, the irradiance I_t also varies. The maximum and minimum values of I_t occur for $\xi = \gamma$ and $\xi = \gamma + \pi/2$, respectively, where $\tan 2\gamma = U_i/Q_i$:

$$I_{\max} = \tfrac{1}{2}(I_i + Q_i \cos 2\gamma + U_i \sin 2\gamma),$$

$$I_{\min} = \tfrac{1}{2}(I_i - Q_i \cos 2\gamma - U_i \sin 2\gamma). \tag{2.88}$$

From (2.88) we obtain the degree of linear polarization

$$\frac{\sqrt{Q_i^2 + U_i^2}}{I_i} = \frac{I_{\max} - I_{\min}}{I_{\max} + I_{\min}}.$$

Therefore, by rotating a linear polarizer in an arbitrary beam and noting the maximum and minimum transmitted irradiance, the degree of linear polarization can be measured regardless of the value of V.

An *ideal linear retarder* divides a given incident electric vector into two linearly polarized components E_1 and E_2, which are mutually orthogonal, and introduces a phase difference $\delta_1 - \delta_2$ between them; there is no diminution of irradiance. Thus, the relation between incident field components and field components transmitted by such a retarder is

$$\begin{pmatrix} E_{\parallel t} \\ E_{\perp t} \end{pmatrix} = \begin{pmatrix} \cos\beta & -\sin\beta \\ \sin\beta & \cos\beta \end{pmatrix}\begin{pmatrix} e^{i\delta_1} & 0 \\ 0 & e^{i\delta_2} \end{pmatrix}\begin{pmatrix} \cos\beta & \sin\beta \\ -\sin\beta & \cos\beta \end{pmatrix}\begin{pmatrix} E_{\parallel i} \\ E_{\perp i} \end{pmatrix}, \tag{2.89}$$

where β is the angle between $\hat{e}_\parallel$ and $\hat{e}_1$ (Fig. 2.17). It is straightforward, but laborious, to show that (2.89) yields the Mueller matrix for an ideal linear

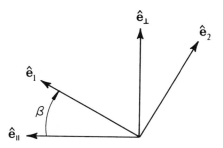

Figure 2.17 $\hat{e}_1$ and $\hat{e}_2$ specify the axes of an ideal linear retarder.

retarder:

$$
\begin{pmatrix}
1 & 0 & 0 & 0 \\
0 & C^2 + S^2\cos\delta & SC(1 - \cos\delta) & -S\sin\delta \\
0 & SC(1 - \cos\delta) & S^2 + C^2\cos\delta & C\sin\delta \\
0 & S\sin\delta & -C\sin\delta & \cos\delta
\end{pmatrix},
\qquad (2.90)
$$

where $C = \cos 2\beta$, $S = \sin 2\beta$, and the *retardance* δ is $\delta_1 - \delta_2$.

The usefulness of the Mueller formulation becomes apparent when we realize that Mueller matrices give us a simple means of determining the state of polarization of a beam transmitted by an optical element for an arbitrarily polarized incident beam. Moreover, if a series of optical elements is interposed in a beam, the combined effect of all these elements may be determined by merely multiplying their associated Mueller matrices. As an example, let us consider how a circular polarizer can be constructed by superposing a linear polarizer and a linear retarder. The beam is first incident on a linear polarizer with horizontal transmission axis ($\xi = 0°$), the Mueller matrix for which is obtained from (2.87):

$$
\frac{1}{2}\begin{pmatrix}
1 & 1 & 0 & 0 \\
1 & 1 & 0 & 0 \\
0 & 0 & 0 & 0 \\
0 & 0 & 0 & 0
\end{pmatrix}.
\qquad (2.91)
$$

The beam transmitted by the polarizer (2.91) is then incident on a retarder with $\delta = 90°$ and $\beta = 45°$, the Mueller matrix for which is obtained from (2.90):

$$
\begin{pmatrix}
1 & 0 & 0 & 0 \\
0 & 0 & 0 & -1 \\
0 & 0 & 1 & 0 \\
0 & 1 & 0 & 0
\end{pmatrix}.
$$

The combined effect of polarizer and retarder is obtained by matrix multiplication:

$$\frac{1}{2}\begin{pmatrix} 1 & 0 & 0 & 0 \\ 0 & 0 & 0 & -1 \\ 0 & 0 & 1 & 0 \\ 0 & 1 & 0 & 0 \end{pmatrix}\begin{pmatrix} 1 & 1 & 0 & 0 \\ 1 & 1 & 0 & 0 \\ 0 & 0 & 0 & 0 \\ 0 & 0 & 0 & 0 \end{pmatrix} = \frac{1}{2}\begin{pmatrix} 1 & 1 & 0 & 0 \\ 0 & 0 & 0 & 0 \\ 0 & 0 & 0 & 0 \\ 1 & 1 & 0 & 0 \end{pmatrix}. \quad (2.92)$$

Thus, if unpolarized light or, indeed, light of arbitrary polarization is incident on the optical system described by the Mueller matrix (2.92), the transmitted light will be 100% right-circularly polarized. Note that matrix multiplication is not commutative: the order of elements in a train must be properly taken into account. Further details about Mueller matrices and experimental means for realizing polarizers, retarders, and other optical elements are found in the excellent book by Shurcliff (1962).

NOTES AND COMMENTS

There are many good books on Fourier transforms. One we have found particularly useful for our purposes is by Champeney (1973).

Toll (1956) examines the logical foundations of causality and the dispersion relations. Goldberger (1960) begins his article with a good historical survey; another discussion within the context of high-energy physics is by Scadron (1979, pp. 326–329), whereas optical properties of solids form the backdrop for Stern's (1963) discussion. And an entire book by Nussenzveig (1972) is devoted to dispersion relations.

Spatial dispersion is the subject of a review article by Rukhadze and Silin (1961) and a book by Agranovich and Ginzburg (1966).

A good review article on optical constants and their measurement is that by Bell (1967). Determination of optical constants from reflectance measurements is treated by Wendlandt and Hecht (1966) and from internal reflection spectroscopy by Harrick (1967). Ellipsometric techniques are discussed at length by Azzam and Bashara (1977).

In Section 2.10 we made an analogy between slabs and particles while cautioning not to push this analogy too far. That caution is necessary is evident from calculations of volumetric absorption by slabs and spheres of the same material (Faxvog and Roessler, 1981).

McMaster (1954) takes a quantum-mechanical approach to the Stokes parameters and polarized light. Two books are devoted entirely to polarization: Shurcliff (1962) and Clarke and Grainger (1971). And a splendid collection of papers on many aspects of polarized light has been edited by Gehrels (1974a). Another collection worth consulting is that compiled by Swindell (1975); it contains several of the classical papers on polarization.

Chapter 3

Absorption and Scattering by an Arbitrary Particle

When a particle is illuminated by a beam of light with specified characteristics, the amount and angular distribution of the light scattered by the particle, as well as the amount absorbed, depends in a detailed way on the nature of the particle, that is, its shape, size, and the materials of which it is composed. This presents us with an almost unlimited number of distinct possibilities. Nevertheless, there are some features common to the phenomena of scattering and absorption by small particles. In this chapter, therefore, our goal is to say as much as possible about such phenomena without invoking any specific particle. This will establish the mathematical and physical framework underlying all the specific problems encountered in later chapters.

3.1 GENERAL FORMULATION OF THE PROBLEM

Our fundamental problem is as follows: Given a particle of specified size, shape and optical properties that is illuminated by an arbitrarily polarized monochromatic wave, determine the electromagnetic field at all points in the particle and at all points of the homogeneous medium in which the particle is embedded. Although we limit our consideration to plane harmonic waves, this is less of a restriction than it might seem at first glance: in Section 2.4 we showed that an arbitrary field can be decomposed into its Fourier components, which are plane waves. Therefore, regardless of the illumination we can obtain the solution to the scattering–absorption problem by superposition.

The field inside the particle is denoted by $(\mathbf{E}_1, \mathbf{H}_1)$; the field $(\mathbf{E}_2, \mathbf{H}_2)$ in the medium surrounding the particle is the superposition of the incident field $(\mathbf{E}_i, \mathbf{H}_i)$ and the scattered field $(\mathbf{E}_s, \mathbf{H}_s)$ (Fig. 3.1):

$$\mathbf{E}_2 = \mathbf{E}_i + \mathbf{E}_s, \qquad \mathbf{H}_2 = \mathbf{H}_i + \mathbf{H}_s,$$

where

$$\mathbf{E}_i = \mathbf{E}_0 \exp(i\mathbf{k} \cdot \mathbf{x} - i\omega t), \qquad \mathbf{H}_i = \mathbf{H}_0 \exp(i\mathbf{k} \cdot \mathbf{x} - i\omega t),$$

and $\mathbf{k}$ is the wave vector appropriate to the surrounding medium. The fields

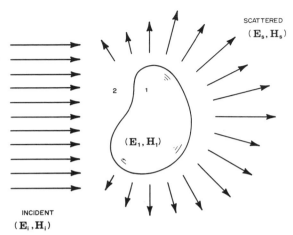

Figure 3.1 The incident field $(\mathbf{E}_i, \mathbf{H}_i)$ gives rise to a field $(\mathbf{E}_1, \mathbf{H}_1)$ inside the particle and a scattered field $(\mathbf{E}_s, \mathbf{H}_s)$ in the medium surrounding the particle.

must satisfy the Maxwell equations

$$\nabla \cdot \mathbf{E} = 0, \tag{3.1}$$

$$\nabla \cdot \mathbf{H} = 0, \tag{3.2}$$

$$\nabla \times \mathbf{E} = i\omega\mu\mathbf{H}, \tag{3.3}$$

$$\nabla \times \mathbf{H} = -i\omega\varepsilon\mathbf{E}, \tag{3.4}$$

at all points where ε and μ are continuous. The curl of (3.3) and (3.4) is

$$\nabla \times (\nabla \times \mathbf{E}) = i\omega\mu\nabla \times \mathbf{H} = \omega^2\varepsilon\mu\mathbf{E},$$

$$\nabla \times (\nabla \times \mathbf{H}) = -i\omega\varepsilon\nabla \times \mathbf{E} = \omega^2\varepsilon\mu\mathbf{H},$$

and if we use the vector identity

$$\nabla \times (\nabla \times \mathbf{A}) = \nabla(\nabla \cdot \mathbf{A}) - \nabla \cdot (\nabla\mathbf{A}) \tag{3.5}$$

we obtain

$$\nabla^2\mathbf{E} + k^2\mathbf{E} = 0, \qquad \nabla^2\mathbf{H} + k^2\mathbf{H} = 0, \tag{3.6}$$

where $k^2 = \omega^2\varepsilon\mu$ and $\nabla^2\mathbf{A} = \nabla \cdot (\nabla\mathbf{A})$. Thus, $\mathbf{E}$ and $\mathbf{H}$ satisfy the *vector wave equation*. Any vector field with zero divergence that satisfies the vector wave equation is an admissible electric field; the associated magnetic field is related to the curl of the electric field through (3.3). Caution: The symbol ∇^2 in (3.6) should be looked upon as shorthand notation for the *vector* operator $\nabla \cdot \nabla$; that is, $\nabla \mathbf{A}$ is a dyadic which when operated on by the divergence operator $\nabla \cdot$ yields a vector. Alternatively, we can consider (3.5) to define $\nabla^2 \mathbf{A}$. It is *not* true that the components of $\mathbf{E}$ separately satisfy the *scalar wave equation*

$$\nabla^2 \psi + k^2 \psi = 0,$$

as a superficial glance at (3.6) might lead one to believe, except in the special case where $\mathbf{E}$ is specified relative to a rectangular Cartesian coordinate system.

3.1.1 Boundary Conditions

The electromagnetic field is required to satisfy the Maxwell equations at points where ε and μ are continuous. However, as one crosses the boundary between particle and medium, there is, in general, a sudden change in these properties. This change occurs over a transition region with thickness of the order of atomic dimensions. From a macroscopic point of view, therefore, there is a discontinuity at the boundary. At such boundary points we impose the following conditions on the fields:

$$\begin{aligned} \left[\mathbf{E}_2(\mathbf{x}) - \mathbf{E}_1(\mathbf{x}) \right] \times \hat{\mathbf{n}} &= 0, \\ \left[\mathbf{H}_2(\mathbf{x}) - \mathbf{H}_1(\mathbf{x}) \right] \times \hat{\mathbf{n}} &= 0, \end{aligned} \qquad \mathbf{x} \text{ on } S, \qquad (3.7)$$

where $\hat{\mathbf{n}}$ is the outward directed normal to the surface S of the particle. The boundary conditions (3.7) are the requirement that the *tangential components* of $\mathbf{E}$ and $\mathbf{H}$ are continuous across a boundary separating media with different properties.

At this point we depart from the traditional derivation of (3.7)—what might be called the "pharmaceutical approach" of constructing pillboxes and loops that straddle the boundary and taking various limits—and give a physical justification of these boundary conditions by appealing to conservation of energy. Consider a closed surface A, with outward normal $\hat{\mathbf{n}}$, which is the boundary between regions 1 and 2 (Fig. 3.2). There are no restrictions on the properties of these regions. The rate at which electromagnetic energy is transferred across a closed surface arbitrarily near A in region 1 (shown by the dashed line in Fig. 3.2) is

$$\int_A \mathbf{S}_1 \cdot \hat{\mathbf{n}} \, dA = \int_A \hat{\mathbf{n}} \cdot (\mathbf{E}_1 \times \mathbf{H}_1) \, dA, \qquad (3.8)$$

where the electromagnetic field $(\mathbf{E}_1, \mathbf{H}_1)$ is not restricted to be time harmonic. Similarly, the rate of electromagnetic energy transfer across a closed surface arbitrarily near A in region 2 (shown by the dotted line in Fig. 3.2) is

$$\int_A \mathbf{S}_2 \cdot \hat{\mathbf{n}} \, dA = \int_A \hat{\mathbf{n}} \cdot (\mathbf{E}_2 \times \mathbf{H}_2) \, dA. \qquad (3.9)$$

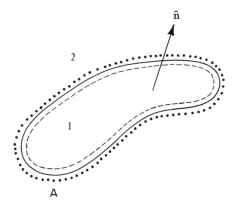

Figure 3.2 Closed surface separating regions 1 and 2.

If the boundary conditions (3.7) are imposed, then $\mathbf{E}_2 \times \hat{\mathbf{n}} = \mathbf{E}_1 \times \hat{\mathbf{n}}$, $\mathbf{H}_2 \times \hat{\mathbf{n}}$ $= \mathbf{H}_1 \times \hat{\mathbf{n}}$, and the integrals (3.8) and (3.9) may be written

$$\int_A \mathbf{S}_1 \cdot \hat{\mathbf{n}} \, dA = \int_A \mathbf{H}_1 \cdot (\hat{\mathbf{n}} \times \mathbf{E}_1) \, dA = \int_A \mathbf{H}_1 \cdot (\hat{\mathbf{n}} \times \mathbf{E}_2) \, dA,$$

$$\int_A \mathbf{S}_2 \cdot \hat{\mathbf{n}} \, dA = \int_A \mathbf{E}_2 \cdot (\mathbf{H}_2 \times \hat{\mathbf{n}}) \, dA = \int_A \mathbf{H}_1 \cdot (\hat{\mathbf{n}} \times \mathbf{E}_2) \, dA,$$

where we have used the permutation rule for the triple scalar product: $\mathbf{A} \cdot (\mathbf{B} \times \mathbf{C}) = \mathbf{B} \cdot (\mathbf{C} \times \mathbf{A}) = \mathbf{C} \cdot (\mathbf{A} \times \mathbf{B})$. Therefore, there are no sources or sinks of electromagnetic energy on A:

$$\int_A \mathbf{S}_1 \cdot \hat{\mathbf{n}} \, dA = \int_A \mathbf{S}_2 \cdot \hat{\mathbf{n}} \, dA.$$

Thus, the requirement that the tangential components of the electromagnetic field are continuous across a boundary of discontinuity is a *sufficient* condition for energy conservation across that boundary.

3.1.2 Superposition

Our fundamental task is to construct solutions to the Maxwell equations (3.1)–(3.4), both inside and outside the particle, which satisfy (3.7) at the boundary between particle and surrounding medium. If the incident electromagnetic field is arbitrary, subject to the restriction that it can be Fourier analyzed into a superposition of plane monochromatic waves (Section 2.4), the solution to the problem of interaction of such a field with a particle can be obtained in principle by superposing fundamental solutions. That this is possible is a consequence of the *linearity* of the Maxwell equations *and* the boundary conditions. That is, if $\mathbf{E}_a$ and $\mathbf{E}_b$ are solutions to the field equations, their sum $\mathbf{E}_a + \mathbf{E}_b$ is also a solution; and if

$$(\mathbf{E}_{a2} - \mathbf{E}_{a1}) \times \hat{\mathbf{n}} = 0, \qquad (\mathbf{E}_{b2} - \mathbf{E}_{b1}) \times \hat{\mathbf{n}} = 0,$$

then

$$(\mathbf{E}_2 - \mathbf{E}_1) \times \hat{\mathbf{n}} = 0,$$

where $\mathbf{E}_2 = \mathbf{E}_{a2} + \mathbf{E}_{b2}$ and $\mathbf{E}_1 = \mathbf{E}_{a1} + \mathbf{E}_{b1}$. This, therefore, is our justification for considering only scattering of plane monochromatic waves. An arbitrarily polarized wave can be expressed as the superposition of two orthogonal polarization states (Section 2.11). Therefore, we need only solve each scattering problem twice (for a given direction of propagation) in order to determine the scattering of an arbitrarily polarized plane wave.

3.2 THE AMPLITUDE SCATTERING MATRIX

Consider an arbitrary particle that is illuminated by a plane harmonic wave (Fig. 3.3). The direction of propagation of the incident light defines the z axis, the *forward direction*. Any point in the particle may be chosen as the origin O of a rectangular Cartesian coordinate system (x, y, z), where the x and y axes are orthogonal to the z axis and to each other but are otherwise arbitrary. The orthonormal basis vectors $\hat{\mathbf{e}}_x, \hat{\mathbf{e}}_y, \hat{\mathbf{e}}_z$ are in the directions of the positive x, y, and z axes. The scattering direction $\hat{\mathbf{e}}_r$ and the forward direction $\hat{\mathbf{e}}_z$ define a plane called the *scattering plane*, which is analogous to the plane of incidence in problems of reflection at an interface (Section 2.7). The scattering plane is uniquely determined by the azimuthal angle ϕ except when $\hat{\mathbf{e}}_r$ is parallel to the z axis. In these two instances ($\hat{\mathbf{e}}_r = \pm \hat{\mathbf{e}}_z$) any plane containing the z axis is a suitable scattering plane. It is convenient to resolve the incident electric field $\mathbf{E}_i$, which lies in the xy plane, into components parallel ($E_{\parallel i}$) and perpendicular ($E_{\perp i}$) to the scattering plane:

$$\mathbf{E}_i = \left(E_{0\parallel} \hat{\mathbf{e}}_{\parallel i} + E_{0\perp} \hat{\mathbf{e}}_{\perp i} \right) \exp(ikz - i\omega t) = E_{\parallel i} \hat{\mathbf{e}}_{\parallel i} + E_{\perp i} \hat{\mathbf{e}}_{\perp i},$$

where $k = 2\pi N_2/\lambda$ is the wave number in the medium surrounding the particle, N_2 is the refractive index, and λ is the wavelength of the incident light *in vacuo*. The orthonormal basis vectors $\hat{\mathbf{e}}_{\parallel i}$ and $\hat{\mathbf{e}}_{\perp i}$, where

$$\hat{\mathbf{e}}_{\perp i} = \sin\phi \hat{\mathbf{e}}_x - \cos\phi \hat{\mathbf{e}}_y, \qquad \hat{\mathbf{e}}_{\parallel i} = \cos\phi \hat{\mathbf{e}}_x + \sin\phi \hat{\mathbf{e}}_y,$$

form a right-handed triad with $\hat{\mathbf{e}}_z$:

$$\hat{\mathbf{e}}_{\perp i} \times \hat{\mathbf{e}}_{\parallel i} = \hat{\mathbf{e}}_z.$$

We also have

$$\hat{\mathbf{e}}_{\perp i} = -\hat{\mathbf{e}}_\phi, \qquad \hat{\mathbf{e}}_{\parallel i} = \sin\theta \hat{\mathbf{e}}_r + \cos\theta \hat{\mathbf{e}}_\theta,$$

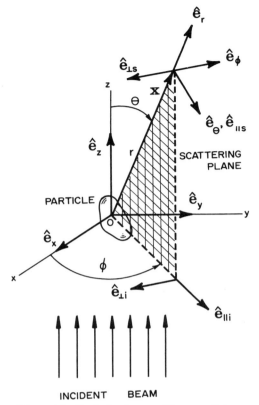

Figure 3.3 Scattering by an arbitrary particle.

where $\hat{e}_r$, $\hat{e}_\theta$, $\hat{e}_\phi$ are the orthonormal basis vectors associated with the spherical polar coordinate system (r, θ, ϕ). If the x and y components of the incident field are denoted by E_{xi} and E_{yi}, then

$$E_{\|i} = \cos\phi E_{xi} + \sin\phi E_{yi},$$

$$E_{\perp i} = \sin\phi E_{xi} - \cos\phi E_{yi}.$$

At sufficiently large distances from the origin $(kr \gg 1)$, in the *far-field region*, the scattered electric field $\mathbf{E}_s$ is approximately *transverse* $(\hat{e}_r \cdot \mathbf{E}_s \simeq 0)$ and has the asymptotic form (see, e.g., Jackson, 1975, p. 748)

$$\mathbf{E}_s \sim \frac{e^{ikr}}{-ikr}\mathbf{A} \qquad kr \gg 1, \tag{3.10}$$

where $\hat{e}_r \cdot \mathbf{A} = 0$. Therefore, the scattered field in the far-field region may be written

$$\mathbf{E}_s = E_{\|s}\hat{\mathbf{e}}_{\|s} + E_{\perp s}\hat{\mathbf{e}}_{\perp s},$$

$$\hat{\mathbf{e}}_{\|s} = \hat{\mathbf{e}}_\theta, \qquad \hat{\mathbf{e}}_{\perp s} = -\hat{\mathbf{e}}_\phi, \qquad \hat{\mathbf{e}}_{\perp s} \times \hat{\mathbf{e}}_{\|s} = \hat{\mathbf{e}}_r. \qquad (3.11)$$

The basis vector $\hat{\mathbf{e}}_{\|s}$ is parallel and $\hat{\mathbf{e}}_{\perp s}$ is perpendicular to the scattering plane. Note, however, that $\mathbf{E}_s$ and $\mathbf{E}_i$ are specified relative to *different* sets of basis vectors. Because of the linearity of the boundary conditions (3.7) the amplitude of the field scattered by an arbitrary particle is a linear function of the amplitude of the incident field. The relation between incident and scattered fields is conveniently written in matrix form

$$\begin{pmatrix} E_{\|s} \\ E_{\perp s} \end{pmatrix} = \frac{e^{ik(r-z)}}{-ikr} \begin{pmatrix} S_2 & S_3 \\ S_4 & S_1 \end{pmatrix} \begin{pmatrix} E_{\|i} \\ E_{\perp i} \end{pmatrix}, \qquad (3.12)$$

where the elements $S_j (j = 1, 2, 3, 4)$ of the *amplitude scattering matrix* depend, in general, on θ, the *scattering angle*, and the azimuthal angle ϕ.

Rarely are the real and imaginary parts of the four amplitude scattering matrix elements measured for all values of θ and ϕ. To do so requires measuring the amplitude *and* phase of the light scattered in all directions for two incident orthogonal polarization states, a measurement impeded by the elusiveness of the latter quantity. Hart and Gray (1964) have described a procedure by which phases might be measured from the interference between light scattered by the particle of interest and a nearby particle with known scattering properties. But few such experiments have been performed, a notable exception being the microwave experiments of Greenberg et al. (1961). However, the amplitude scattering matrix elements are related to quantities the measurement of which poses considerably fewer experimental problems than phases; this will be explored in the following two sections.

3.3 SCATTERING MATRIX

Once we have obtained the electromagnetic fields inside and scattered by the particle, we can determine the Poynting vector at any point. However, we are usually interested only in the Poynting vector at points outside the particle. The time-averaged Poynting vector $\mathbf{S}$ at any point in the medium surrounding the particle can be written as the sum of three terms:

$$\mathbf{S} = \tfrac{1}{2} \text{Re}\{\mathbf{E}_2 \times \mathbf{H}_2^*\} = \mathbf{S}_i + \mathbf{S}_s + \mathbf{S}_{\text{ext}},$$

$$\mathbf{S}_i = \tfrac{1}{2} \text{Re}\{\mathbf{E}_i \times \mathbf{H}_i^*\}, \qquad \mathbf{S}_s = \tfrac{1}{2} \text{Re}\{\mathbf{E}_s \times \mathbf{H}_s^*\}, \qquad (3.13)$$

$$\mathbf{S}_{\text{ext}} = \tfrac{1}{2} \text{Re}\{\mathbf{E}_i \times \mathbf{H}_s^* + \mathbf{E}_s \times \mathbf{H}_i^*\}.$$

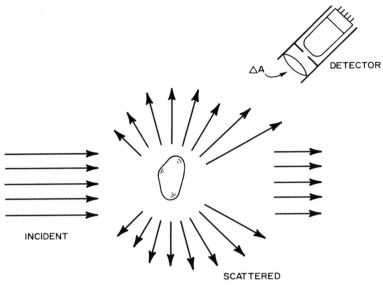

Figure 3.4 The collimated detector responds only to the scattered light.

$\mathbf{S}_i$, the Poynting vector associated with the incident wave, is independent of position if the medium is nonabsorbing; $\mathbf{S}_s$ is the Poynting vector of the scattered field; and we may interpret $\mathbf{S}_{ext}$ as the term that arises because of interaction between the incident and scattered waves.

Suppose that a detector is placed at a distance r from a particle in the far-field region, with its surface ΔA aligned normal to $\hat{\mathbf{e}}_r$ (Fig. 3.4). If the detector is suitably collimated, and if $\hat{\mathbf{e}}_r$ is not too near the forward direction $\hat{\mathbf{e}}_z$, the detector will record a signal proportional to $\mathbf{S}_s \cdot \hat{\mathbf{e}}_r \Delta A$ (ΔA is sufficiently small so that $\mathbf{S}_s$ does not vary greatly over the detector). The detector "sees" only the scattered light provided that it does not "look at" the source of incident light. From (3.10) and (3.13) it follows that

$$\mathbf{S}_s \cdot \hat{\mathbf{e}}_r \Delta A = \frac{k}{2\omega\mu} \frac{|\mathbf{A}|^2}{k^2} \Delta\Omega, \qquad (3.14)$$

where $\Delta\Omega = \Delta A / r^2$ is the solid angle subtended by the detector. Thus, we can obtain $|\mathbf{A}|^2$ as a function of direction, to within a solid angle $\Delta\Omega$, by recording the detector response at various positions on a hemisphere surrounding the particle.

By interposing various polarizers between particle and detector and recording the resulting irradiances in a manner identical to that discussed for a plane wave in Section 2.11, we obtain the Stokes parameters of the light scattered by a particle:

$$I_s = \langle E_{\parallel s} E_{\parallel s}^* + E_{\perp s} E_{\perp s}^* \rangle,$$

$$Q_s = \langle E_{\parallel s} E_{\parallel s}^* - E_{\perp s} E_{\perp s}^* \rangle,$$

$$U_s = \langle E_{\parallel s} E_{\perp s}^* + E_{\perp s} E_{\parallel s}^* \rangle,$$

$$V_s = i\langle E_{\parallel s} E_{\perp s}^* - E_{\perp s} E_{\parallel s}^* \rangle.$$

(3.15)

We again omit the multiplicative factor $k/2\omega\mu$. The relation between incident and scattered Stokes parameters follows from the amplitude scattering matrix (3.12):

$$\begin{pmatrix} I_s \\ Q_s \\ U_s \\ V_s \end{pmatrix} = \frac{1}{k^2 r^2} \begin{pmatrix} S_{11} & S_{12} & S_{13} & S_{14} \\ S_{21} & S_{22} & S_{23} & S_{24} \\ S_{31} & S_{32} & S_{33} & S_{34} \\ S_{41} & S_{42} & S_{43} & S_{44} \end{pmatrix} \begin{pmatrix} I_i \\ Q_i \\ U_i \\ V_i \end{pmatrix}$$

(3.16)

$$S_{11} = \tfrac{1}{2}\left(|S_1|^2 + |S_2|^2 + |S_3|^2 + |S_4|^2\right),$$

$$S_{12} = \tfrac{1}{2}\left(|S_2|^2 - |S_1|^2 + |S_4|^2 - |S_3|^2\right),$$

$$S_{13} = \mathrm{Re}\{S_2 S_3^* + S_1 S_4^*\},$$

$$S_{14} = \mathrm{Im}\{S_2 S_3^* - S_1 S_4^*\},$$

$$S_{21} = \tfrac{1}{2}\left(|S_2|^2 - |S_1|^2 - |S_4|^2 + |S_3|^2\right),$$

$$S_{22} = \tfrac{1}{2}\left(|S_2|^2 + |S_1|^2 - |S_4|^2 - |S_3|^2\right),$$

$$S_{23} = \mathrm{Re}\{S_2 S_3^* - S_1 S_4^*\},$$

$$S_{24} = \mathrm{Im}\{S_2 S_3^* + S_1 S_4^*\},$$

$$S_{31} = \mathrm{Re}\{S_2 S_4^* + S_1 S_3^*\},$$

$$S_{32} = \mathrm{Re}\{S_2 S_4^* - S_1 S_3^*\},$$

$$S_{33} = \mathrm{Re}\{S_1 S_2^* + S_3 S_4^*\},$$

$$S_{34} = \mathrm{Im}\{S_2 S_1^* + S_4 S_3^*\},$$

$$S_{41} = \mathrm{Im}\{S_2^* S_4 + S_3^* S_1\},$$

$$S_{42} = \mathrm{Im}\{S_2^* S_4 - S_3^* S_1\},$$

$$S_{43} = \mathrm{Im}\{S_1 S_2^* - S_3 S_4^*\},$$

$$S_{44} = \mathrm{Re}\{S_1 S_2^* - S_3 S_4^*\}.$$

The 4×4 matrix in (3.16), the *scattering matrix*, is the Mueller matrix for scattering by a single particle; the term *phase matrix* is also used, a particularly inappropriate choice of terminology because the "phase" matrix relates scattered to incident irradiances. The 16 scattering matrix elements for a single particle are not all independent; only seven of them can be independent, corresponding to the four moduli $|S_j|$ ($j = 1, 2, 3, 4$) and the three differences in phase between the S_j. Thus, there must be nine independent relations among the S_{ij}; these are given by Abhyankar and Fymat (1969).

The Stokes parameters of the light scattered by a collection of randomly separated particles are the sum of the Stokes parameters of the light scattered by the individual particles. Therefore, the scattering matrix for such a collection is merely the sum of the individual particle scattering matrices (we assume that the linear dimensions of the volume occupied by the scatterers is small compared with the distance r at which the scattered light is observed). In general, there are 16 nonzero, independent matrix elements, although this number may be reduced because of symmetry. We shall return to this matter of symmetry in later chapters when we consider specific scattering matrices and experimental means for their measurement. For the moment, we consider the most general scattering matrix. The S_{ij} must be independent of ϕ for any particle or collection of particles that is invariant with respect to arbitrary rotation about the z axis.

If unpolarized light of irradiance I_i is incident on one or more particles, the Stokes parameters of the scattered light are

$$\frac{I_s}{I_i} = S_{11}, \qquad \frac{Q_s}{I_i} = S_{21}, \qquad \frac{U_s}{I_i} = S_{31}, \qquad \frac{V_s}{I_i} = S_{41};$$

for convenience we omit the factor $(kr)^{-2}$. Therefore, S_{11} specifies the angular distribution of the scattered light given unpolarized incident light. This scattered light is, in general, partially polarized with degree of polarization

$$\sqrt{\left(S_{21}^2 + S_{31}^2 + S_{41}^2\right)/S_{11}^2} \ .$$

This clearly demonstrates a very general aspect of scattering by particles regardless of their nature: scattering is a mechanism for polarizing light. S_{ij} depends on the scattering direction and, therefore, so does the degree of polarization.

If the incident light is right-circularly polarized, then the irradiance I_R of the scattered light is $S_{11} + S_{14}$ (this notation should not mislead the reader that the scattered light is also right-circularly polarized: it is not, in general). Similarly, the irradiance I_L of the scattered light, given incident left-circularly polarized light, is $S_{11} - S_{14}$. Therefore, S_{14} is readily interpretable in terms of the difference of the irradiances of scattered light for incident right-circularly and left-circularly polarized light:

$$S_{14} = \frac{1}{2} \frac{I_R - I_L}{I_i}.$$

At this point, it is well to remind ourselves that as the scattering direction varies, so does the scattering plane and, as a consequence, the Stokes parameters Q_i and U_i (although $Q_i^2 + U_i^2$ is independent of the scattering plane). If, for a given scattering direction, we consider incident light polarized parallel and perpendicular to the associated scattering plane, it follows that S_{12} is

$$S_{12} = \frac{1}{2} \frac{I_\parallel - I_\perp}{I_i},$$

where $I_\parallel$ and $I_\perp$ are the scattered irradiances for incident light polarized parallel and perpendicular to the scattering plane.

By considering incident light polarized obliquely to the scattering plane ($+45°$ and $-45°$), we obtain a straightforward physical interpretation of the matrix element S_{13}. The remaining matrix elements are a bit more difficult to interpret individually, although various combinations of them are related to *changes* in the state of polarization of incident light. As an example, let us consider the change in degree of polarization of incident light that is completely polarized parallel to a particular scattering plane. The Stokes parameters of the scattered light are $I_s = (S_{11} + S_{12})I_i$, $Q_s = (S_{21} + S_{22})I_i$, $U_s = (S_{31} + S_{32})I_i$, $V_s = (S_{41} + S_{42})I_i$, and the degree of polarization is

$$\frac{\sqrt{(S_{21} + S_{22})^2 + (S_{31} + S_{32})^2 + (S_{41} + S_{42})^2}}{S_{11} + S_{12}}. \tag{3.17}$$

If we are considering scattering by a *single* particle or a collection of *identical* particles (by identical is meant identical in size, shape, composition, and orientation relative to the incident beam), it follows from (3.16) and (3.17) that the scattered light is completely polarized. Although we chose a particular example, this general conclusion is true for arbitrary incident light that is completely polarized. Scattering by a single particle or collection of identical particles does not decrease the *degree* of polarization of 100% polarized incident light. Note, however, that the *nature* of the polarization will, in general, be changed; for example, linearly polarized incident light will be transformed into elliptically polarized light upon scattering. If, on the other hand, the collection is composed of *nonidentical* particles, then (3.17) will be less than (or at most, equal to) 100%. Therefore, scattering by a collection of nonidentical particles results in *depolarization* of incident polarized light. This is another general feature of scattering that is independent of the specific nature of the particles.

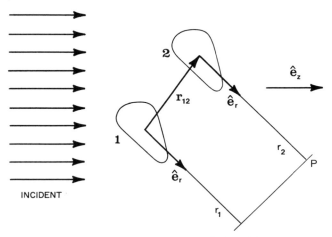

Figure 3.5   Scattering by two identical particles.

Light scattered in the forward direction ($\hat{\mathbf{e}}_r = \hat{\mathbf{e}}_z$) has unique characteristics not possessed by light scattered in all other directions. No matter how small the solid angle $\Delta\Omega$ subtended by the detector, it is not possible to separate the incident beam from the light scattered in the forward direction: the detector unavoidably responds to a superposition of incident and forward scattered fields. There is another singular aspect of the forward scattering direction. Consider a collection of *identical* particles, two of which are shown in Fig. 3.5. For a given scattering direction $\hat{\mathbf{e}}_r$, there is a difference in phase $\Delta\phi$ between the fields $\mathbf{E}_{s1}(\hat{\mathbf{e}}_r)$ and $\mathbf{E}_{s2}(\hat{\mathbf{e}}_r)$ scattered by particles 1 and 2:

$$\mathbf{E}_{s2}(\hat{\mathbf{e}}_r) \simeq \mathbf{E}_{s1}(\hat{\mathbf{e}}_r)\,e^{i\Delta\phi},$$

where $\mathbf{E}_s$ is evaluated in the far-field region at points on the plane P normal to $\hat{\mathbf{e}}_r$, and the phase difference is

$$\Delta\phi = \mathrm{k}\big[\mathbf{r}_{12}\cdot(\hat{\mathbf{e}}_z - \hat{\mathbf{e}}_r)\big]. \tag{3.18}$$

We have also assumed that $r_1 \gg |\mathbf{r}_{12}|$, $r_2 \gg |\mathbf{r}_{12}|$. Except near the forward direction, there is a random distribution of phase differences for light scattered by randomly separated identical particles in a large collection. As we approach the forward direction ($\hat{\mathbf{e}}_r \to \hat{\mathbf{e}}_z$), however, the phase difference (3.18) approaches zero *regardless of the particle separation*. Therefore, scattering near the forward direction is *coherent*. If the particles are not identical, the difference in phase between light scattered by various pairs of particles, does not, in general, vanish in the forward direction, although it is independent of particle separation; the phase difference may, however, depend on the relative orientation of the two particles. It is clear that scattering in or near the forward direction is sufficiently singular to require careful consideration.

3.4 EXTINCTION, SCATTERING, AND ABSORPTION

Suppose that one or more particles are placed in a beam of electromagnetic radiation (Fig. 3.6). The rate at which electromagnetic energy is received by a detector D downstream from the particles is denoted by U. If the particles are removed, the power received by the detector is U_0, where $U_0 > U$. We say that the presence of the particles has resulted in *extinction* of the incident beam. If the medium in which the particles are embedded is nonabsorbing, the difference $U_0 - U$ is accounted for by *absorption* in the particles (i.e., transformation of electromagnetic energy into other forms) and *scattering* by the particles. This extinction depends on the chemical composition of the particles, their size, shape, orientation, the surrounding medium, the number of particles, and the polarization state and frequency of the incident beam. Although the specific details of extinction depend on all these parameters, certain general features are shared in common by all particles.

Let us now consider extinction by a single arbitrary particle embedded in a nonabsorbing medium (not necessarily a vacuum) and illuminated by a plane wave (Fig. 3.7). We construct an imaginary sphere of radius r around the particle; the net rate at which electromagnetic energy crosses the surface A of this sphere is

$$W_a = -\int_A \mathbf{S} \cdot \hat{\mathbf{e}}_r \, dA.$$

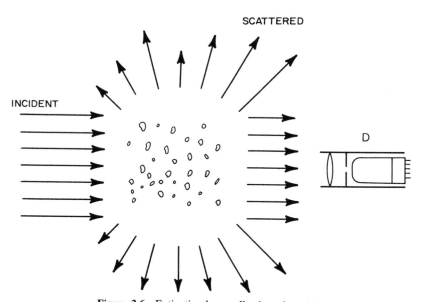

Figure 3.6 Extinction by a collection of particles.

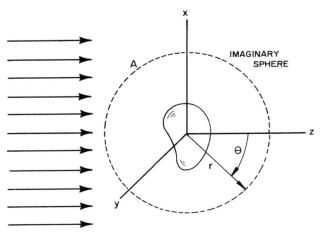

Figure 3.7   Extinction by a single particle.

If $W_a > 0$ (if W_a is negative, energy is being created within the sphere, a possibility we exclude from consideration), energy is absorbed within the sphere. But the medium is nonabsorbing, which implies that W_a is the rate at which energy is absorbed by the particle. Because of (3.13) W_a may be written as the sum of three terms: $W_a = W_i - W_s + W_{ext}$, where

$$W_i = -\int_A \mathbf{S}_i \cdot \hat{\mathbf{e}}_r \, dA, \qquad W_s = \int_A \mathbf{S}_s \cdot \hat{\mathbf{e}}_r \, dA, \qquad W_{ext} = -\int_A \mathbf{S}_{ext} \cdot \hat{\mathbf{e}}_r \, dA.$$

$$(3.19)$$

W_i vanishes identically for a nonabsorbing medium; W_s is the rate at which energy is scattered across the surface A. Therefore, W_{ext} is just the sum of the energy absorption rate and the energy scattering rate:

$$W_{ext} = W_a + W_s. \qquad (3.20)$$

For convenience we take the incident electric field $\mathbf{E}_i = E\hat{\mathbf{e}}_x$ to be x-polarized. Because the medium is nonabsorbing, W_a is independent of the radius r of the imaginary sphere. Therefore, we may choose r sufficiently large such that we are in the far-field region where

$$\mathbf{E}_s \sim \frac{e^{ik(r-z)}}{-ikr} \mathbf{X} E, \qquad \mathbf{H}_s \sim \frac{k}{\omega\mu} \hat{\mathbf{e}}_r \times \mathbf{E}_s, \qquad (3.21)$$

and $\hat{\mathbf{e}}_r \cdot \mathbf{X} = 0$. As a reminder that the incident light is x-polarized we use the symbol $\mathbf{X}$ for the *vector scattering amplitude*, which is related to the (scalar) amplitude scattering matrix elements S_j as follows:

$$\mathbf{X} = (S_2\cos\phi + S_3\sin\phi)\hat{\mathbf{e}}_{\parallel s} + (S_4\cos\phi + S_1\sin\phi)\hat{\mathbf{e}}_{\perp s}. \tag{3.22}$$

After a considerable amount of algebraic manipulation we obtain

$$W_{ext} = \frac{-k}{2\omega\mu}|E|^2 \text{Re}\left\{\frac{e^{-ikr}}{ikr}\int_A e^{ikz}\hat{\mathbf{e}}_x \cdot \mathbf{X}^* \, dA\right.$$

$$-\frac{e^{ikr}}{ikr}\int_A e^{-ikz}\cos\theta\hat{\mathbf{e}}_x \cdot \mathbf{X} \, dA$$

$$\left. +\frac{e^{ikr}}{ikr}\int_A e^{-ikz}\sin\theta\cos\phi\hat{\mathbf{e}}_z \cdot \mathbf{X} \, dA\right\}. \tag{3.23}$$

Equation (3.23) contains integrals of the form

$$\int_{-1}^{1} e^{ikr\mu}f(\mu) \, d\mu,$$

where $\mu = \cos\theta$, which can be integrated by parts to yield

$$\frac{e^{ikr}f(1) - e^{-ikr}f(-1)}{ikr} + O\left(\frac{1}{k^2r^2}\right),$$

provided that $df/d\mu$ is bounded. The limiting value of W_{ext} as $kr \to \infty$ is therefore

$$W_{ext} = I_i\frac{4\pi}{k^2}\text{Re}\{(\mathbf{X}\cdot\hat{\mathbf{e}}_x)_{\theta=0}\},$$

where I_i is the incident irradiance. The ratio of W_{ext} to I_i is a quantity with dimensions of area:

$$C_{ext} = \frac{W_{ext}}{I_i} = \frac{4\pi}{k^2}\text{Re}\{(\mathbf{X}\cdot\hat{\mathbf{e}}_x)_{\theta=0}\}. \tag{3.24}$$

It follows from (3.20) that the *extinction cross section* C_{ext} may be written as the sum of the *absorption cross section* C_{abs} and the *scattering cross section* C_{sca}:

$$C_{ext} = C_{abs} + C_{sca}, \tag{3.25}$$

where $C_{abs} = W_{abs}/I_i$ and $C_{sca} = W_s/I_i$. From (3.19) and (3.21) we have

$$C_{sca} = \int_0^{2\pi}\int_0^{\pi}\frac{|\mathbf{X}|^2}{k^2}\sin\theta \, d\theta \, d\phi = \int_{4\pi}\frac{|\mathbf{X}|^2}{k^2} \, d\Omega. \tag{3.26}$$

The quantity $|\mathbf{X}|^2/k^2$ is sometimes called the *differential scattering cross section*, a familiar term in atomic and nuclear physics, and denoted symbolically by $dC_{sca}/d\Omega$; this should not be interpreted as the derivative of a function of Ω: the differential scattering cross section is *formally* written as a derivative merely as an aid to the memory. Physically, $dC_{sca}/d\Omega$ specifies the angular distribution of the scattered light: the amount of light (for unit incident irradiance) scattered into a unit solid angle about a given direction. In light scattering theory one commonly encounters the term *phase function*, defined as $|\mathbf{X}|^2/k^2 C_{sca}$ and denoted by the symbol p; it is normalized:

$$\int_{4\pi} p \, d\Omega = 1.$$

We previously voiced our objection to the term phase used to designate irradiances. A less commonly encountered, although perhaps better term for the phase function is the *scattering diagram*.

The average cosine of the scattering angle, or the *asymmetry parameter g* is

$$g = \langle \cos \theta \rangle = \int_{4\pi} p \cos \theta \, d\Omega.$$

For a particle that scatters light *isotropically* (i.e., the same in all directions), g vanishes; g also vanishes if the scattering is symmetric about a scattering angle of $90°$. If the particle scatters more light toward the forward direction ($\theta = 0°$), g is positive; g is negative if the scattering is directed more toward the back direction ($\theta = 180°$).

We may define *efficiencies* (or efficiency factors) for extinction, scattering, and absorption:

$$Q_{ext} = \frac{C_{ext}}{G}, \qquad Q_{sca} = \frac{C_{sca}}{G}, \qquad Q_{abs} = \frac{C_{abs}}{G},$$

where G is the particle cross-sectional area projected onto a plane perpendicular to the incident beam (e.g., $G = \pi a^2$ for a sphere of radius a). The word "efficiency," together with our intuitive notions molded by geometrical optics, might lead us to believe that extinction efficiencies can never be greater than unity. Indeed, *if* geometrical optics were a completely trustworthy guide into the world of small particles, the extinction efficiency of *all* particles would be identically equal to unity: all rays incident on a particle are either absorbed or deflected by reflection and refraction. In later chapters we shall see that there are very many particles of a rather common sort which can scatter and absorb *more* light, often much more, than is geometrically incident upon them. So the wisest course is to look on the efficiencies as merely dimensionless cross sections and not hobble our thinking with imagined constraints on the values they can take.

The expressions for C_{ext} and C_{sca} were derived under the assumption of x-polarized incident light. It is clear, however, that the form of these expressions is the same for arbitrary linearly polarized incident light: we need merely reinterpret what is meant by the x direction. We must keep in mind, however, that $\mathbf{X}$ depends on the direction of polarization $\hat{\mathbf{e}}_x$.

If the incident field $\mathbf{E}_i = E_x\hat{\mathbf{e}}_x + E_y\hat{\mathbf{e}}_y$ is arbitrarily polarized, the expressions for the cross sections are

$$C_{ext} = \frac{4\pi}{k^2|\mathbf{E}_i|^2}\, \mathrm{Re}\{(\mathbf{E}_i^* \cdot \mathbf{T})_{\theta=0}\},$$

$$C_{sca} = \int_{4\pi} \frac{|\mathbf{T}|^2}{k^2|\mathbf{E}_i|^2}\, d\Omega,$$

where $\mathbf{T} = E_x\mathbf{X} + E_y\mathbf{Y}$ and $\mathbf{Y}$ is the vector scattering amplitude for incident y-polarized light. For incident *unpolarized* light ($\langle E_x E_x^* \rangle = \langle E_y E_y^* \rangle$, $\langle E_x E_y^* \rangle = \langle E_x^* E_y \rangle = 0$), these expressions yield

$$C_{ext} = \tfrac{1}{2}\big(C_{ext,\,x} + C_{ext,\,y}\big), \qquad C_{sca} = \tfrac{1}{2}\big(C_{sca,\,x} + C_{sca,\,y}\big),$$

where subscripts x and y denote cross sections for incident x-polarized and y-polarized light.

Equation (3.24) is one particular form of the *optical theorem*, the 100-year history of which has been related by Newton (1976). This theorem, which is common to all kinds of seemingly disparate scattering phenomena involving acoustic waves, electromagnetic waves, and elementary particles, expresses a very curious fact: extinction depends only on the scattering amplitude *in the forward direction*. Yet extinction is the combined effect of absorption in the particle and scattering *in all directions* by the particle. To explain this, we must examine in more detail the measurement of extinction. In so doing, we shall rely heavily on a physically intuitive derivation of the optical theorem given by van de Hulst (1949).

Consider a single arbitrary particle interposed between a source of light (taken to be x-polarized) and a detector D (Fig. 3.6). The power U incident on the detector is

$$U = \iint_D \mathbf{S}_i \cdot \hat{\mathbf{e}}_z\, dx\, dy + \iint_D \mathbf{S}_s \cdot \hat{\mathbf{e}}_z\, dx\, dy + \iint_D \mathbf{S}_{ext} \cdot \hat{\mathbf{e}}_z\, dx\, dy$$

$$= U_i + U_s + U_{ext}, \tag{3.27}$$

where integration is taken over the area of the detector. The first term in (3.27) is just $U_i = I_i A(D)$, where I_i is the incident irradiance and $A(D)$ is the area of the detector. We take the distance z between particle and detector to be sufficiently large ($kz \gg 1$) so that the detector is in the far-field region:

$$U_s = I_i \iint_D \frac{|\mathbf{X}|^2}{(kr)^2} \cos \theta \, dx \, dy. \tag{3.28}$$

If $R/z \ll 1$, where R is the maximum linear dimension of the detector, then $|\mathbf{X}|^2$, $\cos \theta$, and r are approximately constant on D, and (3.28) is

$$U_s \simeq I_i \frac{|\mathbf{X}|^2_{\theta=0}}{k^2} \Omega(D), \tag{3.29}$$

where $\Omega(D) \simeq A(D)/z^2$ is the solid angle subtended by the detector. The third term U_{ext} is

$$
\begin{aligned}
U_{\text{ext}} = I_i \text{Re} \Bigg\{ & \iint_D \frac{e^{-ik(r-z)}}{ikr} \cos \theta (\hat{\mathbf{e}}_x \cdot \mathbf{X}^*) \, dx \, dy \\
& - \iint_D \frac{e^{-ik(r-z)}}{ikr} \sin \theta \cos \phi (\hat{\mathbf{e}}_z \cdot \mathbf{X}^*) \, dx \, dy \\
& - \iint_D \frac{e^{ik(r-z)}}{ikr} (\hat{\mathbf{e}}_x \cdot \mathbf{X}) \, dx \, dy \Bigg\}.
\end{aligned} \tag{3.30}
$$

Equation (3.30) contains integrals of the form

$$J = \iint_D e^{ikzf(x, y)} g(x, y) \, dx \, dy, \tag{3.31}$$

the asymptotic behavior of which have been investigated extensively by Jones and Kline (1958) using the *method of stationary phase*. The value of J is determined by the behavior of f in the neighborhood of certain *critical points* interior to and on the boundary of D. The integrals in (3.30) are of the form (3.31) with $f = r/z - 1$. The only critical point in the interior of D is at $x = 0$, $y = 0$, where f is stationary ($\partial f/\partial x = \partial f/\partial y = 0$). If the detector is chosen so that there are no critical points on the boundary of D, then

$$J = \frac{2\pi i z}{k} g(0,0) + O\left(\frac{1}{k^2 z^2}\right). \tag{3.32}$$

In particular, a circular boundary (centered at $x = 0$, $y = 0$) for D is excluded. We also require that $kR^2/z \gg 4\pi$, which ensures that the domain of integration includes a large number of maxima and minima of the oscillatory function $\exp[ik(r - z)]$. If we use (3.32), then (3.30) becomes

$$U_{\text{ext}} = -I_i C_{\text{ext}},$$

for sufficiently large kz. Therefore, the power received by the detector is

$$U = I_i \left[A(D) - C_{ext} + \frac{|\mathbf{X}|^2_{\theta=0}}{k^2} \Omega(D) \right]. \tag{3.33}$$

The third quantity in brackets in (3.33) is the amount of energy scattered into a solid angle $\Omega(D)$ centered about the forward direction; if this solid angle is sufficiently small, consistent with the requirement that $kR^2/z \gg 4\pi$, then

$$U = I_i [A(D) - C_{ext}]. \tag{3.34}$$

Therefore, C_{ext} is a well-defined observable quantity: we measure U with and without the particle interposed between source and detector. Because C_{ext} is inherently positive, the effect of the particle is to reduce the detector area by C_{ext}; this, then, is the interpretation of C_{ext} as an area. In the language of geometrical optics we would say that the particle "casts a shadow" of area C_{ext}. However, as stated previously, this "shadow" can be considerably greater—or much less—than the particle's geometrical shadow. We note from (3.33) that C_{ext} is the *maximum* observable extinction. The scattering term $\Omega(D)|\mathbf{X}|^2_{\theta=0}/k^2$ cannot be greater than C_{sca} and is positive; therefore, the observed extinction C'_{ext} lies within the limits

$$C_{abs} \leq C'_{ext} \leq C_{ext}.$$

The full extinction C_{ext} will be observed only if the detector subtends a sufficiently small solid angle. As the detector is moved closer to the particle, however, the observed extinction will decrease. We shall see when we consider specific examples that light scattered by particles much larger than the wavelength of the incident light tends to be concentrated around the forward direction. Therefore, the larger the particle, the more difficult it is to exclude scattered light from the detector.

From (3.13) and (3.27) we have

$$U_{ext} = \iint\limits_{D} \tfrac{1}{2} \operatorname{Re}\{\mathbf{E}_i \times \mathbf{H}^*_s + \mathbf{E}_s \times \mathbf{H}^*_i\} \cdot \hat{\mathbf{e}}_r \, dA. \tag{3.35}$$

It is obvious from the form of the integrand in (3.35) that it is a manifestation of interference between the incident and forward scattered light. Conservation of energy then requires that the light removed from the incident beam by interference is accounted for by scattering in all directions and absorption in the particle.

We derived C_{ext} by two different methods. The first, integrating the Poynting vector over an imaginary sphere around the particle, emphasized the conservation of energy aspect of extinction: extinction = scattering + absorption. The second, focusing attention on what is measured in a hypothetical extinction experiment, emphasized the interference aspect of extinction: extinction = interference between incident and forward scattered light.

Up to this point we have considered only extinction by a single particle. However, the vast majority of extinction measurements involve collections of very many particles. Let us now consider such a collection, which is confined to a finite volume, the *scattering volume*. The total Poynting vector is

$$\mathbf{S} = \tfrac{1}{2} \operatorname{Re}\left\{ \mathbf{E}_i \times \mathbf{H}_i^* + \sum_j (\mathbf{E}_i \times \mathbf{H}_{sj}^* + \mathbf{E}_{sj} \times \mathbf{H}_i^*) + \sum_j \sum_k \mathbf{E}_{sj} \times \mathbf{H}_{sk}^* \right\},$$

where $(\mathbf{E}_{sj}, \mathbf{H}_{sj})$ is the electromagnetic field scattered by the jth particle. As before, we construct an imaginary sphere centered on an arbitrary point taken as origin in the scattering volume, the dimensions of which are small compared with the sphere radius r. The electric field scattered by the jth particle is

$$\mathbf{E}_{sj} \sim \frac{e^{ik(r-z)}}{-ikr} \mathbf{X}_j e^{i\delta_j} E,$$

where the incident field is taken to be x-polarized. The phase δ_j is approximately $(r \gg \xi_j)$

$$\delta_j \simeq k\left[(\hat{\mathbf{e}}_z - \hat{\mathbf{e}}_r) \cdot \boldsymbol{\xi}_j + \frac{\xi_j^2}{2r} \right],$$

where $\boldsymbol{\xi}_j$ is the position vector of the jth particle relative to the origin and $\hat{\mathbf{e}}_r = \mathbf{r}/r$ is the scattering direction. If we integrate $\mathbf{S}$ over the surface of the sphere, we obtain

$$C_{\text{ext}} = \frac{W_{\text{ext}}}{I_i} = \sum_j C_{\text{ext}, j},$$

$$C_{\text{ext}, j} = C_{\text{abs}, j} + C_{\text{sca}, j},$$

provided that $\delta_j(\theta = 0^\circ) = k\xi_j^2/2r \ll 1$ and that

$$\sum_j \int_A \frac{|\mathbf{X}_j|^2}{k^2 r^2} \, dA \gg \left| \sum_j \sum_{\substack{k \\ j \neq k}} \int_A \frac{\mathbf{X}_j \cdot \mathbf{X}_k^*}{k^2 r^2} e^{i(\delta_j - \delta_k)} \, dA \right|, \qquad (3.36)$$

(i.e., the scattering is *incoherent*); $C_{\text{ext}, j}$, $C_{\text{abs}, j}$, and $C_{\text{sca}, j}$ are the single-particle cross sections. It is difficult to give precise criteria under which (3.36) is satisfied. It is necessary, however, that the separations between the particles be uncorrelated during the time required to make a measurement.

If, in a measurement of extinction and scattering by a collection of small particles, a converging lens is placed in front of a detector which lies in the focal plane of the lens, r becomes effectively infinite. Therefore, under conditions that are likely to be frequently met in practice, the cross sections of a collection of particles are additive.

3.4.1 Extinction by a Slab of Particles

As a final example of extinction by a collection of particles let us consider a semi-infinite region $0 \le z \le h$, $-\infty < x < \infty$, $-\infty < y < \infty$, throughout which particles are more or less uniformly distributed (Fig. 3.8). The field $\mathbf{E}_t$ at the point P is the sum of the incident field $\mathbf{E}_i = E_0 e^{ikz}\hat{\mathbf{e}}_x$ and the fields scattered by the particles:

$$\mathbf{E}_t = \mathbf{E}_i + \sum_j \mathbf{E}_{sj}, \tag{3.37}$$

where the contribution to $\mathbf{E}_t$ from the particle with coordinates (x_j, y_j, z_j) is

$$\mathbf{E}_{sj} = \frac{e^{ikR_j}}{-ikR_j}\mathbf{X}_j(\hat{\mathbf{e}}_j) E_0 e^{ikz_j}, \qquad R_j = |\mathbf{R}_j|,$$

$$\hat{\mathbf{e}}_j = \frac{\mathbf{R}_j}{R_j}, \qquad \mathbf{R}_j = -x_j\hat{\mathbf{e}}_x - y_j\hat{\mathbf{e}}_y + (d - z)\hat{\mathbf{e}}_z.$$

We may assume without appreciable loss of generality that the particles are identical $(\mathbf{X}_j = \mathbf{X})$. A more important assumption is that $\mathfrak{N}$, the number of particles per unit volume, is sufficiently large such that the summation in (3.37) may be replaced by integration:

$$\sum_j \rightarrow \int\int\int \mathfrak{N}\, dx\, dy\, dz.$$

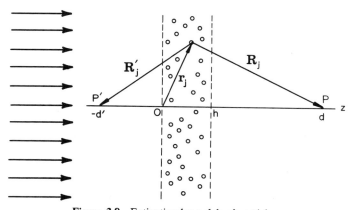

Figure 3.8 Extinction by a slab of particles.

With this assumption, the discrete variables x_j, y_j, z_j are replaced by the continuous variables x, y, z, and

$$\sum_j \mathbf{E}_{sj} \simeq E_0 \int_0^h dz \int_{-\infty}^{\infty} \int_{-\infty}^{\infty} \frac{e^{ik(R+z)}}{-ikR} \mathbf{X}(\hat{\mathbf{e}}) \, \mathfrak{N} \, dx \, dy, \qquad (3.38)$$

where $R = \sqrt{x^2 + y^2 + (d - z)^2}$ and $\hat{\mathbf{e}} = \mathbf{R}/R$. The integral over x and y in (3.38) can be evaluated in a straightforward manner by the method of stationary phase; although the limits of this integral are infinite, its value is independent of the lateral extent of the slab provided that it is large compared with $\sqrt{4\pi d/k}$. After performing the necessary integrations we obtain

$$\mathbf{E}_t = E_0 e^{ikd} \left\{ \left[1 - \frac{2\pi \mathfrak{N}}{k^2} (\mathbf{X} \cdot \hat{\mathbf{e}}_x)_{\theta=0} h \right] \hat{\mathbf{e}}_x - \frac{2\pi \mathfrak{N}}{k^2} (\mathbf{X} \cdot \hat{\mathbf{e}}_y)_{\theta=0} h \hat{\mathbf{e}}_y \right\}.$$

Although the incident light is x-polarized, the transmitted light has a y component if $(\mathbf{X} \cdot \hat{\mathbf{e}}_y)_{\theta=0} \neq 0$; that is, the direction of vibration of the incident beam is, in general, *rotated* on transmission through the slab. If we assume that there is no rotation and, moreover, if $|2\pi \mathfrak{N} k^{-2} h(\mathbf{X} \cdot \hat{\mathbf{e}}_x)_{\theta=0}| \ll 1$, we may write

$$\mathbf{E}_t = \mathbf{E}_i \exp\left[-\frac{2\pi \mathfrak{N} h}{k^2} (\mathbf{X} \cdot \hat{\mathbf{e}}_x)_{\theta=0} \right]. \qquad (3.39)$$

The transmission coefficient of a *homogeneous* slab of thickness h and refractive index $\tilde{N}$, which is embedded in a homogeneous medium with refractive index N, where $\tilde{N} \simeq N$, is given by (2.73):

$$\tilde{t}_{slab} \simeq e^{i(\tilde{k} - k)h}, \qquad (3.40)$$

where $\tilde{k} = 2\pi \tilde{N}/\lambda$. If we compare (3.39) and (3.40), we note that the slab of particles is equivalent, at least as far as transmission is concerned, to the homogeneous slab if

$$\frac{\tilde{N}}{N} = 1 + i \frac{2\pi \mathfrak{N}}{k^3} (\mathbf{X} \cdot \hat{\mathbf{e}}_x)_{\theta=0}. \qquad (3.41)$$

Therefore, within limits, we may interpret $\tilde{N}$ in (3.41) as the effective refractive index of the slab of particles. This leads naturally to the question: To what extent is $\tilde{N}$ similar to the refractive index of a homogeneous medium? For example, under what conditions, if any, will substitution of $\tilde{N}$ into the expression for the reflection coefficient of a homogeneous slab yield physically correct results? We can answer the latter and more specific of these two questions by calculating the field $\mathbf{E}_r$ at the point P' (Fig. 3.8), which is the sum of the individual fields scattered by the particles:

$$\mathbf{E}_r = \sum_j \frac{e^{ikR_j}}{-ikR_j} \mathbf{X}_j(\hat{\mathbf{e}}_j) E_0 e^{ikz_j}, \tag{3.42}$$

where $\mathbf{R}_j = -[x_j \hat{\mathbf{e}}_x + y_j \hat{\mathbf{e}}_y + (d + z_j)\hat{\mathbf{e}}_z]$. Again, we assume identical particles, and the summation (3.42) is approximated by an integral, which can be evaluated by the method of stationary phase:

$$\mathbf{E}_r = -E_0 e^{ikd}(1 - e^{i2kh}) \frac{i\pi \mathfrak{N}}{k^3} \mathbf{X}_{\theta = 180}. \tag{3.43}$$

The reflection coefficient for a homogeneous slab with refractive index $\tilde{N} \simeq N$ is, from (2.72),

$$\tilde{r}_{\text{slab}} \simeq \frac{N - \tilde{N}}{2N}(1 - e^{i2\tilde{k}h}). \tag{3.44}$$

Thus, if we assume that $(\mathbf{X} \cdot \hat{\mathbf{e}}_y)_{\theta = 180} = 0$, then (3.41) substituted into (3.44) is consistent with (3.43) provided that

$$\mathbf{X}_{\theta = 0} = \mathbf{X}_{\theta = 180}. \tag{3.45}$$

As we shall see in later chapters, (3.45) is satisfied for particles small compared with the wavelength. For a collection of such particles, therefore, the concept of an effective refractive index is meaningful at least as far as transmission and reflection are concerned. However, even if the particles are small compared with the wavelength, $\tilde{N}$ should not be interpreted too literally as a refractive index on the same footing as the refractive index of a homogeneous medium. For example, attenuation in a strictly homogeneous medium is the result of absorption, which is accounted for quantitatively by the imaginary part of the refractive index. In a particulate medium, however, attenuation may be wholly or in part the result of scattering. Even if the particles are nonabsorbing, the imaginary part of the effective refractive index (3.41) can be nonzero.

From (3.39) and the optical theorem (3.24) it follows that the irradiance is attenuated according to $I_t = I_i \exp(-\alpha_{\text{ext}} h)$ as the incident beam traverses the slab of particles, where the *attenuation coefficient* α_{ext} is

$$\alpha_{\text{ext}} = \mathfrak{N} C_{\text{ext}} = \mathfrak{N} C_{\text{abs}} + \mathfrak{N} C_{\text{sca}}. \tag{3.46}$$

Although we assumed identical particles, this was done to avoid a cluttered notation, and it is not a restriction on the validity of our analysis; the results above are readily generalized to a mixture of different particles. For example, the attenuation coefficient of such a mixture is

$$\alpha_{\text{ext}} = \sum_j \mathfrak{N}_j C_{\text{ext}, j},$$

where $\mathfrak{N}_j$ is the number of particles of type j per unit volume and $C_{\mathrm{ext},\,j}$ is the corresponding extinction cross section.

Underlying (3.39), and hence exponential attenuation of irradiance in particulate media, is the requirement that $\alpha_{\mathrm{ext}} h \ll 1$. This condition may be relaxed somewhat if the scattering contribution to total attenuation is small (i.e., $\mathfrak{N} C_{\mathrm{sca}} h \ll 1$). To justify this assertion fully would take us somewhat afield into the theory of radiative transfer. But we can give a brief heuristic argument as follows. An amount of light dI is removed from a beam propagating in the z direction through an infinitesimal distance between z and $z + dz$ in a slab of particles:

$$dI = -\alpha_{\mathrm{ext}} I\, dz, \tag{3.47}$$

where I is the beam irradiance at z. However, light can get back into the beam by multiple scattering; that is, light scattered at other positions in the slab may ultimately contribute to the irradiance at z. Scattered light, in contradistinction to absorbed light, is not irretrievably lost from the system—it merely changes direction and is lost from a beam propagating in a particular direction—but contributes to other directions. Clearly, the greater the scattering cross section, number density of particles, and slab thickness h, the greater will be the multiple scattering contribution to the irradiance at z. Thus, if $\mathfrak{N} C_{\mathrm{sca}} h$ is sufficiently small, we may ignore multiple scattering and (3.47) can be integrated to yield $I_t = I_i \exp(-\alpha_{\mathrm{ext}} h)$.

Sometimes it is of interest to compare attenuation of light by a given material in the bulk with that in the finely divided, or particulate, states. In order for the comparison to be fair, however, we have to consider equal masses or, equivalently, equal volumes of material in the two states. If $\mathfrak{N}$ is the particle number density, $1/\mathfrak{N}$ is the average volume allocated to a single particle, and the volume fraction f of particles in the collection is $\mathfrak{N} v$, where v is the volume of a single particle. Thus, we can write the attenuation coefficient (3.46) as $\alpha_{\mathrm{ext}} = f C_{\mathrm{ext}}/v$. Imagine now that the particles are compressed into a homogeneous slab ($f = 1$) without, however, losing their individual identities and properties. We shall call the resulting attenuation coefficient, which is the extinction cross section per unit particle volume, the *volume attenuation coefficient* α_v:

$$\alpha_v = \frac{C_{\mathrm{ext}}}{v}. \tag{3.48}$$

The *mass attenuation coefficient* α_m, defined as the extinction cross section per unit particle mass, is related to the volume attenuation coefficient by

$$\alpha_m = \frac{\alpha_v}{\rho},$$

where ρ is the density of the particle. If any quantity deserves to be called an extinction "efficiency," it is the extinction cross section per unit volume (or mass) rather than the extinction cross section per unit area. For it is the former quantity that tells us how effective a *fixed mass* of particles is in removing light from a beam. Suppose that we set on a chunk of material with a hammer and smash it to bits. What size should the bits be to most effectively extinguish light of a given wavelength? To answer this question, it is clear that we should plot C_{ext}/v as a function of size rather than, as is traditionally done, Q_{ext}. Sacrosanct though it may be, Q_{ext} conveys less physical information than C_{ext}/v, and we shall often present the latter rather than the former.

A set of measurements of I_t/I_i over some range of wavelengths for a homogeneous slab of material is called the *transmission*, or *absorption*, *spectrum* of the material. We may look on this spectrum as the fingerprints of the material. However, it is possible to smudge these fingerprints drastically. If, for example, we measure the transmission spectrum of a given homogeneous material and then divide it by some means into a collection of small particles, the transmission spectrum of the particulate medium will often bear little resemblance to that of the bulk parent material. The chemical composition is the same in both instances, but the state of aggregation has changed. An obvious source of difference between the two spectra is scattering: if fluctuations are ignored, the homogeneous slab does not scatter light, whereas the transmission spectrum of the particulate medium may be primarily the result of scattering. But the difference goes deeper than this: even if one could correct for scattering (e.g., by suitably collecting the scattered light), large differences might still exist. Thus, the gross optical properties (e.g., reflection and transmission) of a given material can and do differ appreciably depending on its state of aggregation. We shall encounter many examples of this in later chapters.

NOTES AND COMMENTS

Our derivation of (3.24) is similar to that of Jones (1955).

Chapter 4

Absorption and Scattering by a Sphere

Perhaps the most important exactly soluble problem in the theory of absorption and scattering by small particles is that for a sphere of arbitrary radius and refractive index. Although the formal solution to this problem has been available for many years, only since the advent of large digital computers has it been a practical means for detailed computations. In 1908, Gustav Mie developed the theory in an effort to understand the varied colors in absorption and scattering exhibited by small colloidal particles of gold suspended in water. About the same time Peter Debye considered the problem of the radiation pressure exerted on small particles in space. Debye's work, which was the subject of his doctoral dissertation, is one of the first applications of the theory to an astrophysical problem. Neither Mie nor Debye was the first to construct a solution to the sphere problem; however, establishing who precisely was the first is not an easy task, although Lorenz is a strong contender for this honor. Without trying to establish historical precedents, we shall accept the most common term, the Mie theory. A good concise but thorough treatment of the history of the sphere problem is given in Kerker (1969, pp. 54–59).

The mathematical basis of the Mie theory is the subject of this chapter. Expressions for absorption and scattering cross sections and angle-dependent scattering functions are derived; reference is then made to the computer program in Appendix A, which provides for numerical calculations of these quantities. This is the point of departure for a host of applications in several fields of applied science, which are covered in more detail in Part 3. The mathematics, divorced from physical phenomena, can be somewhat boring. For this reason, a few illustrative examples are sprinkled throughout the chapter. These are just appetizers to help maintain the reader's interest; a fuller meal will be served in Part 3.

Whereas the mathematics of the Mie theory is straightforward, if somewhat cumbersome, the physics of the interaction of an electromagnetic wave with a sphere is extremely complicated. It is a relatively easy matter to write the infinite series expansions of the electromagnetic fields at all points of space. It is an even easier matter these days to produce great reams of output from Mie computations. A more difficult task, however, is to visualize the fields, to categorize the significant electromagnetic modes inside and outside the sphere, and to acquire some intuitive feeling for how a sphere of given size and optical properties absorbs and scatters light.

On the one hand, there are those who scoff at the use of the Mie theory to describe any properties of nonspherical particles, the type of particles that are likely to inhabit planetary atmospheres and the interstellar medium; on the other hand, there are those who unquestioningly use Mie theory for any and every aspect of light interaction with such particles. Neither attitude is enlightened. The Mie theory, limited though it may be, does provide a first-order description of optical effects in nonspherical particles, and it correctly describes many small-particle effects that are not intuitively obvious. For example, we shall see in Chapter 11, where we consider nonspherical particles, that the sphere solution is an excellent guide to the changes that occur when absorption increases or size dispersion changes. Therefore, our approach will be to explore fully the Mie theory and the optical effects it describes and then, in succeeding chapters, to critically examine its failings in dealing with nonspherical particles.

4.1 SOLUTIONS TO THE VECTOR WAVE EQUATIONS

We showed in Chapter 3 that a physically realizable time-harmonic electromagnetic field $(\mathbf{E}, \mathbf{H})$ in a linear, isotropic, homogeneous medium must satisfy the wave equation

$$\nabla^2 \mathbf{E} + k^2 \mathbf{E} = 0, \qquad \nabla^2 \mathbf{H} + k^2 \mathbf{H} = 0,$$

where $k^2 = \omega^2 \varepsilon \mu$, and be divergence-free

$$\nabla \cdot \mathbf{E} = 0, \qquad \nabla \cdot \mathbf{H} = 0.$$

In addition, $\mathbf{E}$ and $\mathbf{H}$ are not independent:

$$\nabla \times \mathbf{E} = i\omega\mu\mathbf{H}, \qquad \nabla \times \mathbf{H} = -i\omega\varepsilon\mathbf{E}.$$

Suppose that, given a *scalar* function ψ and an arbitrary *constant* vector $\mathbf{c}$, we construct a *vector* function $\mathbf{M}$:

$$\mathbf{M} = \nabla \times (\mathbf{c}\psi).$$

The divergence of the curl of any vector function vanishes:

$$\nabla \cdot \mathbf{M} = 0.$$

If we use the vector identities

$$\nabla \times (\mathbf{A} \times \mathbf{B}) = \mathbf{A}(\nabla \cdot \mathbf{B}) - \mathbf{B}(\nabla \cdot \mathbf{A}) + (\mathbf{B} \cdot \nabla)\mathbf{A} - (\mathbf{A} \cdot \nabla)\mathbf{B},$$

$$\nabla(\mathbf{A} \cdot \mathbf{B}) = \mathbf{A} \times (\nabla \times \mathbf{B}) + \mathbf{B} \times (\nabla \times \mathbf{A}) + (\mathbf{B} \cdot \nabla)\mathbf{A} + (\mathbf{A} \cdot \nabla)\mathbf{B},$$

we obtain

$$\nabla^2 \mathbf{M} + k^2 \mathbf{M} = \nabla \times \left[\mathbf{c} (\nabla^2 \psi + k^2 \psi) \right].$$

Therefore, $\mathbf{M}$ satisfies the *vector* wave equation if ψ is a solution to the *scalar* wave equation

$$\nabla^2 \psi + k^2 \psi = 0.$$

We may also write $\mathbf{M} = -\mathbf{c} \times \nabla \psi$, which shows that $\mathbf{M}$ is perpendicular to $\mathbf{c}$. Let us construct from $\mathbf{M}$ another vector function

$$\mathbf{N} = \frac{\nabla \times \mathbf{M}}{k}$$

with zero divergence, which also satisfies the vector wave equation

$$\nabla^2 \mathbf{N} + k^2 \mathbf{N} = 0.$$

We also have

$$\nabla \times \mathbf{N} = k\mathbf{M}.$$

Therefore, $\mathbf{M}$ and $\mathbf{N}$ have all the required properties of an electromagnetic field: they satisfy the vector wave equation, they are divergence-free, the curl of $\mathbf{M}$ is proportional to $\mathbf{N}$, and the curl of $\mathbf{N}$ is proportional to $\mathbf{M}$. Thus, the problem of finding solutions to the field equations reduces to the comparatively simpler problem of finding solutions to the scalar wave equation. We shall call the scalar function ψ a *generating function* for the *vector harmonics* $\mathbf{M}$ and $\mathbf{N}$; the vector $\mathbf{c}$ is sometimes called the *guiding* or *pilot* vector.

The choice of generating functions is dictated by whatever symmetry may exist in the problem. In this chapter we are interested in scattering by a sphere; therefore, we choose functions ψ that satisfy the wave equation in spherical polar coordinates r, θ, ϕ (Fig. 4.1). The choice of pilot vector is somewhat less obvious. We could choose some arbitrary vector $\mathbf{c}$. However, if we take

$$\mathbf{M} = \nabla \times (\mathbf{r}\psi), \tag{4.1}$$

where $\mathbf{r}$ is the radius vector, then $\mathbf{M}$ is a solution to the vector wave equation in *spherical polar coordinates*. In problems involving spherical symmetry, therefore, we shall take $\mathbf{M}$ given in (4.1) and the associated $\mathbf{N}$ as our fundamental solutions to the field equations. Note that $\mathbf{M}$ is everywhere tangential to any sphere $|\mathbf{r}| = $ constant (i.e., $\mathbf{r} \cdot \mathbf{M} = 0$).

The scalar wave equation in spherical polar coordinates is

$$\frac{1}{r^2} \frac{\partial}{\partial r} \left(r^2 \frac{\partial \psi}{\partial r} \right) + \frac{1}{r^2 \sin \theta} \frac{\partial}{\partial \theta} \left(\sin \theta \frac{\partial \psi}{\partial \theta} \right) + \frac{1}{r^2 \sin \theta} \frac{\partial^2 \psi}{\partial \phi^2} + k^2 \psi = 0. \tag{4.2}$$

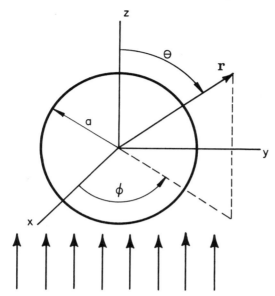

Figure 4.1 Spherical polar coordinate system centered on a spherical particle of radius a.

We seek particular solutions to (4.2) of the form

$$\psi(r, \theta, \phi) = R(r)\Theta(\theta)\Phi(\phi),$$

which when substituted into (4.2) yield the three separated equations:

$$\frac{d^2\Phi}{d\phi^2} + m^2\Phi = 0, \tag{4.3}$$

$$\frac{1}{\sin\theta}\frac{d}{d\theta}\left(\sin\theta\frac{d\Theta}{d\theta}\right) + \left[n(n+1) - \frac{m^2}{\sin^2\theta}\right]\Theta = 0, \tag{4.4}$$

$$\frac{d}{dr}\left(r^2\frac{dR}{dr}\right) + \left[k^2r^2 - n(n+1)\right]R = 0, \tag{4.5}$$

where the *separation constants* m and n are determined by subsidiary conditions that ψ must satisfy. We first note that if, for a given m, Φ_m is a solution to (4.3), then Φ_{-m} is not a linearly independent solution. The linearly independent solutions are

$$\Phi_e = \cos m\phi, \qquad \Phi_o = \sin m\phi,$$

where subscripts e and o denote *even* and *odd*. We require that ψ be a *single-valued* function of the azimuthal angle ϕ:

$$\lim_{\nu \to 2\pi} \psi(\phi + \nu) = \psi(\phi) \tag{4.6}$$

for all ϕ except, possibly, at points on the boundary between regions with different properties. However, we need not concern ourselves with such boundary points; we are only interested in solutions to the scalar wave equation at interior points of homogeneous regions. Condition (4.6) then requires m to be an integer or zero; positive values of m are sufficient to generate all the linearly independent solutions to (4.3).

The solutions to (4.4) that are finite at $\theta = 0$ and $\theta = \pi$ are the *associated Legendre functions* of the first kind $P_n^m(\cos\theta)$ of degree n and order m, where $n = m, m + 1, \ldots$ (see, e.g., Courant and Hilbert, 1953, pp. 326, 327). These functions are orthogonal:

$$\int_{-1}^{1} P_n^m(\mu) P_{n'}^m(\mu)\, d\mu = \delta_{n'n} \frac{2}{2n + 1} \frac{(n + m)!}{(n - m)!}, \tag{4.7}$$

where $\mu = \cos\theta$ and $\delta_{n'n}$, the Kronecker delta, is unity if $n = n'$ and zero otherwise. When $m = 0$ the associated Legendre functions are the *Legendre polynomials*, which are denoted by P_n.

If we introduce the dimensionless variable $\rho = kr$ and define the function $Z = R\sqrt{\rho}$, (4.5) becomes

$$\rho \frac{d}{d\rho}\left(\rho \frac{dZ}{d\rho}\right) + \left[\rho^2 - (n + \tfrac{1}{2})^2\right] Z = 0. \tag{4.8}$$

The linearly independent solutions to (4.8) are the *Bessel functions* of first and second kind J_ν and Y_ν (the symbol N_ν is often used instead of Y_ν), where the *order* $\nu = n + \tfrac{1}{2}$ is half-integral. Therefore, the linearly independent solutions to (4.5) are the *spherical Bessel functions*

$$j_n(\rho) = \sqrt{\frac{\pi}{2\rho}}\, J_{n+1/2}(\rho), \tag{4.9}$$

$$y_n(\rho) = \sqrt{\frac{\pi}{2\rho}}\, Y_{n+1/2}(\rho), \tag{4.10}$$

where the constant factor $\sqrt{\pi/2}$ is introduced for convenience. The spherical Bessel functions satisfy the recurrence relations

$$z_{n-1}(\rho) + z_{n+1}(\rho) = \frac{2n + 1}{\rho} z_n(\rho), \tag{4.11}$$

$$(2n + 1)\frac{d}{d\rho} z_n(\rho) = n z_{n-1}(\rho) - (n + 1) z_{n+1}(\rho), \tag{4.12}$$

where z_n is either j_n or y_n. From the first two orders

$$j_0(\rho) = \frac{\sin \rho}{\rho}, \qquad j_1(\rho) = \frac{\sin \rho}{\rho^2} - \frac{\cos \rho}{\rho},$$

$$y_0(\rho) = -\frac{\cos \rho}{\rho}, \qquad y_1(\rho) = -\frac{\cos \rho}{\rho^2} - \frac{\sin \rho}{\rho},$$

higher-order functions can be generated by recurrence. Note that for all orders n, $y_n(kr)$ becomes infinite as r approaches the origin. In Fig. 4.2 we show $j_n(x)$ and $y_n(x)(n = 0, 1, 2, 3)$ for real values of x, although the spherical Bessel functions are not restricted to real arguments.

Any linear combination of j_n and y_n is also a solution to (4.5). If the mood were to strike us, therefore, we could just as well take as fundamental solutions to (4.5) any two linearly independent combinations. Two such combinations deserve special attention, the *spherical Bessel functions of the third kind* (sometimes called spherical Hankel functions):

$$h_n^{(1)}(\rho) = j_n(\rho) + iy_n(\rho), \tag{4.13}$$

$$h_n^{(2)}(\rho) = j_n(\rho) - iy_n(\rho). \tag{4.14}$$

We hasten to add that we have introduced even more Bessel functions neither for "completeness" nor for the further aggrandizement of Friedrich Wilhelm Bessel (1784–1846), who, with a veritable zoo of functions to his credit, not to mention infinite series, a revered inequality, an interpolation scheme, and various other mathematical artifacts, needs no publicity; as we shall see, (4.13) and (4.14) will save some labor, a sufficient reason for admitting more functions into our larder.

We have now done enough work to construct generating functions that satisfy the scalar wave equation in spherical polar coordinates:

$$\psi_{emn} = \cos m\phi P_n^m(\cos \theta) z_n(kr), \tag{4.15}$$

$$\psi_{omn} = \sin m\phi P_n^m(\cos \theta) z_n(kr), \tag{4.16}$$

where z_n is any of the four spherical Bessel functions j_n, y_n, $h_n^{(1)}$, or $h_n^{(2)}$. Moreover, because of the *completeness* of the functions $\cos m\phi$, $\sin m\phi$, $P_n^m(\cos \theta)$, $z_n(kr)$, any function that satisfies the scalar wave equation in spherical polar coordinates may be expanded as an infinite series in the functions (4.15) and (4.16). The vector spherical harmonics generated by ψ_{emn} and ψ_{omn} are

$$\mathbf{M}_{emn} = \nabla \times (\mathbf{r}\psi_{emn}), \qquad \mathbf{M}_{omn} = \nabla \times (\mathbf{r}\psi_{omn}),$$

$$\mathbf{N}_{emn} = \frac{\nabla \times \mathbf{M}_{emn}}{k}, \qquad \mathbf{N}_{omn} = \frac{\nabla \times \mathbf{M}_{omn}}{k},$$

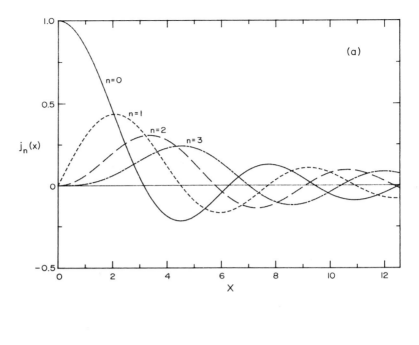

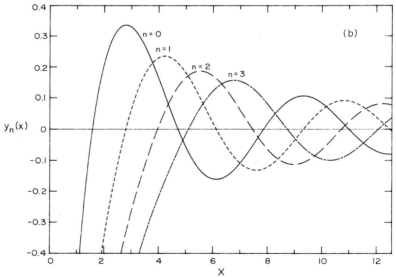

Figure 4.2 Spherical Bessel functions of the first (a) and second (b) kind.

which, in component form, may be written

$$\mathbf{M}_{emn} = \frac{-m}{\sin\theta}\sin m\phi\, P_n^m(\cos\theta)z_n(\rho)\hat{\mathbf{e}}_\theta$$

$$-\cos m\phi\frac{dP_n^m(\cos\theta)}{d\theta}z_n(\rho)\hat{\mathbf{e}}_\phi, \tag{4.17}$$

$$\mathbf{M}_{omn} = \frac{m}{\sin\theta}\cos m\phi\, P_n^m(\cos\theta)z_n(\rho)\hat{\mathbf{e}}_\theta$$

$$-\sin m\phi\frac{dP_n^m(\cos\theta)}{d\theta}z_n(\rho)\hat{\mathbf{e}}_\phi, \tag{4.18}$$

$$\mathbf{N}_{emn} = \frac{z_n(\rho)}{\rho}\cos m\phi\, n(n+1)P_n^m(\cos\theta)\hat{\mathbf{e}}_r$$

$$+\cos m\phi\frac{dP_n^m(\cos\theta)}{d\theta}\frac{1}{\rho}\frac{d}{d\rho}\big[\rho z_n(\rho)\big]\hat{\mathbf{e}}_\theta$$

$$-m\sin m\phi\frac{P_n^m(\cos\theta)}{\sin\theta}\frac{1}{\rho}\frac{d}{d\rho}\big[\rho z_n(\rho)\big]\hat{\mathbf{e}}_\phi, \tag{4.19}$$

$$\mathbf{N}_{omn} = \frac{z_n(\rho)}{\rho}\sin m\phi\, n(n+1)P_n^m(\cos\theta)\hat{\mathbf{e}}_r$$

$$+\sin m\phi\frac{dP_n^m(\cos\theta)}{d\theta}\frac{1}{\rho}\frac{d}{d\rho}\big[\rho z_n(\rho)\big]\hat{\mathbf{e}}_\theta$$

$$+m\cos m\phi\frac{P_n^m(\cos\theta)}{\sin\theta}\frac{1}{\rho}\frac{d}{d\rho}\big[\rho z_n(\rho)\big]\hat{\mathbf{e}}_\phi, \tag{4.20}$$

where the r-component of $\mathbf{N}_{mn}$ has been simplified by using the fact that P_n^m satisfies (4.4). Any solution to the field equations can now be expanded in an infinite series of the functions (4.17)–(4.20). Thus, armed with vector harmonics, we are ready to attack the problem of scattering by an arbitrary sphere.

4.2 EXPANSION OF A PLANE WAVE IN VECTOR SPHERICAL HARMONICS

Expansion of a plane wave in vector spherical harmonics is a lengthy, although straightforward, procedure. In this section we outline how one goes about determining the coefficients in such an expansion.

The problem with which we are concerned is scattering of a plane x-polarized wave, written in spherical polar coordinates as

$$\mathbf{E}_i = E_0 e^{ikr\cos\theta}\hat{\mathbf{e}}_x, \tag{4.21}$$

where

$$\hat{\mathbf{e}}_x = \sin\theta\cos\phi\,\hat{\mathbf{e}}_r + \cos\theta\cos\phi\,\hat{\mathbf{e}}_\theta - \sin\phi\,\hat{\mathbf{e}}_\phi, \tag{4.22}$$

by an arbitrary sphere. The first step toward the solution to this problem is expanding (4.21) in vector spherical harmonics:

$$\mathbf{E}_i = \sum_{m=0}^{\infty} \sum_{n=m}^{\infty} \left(B_{emn}\mathbf{M}_{emn} + B_{omn}\mathbf{M}_{omn} \right.$$

$$\left. + A_{emn}\mathbf{N}_{emn} + A_{omn}\mathbf{N}_{omn} \right). \tag{4.23}$$

Because $\sin m\phi$ is orthogonal to $\cos m'\phi$ for all m and m' it follows that $\mathbf{M}_{emn}$ and $\mathbf{M}_{omn}$ are orthogonal in the sense that

$$\int_0^{2\pi}\int_0^{\pi}\mathbf{M}_{em'n'}\cdot\mathbf{M}_{omn}\sin\theta\,d\theta\,d\phi = 0 \qquad (\text{all } m, m', n, n').$$

Similarly, $(\mathbf{N}_{omn}, \mathbf{N}_{emn})$, $(\mathbf{M}_{omn}, \mathbf{N}_{omn})$ and $(\mathbf{M}_{emn}, \mathbf{N}_{emn})$ are mutually orthogonal sets of functions. The orthogonality properties of $\cos m\phi$ and $\sin m\phi$ imply that all vector harmonics of different order m are mutually orthogonal.

To prove that the functions $(\mathbf{M}_{emn}, \mathbf{N}_{omn})$ and $(\mathbf{N}_{emn}, \mathbf{M}_{omn})$ are orthogonal, we must show that the integral

$$m\int_0^{\pi}\left(P_n^m\frac{dP_{n'}^m}{d\theta} + P_{n'}^m\frac{dP_n^m}{d\theta} \right)d\theta = P_n^m P_{n'}^m|_0^{\pi} \tag{4.24}$$

vanishes for all n and n'. The associated Legendre function P_n^m is related to the mth derivative of the corresponding Legendre polynomial P_n,

$$P_n^m(\mu) = \left(1 - \mu^2\right)^{m/2}\frac{d^m P_n(\mu)}{d\mu^m}, \tag{4.25}$$

where $\mu = \cos\theta$, from which it follows that P_n^m vanishes for $\theta = 0$ and $\theta = \pi$ except when $m = 0$. Therefore, (4.24) vanishes for all m, n, and n'.

The proof of the remaining orthogonality relations

$$\int_0^{2\pi}\int_0^{\pi}\mathbf{M}_{emn}\cdot\mathbf{M}_{emn'}\sin\theta\,d\theta\,d\phi = \int_0^{2\pi}\int_0^{\pi}\mathbf{M}_{omn}\cdot\mathbf{M}_{omn'}\sin\theta\,d\theta\,d\phi = 0,$$

$$\int_0^{2\pi}\int_0^{\pi}\mathbf{N}_{emn}\cdot\mathbf{N}_{emn'}\sin\theta\,d\theta\,d\phi = \int_0^{2\pi}\int_0^{\pi}\mathbf{N}_{omn}\cdot\mathbf{N}_{omn'}\sin\theta\,d\theta\,d\phi = 0,$$

when $n \neq n'$ and $m \neq 0$, requires showing that

$$\int_0^{\pi}\left(\frac{dP_n^m}{d\theta}\frac{dP_{n'}^m}{d\theta} + m^2\frac{P_n^m P_{n'}^m}{\sin^2\theta} \right)\sin\theta\,d\theta = 0. \tag{4.26}$$

Because both P_n^m and $P_{n'}^m$ satisfy (4.4), we have, after a bit of manipulation

$$2\sin\theta\left(\frac{dP_n^m}{d\theta}\frac{dP_{n'}^m}{d\theta} + m^2\frac{P_n^m P_{n'}^m}{\sin^2\theta} \right) = 2n(n+1)P_n^m P_{n'}^m\sin\theta$$

$$+ \frac{d}{d\theta}\left(\sin\theta\frac{dP_{n'}^m}{d\theta}P_n^m + \sin\theta\frac{dP_n^m}{d\theta}P_{n'}^m \right), \tag{4.27}$$

from which, together with the orthogonality relations for the P_n^m, (4.26) readily follows. When $m = 0$, $\mathbf{N}_{omn}$ and $\mathbf{M}_{omn}$ vanish; the orthogonality of the $\mathbf{M}_{emn}$ and the $\mathbf{N}_{emn}$ when $m = 0$ also follows from (4.26) and (4.27).

The orthogonality of all the vector spherical harmonics, which was established in the preceding section, implies that the coefficients in the expansion (4.23) are of the form

$$B_{emn} = \frac{\int_0^{2\pi}\int_0^{\pi}\mathbf{E}_i\cdot\mathbf{M}_{emn}\sin\theta\,d\theta\,d\phi}{\int_0^{2\pi}\int_0^{\pi}|\mathbf{M}_{emn}|^2\sin\theta\,d\theta\,d\phi},$$

with similar expressions for B_{omn}, A_{emn}, and A_{omn}. It follows from (4.17), (4.20), and (4.22), together with the orthogonality of the sine and cosine, that $B_{emn} = A_{omn} = 0$ for all m and n. Moreover, the remaining coefficients vanish unless $m = 1$ for the same reason. The incident field is finite at the origin, which requires that $j_n(kr)$ is the appropriate spherical Bessel function in the generating functions ψ_{o1n} and ψ_{e1n}; we reject y_n because of its misbehavior at the origin. We shall append the superscript (1) to vector spherical harmonics for which the radial dependence of the generating functions is specified by j_n. Thus, the expansion for $\mathbf{E}_i$ has the form

$$\mathbf{E}_i = \sum_{n=1}^{\infty}\left(B_{o1n}\mathbf{M}_{o1n}^{(1)} + A_{e1n}\mathbf{N}_{e1n}^{(1)}\right). \tag{4.28}$$

The integral in the denominator of the expression for B_{o1n} can readily be evaluated from (4.27); the numerator, however, contains the integral

$$\int_0^{\pi}\frac{d}{d\theta}\left(\sin\theta P_n^1\right)e^{i\rho\cos\theta}\,d\theta. \tag{4.29}$$

From (4.25) we have

$$P_n^1 = -\frac{dP_n}{d\theta}, \tag{4.30}$$

where the Legendre polynomials of degree n satisfy (4.4):

$$\frac{d}{d\theta}\left(\sin\theta\frac{dP_n}{d\theta}\right) = -n(n+1)P_n\sin\theta. \tag{4.31}$$

Thus, (4.29) is proportional to

$$\int_0^{\pi}e^{i\rho\cos\theta}P_n\sin\theta\,d\theta.$$

The final step is Gegenbauer's generalization of Poisson's integral (Watson,

1958, p. 50):

$$j_n(\rho) = \frac{i^{-n}}{2} \int_0^\pi e^{i\rho \cos \theta} P_n \sin \theta \, d\theta. \tag{4.32}$$

Without further fanfare, therefore, we arrive at the expansion coefficients

$$B_{o1n} = i^n E_0 \frac{2n + 1}{n(n + 1)}. \tag{4.33}$$

The expansion coefficients A_{emn} are somewhat less tractable. For example, we are faced with the integral

$$\int_0^\pi P_n^1 \sin \theta e^{i\rho \cos \theta} \sin \theta \, d\theta, \tag{4.34}$$

which may be integrated by parts to yield

$$\frac{2n(n + 1) j_n(\rho) i^n}{i\rho},$$

where we have also used (4.30), (4.31), and (4.32). The nastiest integral of the lot, however, is

$$\int_0^\pi \left(\cos \theta \frac{dP_n^1}{d\theta} + \frac{P_n^1}{\sin \theta} \right) e^{i\rho \cos \theta} \sin \theta \, d\theta, \tag{4.35}$$

which may be brought to earth by first multiplying (4.32) by ρ and then differentiating the resulting expression with respect to ρ. After a good bit of algebra we obtain

$$\frac{2n(n + 1)i^n}{\rho} \frac{d}{d\rho}(\rho j_n)$$

for (4.35). The expansion coefficients then follow straightforwardly:

$$A_{e1n} = -iE_0 i^n \frac{2n + 1}{n(n + 1)}. \tag{4.36}$$

The desired expansion of a plane wave in spherical harmonics

$$\mathbf{E}_i = E_0 \sum_{n=1}^{\infty} i^n \frac{2n + 1}{n(n + 1)} \left(\mathbf{M}_{o1n}^{(1)} - i\mathbf{N}_{e1n}^{(1)} \right) \tag{4.37}$$

was not achieved without difficulty. This is undoubtedly the result of the unwillingness of a plane wave to wear a guise in which it feels uncomfortable; expanding a plane wave in spherical wave functions is somewhat like trying to force a square peg into a round hole. However, the reader who has painstakingly followed the derivation of (4.37), and thereby acquired virtue through suffering, may derive some comfort from the knowledge that it is relatively clear sailing from here on.

4.3 THE INTERNAL AND SCATTERED FIELDS

Suppose that a plane x-polarized wave is incident on a homogeneous, isotropic sphere of radius a (Fig. 4.1). As we showed in the preceding section, the incident electric field may be expanded in an infinite series of vector spherical harmonics. The corresponding incident magnetic field is obtained from the curl of (4.37):

$$\mathbf{H}_i = \frac{-\mathbf{k}}{\omega\mu} E_o \sum_{n=1}^{\infty} i^n \frac{2n+1}{n(n+1)} \left(\mathbf{M}_{e1n}^{(1)} + i\mathbf{N}_{o1n}^{(1)} \right). \tag{4.38}$$

We may also expand the scattered electromagnetic field $(\mathbf{E}_s, \mathbf{H}_s)$ and the field $(\mathbf{E}_1, \mathbf{H}_1)$ inside the sphere in vector spherical harmonics. At the boundary between the sphere and the surrounding medium we impose the conditions (3.7):

$$(\mathbf{E}_i + \mathbf{E}_s - \mathbf{E}_1) \times \hat{\mathbf{e}}_r = (\mathbf{H}_i + \mathbf{H}_s - \mathbf{H}_1) \times \hat{\mathbf{e}}_r = 0. \tag{4.39}$$

The boundary conditions (4.39), the orthogonality of the vector harmonics, and the form of the expansion of the incident field dictate the form of the expansions for the scattered field and the field inside the sphere: the coefficients in these expansions vanish for all $m \neq 1$. Finiteness at the origin requires that we take $j_n(\mathbf{k}_1 r)$, where $\mathbf{k}_1$ is the wave number in the sphere, as the appropriate spherical Bessel functions in the generating functions for the vector harmonics inside the sphere. Thus, the expansion of the field $(\mathbf{E}_1, \mathbf{H}_1)$ is

$$\mathbf{E}_1 = \sum_{n=1}^{\infty} E_n \left(c_n \mathbf{M}_{o1n}^{(1)} - i d_n \mathbf{N}_{e1n}^{(1)} \right),$$

$$\mathbf{H}_1 = \frac{-\mathbf{k}_1}{\omega\mu_1} \sum_{n=1}^{\infty} E_n \left(d_n \mathbf{M}_{e1n}^{(1)} + i c_n \mathbf{N}_{o1n}^{(1)} \right), \tag{4.40}$$

where $E_n = i^n E_0 (2n+1)/n(n+1)$ and μ_1 is the permeability of the sphere.

In the region outside the sphere j_n and y_n are well behaved; therefore, the expansion of the scattered field involves both of these functions. However, it is convenient if we now switch our allegiance to the spherical Hankel functions $h_n^{(1)}$ and $h_n^{(2)}$. We can show that only one of these functions is required by considering the asymptotic expansions of the Hankel functions of order ν for large values of $|\rho|$ (Watson, 1958, p. 198):

$$H_\nu^{(1)}(\rho) \sim \sqrt{\frac{2}{\pi\rho}} e^{i[\rho - \nu\pi/2 - \pi/4]} \sum_{m=0}^{\infty} \frac{(-1)^m (\nu, m)}{(2i\rho)^m},$$

$$H_\nu^{(2)}(\rho) \sim \sqrt{\frac{2}{\pi\rho}} e^{-i[\rho - \nu\pi/2 - \pi/4]} \sum_{m=0}^{\infty} \frac{(\nu, m)}{(2i\rho)^m}, \tag{4.41}$$

where $(\nu, m) = \Gamma(\nu + m + 1/2)/m!\Gamma(\nu - m + 1/2)$ and Γ is the gamma function; $\Gamma(n + 1) = n!$ if n is a nonnegative integer. It follows from (4.41) that the spherical Hankel functions are asymptotically given by

$$h_n^{(1)}(\mathbf{k}r) \sim \frac{(-i)^n e^{ikr}}{ikr}, \qquad (4.42)$$

$$\qquad\qquad\qquad kr \gg n^2$$

$$h_n^{(2)}(\mathbf{k}r) \sim -\frac{i^n e^{-ikr}}{ikr}. \qquad (4.43)$$

The first of these asymptotic expressions corresponds to an *outgoing* spherical wave; the second corresponds to an *incoming* spherical wave. If, on physical grounds, the scattered field is to be an outgoing wave at large distances from the particle, then only $h_n^{(1)}$ should be used in the generating functions. When we consider the scattered field at large distances we shall also need the asymptotic expression for the derivative of $h_n^{(1)}$; it follows from the identity

$$\frac{d}{d\rho} z_n = \frac{n z_{n-1} - (n + 1) z_{n+1}}{2n + 1}$$

and (4.42) that

$$\frac{dh_n^{(1)}}{d\rho} \sim \frac{(-i)^n e^{i\rho}}{\rho} \qquad (\rho \gg n^2). \qquad (4.44)$$

The expansion of the scattered field is therefore

$$\mathbf{E}_s = \sum_{n=1}^{\infty} E_n \left(i a_n \mathbf{N}_{e1n}^{(3)} - b_n \mathbf{M}_{o1n}^{(3)} \right),$$

$$\qquad\qquad\qquad (4.45)$$

$$\mathbf{H}_s = \frac{\mathbf{k}}{\omega\mu} \sum_{n=1}^{\infty} E_n \left(i b_n \mathbf{N}_{o1n}^{(3)} + a_n \mathbf{M}_{e1n}^{(3)} \right),$$

where we append the superscript (3) to vector spherical harmonics for which the radial dependence of the generating functions is specified by $h_n^{(1)}$.

4.3.1   Angle-Dependent Functions

It is now convenient to define the functions

$$\pi_n = \frac{P_n^1}{\sin\theta}, \qquad \tau_n = \frac{dP_n^1}{d\theta}. \qquad (4.46)$$

The angle-dependent functions π_n and τ_n appear to pose no particular computational problems—at least no one has complained about their misbehavior in print—and can be computed by upward recurrence from the relations

$$\pi_n = \frac{2n-1}{n-1}\mu\pi_{n-1} - \frac{n}{n-1}\pi_{n-2},$$

$$\tau_n = n\mu\pi_n - (n+1)\pi_{n-1}, \tag{4.47}$$

where $\mu = \cos\theta$, beginning with $\pi_0 = 0$ and $\pi_1 = 1$; π_n and τ_n are alternately even and odd functions of μ:

$$\pi_n(-\mu) = (-1)^{n-1}\pi_n(\mu), \qquad \tau_n(-\mu) = (-1)^n\tau_n(\mu). \tag{4.48}$$

Although π_n and τ_n are neither mutually orthogonal nor orthogonal to each other, it follows from (4.26) and (4.27) that $\pi_n + \tau_n$, as well as $\pi_n - \tau_n$, are orthogonal sets of functions:

$$\int_0^\pi (\tau_n + \pi_n)(\tau_m + \pi_m)\sin\theta\, d\theta = \int_0^\pi (\tau_n - \pi_n)(\tau_m - \pi_m)\sin\theta\, d\theta = 0$$

$$(m \neq n). \quad (4.49)$$

We can now write the vector spherical harmonics (4.17)–(4.20) (with $m = 1$) in the expansions of the internal field(4.40) and the scattered field (4.45) in a more concise form:

$$\mathbf{M}_{o1n} = \cos\phi\,\pi_n(\cos\theta)z_n(\rho)\hat{\mathbf{e}}_\theta - \sin\phi\,\tau_n(\cos\theta)z_n(\rho)\hat{\mathbf{e}}_\phi,$$

$$\mathbf{M}_{e1n} = -\sin\phi\,\pi_n(\cos\theta)z_n(\rho)\hat{\mathbf{e}}_\theta - \cos\phi\,\tau_n(\cos\theta)z_n(\rho)\hat{\mathbf{e}}_\phi,$$

$$\mathbf{N}_{o1n} = \sin\phi\,n(n+1)\sin\theta\,\pi_n(\cos\theta)\frac{z_n(\rho)}{\rho}\hat{\mathbf{e}}_r$$

$$+ \sin\phi\,\tau_n(\cos\theta)\frac{[\rho z_n(\rho)]'}{\rho}\hat{\mathbf{e}}_\theta + \cos\phi\,\pi_n(\cos\theta)\frac{[\rho z_n(\rho)]'}{\rho}\hat{\mathbf{e}}_\phi$$

$$\mathbf{N}_{e1n} = \cos\phi\,n(n+1)\sin\theta\,\pi_n(\cos\theta)\frac{z_n(\rho)}{\rho}\hat{\mathbf{e}}_r$$

$$+ \cos\phi\,\tau_n(\cos\theta)\frac{[\rho z_n(\rho)]'}{\rho}\hat{\mathbf{e}}_\theta - \sin\phi\,\pi_n(\cos\theta)\frac{[\rho z_n(\rho)]'}{\rho}\hat{\mathbf{e}}_\phi.$$

$$(4.50)$$

Superscripts will be appended to the functions $\mathbf{M}$ and $\mathbf{N}$ to denote the kind of spherical Bessel function z_n: (1) denotes $j_n(k_1 r)$ and (3) denotes $h_n^{(1)}(kr)$. As noted previously, **M** has no radial component, and for sufficiently large kr the radial component of **N** for the scattered field is negligible compared with the transverse component.

We have shown in Fig. 4.2 how the functions j_n and y_n behave, and the functions $\sin\phi$, $\cos\phi$ are well known. Thus, it only remains for us to show the behavior of the functions π_n and τ_n, which determine the θ dependence of the fields. Polar plots of π_n and τ_n for $n = 1\text{–}5$ are shown in Fig. 4.3; these plots are more pleasing to the eye if we allow θ to range from 0 to 360°. Note

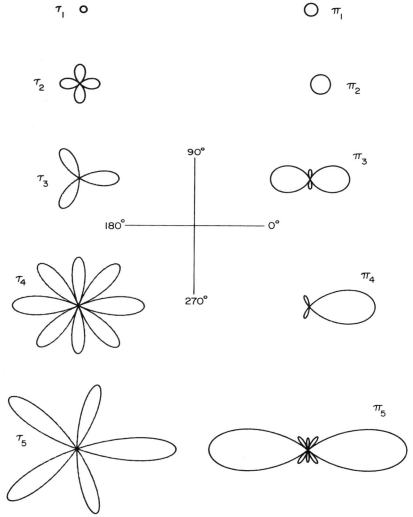

Figure 4.3 Polar plots of the first five angle-dependent functions π_n and τ_n. Both functions are plotted to the same scale.

that these functions (except π_1, which is constant) take on both positive and negative values; for example, τ_2 is positive from 0 to 45°, negative from 45 to 135°, and positive from 135 to 180°. As n increases, the number of lobes increases, with the result that the forward-directed lobe becomes narrower (i.e., the first zero occurs at smaller angles). The absence of a back-directed lobe in the polar plots of π_n and τ_n indicates that they are negative for backward directions; for example, τ_3 is negative for θ between about 149 and 180°. All the functions have forward-directed lobes (i.e., are positive in the forward direction), but the backward lobes disappear for alternate values of n. As we shall see, the larger the sphere, the more high-order functions π_n and τ_n are incorporated in the scattering diagram. Because of the behavior of these functions, therefore, the larger the sphere, the more heavily forward scattering directions are weighted compared with backscattering directions (alternate values of π_n or τ_n tend to cancel in backscattering directions), and the narrower the forward scattered peak.

4.3.2   Field Patterns: Normal Modes

The scattered electromagnetic field has been written as an infinite series in the vector spherical harmonics $\mathbf{M}_n$ and $\mathbf{N}_n$, the electromagnetic *normal modes* of the spherical particle. In the following section we shall discuss the conditions under which a single normal mode might be excited; in general, however, the scattered field is a superposition of normal modes, each weighted by the appropriate coefficient a_n or b_n. Diagrams from Mie's 1908 paper, which show the electric field lines corresponding to the transverse components of the first four normal modes, are given in Fig. 4.4. These diagrams have been reproduced in, among other places, Stratton (1941, p. 567), where so little discussion is given that they are easily misinterpreted, and in Tricker (1970, p. 226), where an excellent discussion of them is found. The field lines are shown on the surface of an imaginary sphere concentric with, but at a distance from, the particle. For each n there are two distinct types of modes: one for which there is no radial magnetic field component, called *transverse magnetic modes*, and another for which there is no radial electric field component, called *transverse electric modes*. As an aid in translating the confusing terminology often found for these modes, we have included in Fig. 4.4 other terms sometimes used, such as *electric type* or *E-waves* for the transverse magnetic modes, and *magnetic type* or *H-waves* for the transverse electric modes. Below each diagram we indicate the corresponding vector spherical harmonic together with the appropriate scattering coefficient. Although we have shown only the electric field patterns, the magnetic field patterns are readily obtained by rotation through an azimuthal angle of 90°; this follows from the relations [see (4.50)]

$$\mathbf{M}_{oln}\left(\phi + \tfrac{1}{2}\pi\right) = \mathbf{M}_{eln}(\phi); \qquad \mathbf{N}_{oln}\left(\phi + \tfrac{1}{2}\pi\right) = \mathbf{N}_{eln}(\phi)$$

and the expansions (4.45).

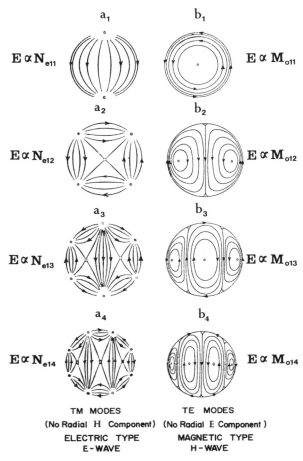

Figure 4.4   Electric field patterns: normal modes (Mie, 1908).

At first glance it may be confusing to see what appear to be free charges outside the particle, that is, points where the field lines appear to converge toward or diverge from. There clearly should be no free charges because each diagram represents field lines on the surface of an imaginary sphere in the medium surrounding the particle, which we may take to be free space. These apparent charge points are positions on the imaginary sphere at which the transverse field vanishes, and radial fields cannot be represented on a spherical surface. This can be made clearer by considering the radial component of the field for a particular mode. We have chosen the a_1 mode, which has particular importance later in the book; this is the field radiated by an oscillating electric dipole. Therefore, we can refer to the dipole radiation pattern for insight into the patterns shown in Fig. 4.4. Field lines in the xy plane ($\theta = \pi/2$) corre-

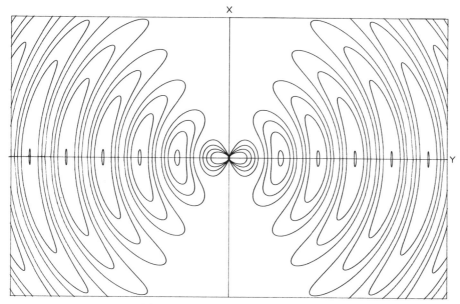

Figure 4.5 Field of a radiating dipole.

sponding to $i\mathbf{N}_{e11}^{(3)}$ are shown in Fig. 4.5. Note that, for a given distance r, the ϕ component of the field vanishes as we approach the x axis; along this axis the field is entirely radial. This, therefore, is why the field lines of the a_1 diagram in Fig. 4.4 vanish near the poles. Similar effects occur in each of the more complicated diagrams of Fig. 4.4.

4.3.3 Scattering Coefficients

We have arrived at the point where further understanding of scattering and absorption by a sphere is difficult to acquire without some numerical examples. What is needed now is some flesh to cover the dry bones of the formal theory; we should like to know how the various observable quantities vary with the size and optical properties of the sphere and the nature of the surrounding medium. To do so the first step is to obtain explicit expressions for the *scattering coefficients* a_n and b_n.

For a given n there are four unknown coefficients a_n, b_n, c_n, and d_n; thus, we need four independent equations, which are obtained from the boundary conditions (4.39) in component form:

$$E_{i\theta} + E_{s\theta} = E_{1\theta}, \qquad E_{i\phi} + E_{s\phi} = E_{1\phi},$$

$$H_{i\theta} + H_{s\theta} = H_{1\theta}, \qquad H_{i\phi} + H_{s\phi} = H_{1\phi}, \qquad r = a.$$

From the orthogonality of $\sin \phi$ and $\cos \phi$, the relations (4.49), the boundary conditions above, together with the expansions (4.37), (4.38), (4.40), (4.45), and the expressions (4.50) for the vector harmonics, we eventually obtain four linear equations in the expansion coefficients:

$$j_n(mx)c_n + h_n^{(1)}(x)b_n = j_n(x),$$

$$\mu[mxj_n(mx)]'c_n + \mu_1[xh_n^{(1)}(x)]'b_n = \mu_1[xj_n(x)]',$$

$$\mu mj_n(mx)d_n + \mu_1 h_n^{(1)}(x)a_n = \mu_1 j_n(x),$$

$$[mxj_n(mx)]'d_n + m[xh_n^{(1)}(x)]'a_n = m[xj_n(x)]',$$

(4.51)

where the prime indicates differentiation with respect to the argument in parentheses and the *size parameter x* and the *relative refractive index m* are

$$x = \mathrm{k}a = \frac{2\pi Na}{\lambda}, \qquad m = \frac{\mathrm{k}_1}{\mathrm{k}} = \frac{N_1}{N}.$$

N_1 and N are the refractive indices of particle and medium, respectively. The four simultaneous linear equations (4.51) are easily solved for the coefficients of the field inside the particle

$$c_n = \frac{\mu_1 j_n(x)[xh_n^{(1)}(x)]' - \mu_1 h_n^{(1)}(x)[xj_n(x)]'}{\mu_1 j_n(mx)[xh_n^{(1)}(x)]' - \mu h_n^{(1)}(x)[mxj_n(mx)]'},$$

$$d_n = \frac{\mu_1 mj_n(x)[xh_n^{(1)}(x)]' - \mu_1 mh_n^{(1)}(x)[xj_n(x)]'}{\mu m^2 j_n(mx)[xh_n^{(1)}(x)]' - \mu_1 h_n^{(1)}(x)[mxj_n(mx)]'},$$

(4.52)

and the scattering coefficients

$$a_n = \frac{\mu m^2 j_n(mx)[xj_n(x)]' - \mu_1 j_n(x)[mxj_n(mx)]'}{\mu m^2 j_n(mx)[xh_n^{(1)}(x)]' - \mu_1 h_n^{(1)}(x)[mxj_n(mx)]'},$$

$$b_n = \frac{\mu_1 j_n(mx)[xj_n(x)]' - \mu j_n(x)[mxj_n(mx)]'}{\mu_1 j_n(mx)[xh_n^{(1)}(x)]' - \mu h_n^{(1)}(x)[mxj_n(mx)]'},$$

(4.53)

Note that the denominators of c_n and b_n are identical as are those of a_n and d_n. If for a particular n the frequency (or radius) is such that one of these denominators is very small, the corresponding normal mode will dominate the scattered field. The a_n mode is dominant if the condition

$$\frac{[xh_n^{(1)}(x)]'}{h_n^{(1)}(x)} = \frac{\mu_1[mxj_n(mx)]'}{\mu m^2 j_n(mx)},$$

(4.54)

is approximately satisfied; similarly, the b_n mode is dominant if

$$\frac{\left[xh_n^{(1)}(x)\right]'}{h_n^{(1)}(x)} = \frac{\mu\left[mxj_n(mx)\right]'}{\mu_1 j_n(mx)} \tag{4.55}$$

is approximately satisfied. In general, of course, the scattered field is a superposition of normal modes.

The frequencies for which (4.54) and (4.55) are exactly satisfied, the so-called *natural* frequencies of the sphere (Stratton, 1941, p. 554), are *complex*, and the associated modes are sometimes said to be *virtual*. If the imaginary parts of these complex frequencies are small compared with the real parts, the latter correspond approximately to the real frequencies of incident electromagnetic waves which excite the various electromagnetic modes. Fuchs and Kliewer (1968) have thoroughly investigated the virtual modes of an ionic sphere with realistic frequency-dependent optical constants; they found that the modes fell naturally into three classes: low-frequency modes, high-frequency modes, and surface modes. In subsequent chapters we shall have more to say about electromagnetic modes in small particles, particularly in Chapter 12, where surface modes will be discussed at length.

The scattering coefficients (4.53) can be simplified somewhat by introducing the *Riccati–Bessel functions*:

$$\psi_n(\rho) = \rho j_n(\rho), \qquad \xi_n(\rho) = \rho h_n^{(1)}(\rho).$$

If we take the permeability of the particle and the surrounding medium to be the same, then

$$a_n = \frac{m\psi_n(mx)\psi_n'(x) - \psi_n(x)\psi_n'(mx)}{m\psi_n(mx)\xi_n'(x) - \xi_n(x)\psi_n'(mx)}, \tag{4.56}$$

$$b_n = \frac{\psi_n(mx)\psi_n'(x) - m\psi_n(x)\psi_n'(mx)}{\psi_n(mx)\xi_n'(x) - m\xi_n(x)\psi_n'(mx)}. \tag{4.57}$$

Note that a_n and b_n vanish as m approaches unity; this is as it should be: when the particle disappears, so does the scattered field.

As far as notation for the scattering coefficients is concerned, we have followed as much as possible van de Hulst (1957) and Kerker (1969), with the exception of the opposite sign convention for the time-harmonic factor $\exp(-i\omega t)$. Kerker (1969, p. 60) gives a table comparing the notation of various authors who have written on the theory of scattering by a sphere.

4.4   CROSS SECTIONS AND MATRIX ELEMENTS

Although we considered only scattering of x-polarized light in the preceding section, the scattered field for arbitrary linearly polarized incident light, and hence any polarization state, follows from the symmetry of the particle. For example, the scattered electric fields for equal-amplitude incident x-polarized and y-polarized plane waves are related by

$$\mathbf{E}_s(\phi; x\text{-polarized}) = \mathbf{E}_s\left(\phi + \frac{\pi}{2}; y\text{-polarized}\right).$$

Thus, if we have in hand the scattering coefficients a_n and b_n, we can determine all the measurable quantities associated with scattering and absorption, such as cross sections and scattering matrix elements.

4.4.1 Cross Sections

We could obtain cross sections for a sphere by appealing to the expressions for an arbitrary particle that were derived in Section 3.4 by calculating the net rate W_a at which electromagnetic energy crosses the surface of an imaginary sphere centered on the particle. If the surrounding medium is nonabsorbing, W_a is independent of the radius of this imaginary sphere, which for convenience was chosen to be sufficiently large that the far-field approximation for the electromagnetic field could be used. However, it is possible to derive expressions for the cross sections of a spherical particle exactly, something that seems to have been overlooked by previous authors. Therefore, it seems worthwhile to provide such a derivation. In so doing, we shall show some of the mathematical properties of the spherical Bessel functions; we may also acquire a bit more confidence in the optical theorem.

As before, we write W_a as $W_{\text{ext}} - W_s$, where

$$W_{\text{ext}} = \tfrac{1}{2}\,\text{Re}\int_0^{2\pi}\int_0^{\pi}\left(E_{i\phi}H_{s\theta}^* - E_{i\theta}H_{s\phi}^* - E_{s\theta}H_{i\phi}^* + E_{s\phi}H_{i\theta}^*\right)r^2\sin\theta\,d\theta\,d\phi,$$

$$W_s = \tfrac{1}{2}\,\text{Re}\int_0^{2\pi}\int_0^{\pi}\left(E_{s\theta}H_{s\phi}^* - E_{s\phi}H_{s\theta}^*\right)r^2\sin\theta\,d\theta\,d\phi, \tag{4.58}$$

and the radius $r \geq a$ of the imaginary sphere is arbitrary. On physical grounds we know that W_{ext} and W_s are independent of the polarization state of the incident light. Therefore, in evaluating the integrals (4.58) we may take the incident light to be x-polarized:

$$E_{i\theta} = \frac{\cos\phi}{\rho}\sum_{n=1}^{\infty} E_n(\psi_n\pi_n - i\psi_n'\tau_n), \qquad H_{i\theta} = \frac{k}{\omega\mu}\tan\phi\,E_{i\theta},$$

$$E_{i\phi} = \frac{\sin\phi}{\rho}\sum_{n=1}^{\infty} E_n(i\psi_n'\pi_n - \psi_n\tau_n), \qquad H_{i\phi} = \frac{-k}{\omega\mu}\cot\phi\,E_{i\phi},$$

where $\rho = kr$. The corresponding scattered field is

$$E_{s\theta} = \frac{\cos\phi}{\rho} \sum_{n=1}^{\infty} E_n(ia_n\xi_n'\tau_n - b_n\xi_n\pi_n),$$

$$E_{s\phi} = \frac{\sin\phi}{\rho} \sum_{n=1}^{\infty} E_n(b_n\xi_n\tau_n - ia_n\xi_n'\pi_n),$$

$$H_{s\theta} = \frac{k}{\omega\mu} \frac{\sin\phi}{\rho} \sum_{n=1}^{\infty} E_n(ib_n\xi_n'\tau_n - a_n\xi_n\pi_n),$$

$$H_{s\phi} = \frac{k}{\omega\mu} \frac{\cos\phi}{\rho} \sum_{n=1}^{\infty} E_n(ib_n\xi_n'\pi_n - a_n\xi_n\tau_n).$$

(4.59)

If we assume that the series expansions (4.59) may be substituted in the integral for W_s and the resulting product series integrated term by term, we obtain

$$W_s = \frac{\pi|E_0|^2}{k\omega\mu} \sum_{n=1}^{\infty} (2n + 1)\mathrm{Re}\{g_n\}(|a_n|^2 + |b_n|^2),$$

where we have used (4.24) and the relation

$$\int_0^{\pi} (\pi_n\pi_m + \tau_n\tau_m)\sin\theta\, d\theta = \delta_{nm}\frac{2n^2(n + 1)^2}{2n + 1},$$

which follows from (4.27). The quantity g_n, defined as $-i\xi_n^*\xi_n'$, may be written in the form

$$g_n = (\chi_n^*\psi_n' - \psi_n^*\chi_n') - i(\psi_n^*\psi_n' + \chi_n^*\chi_n'),$$

where the Riccati–Bessel function χ_n is $-\rho y_n(\rho)$ and, therefore, $\xi_n = \psi_n - i\chi_n$. The functions ψ_n and χ_n are real for real argument; therefore, if we use the Wronskian (Antosiewicz, 1964)

$$\chi_n\psi_n' - \psi_n\chi_n' = 1,$$

(4.60)

it follows that the scattering cross section is

$$C_{\mathrm{sca}} = \frac{W_s}{I_i} = \frac{2\pi}{k^2} \sum_{n=1}^{\infty} (2n + 1)(|a_n|^2 + |b_n|^2).$$

(4.61)

Similarly, the extinction cross section is

$$C_{\mathrm{ext}} = \frac{W_{\mathrm{ext}}}{I_i} = \frac{2\pi}{k^2} \sum_{n=1}^{\infty} (2n + 1)\mathrm{Re}\{a_n + b_n\},$$

(4.62)

where, as in the derivation of the scattering cross section, the key step is the relation (4.60).

4.4.2 Examples of Extinction: Interference and Ripple Structure; Reddening

We now pause in the mathematical development to consider a few examples of extinction. A more complete discussion of extinction will be given in Chapter 11, which is devoted exclusively to this subject. The computational methods used to generate these examples will be discussed later in this chapter. For our brief look at extinction we have chosen water droplets in air; the wavelength-dependent optical constants—from radio to ultraviolet—that were used in these calculations are given in Chapter 10. Calculated extinction curves for three different radii are shown in Fig. 4.6, where extinction efficiency $Q_{ext} = C_{ext}/\pi a^2$ is plotted as a function of inverse wavelength $1/\lambda$. This somewhat unconventional method of displaying extinction may cause some readers to reel in horror, particularly when it is noted that the curves in Fig. 4.6 show marked deviations from those more commonly encountered; extinction efficiencies are usually shown as functions of x for a *fixed* refractive index m, a practice hallowed by tradition. Although the traditional method of displaying extinction is not necessarily incorrect, it is often misleading: x and m are *mathematically* independent variables but they may not be *physically* independent. This elementary fact is often lost sight of when x is considered to be merely a dimensionless variable that is indifferent to whether it changes because of varying wavelength or radius. For if the wavelength varies, so must m: no material substance has optical constants independent of wavelength except over a narrow range. Unfortunately, in some areas to which light scattering theory has been applied, full realization of this has only slowly dawned; the result has been spurious conclusions based on faulty reasoning. One of the central themes of this book is that full understanding of light scattering and absorption by particles requires understanding optical properties of bulk matter. The reason for the traditional method of displaying extinction has more to do with convenience than with fidelity to physical reality: it is relatively easy to calculate Q_{ext} as a function of x for fixed m. The curves shown in Fig. 4.6, however, require considerably more effort than is usual: at each of the many wavelengths for which computations are done, the correct optical properties must be used. The effort required to compute Q_{ext} *per se* is greatly overshadowed by that entailed in compiling optical constants from many sources and suitably interpolating between measured data points. The reward for this effort, however, is a more physically accurate picture of extinction.

In the region where water is weakly absorbing (between about 0.5 and 5 μm^{-1}) the extinction curve for a 1.0 μm droplet has several features: (1) a series of regularly spaced broad maxima and minima called the *interference structure*, which oscillates approximately about the value 2; (2) irregular fine

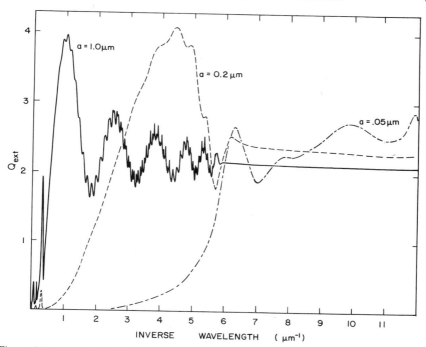

Figure 4.6 Extinction efficiencies for water droplets in air; plotting increment = 0.01 $\mu\mathrm{m}^{-1}$.

structure called the *ripple structure*; and (3) monotonically increasing extinction with decreasing wavelength for $a < \lambda$. We shall briefly consider each of these in turn.

For large ($\gg n^2)x$ and $|mx|$ the *numerator* of a_n is approximately

$$\frac{(m-1)\sin[x-n\pi/2]\cos[mx-n\pi/2]+m\sin[x(m-1)]}{x}. \qquad (4.63)$$

To the same degree of approximation the numerator of b_n is

$$\frac{(1-m)\sin[x-n\pi/2]\cos[mx-n\pi/2]+\sin[x(m-1)]}{mx}. \qquad (4.64)$$

To obtain (4.63) and (4.64) from (4.56) and (4.57) we used the asymptotic relation

$$\psi_n(\rho) \sim \sin(\rho - n\pi/2) \qquad (|\rho| \gg n^2). \qquad (4.65)$$

Note that the numerator of both a_n and b_n contains the common term $\sin[x(m-1)]$, which is independent of n; thus, we expect the extinction cross section maxima to be approximately determined by the maxima of this function, which occur for $x(m - 1) = (2p + 1)\pi/2$, where p is an integer. Therefore, the separation $\Delta(1/\lambda)$ between the maxima of $\sin[x(m - 1)]$, over a wavelength region for which m is approximately constant and real, is $1/2a(m - 1)$. For water at or near visible wavelengths, m may be taken to be about 1.33; thus, we expect the cross-section maxima for a 1.0-μm-radius droplet to be separated by about 1.5 μm^{-1}. That this is indeed so is apparent from Fig. 4.6. The origin of the term "interference structure" applied to these broad extinction peaks lies in the interpretation of extinction as interference between the incident and forward-scattered light (see Section 3.4). For if we adopt the viewpoint of elementary optics, the phase difference $\Delta\phi$ between a ray that traverses a large transparent sphere without deviation (i.e., the forward-scattered or central ray) and a ray that traverses the same physical path outside the sphere is

$$\Delta\phi = \frac{2\pi}{\lambda}2a(N_1 - N) = 2x(m - 1).$$

The condition for destructive interference between these two rays is $\Delta\phi = (2p + 1)\pi$ or, equivalently, $x(m - 1) = (2p + 1)\pi/2$, which is the same condition as that obtained by examining the numerators of a_n and b_n.

We shall defer detailed discussion of the ripple structure, which is considerably more complicated both mathematically and physically than the interference structure, until Chapter 11. Suffice it to say for the moment that the ripple structure has its origins in the roots of the transcendental equations (4.54) and (4.55), the conditions under which the *denominators* of the scattering coefficients vanish.

Both the interference structure and the ripple structure are strongly damped when absorption becomes large, as it does in water if $1/\lambda$ is greater than about 6 μm^{-1}; this is analogous to damping of interference bands in the transmission spectrum of a slab (see Fig. 2.8). If the droplet is small compared with the wavelength, then peaks in the *bulk* absorption spectrum are seen in the *particle* extinction spectrum; for example, the extinction peaks in Fig. 4.6 at about 6 μm^{-1} for a 0.05-μm-radius droplet and at about 0.3 μm^{-1} for a 1.0-μm droplet are neither interference nor ripple structure but bulk absorption peaks. This illustrates the fact that absorption dominates over scattering for small a/λ if there is any appreciable bulk absorption.

A familiar phenomenon is *reddening* of white light on passing through a collection of very small particles. This can be demonstrated easily by putting a few drops of milk into a container of pure water: a collimated beam of white light takes on a reddish tint after transmission through this suspension because the shorter-wavelength blue light is extinguished more effectively than the longer-wavelength red light. Rising extinction toward shorter wavelengths is a general characteristic of nonabsorbing particles small compared with the wavelength; this is exhibited in the extinction curves in Fig. 4.6 for the two smaller particles. Everyone is familiar with such effects through the beautiful red and orange hues of sunset skies, which are partly the result of molecular scattering. Small particles can enhance reddening of the sunset. Periods of strong volcanic activity have been known to increase the beauty of sunset colors for more than a year because of particles in the atmosphere; high levels of particulate air pollution tend to increase the sunset reddening.

Reddening because of extinction by small particles is certainly not limited to the terrestrial environment. Dust particles between stars extinguish blue light more efficiently than red light; starlight transmitted through this dust, therefore, is reddened. This effect is so reliable and uniform when averaged over many thousands of light years that it can be used to measure distances to stars in our galaxy. A "highly reddened star," as the jargon goes, is one that has a large quantity of interstellar dust between it and the observer. It is obvious from Fig. 4.6 that extinction is quite size dependent; for this reason, extinction has occasionally been used to size particles. In fact, this size dependence provides us with our best evidence that interstellar dust grains are predominantly submicron. In the laboratory, however, other types of measurements, such as angular scattering, are usually preferable for particle sizing.

Reddening occurs for collections of particles regardless of their size distribution provided that they are small compared with the wavelength. The opposite spectral effect, "bluing," can be seen on the high-frequency side of the extinction peaks in Fig. 4.6. Such bluing *is* highly dependent on the size distribution and tends to vanish, as do the other characteristics of the interference structure, as the dispersion of particle radii increases. Thus, bluing of sunlight by particles in the atmosphere is quite rare although not unheard of: it happens "once in a blue moon." This saying evidently originates from the fact that there have been a few times recorded in history when the sun and the moon were observed to be blue, such as after giant eruptions of the volcano Krakatoa and following huge forest fires in Canada. According to the conventional explanation, the conditions necessary for this anomalous extinction, which include a narrow range of particle sizes, are rarely met.

4.4.3 The Extinction Paradox; Scalar Diffraction Theory

We noted in the preceding section that Q_{ext} appears to approach the limiting value 2 as the size parameter increases:

$$\lim_{x \to \infty} Q_{ext}(x, m) = 2,$$

which is *twice* as large as that predicted by geometrical optics. Yet geometrical optics is considered to be a good approximation if all dimensions are much larger than the wavelength. Moreover, $Q_{ext} = 2$ contradicts "common sense": we do not expect a large object to remove twice the energy that is incident on it. This perhaps puzzling result is called the *extinction paradox*, which we shall try to resolve in the following paragraphs.

Although geometrical optics is a good approximation to the exact wave theory for large objects, no matter how large an object is it still has an *edge* in the neighborhood of which geometrical optics fails to be valid. Therefore, let us analyze extinction by a large sphere of radius a using a combination of geometrical optics and scalar diffraction theory.

The fundamental problem in scalar diffraction theory is to determine the value of a scalar wave ψ at a point P given the value of ψ and $\nabla\psi$ on a closed surface S surrounding P. A concise and lucid treatment of this theory is given by Wangsness (1963), who shows that

$$\psi(P) = \frac{1}{4\pi} \int_S \left\{ \frac{e^{ikR}}{R} \nabla\psi - \psi\nabla\left(\frac{e^{ikR}}{R}\right) \right\} \cdot \hat{\mathbf{n}} \, dA, \tag{4.66}$$

where R is the distance from P to a point on S and $\hat{\mathbf{n}}$ is the outward directed normal to S. Equation (4.66) is not restricted to electromagnetic waves but applies equally well to any scalar quantity that satisfies the wave equation $\nabla^2\psi + k^2\psi = 0$.

An amount of energy $I_i\pi a^2$ is removed from a beam with irradiance I_i as a result of reflection, refraction, and absorption of the rays that are incident on the sphere; that is, every ray is either absorbed or changes its direction and is therefore counted as having been removed from the incident beam. An *opaque disk* of radius a also removes an amount of energy $I_i\pi a^2$, and to the extent that scalar diffraction theory is valid, a sphere and an opaque disk have the same diffraction pattern. Therefore, for purposes of this analysis, we may replace the sphere by an opaque disk.

Although we shall be interested primarily in diffraction by an opaque circular disk, no extra labor is entailed if the shape of the planar obstacle is unrestricted at this stage of our argument (Fig. 4.7a). It is more convenient to consider diffraction by a planar *aperture*, with the same shape and dimensions as the obstacle, in an otherwise opaque screen (Fig. 4.7b). If $\psi(P)$ is the value of the wave function at P when the aperture is in place, we can invoke *Babinet's principle*,

$$\bar{\psi}(P) + \psi(P) = \psi_0(P), \tag{4.67}$$

to obtain the wave function $\bar{\psi}$ when the obstacle is in place, where ψ_0 is the unimpeded (incident) wave function, which we take to be a plane wave $E_0\exp(ikz)$.

To evaluate the integral in (4.66) we need to know ψ and its gradient over the surface S. It is physically plausible to assume that the only contribution to $\psi(P)$ comes from the aperture $\mathcal{C}$, over which ψ may be approximated by the incident wave function ψ_0. With this assumption (4.66) becomes

$$\psi(P) = \frac{-ikE_0}{4\pi} \int_{\mathcal{C}} \frac{e^{ikR}}{R}(1 + \hat{\mathbf{e}}_R \cdot \hat{\mathbf{e}}_z) \, dA, \tag{4.68}$$

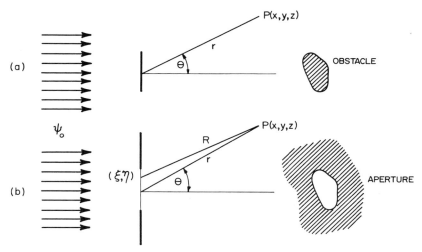

Figure 4.7 (*a*) Diffraction by an opaque planar obstacle. (*b*) Diffraction by an aperture with the same shape as the obstacle.

where we have also taken $kR \gg 1$; that is, the point P is at a distance from the aperture large compared with the wavelength. The unit vector $\hat{\mathbf{e}}_R$ is directed along the line from P to a point on the aperture with coordinates (ξ, η), and the distance R is

$$R = \sqrt{r^2 - 2x\xi - 2y\eta + \xi^2 + \eta^2}. \tag{4.69}$$

If the linear dimensions of the aperture are small compared with r, we can expand (4.69) in powers of ξ/r and η/r; *Fraunhofer diffraction* results if we terminate this expansion at linear terms:

$$\psi(P) = E_0 \frac{e^{ikr}}{ikr} S(\theta, \phi),$$

$$\tag{4.70}$$

$$S(\theta, \phi) = \frac{k^2}{4\pi} \int_{@} e^{-ik \sin \theta (\xi \cos \phi + \eta \sin \phi)} (1 + \cos \theta) \, d\xi \, d\eta.$$

Therefore, with the obstacle in place, we have from (4.67) and (4.70),

$$\bar{\psi}(P) = E_0 e^{ikz} - E_0 \frac{e^{ikr}}{ikr} S(\theta, \phi). \tag{4.71}$$

The first term in (4.71) is just the incident wave; the second term is the wave scattered (diffracted) by the obstacle. The optical theorem for scalar waves is formally identical to that for vector waves, (3.24); thus, the extinction cross section is

$$C_{\text{ext}} = \frac{4\pi}{k^2} \operatorname{Re}\{S(\theta = 0)\} = 2G, \qquad (4.72)$$

where G is the area of the obstacle. Therefore, the extinction cross section of the obstacle is twice its geometrical cross section: all the energy incident on the opaque obstacle, an amount equal to $I_i G$, is absorbed; in addition, an equal amount of energy $W_s = I_i C_{\text{sca}}$, where

$$C_{\text{sca}} = \int_0^{2\pi} \int_0^{\pi} \frac{|S(\theta, \phi)|^2}{k^2} \sin\theta \, d\theta \, d\phi = G,$$

is scattered (diffracted) by the obstacle. Roughly speaking, we may say that the incident wave is influenced beyond the physical boundaries of the obstacle: the edge deflects rays in its neighborhood, rays that, from the viewpoint of geometrical optics, would have passed unimpeded. These rays, regardless of how small the angle through which they are deflected, are counted as having been removed from the incident beam and therefore contribute to the total extinction. To the extent that replacing a particle very much larger than the wavelength by an opaque planar obstacle with the same projected area is a valid approximation, the same interpretation elucidates why the extinction cross section of such a particle is twice its geometrical cross section.

We have yet to explain why $C_{\text{ext}} = 2G$ is not necessarily observed. To do so it will help if we consider a specific example. The scattering amplitude for a circular disk is independent of the azimuthal angle ϕ:

$$S(\theta) = \frac{k^2}{4\pi} \int_{\mathcal{C}} e^{-ik\xi \sin\theta} (1 + \cos\theta) \, d\xi \, d\eta. \qquad (4.73)$$

We can evaluate the integral in (4.73) by transforming to plane polar coordinates and using the integral representation of the Bessel function J_0,

$$J_0(z) = \frac{1}{\pi} \int_0^{\pi} e^{iz \cos\psi} d\psi$$

together with the identity $d(zJ_1)/dz = zJ_0$; the result is

$$S(\theta) = x^2 \frac{(1 + \cos\theta)}{2} \frac{J_1(x \sin\theta)}{x \sin\theta}.$$

The size parameter is very large and $J_1(x \sin\theta)/x \sin\theta$ is negligibly small for $x \sin\theta$ greater than about 10; therefore, the factor $(1 + \cos\theta)/2$ is unity to a very good approximation over the angular region of interest. In Fig. 4.8 we show the scattering diagram normalized to the forward direction as a function of $x \sin\theta$. Note that almost all the scattered light is confined within a cone of

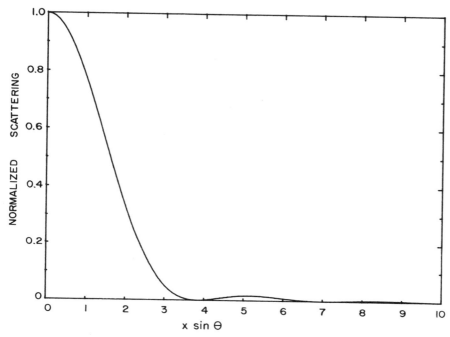

Figure 4.8   Scattering diagram for diffraction by a circular disk.

half-angle $\theta \simeq 10/x$. If a detector is to record the full extinction by a large spherical object, its acceptance angle must be much less than this, say $\theta_{acc} < 1/2x$. Thus, in any measurement of extinction by particles much larger than the wavelength, the possible effect of instrument geometry should be considered carefully. Failure to do so may result in spurious disagreement between theory and experiment, as well as lack of experimental reproducibility.

4.4.4   Scattering Matrix

We assume that the series expansion (4.45) of the scattered field is uniformly convergent. Therefore, we can terminate the series after n_c terms and the resulting error will be arbitrarily small for all kr if n_c is sufficiently large. If, in addition, $kr \gg n_c^2$, we may substitute the asymptotic expressions (4.42) and (4.44) in the truncated series; the resulting transverse components of the scattered electric field are

$$E_{s\theta} \sim E_0 \frac{e^{ikr}}{-ikr} \cos \phi \, S_2(\cos \theta),$$

$$E_{s\phi} \sim -E_0 \frac{e^{ikr}}{-ikr} \sin \phi \, S_1(\cos \theta),$$

where

$$S_1 = \sum_n \frac{2n+1}{n(n+1)} (a_n \pi_n + b_n \tau_n),$$

$$S_2 = \sum_n \frac{2n+1}{n(n+1)} (a_n \tau_n + b_n \pi_n),$$

(4.74)

and the series are terminated after n_c terms. The relation between incident and scattered field amplitudes is therefore

$$\begin{pmatrix} E_{\|s} \\ E_{\perp s} \end{pmatrix} = \frac{e^{ik(r-z)}}{-ikr} \begin{pmatrix} S_2 & 0 \\ 0 & S_1 \end{pmatrix} \begin{pmatrix} E_{\|i} \\ E_{\perp i} \end{pmatrix}.$$

(4.75)

We can show from (4.25) that

$$\pi_n(1) = \tau_n(1) = \left. \frac{dP_n}{d\mu} \right|_{\mu=1}.$$

But P_n satisfies the differential equation (4.4), from which, together with $P_n(1) = 1$, it follows that

$$\pi_n(1) = \tau_n(1) = \frac{n(n+1)}{2}.$$

Thus, in the forward direction ($\theta = 0°$)

$$S_2(0°) = S_1(0°) = S(0°) = \tfrac{1}{2} \sum_n (2n+1)(a_n + b_n),$$

which when substituted in the optical theorem (3.24) yields the extinction cross section (4.62):

$$C_{ext} = \frac{4\pi}{k^2} \, \text{Re}\{S(0°)\}.$$

(4.76)

The relation between incident and scattered Stokes parameters follows from (4.75):

$$\begin{pmatrix} I_s \\ Q_s \\ U_s \\ V_s \end{pmatrix} = \frac{1}{k^2 r^2} \begin{pmatrix} S_{11} & S_{12} & 0 & 0 \\ S_{12} & S_{11} & 0 & 0 \\ 0 & 0 & S_{33} & S_{34} \\ 0 & 0 & -S_{34} & S_{33} \end{pmatrix} \begin{pmatrix} I_i \\ Q_i \\ U_i \\ V_i \end{pmatrix},$$

(4.77)

$$S_{11} = \tfrac{1}{2}(|S_2|^2 + |S_1|^2), \qquad S_{12} = \tfrac{1}{2}(|S_2|^2 - |S_1|^2),$$

$$S_{33} = \tfrac{1}{2}(S_2^* S_1 + S_2 S_1^*), \qquad S_{34} = \frac{i}{2}(S_1 S_2^* - S_2 S_1^*).$$

Only three of these four matrix elements are independent: $S_{11}^2 = S_{12}^2 + S_{33}^2 + S_{34}^2$.

If the incident light is 100% polarized *parallel* to a particular scattering plane (it makes no difference which scattering plane), the Stokes parameters of the scattered light are

$$I_s = (S_{11} + S_{12})I_i, \qquad Q_s = I_s, \qquad U_s = V_s = 0,$$

where we have omitted the factor $1/k^2r^2$. Thus, the scattered light is also 100% polarized parallel to the scattering plane. We denote by $i_{\parallel}$ the scattered irradiance per unit incident irradiance given that the incident light is polarized parallel to the scattering plane:

$$i_{\parallel} = S_{11} + S_{12} = |S_2|^2.$$

If the incident light is polarized *perpendicular* to the scattering plane, the Stokes parameters of the scattered light are

$$I_s = (S_{11} - S_{12})I_i, \qquad Q_s = -I_s, \qquad U_s = V_s = 0.$$

Thus, the scattered light is also polarized perpendicular to the scattering plane. We denote by $i_{\perp}$ the scattered irradiance per unit incident irradiance given that the incident light is polarized perpendicular to the scattering plane:

$$i_{\perp} = S_{11} - S_{12} = |S_1|^2.$$

If the incident light is *unpolarized*, the Stokes parameters of the scattered light are

$$I_s = S_{11}I_i, \qquad Q_s = S_{12}I_i, \qquad U_s = V_s = 0.$$

The ratio

$$P = -\frac{S_{12}}{S_{11}} = \frac{i_{\perp} - i_{\parallel}}{i_{\perp} + i_{\parallel}} \tag{4.78}$$

is such that $|P| \leq 1$; if P is *positive*, the scattered light is partially polarized *perpendicular* to the scattering plane; if P is *negative*, the scattered light is partially polarized *parallel* to the scattering plane; the degree of polarization is $|P|$. Regardless of the size and composition of the sphere, $P(0°) = P(180°) = 0$.

If the incident light is *obliquely* polarized at an angle of 45° to the scattering plane, the scattered light will, in general, be *elliptically* polarized, although the azimuth of the vibration ellipse need not be 45°. The amount of rotation of the azimuth, as well as the ellipticity, depends not only on the particle characteristics but also on the direction in which the light is scattered.

Table 4.1   Scattering Coefficients for a Water Droplet in Air with Size Parameter $x = 3$ and Complex Refractive Index $m = 1.33 + i10^{-8}$

n	$\dfrac{2n+1}{n(n+1)}$	a_n	b_n
1	$\frac{3}{2}$	$5.1631 \times 10^{-1} - i4.9973 \times 10^{-1}$	$7.3767 \times 10^{-1} - i4.3990 \times 10^{-1}$
2	$\frac{5}{6}$	$3.4192 \times 10^{-1} - i4.7435 \times 10^{-1}$	$4.0079 \times 10^{-1} - i4.9006 \times 10^{-1}$
3	$\frac{7}{12}$	$4.8467 \times 10^{-2} - i2.1475 \times 10^{-1}$	$9.3553 \times 10^{-3} - i9.6269 \times 10^{-2}$
4	$\frac{9}{20}$	$1.0346 \times 10^{-3} - i3.2148 \times 10^{-2}$	$6.8810 \times 10^{-5} - i8.2949 \times 10^{-3}$
5	$\frac{11}{30}$	$9.0375 \times 10^{-6} - i3.0062 \times 10^{-3}$	$2.8309 \times 10^{-7} - i5.3204 \times 10^{-4}$

4.4.5   An Example of Angle-Dependent Scattering

As an example of angle-dependent scattering by a sphere we have chosen a water droplet with size parameter $x = 3$ illuminated by visible light of wavelength 0.55 μm. At this wavelength the complex refractive index of water is $1.33 + i10^{-8}$; $x = 3$ corresponds to a droplet radius of about 0.26 μm. The first five scattering coefficients for this particle are given in Table 4.1, from which it is clear that the first two or three functions π_n and τ_n determine the angular dependence of the scattering.

The results of computations using the program of Appendix A are shown in Fig. 4.9: linear polar plots of $i_\perp$ and $i_\parallel$ in part a; the logarithms of $i_\perp$ and $i_\parallel$ in b; and the polarization (4.78) in c; in all three sets of curves the independent

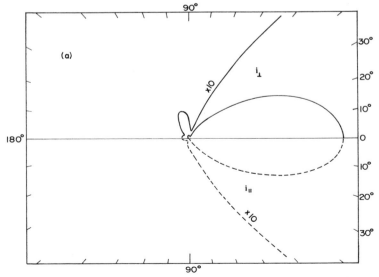

Figure 4.9   Scattering by a sphere with $x = 3$ and $m = 1.33 + i10^{-8}$.

variable is the scattering angle θ. Perhaps the most important point to note is that the scattering is highly peaked in the forward direction. This is seen most strikingly in the linear polar plot of part *a*. The small scattering lobes for $\theta > 90°$ are almost imperceptible compared with the strong forward-scattering lobes; indeed, for the backscattering lobes to be seen at all requires that we magnify the polar plots by a factor of 10. The scattered irradiance in the forward direction is more than 100 times greater than that in the backward direction; such directional asymmetry becomes even more pronounced as the size parameter increases, to the point that it is of little value to display scattering diagrams in a linear fashion. Our intent in showing this one polar plot is to emphasize the predominance of forward scattering even for rather small spheres—a 0.26-μm water droplet is so small as to be unheard of in clouds.

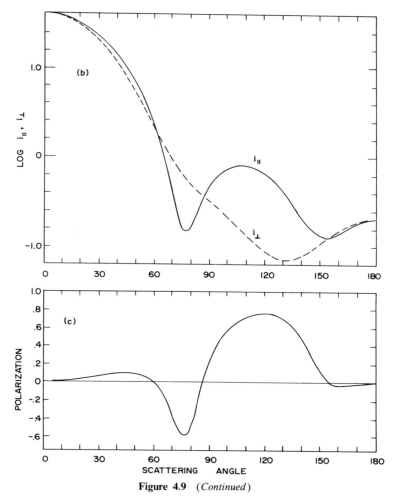

Figure 4.9 (*Continued*)

We encounter the consequences of strong forward scattering almost every day. An evening drive toward a bright setting sun can be a blinding experience, even if the direct sunlight is blocked by the sun visor, because of intense forward scattering by particles in the atmosphere and on the windshield. This is easily remedied by driving in the opposite direction—180° scattering is orders of magnitude less intense—but this solution usually has little practical appeal. In a similar way, night driving in fog or with a dirty windshield can be difficult: light from oncoming automobile headlights is scattered in the forward direction by fog droplets or particles to produce bothersome glare.

4.4.6   Integrated Extinction: A Sum Rule

We showed in Section 2.3 that the real and imaginary parts of the electric susceptibility are connected by the dispersion relations (2.36) and (2.37). This followed as a consequence of the linear causal relation between the electric field and polarization together with the vanishing of $\chi(\omega)$ in the limit of infinite frequency ω. We also stated that, in general, similar relations are expected to hold for any frequency-dependent function that connects an output with an input in a linear causal way. An example is the amplitude scattering matrix (4.75): the scattered field is linearly related to the incident field. Moreover, this relation must be causal: the scattered field cannot precede in time the incident field that excited it. Therefore, the matrix elements should satisfy dispersion relations. In particular, this is true for the forward direction $\theta = 0°$. But $S(0°, \omega)$ does not have the required asymptotic behavior: it is clear from the diffraction theory approximation (4.73) that for sufficiently large frequencies, $S(0°, \omega)$ is proportional to ω^2. Nevertheless, only minor fiddling with S makes it behave properly: the function

$$F(\omega) = \frac{S(0°, \omega)}{\omega^2} - \frac{a^2}{2c^2}$$

vanishes in the limit of infinite frequency; c is the speed of light *in vacuo* and we have taken the sphere to be in free space for convenience. When the domain of definition is extended to the complex $\tilde{\omega}$ plane, F is analytic in its upper half because of the analyticity of S; F also satisfies the same crossing condition as χ: $F(-\omega) = F^*(\omega)$. Therefore, the real and imaginary parts of $F = F' + iF''$ are connected by the integral relations (2.36) and (2.37):

$$F'(\omega) = \frac{2}{\pi} P \int_0^\infty \frac{\Omega F''(\Omega)}{\Omega^2 - \omega^2} d\Omega, \qquad F''(\omega) = \frac{-2\omega}{\pi} P \int_0^\infty \frac{F'(\Omega)}{\Omega^2 - \omega^2} d\Omega.$$

From these relations, the optical theorem (4.76), and (2.51) it follows that

$$2\pi^2 c^2 \frac{\operatorname{Im}\{S(0°, \omega)\}}{\omega^3} = -P \int_0^\infty \frac{C_{\text{ext}}(\Omega)}{\Omega^2 - \omega^2} d\Omega. \tag{4.79}$$

It is easy to show from (5.4) that for sufficiently small frequencies the forward scattering matrix element is given by

$$S(0^0, \omega) = -i\frac{\omega^3}{c^3}\frac{\epsilon(\omega) - 1}{\epsilon(\omega) + 2},$$

where the dielectric function $\epsilon = m^2$ is the permittivity of the particle relative to that of free space. Therefore, the limit of (4.79) as ω approaches zero is

$$\frac{2\pi^2 a^3}{c} \lim_{\omega \to 0} \text{Im}\left\{ i\frac{\epsilon(\omega) - 1}{\epsilon(\omega) + 2} \right\} = \int_0^\infty \frac{C_{\text{ext}}(\Omega)}{\Omega^2} d\Omega. \tag{4.80}$$

This sum rule for extinction is written more compactly if we transform the integration variable from frequency to wavelength and assume that the static dielectric function is real and finite:

$$\int_0^\infty C_{\text{ext}}(\lambda) \, d\lambda = 4\pi^3 a^3 \frac{\epsilon(0) - 1}{\epsilon(0) + 2}. \tag{4.81}$$

Some rather remarkable conclusions follow from (4.81): although the dependence of C_{ext} on particle radius at a given wavelength may be quite complicated, the integrated extinction is merely proportional to the particle volume. Moreover, the optical properties of the particle enter into the integrated extinction only through the *static* dielectric function; the greater it is, the greater will be the integrated extinction. Thus, regardless of particle composition, we have an upper limit to integrated extinction:

$$\int_0^\infty C_{\text{ext}}(\lambda) \, d\lambda \leq 4\pi^3 a^3.$$

The sum rule (4.81) for extinction was first obtained by Purcell (1969) in a paper which we belive has not received the attention it deserves. Our path to this sum rule is different from that of Purcell's but we obtain essentially the same results. Purcell did not restrict himself to spherical particles but considered the more general case of spheroids. Regardless of the shape of the particle, however, it is plausible on physical grounds that integrated extinction should be proportional to the volume of an arbitrary particle, where the proportionality factor depends on its shape and static dielectric function.

4.4.7   Finite Beam Width

The expressions for the field scattered by a sphere were obtained under the assumption that the beam is infinite in lateral extent; such beams, however, are difficult to produce in the laboratory. Nevertheless, it is physically plausible that scattering and absorption by any particle will be independent of the extent of the beam provided that it is large compared with the particle size; that is, the particle is completely bathed in the incident light. Our physical intuition is bolstered by the analysis of Tsai and Pogorzelski (1975), who obtained exact expressions for the field scattered by a sphere when the incident beam is cylindrically symmetric with a finite cross section. Their calculations of the angular dependence of the light scattered by a conducting sphere show no difference between infinite and finite beams provided that the beam radius is about 10 times larger than the sphere radius. In most scattering experiments, even those using highly collimated laser beams, this condition will certainly be satisfied. Thus, we are usually justified in ignoring finite beam width.

4.4.8 Charged Sphere

We have also assumed that the particle carries no net surface charge. This assumption, although not explicitly stated, is implicit in the boundary conditions (3.7). However, naturally occurring charged particles are not uncommon: water droplets formed in ocean sprays, water droplets and ice crystals in thunderstorms, drifting snow, and dust can be electrically charged; it is also believed that interstellar grains are charged (Spitzer, 1948). Thus, we are naturally led to the question: Does a particle that carries a physically realistic net charge scatter electromagnetic waves in any manner that is observably different from that of an identical uncharged particle? In an attempt to answer this question, Bohren and Hunt (1977) considered the problem of scattering by a sphere in which the conditions

$$(\mathbf{E}_2 - \mathbf{E}_1) \times \hat{\mathbf{e}}_r = 0, \qquad (\mathbf{H}_2 - \mathbf{H}_1) \times \hat{\mathbf{e}}_r = \mathbf{K},$$

where $\mathbf{K}$ is the surface current density of excess surface charge, were imposed at the boundary between particle and surrounding medium. If it is assumed that $\mathbf{K} = \sigma_s \mathbf{E}_{1t}$, where σ_s is a phenomenological surface conductivity and $\mathbf{E}_{1t}$ is the tangential field at the surface of the sphere, the coefficients of the scattered field can be obtained in a manner similar to that for an uncharged sphere. Obtaining the mathematical *form* of the scattering coefficients is easy enough; the difficulties arise when one tries to make *quantitative* conclusions in the absence of either measured values of σ_s or suitable microscopic theories. On the basis of a simple microscopic theory of free excess surface charges, Bohren and Hunt concluded that surface charges on metallic particles small compared with the wavelength do not appreciably affect the extinction cross section. However, a full understanding of scattering by charged particles awaits a satisfactory treatment of σ_s. In the interim, we shall assume, in the absence of evidence to the contrary, that surface charges on a particle only slightly perturb its scattering and absorbing properties.

4.5 ASYMMETRY PARAMETER AND RADIATION PRESSURE

The asymmetry parameter, which was defined in Section 3.4 as the average cosine of the scattering angle, depends, in general, on the polarization state of the incident light. However, the asymmetry parameter for a spherical particle is clearly independent of polarization and is given by

$$k^2 C_{\text{sca}} \langle \cos \theta \rangle = \pi \int_{-1}^{1} (|S_1|^2 + |S_2|^2)\mu \, d\mu,$$

where $\mu = \cos \theta$ and

$$|S_1|^2 + |S_2|^2 = \sum_n \sum_m \frac{2n + 1}{n(n + 1)} \frac{2m + 1}{m(m + 1)} \left[(a_n a_m^* + b_n b_m^*)(\tau_n \tau_m + \pi_n \pi_m) \right.$$

$$\left. + (a_n b_m^* + a_m^* b_n)(\tau_n \pi_m + \pi_n \tau_m) \right].$$

Therefore, to obtain the asymmetry parameter, we must evaluate the integrals

$$T_{nm}^{(1)} = \int_{-1}^{1} (\tau_n \tau_m + \pi_n \pi_m)\mu \, d\mu, \qquad T_{nm}^{(2)} = \int_{-1}^{1} (\tau_n \pi_m + \pi_n \tau_m)\mu \, d\mu.$$

The first of these integrals can be written in the form

$$T_{nm}^{(1)} = \int_{0}^{\pi} \left(\frac{dP_m^1}{d\theta} P_n^1 + P_m^1 \frac{dP_n^1}{d\theta} \right) \cos \theta \, d\theta$$

and integrated by parts to yield

$$T_{nm}^{(1)} = \delta_{nm} \frac{2n(n + 1)}{2n + 1},$$

where we have used the orthogonality of the P_n^1 (4.7). It follows from (4.48) that $\tau_n \tau_m + \pi_n \pi_m$ is an even function of μ if $m + n$ is even; therefore, $T_{nm}^{(2)}$ vanishes unless $m = n \pm p$, where $p = 1, 3, \ldots$. By using the recurrence relations

$$\tau_n = n \pi_{n+1} - (n + 1)\mu \pi_n, \qquad \tau_n = n\mu \pi_n - (n + 1)\pi_{n-1}$$

we can show that

$$\mu \tau_n = \frac{\pi_n}{n(n + 1)} + \frac{n^2 \tau_{n+1}}{(n + 1)(2n + 1)} + \frac{(n + 1)^2}{n(2n + 1)} \tau_{n-1},$$

$$\mu \pi_n = \frac{\tau_n}{n(n + 1)} + \frac{n^2 \pi_{n+1}}{(n + 1)(2n + 1)} + \frac{(n + 1)^2}{n(2n + 1)} \pi_{n-1},$$

from which, together with (4.24) and (4.26), it follows that

$$
T_{nm}^{(2)} = \begin{bmatrix} \dfrac{2n^2(n+1)(n+2)^2}{(2n+1)(2n+3)} & \text{if } m = n+1 \\[3mm] \dfrac{2n(n+1)^2(n-1)^2}{(2n+1)(2n-1)} & \text{if } m = n-1 \\[3mm] 0 & \text{if } m \neq n \pm 1 \end{bmatrix}
$$

Therefore, the asymmetry parameter is given by

$$
Q_{\text{sca}}\langle \cos\theta \rangle = \frac{4}{x^2} \left[\sum_n \frac{n(n+2)}{n+1} \, \text{Re}\{a_n a_{n+1}^* + b_n b_{n+1}^*\} \right.
$$

$$
\left. + \sum_n \frac{2n+1}{n(n+1)} \, \text{Re}\{a_n b_n^*\} \right].
$$

In addition to energy, light carries *momentum*; therefore, a beam that interacts with a particle will exert a force on the particle, called *radiation pressure*. The momentum flux of a plane, homogeneous wave with phase velocity v is S/v. If we now adopt the viewpoint that a beam of light consists of a stream of photons, it is physically reasonable to assert that the photons absorbed by the particle transfer all their momentum to the particle and therefore exert a force in the direction of propagation. If we interpret C_{abs} as the effective area for absorption, the momentum transfer to the particle is proportional to $I_i C_{\text{abs}}$, where I_i is the irradiance of the incident beam. Now let us interpret C_{sca} as the effective area for scattering. The photons incident on this area are elastically scattered through some distribution of angles θ, and the net rate of momentum transfer in the direction of propagation is therefore proportional to $I_i C_{\text{sca}}(1 - \langle \cos\theta \rangle)$. Thus, the total rate of momentum transfer to the particle is proportional to $I_i(C_{\text{ext}} - C_{\text{sca}}\langle \cos\theta \rangle)$, and we may define the *efficiency for radiation pressure* Q_{pr} as

$$
Q_{\text{pr}} = Q_{\text{ext}} - Q_{\text{sca}}\langle \cos\theta \rangle.
$$

The derivation above of the efficiency for radiation pressure is heuristic; a rigorous derivation of this result, which was first obtained by Debye (1909), entails integrating the stress tensor of the electromagnetic field over a spherical surface surrounding the particle.

4.6   RADAR BACKSCATTERING CROSS SECTION

McDonald (1962) has described the definition of the *radar backscattering cross section* as "intrinsically awkward"; we heartily agree that it is awkward, but we suggest that it is less intrinsically than unnecessarily so. There are several definitions extant, but none with the power to elicit a clear image of just what is meant *physically* by the concept. Perhaps the clearest statement of the definition has been given by Battan (1973, p. 30), whom we paraphrase here. Consider an arbitrary particle illuminated by a beam with irradiance I_i, which is taken to be x-polarized. It is clear from the steps leading to (3.26) that the quantity $I_i|\mathbf{X}(\theta, \phi)|^2/k^2$ is the amount of energy scattered into a unit solid angle about a particular direction (θ, ϕ), where $\mathbf{X}$ is the vector scattering amplitude for the particle. Now consider a hypothetical *isotropic* scatterer illuminated by the same beam, where the vector scattering amplitude $\mathbf{X}_{iso}$ is independent of direction and is taken to be equal to the scattering amplitude in the backscattering direction $(\theta = 180°)$ for the particle of interest: $\mathbf{X}_{iso} = \mathbf{X}(180°)$. The total energy W_{sca} scattered in all directions by the hypothetical particle is therefore

$$W_{sca} = \frac{I_i 4\pi |\mathbf{X}_{iso}|^2}{k^2} = \frac{I_i 4\pi |\mathbf{X}(180°)|^2}{k^2}.$$

The backscattering cross section σ_b is then *defined* by

$$I_i \sigma_b = W_{sca} = \frac{I_i 4\pi |\mathbf{X}(180°)|^2}{k^2},$$

$$\sigma_b = \frac{4\pi |\mathbf{X}(180°)|^2}{k^2}. \tag{4.82}$$

It is the presence of the factor 4π in (4.82) that is the obstacle to interpreting σ_b; were it not for this factor, σ_b would merely be the differential scattering cross section for scattering into a unit solid angle around the backscattering direction. In fact, it is obvious from (3.14) and (3.21) that the signal received by a detector subtending a solid angle $\Delta\Omega$ at the particle is proportional to $I_i \Delta\Omega |\mathbf{X}(\theta, \phi)|^2/k^2$ for *all* scattering angles. In addition to causing problems of interpretation, the historical definition also leads to a needless paradox: the backscattering cross section for a sphere small compared with the wavelength is *greater* than the total scattering cross section (Section 5.1). This implies, at first glance, that a part is greater than the whole!

The traditional definition of the radar backscattering cross section can be stated clearly in a few words: it is just 4π times what it ought to be. We therefore counsel the reader to mentally delete the factor 4π in (4.82) and only reintroduce it as a sop to convention when necessary.

For a sphere, we have from (3.22), (4.48), and (4.74)

$$|\mathbf{X}(180°)|^2 = |S_2(180°)|^2 \cos^2\phi + |S_1(180°)|^2 \sin^2\phi,$$

$$S_2(180°) = -S_1(180°) = \tfrac{1}{2} \sum_n (2n+1)(-1)^n (a_n - b_n).$$

Therefore, the *efficiency for backscattering* Q_b is

$$Q_b = \frac{\sigma_b}{\pi a^2} = \frac{1}{x^2}\left|\sum_n (2n + 1)(-1)^n (a_n - b_n)\right|^2.$$

McDonald (1962) gave a physical derivation of the limiting value of Q_b as $x \to \infty$, which, because of its appealing simplicity, is worth repeating here. Consider a sphere of radius a, which is taken to be large compared with the wavelength so that geometrical optics is a good approximation (see Chapter 7 for more details). The sphere is sufficiently absorbing so that all rays that are not reflected at the first interface are absorbed within the sphere; thus, scattering (excluding the forward diffraction peak, which does not contribute to the backscattering cross section) is entirely the result of reflection. Consider all those reflected rays that are confined within a set of directions defined by a cone of half-angle 2Θ about the backward direction, where $\Theta \ll 1$ but is shown on an enlarged scale in Fig. 4.10. All the energy ΔW_{sca} scattered into the solid angle $\Delta\Omega = \pi(2\Theta)^2$ results from reflection of incident rays with angles of incidence between 0 and Θ. Because Θ is small, this scattered energy is approximately $\Delta W_{sca} = I_i \pi a^2 \Theta^2 R(0°)$, where $R(0°)$ is the reflectance at normal incidence (2.58). Therefore, the backscattering cross section is given by

$$I_i \sigma_b = \frac{4\pi \Delta W_{sca}}{\Delta\Omega} = \pi a^2 R(0°) I_i,$$

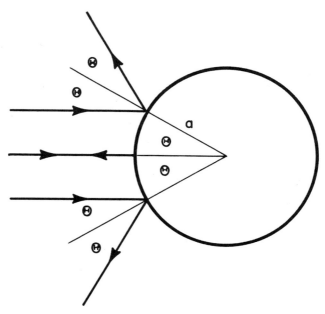

Figure 4.10 Backscattering by a large sphere in the geometrical optics approximation.

and the backscattering efficiency has the limiting value

$$\lim_{x \to \infty} Q_b = R(0^\circ),$$ (4.83)

a surprisingly simple result which, because of the "peculiar nature" of the Q_b definition, is "less than obvious." Deirmendjian's calculations (1969, p. 40) of the backscattering efficiency for large metallic spheres supports this limiting value for Q_b. We emphasize that (4.83) is expected to be correct only if the sphere is absorbing so that internally reflected rays do not contribute to the backscattering. This in turn implies that for given x (provided, of course, that $x \gg 1$), $R(0^\circ)$ will be a better approximation to the exact value the greater the absorption coefficient; the computations of Herman and Battan (1961) are consistent with this assertion.

Some of the more interesting applications of radar backscattering are given in *Radar Ornithology* by Eastwood (1967), in which one can find measured backscattering cross sections at 3-cm wavelength for pigeons, starlings, and house sparrows, together with calculations for "equivalent" spherical birds composed of water. It seems that Mie theory is sufficiently broad to embrace an unexpectedly large variety of objects.

4.7   THERMAL EMISSION

At temperatures above absolute zero, particles can emit as well as absorb and scatter electromagnetic radiation. Emission does not strictly fall within the bounds imposed in the first chapter; it is more akin to such phenomena as luminescence than to elastic scattering. However, because of the relation between emission and absorption, and because emission can be an important cooling mechanism for particles, it seems appropriate to discuss, at least briefly, thermal emission by a sphere.

Consider an enclosure of dimensions large compared with any wavelengths under consideration, which is opaque but otherwise arbitrary in shape and composition (Fig. 4.11). If the enclosure is maintained at a constant absolute temperature T, the equilibrium radiation field will be isotropic, homogeneous, and unpolarized (see Reif, 1965, p. 373 *et seq.* for a good discussion of equilibrium radiation in an enclosure). At any point the amount of radiant energy per unit frequency interval, confined to a unit solid angle about any direction, which crosses a unit area normal to this direction in unit time is given by the Planck function

$$\mathcal{P}_e = \frac{\hbar \omega^3}{4\pi^3 c^2} \frac{1}{\exp(\hbar\omega/k_B T) - 1},$$

where $\hbar = h/2\pi$, h is Planck's constant, c is the speed of light *in vacuo*, and k_B is Boltzmann's constant. $\mathcal{P}_e$ (or a similar expression) is sometimes referred to as

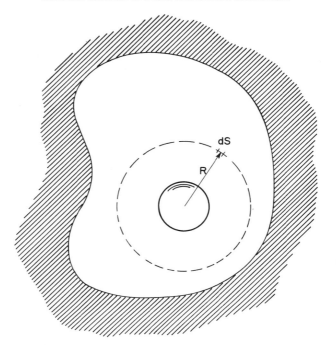

Figure 4.11 An enclosure, spherical particle, and radiation field in thermodynamic equilibrium.

the "blackbody" radiation distribution function. But the physical significance of the Planck function is that it is the distribution of radiation in equilibrium with matter; it does not require for its realization the existence of hypothetical "perfectly black" bodies. Indeed, as we shall see, rigid adherence to the notion of a perfect absorber leads to a needless paradox for small particles.

If a spherical particle is placed in the enclosure, then in equilibrium the distribution of radiation is unchanged. We may imagine that surrounding the particle there is a spherical surface of radius R, where R is much larger than the particle radius a, each element dS of which is the source of a nearly plane wave that illuminates the particle with irradiance $\mathcal{P}_e \, dS/R^2$ (Fig. 4.11). Therefore, the amount of energy absorbed per unit time by the particle is

$$\int_0^\infty \int_S \mathcal{P}_e C_{\text{abs}} \frac{dS}{R^2} d\omega = 4\pi \int_0^\infty \mathcal{P}_e C_{\text{abs}} \, d\omega.$$

In equilibrium the total power emitted by the particle must be equal to that absorbed:

$$\int_0^\infty W_e \, d\omega = 4\pi \int_0^\infty \mathcal{P}_e C_{\text{abs}} \, d\omega. \tag{4.84}$$

W_e is the power in a unit frequency interval, which by symmetry is emitted uniformly in all directions. We define the *emissivity* e as the ratio of the power emitted by the particle to the power emitted by a particle that emits according to the Planck function:

$$e = \frac{W_e}{4\pi^2 a^2 \mathcal{P}_e}.$$ (4.85)

It follows from (4.84) and (4.85) that

$$\int_0^\infty \mathcal{P}_e (Q_{abs} - e)\, d\omega = 0.$$ (4.86)

A *sufficient* condition that (4.86) be satisfied is

$$Q_{abs} = e,$$ (4.87)

which may be interpreted as Kirchhoff's law for emission and absorption by an arbitrary spherical particle. However, it is not *necessary* that (4.87) hold in order for the energy of the particle to be conserved: an excess of emission over absorption at some frequencies could be balanced by an excess of absorption over emission at other frequencies. To show that $Q_{abs} = e$ is also necessary requires invoking the principle of detailed balance, a thorough discussion of which is beyond the scope of this book. Briefly stated, however, detailed balance is a consequence of time-reversal symmetry and requires that the probability of any process be equal to the probability of the reverse process (Reif, 1965, p. 384). A lengthy and rigorous derivation of (4.87) has been given by Kattawar and Eisner (1970), who solved the field equations for a homogeneous isothermal sphere with a fluctuating electric polarization.

Equation (4.87) was obtained under the assumption of strict thermodynamic equilibrium between the particle and the surrounding radiation field; that is, the particle at temperature T is embedded in a radiation field characterized by the same temperature. However, we are almost invariably interested in applying (4.87) to particles that are not in thermodynamic equilibrium with the surrounding radiation. For example, if the only mechanisms for energy transfer are radiative, then a particle illuminated by the sun or another star will come to constant temperature when emission balances absorption; but the particle's steady temperature will not, in general, be the same as that of the star. The validity of Kirchhoff's law for a body in a nonequilibrium environment has been the subject of some controversy. However, from the review by Baltes (1976) and the papers cited therein, it appears that questions about the validity of Kirchhoff's law are merely the result of different definitions of emission and absorption, and we are justified in using (4.87) for particles under arbitrary illumination.

We shall occasionally encounter spherical particles with absorption efficiencies greater than 1, sometimes much greater (see, e.g., Chapter 12). But if

Q_{abs} can be greater than 1, the emissivity can be greater than 1, which treads heavily on deep-seated prejudices about the upper limit a proper emissivity can assume; at first glance, an emissivity greater than 1 implies that the particle emits more than a "perfectly black particle." But what is a perfectly black particle? The standard definition of a perfect blackbody is that it absorbs all the light that is *incident on it*. The key phrase here is italicized; the notion of light geometrically incident on a body is a concept from geometrical optics, which fails to be valid for particles with dimensions comparable to or less than the wavelength. This was recognized by Planck (1913), who stated that "throughout the following discussion it will be assumed that the... radii of curvature of all surfaces under consideration are large compared with the wavelengths of the rays considered." According to Baltes (1976), Kirchhoff was also well aware of the restrictions on his derivations. Unfortunately, as so often happens in physics, each successive author in a chain extending from the source of a theory tends to omit more of the fine print underlying its validity. When a "paradox" is inevitably uncovered, brickbats are unfairly hurled at the theory when their proper target is those who uncritically use it in a state of blissful ignorance about its limitations.

We shall show in Section 7.1 that the absorption efficiency, and hence the emissivity, of a sufficiently large absorbing sphere is not greater than 1. Thus, when the sphere radius is much larger than the wavelength, the definition of particle emissivity (4.85) is consonant with elementary notions about the emissivity of a body. It is also interesting to note that if particles with individual emissivities greater than 1 are strewn onto a large substrate, the resulting emissivity of the composite system is not greater than 1.

4.8 COMPUTATION OF SCATTERING COEFFICIENTS AND CROSS SECTIONS

To obtain quantitative results from the Mie theory it might seem that we are faced with a straightforward task: we need merely calculate the scattering coefficients a_n and b_n together with the angular functions π_n and τ_n and sum the series (4.61) and (4.62) for the cross sections and (4.74) for the amplitude scattering matrix elements. However, the number of terms required for convergence can be quite large: a rough rule of thumb is that about x terms are sufficient (see, e.g., Table 4.1). Thus, if we were interested in investigating the rainbow—a visible scattering phenomenon familiar to all except perhaps inhabitants of the Atacama desert—we would need to sum about 12,000 terms assuming a water droplet radius of 1 mm. Such a calculation clearly requires more than just patience, pencil, pad of paper, and pocket calculator. Even for smaller particles the number of calculations can be painfully large. Indeed, until the advent of high-speed digital computers, scattering calculations were laborious, boring, and time consuming; and the literature on scattering as recently as a decade ago tended to be dominated by papers presenting numerical results for special cases. Although computers can greatly reduce the time required to sum series, there are problems inherent in the computation of the scattering coefficients themselves; a_n and b_n are rather complicated functions of the spherical Bessel functions and their derivatives, the arguments of which are, in general, complex. Fortunately, the Bessel functions satisfy simple recurrence relations, (4.11) and (4.12), and, moreover, the first few orders are elementary trigonometric functions. We might therefore be tempted to assume that we could bootstrap our way forward by calculating Bessel functions of arbitrary order from the functions of the two preceding orders beginning with $n = 2$. Such would indeed be possible with a perfect computer at hand. Perfection is not of this world, however, and the *roundoff error* associated with the unavoidable representation of a number with an infinite number of digits by one with a finite number can accumulate in such a way as to yield incorrect results. There is not a unanimity of opinion about the conditions under which roundoff error accumulation can be a problem; this is most likely a consequence of different word lengths at various computer facilities, the fact that most people are usually interested in a limited range of sizes and optical properties, together with the unfortunate human tendency to generalize too readily on the basis of limited experience. However, there does seem to be common agreement that the *form* (4.56) and (4.57) of the scattering coefficients is not the one best suited for computations.

Aden (1951) was apparently the first to introduce the *logarithmic derivative*

$$D_n(\rho) = \frac{d}{d\rho} \ln \psi_n(\rho)$$

in the context of computing scattering coefficients for a sphere. We may therefore recast (4.56) and (4.57) in the form

$$a_n = \frac{[D_n(mx)/m + n/x]\psi_n(x) - \psi_{n-1}(x)}{[D_n(mx)/m + n/x]\xi_n(x) - \xi_{n-1}(x)},$$

$$b_n = \frac{[mD_n(mx) + n/x]\psi_n(x) - \psi_{n-1}(x)}{[mD_n(mx) + n/x]\xi_n(x) - \xi_{n-1}(x)},$$

$$(4.88)$$

where we have used the recurrence relations

$$\psi_n'(x) = \psi_{n-1}(x) - \frac{n\psi_n(x)}{x}, \qquad \xi_n'(x) = \xi_{n-1}(x) - \frac{n\xi_n(x)}{x}$$

to eliminate ψ_n' and ξ_n'. Equations (4.88) are just *one* out of many possible ways of rewriting the scattering coefficients in a form more suitable for computation. The logarithmic derivative satisfies the recurrence relation

$$D_{n-1} = \frac{n}{\rho} - \frac{1}{D_n + n/\rho} \qquad (4.89)$$

as a consequence of the recurrence relations (4.11) and (4.12).

There are two possible schemes for calculating $D_n(mx)$ in (4.88): *upward recurrence* (higher orders are generated from lower orders) and *downward recurrence* (lower orders are generated from higher orders). Kattawar and Plass (1967) have shown that D_n is numerically stable with respect to downward recurrence; that is, if e_n is the error in D_n, then the error in D_{n-1} generated from (4.89) is such that $|e_{n-1}| \ll |e_n|$. Thus, beginning with an estimate for D_n, where n is larger than the number of terms required for convergence, successively more accurate lower-order logarithmic derivatives can be generated by downward recurrence. The downward stability of D_n is a consequence of the downward stability of the spherical Bessel functions j_n (Abramowitz and Stegun, 1964, p. xiii). But j_n is just one of two linearly independent solutions to the second-order differential equation (4.8); the other solution y_n satisfies the same recurrence relations as j_n but is numerically stable with respect to *upward* recurrence. If one begins with j_0 and j_1, it is possible to calculate accurate values of successive j_n by upward recurrence—up to a point. Where that point is depends on the precision of the computer; no matter how great the precision, however, an upward recurrence scheme for j_n must eventually lead to grief. For it is y_n that is stable by upward recurrence, and any such scheme must eventually find its way onto the stable solution regardless of how recurrence is initialized.

In the light of the preceding paragraph it appears that the most conservative method for computing scattering coefficients is to compute D_n and j_n by downward recurrence and y_n by upward recurrence (recall that $\psi_n = \rho j_n$ and $\xi_n = \rho j_n + i\rho y_n$). This does not seem to be necessary, however, provided that one does not demand more scattering coefficients than are necessary for convergence of the cross sections. One occasionally encounters the assertion that computations of scattering coefficients can always be done using only upward recurrence. This may indeed be true for weakly or moderately absorbing spheres. However, following Dave (1968), we have convinced ourselves that an otherwise satisfactory upward recurrence program will generate negative cross sections for sufficiently high absorption ($k_1 x > 80$ seems to be a reasonable criterion). Therefore, the calculations in this book were obtained with a program in which $D_n(mx)$ is calculated by downward recurrence, $\xi_n(x)$ and $\psi_n(x)$ by upward recurrence. We do not claim that this program, the details of which are given in Appendix A, is the most accurate or the fastest way of calculating scattering coefficients; it has met our modest computational needs satisfactorily. The disadvantage of downward recurrence is that all the scattering coefficients must be computed and stored before summing the series for the matrix elements and cross sections. With upward recurrence the scattering coefficients are computed and successively added to the partial sums of the series; as a consequence, the storage requirements are usually much less than for downward recurrence. As computers grow ever larger, however, storage problems become less acute.

Wiscombe (1979, 1980) has recently discussed very thoroughly many of the problems encountered in computing scattering coefficients and has suggested techniques for greatly increasing the speed of such computations.

Caution: Even if D_n and j_n are calculated by downward recurrence and y_n by upward recurrence with as much precision as can be squeezed out of the largest computer available, there may be traps for the unwary in unexplored regions of $m-x$ space.

NOTES AND COMMENTS

In an interesting historical article, Logan (1965) cites many of the early papers on the problem of scattering by a sphere.

The most detailed observations of extinction by the atmospheric particles responsible for the blue suns and moons widely observed in September 1950 were made by Wilson (1951). Subsequently, his and other data were thoroughly analyzed by Penndorf (1953). More recently, Porch et al. (1973) concluded on the basis of observations made at remote locations that bluing is a rather common property of the background (i.e., nonurban) aerosol. An elementary discussion of the blue moon, together with instructions on how to demonstrate one with cigarette smoke, was given by Bohren and Brown (1981).

Brillouin (1949) discussed the extinction paradox in much more detail than we have done in this chapter.

Box and McKellar (1978) derived the sum rule (4.81) under the assumption of a constant refractive index and within the framework of the anomalous diffraction approximation of van de Hulst (1957, Chap. 11).

Chapter 5

Particles Small Compared with the Wavelength

If we were interested only in scattering and absorption by spheres, we would need go no further than the Mie theory. There is, strictly speaking, no need for approximations because we have the exact theory in hand. Given enough time, a suitable computer program will generate cross sections and scattering matrix elements for an arbitrary sphere. But physics is—or should be—more than just a semi-infinite strip of computer output: we need not denude vast tracts of forest in order to obtain some insight into scattering and absorption by small particles. In fact, great reams of calculations often serve only to obscure from view the basic physics, which can be quite simple. Therefore, it is worthwhile to consider approximate expressions, which are valid in certain limiting cases, in the hope that we may acquire some insight. Aside from the immediate applicability of these expressions to back-of-the-envelope calculations without worrying about convergence of series, misbehavior of Bessel functions, and significant digits, they point the way toward approximate methods to be used to tackle problems for which there is no exact theory.

5.1 SPHERE SMALL COMPARED WITH THE WAVELENGTH

The power series expansions of the spherical Bessel functions are (Antosiewicz, 1964)

$$j_n(\rho) = \frac{\rho^n}{1 \cdot 3 \cdot 5 \cdots (2n+1)} \left[1 - \frac{\frac{1}{2}\rho^2}{1!(2n+3)} \right.$$
$$\left. + \frac{\left(\frac{1}{2}\rho^2\right)^2}{2!(2n+3)(2n+5)} - \cdots \right], \quad (5.1)$$

$$y_n(\rho) = -\frac{1 \cdot 3 \cdot 5 \cdots (2n-1)}{\rho^{n+1}} \left[1 - \frac{\frac{1}{2}\rho^2}{1!(1-2n)} \right.$$
$$\left. + \frac{\left(\frac{1}{2}\rho^2\right)^2}{2!(1-2n)(3-2n)} - \cdots \right]. \quad (5.2)$$

130

Let us expand the various functions in the scattering coefficients a_n and b_n in power series and retain only the first few terms. From (5.1) and (5.2) we have

$$\psi_1(\rho) \simeq \frac{\rho^2}{3} - \frac{\rho^4}{30}, \qquad\qquad \psi_1'(\rho) \simeq \frac{2\rho}{3} - \frac{2\rho^3}{15},$$

$$\xi_1(\rho) \simeq -\frac{i}{\rho} - \frac{i\rho}{2} + \frac{\rho^2}{3}, \qquad\qquad \xi_1'(\rho) \simeq \frac{i}{\rho^2} - \frac{i}{2} + \frac{2\rho}{3}, \qquad (5.3)$$

$$\psi_2(\rho) \simeq \frac{\rho^3}{15}, \qquad\qquad \psi_2'(\rho) \simeq \frac{\rho^2}{5},$$

$$\xi_2(\rho) \simeq -\frac{i3}{\rho^2}, \qquad\qquad \xi_2'(\rho) \simeq \frac{i6}{\rho^3}.$$

We have retained a sufficient number of terms in the expansions (5.3) to ensure that the scattering coefficients are accurate to terms of order x^6. The first four coefficients so obtained are

$$a_1 = -\frac{i2x^3}{3}\frac{m^2-1}{m^2+2} - \frac{i2x^5}{5}\frac{(m^2-2)(m^2-1)}{(m^2+2)^2}$$

$$+ \frac{4x^6}{9}\left(\frac{m^2-1}{m^2+2}\right)^2 + O(x^7),$$

$$b_1 = -\frac{ix^5}{45}(m^2-1) + O(x^7),$$

$$a_2 = -\frac{ix^5}{15}\frac{m^2-1}{2m^2+3} + O(x^7),$$

$$b_2 = O(x^7),$$

where we have taken the permeability of the sphere to be equal to that of the surrounding medium. The expansions for higher-order scattering coefficients involve terms of order x^7 and higher. If $|m|x \ll 1$, then $|b_1| \ll |a_1|$; with this assumption the amplitude scattering matrix elements to terms of order x^3 are

$$S_1 = \tfrac{3}{2}a_1, \qquad S_2 = \tfrac{3}{2}a_1\cos\theta,$$

$$a_1 = -\frac{i2x^3}{3}\frac{m^2-1}{m^2+2}. \qquad (5.4)$$

The corresponding scattering matrix, accurate to terms of order x^6, is

$$\frac{9|a_1|^2}{4k^2r^2}\begin{pmatrix} \frac{1}{2}(1+\cos^2\theta) & \frac{1}{2}(\cos^2\theta-1) & 0 & 0 \\ \frac{1}{2}(\cos^2\theta-1) & \frac{1}{2}(1+\cos^2\theta) & 0 & 0 \\ 0 & 0 & \cos\theta & 0 \\ 0 & 0 & 0 & \cos\theta \end{pmatrix}. \tag{5.5}$$

If the incident light is unpolarized with irradiance I_i, the scattered irradiance I_s is

$$I_s = \frac{8\pi^4 N a^6}{\lambda^4 r^2}\left|\frac{m^2-1}{m^2+2}\right|^2 (1+\cos^2\theta)I_i. \tag{5.6}$$

Thus, if the quantity $|(m^2-1)/(m^2+2)|^2$ is weakly dependent on wavelength (this is *not* always true), the irradiance scattered by a sphere small compared with the wavelength or, indeed, *any* sufficiently small particle regardless of its shape, is proportional to $1/\lambda^4$. Such scattering is often referred to as Rayleigh scattering, and we are quite content to adopt this term in all that follows. However, there are those who object to associating Rayleigh's name with small particles if they absorb as well as scatter light: it seems that Rayleigh did not explicitly consider absorbing particles. Nevertheless, we shall attach the name "Rayleigh" to small-particle scattering for convenience even though the term may lack strict historical accuracy. We have not followed the historical path to the $1/\lambda^4$ scattering law but, rather, have considered the limiting case of the Mie theory. However, Rayleigh's (1871) original derivation was simplicity itself, and it is worth reproducing here to show how much can be obtained from such a small expenditure of effort.

Having disposed of the polarization, let us now consider how the intensity of the scattered light varies from one part of the spectrum to another, still supposing that all the particles are many times smaller than the wavelength even of violet light. The whole question admits of analytical treatment; but before entering upon that, it may be worthwhile to show how the principal result may be anticipated from a consideration of the *dimensions* of the quantities concerned.

The object is to compare the intensities of the incident and scattered rays; for these will clearly be proportional. The number (i) expressing the ratio of the two amplitudes is a function of the following quantities: T, the volume of the disturbing particle; r, the distance of the point under consideration from it; λ, the wavelength; b, the velocity of propagation of light; D and D', the original and altered densities [the density parameter of the ether, which was still in vogue in Rayleigh's day, corresponds to the dielectric function (Twersky, 1964)]: of which the first three depend only on space, the fourth on space and time, while the fifth and sixth introduce the consideration of mass. Other elements of the problem there are none, except mere numbers and angles, which do not depend on the fundamental measurements of space, time, and mass. Since the ratio i, whose expression we seek, is of no dimension in mass, it follows at once that D and D' only occur under the form $D : D'$, which is a simple number and may therefore be omitted. It remains to find how i varies with T, r, λ, and b.

Now of these quantities, b is the only one depending on time; and therefore, as i is of no dimensions in time, b cannot occur in its expression. We are left, then, with T, r, and λ; and from what we know of the dynamics of the question, we may be sure that i varies directly as T and inversely as r, and must therefore be proportional to $T \div \lambda^2 r$, T being of three dimensions in space. In passing from one part of the spectrum to another λ is the only quantity which varies, and we have the important law:

When light is scattered by particles which are very small compared with any of the wavelengths, the ratio of the amplitudes of the vibrations of the scattered and incident light varies inversely as the square of the wavelength and the intensity of the lights themselves as the inverse fourth power.

Equation (5.6) applies to incident unpolarized light; it is important to remember that the angular distribution of the scattered light depends on the polarization of the incident light:

$$i_{\parallel} = \frac{9|a_1|^2}{4k^2r^2}\cos^2\theta \qquad \text{\textit{Incident light polarized parallel to the scattering plane.}}$$

$$i_{\perp} = \frac{9|a_1|^2}{4k^2r^2} \qquad \text{\textit{Incident light polarized perpendicular to the scattering plane.}}$$

$$i = \tfrac{1}{2}(i_{\parallel} + i_{\perp}) \qquad \text{\textit{Unpolarized incident light.}}$$

The angular distribution of the scattered light (normalized to the forward direction) for incident light polarized parallel and perpendicular to the scattering plane and unpolarized is shown in Fig. 5.1; both linear and polar plots are given.

If the incident light is 100% polarized, the scattered light will be similarly polarized. However, because light of two different polarization states is scattered differently, the scattered light will be partially polarized if the incident light is unpolarized. From (4.78) we have

$$P = \frac{1 - \cos^2\theta}{1 + \cos^2\theta},$$

where the degree of polarization of the scattered light, given incident unpolarized light, is $|P|$ (Fig. 5.2). P is always positive; therefore, the scattered light is partially polarized *perpendicular* to the scattering plane. If a sufficiently small sphere is illuminated by unpolarized light, the scattered light is 100%

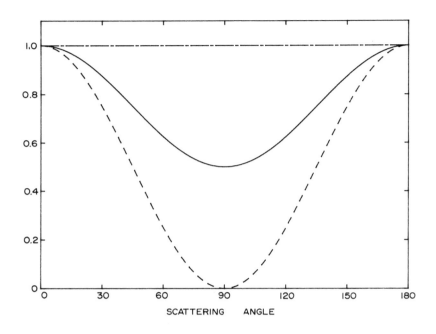

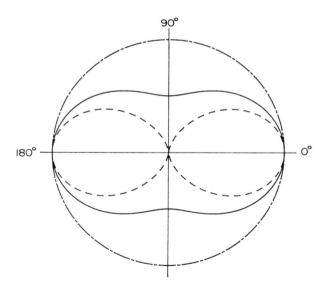

Figure 5.1 Angular distribution (normalized) of the light scattered by a sphere small compared with the wavelength: incident light polarized parallel (– – – –) and perpendicular (–·–·–) to the scattering plane; (––––) unpolarized incident light.

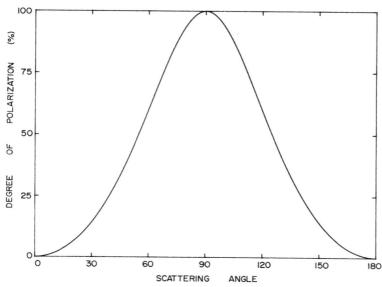

Figure 5.2 Degree of polarization of light scattered by a sphere small compared with the wavelength for incident unpolarized light.

polarized at a scattering angle of 90°. This is analogous to complete polarization upon reflection of unpolarized light incident on a plane interface at the Brewster angle (Section 2.7).

Note that P is independent of particle size, as are the functional forms of $i_{\parallel}(\theta)$ and $i_{\perp}(\theta)$; the *absolute* scattered irradiance depends on size (the volume squared), but it is difficult to measure absolute irradiances. Therefore, the radius of small spheres cannot readily be determined from scattering measurements; in this sense, all small spheres are equal.

The extinction, scattering, and radar backscattering efficiencies are (to terms of order x^4)

$$Q_{\text{ext}} = 4x \,\text{Im}\left\{ \frac{m^2 - 1}{m^2 + 2}\left[1 + \frac{x^2}{15}\left(\frac{m^2 - 1}{m^2 + 2} \right) \frac{m^4 + 27m^2 + 38}{2m^2 + 3} \right]\right\}$$

$$+ \frac{8}{3} x^4 \text{Re}\left\{ \left(\frac{m^2 - 1}{m^2 + 2} \right)^2 \right\}, \tag{5.7}$$

$$Q_{\text{sca}} = \frac{8}{3} x^4 \left| \frac{m^2 - 1}{m^2 + 2} \right|^2, \tag{5.8}$$

$$Q_b = 4x^4 \left| \frac{m^2 - 1}{m^2 + 2} \right|^2, \tag{5.9}$$

and the absorption efficiency Q_{abs} is $Q_{ext} - Q_{sca}$. If $|m|x \ll 1$, the coefficient (in brackets) of $(m^2 - 1)/(m^2 + 2)$ in the first term of (5.7) is approximately unity; with this restriction, the absorption efficiency is

$$Q_{abs} = 4x \operatorname{Im}\left\{ \frac{m^2 - 1}{m^2 + 2} \right\}\left[1 + \frac{4x^3}{3} \operatorname{Im}\left(\frac{m^2 - 1}{m^2 + 2} \right) \right]. \qquad (5.10)$$

Therefore, if $(4x^3/3)\operatorname{Im}\{(m^2 - 1)/(m^2 + 2)\} \ll 1$, a condition that will be satisfied for sufficiently small x, the absorption efficiency is approximately

$$Q_{abs} = 4x \operatorname{Im}\left\{ \frac{m^2 - 1}{m^2 + 2} \right\}. \qquad (5.11)$$

To the extent that (5.11) is a good approximation, the absorption cross section $C_{abs} = \pi a^2 Q_{abs}$ is proportional to the *volume* of the particle.

Equations (5.8), (5.11), and the scattering matrix (5.5) are widely—and wildly—quoted in the literature. However, some care must be exercised if they are to be used in calculations. Kerker et al. (1978) have investigated the validity of Rayleigh theory by calculating the scattered irradiance $i_{\parallel}(0°)$ according to the exact and approximate theories for size parameters between 0.01 and 0.11 and a range of real and imaginary parts of m. Their results clearly show that, for given x, the accuracy of the Rayleigh theory decreases as $|m|$ is increased.

If $(m^2 - 1)/(m^2 + 2)$ is a weak function of wavelength over some interval (this is not true, for example, for *metallic* particles), then for sufficiently small particles

$$Q_{abs} \propto \frac{1}{\lambda}, \qquad Q_{sca} \propto \frac{1}{\lambda^4}.$$

If extinction is dominated by absorption, the extinction spectrum will vary as $1/\lambda$; if extinction is dominated by scattering, the extinction spectrum will vary as $1/\lambda^4$. In either case, and in intermediate cases as well, shorter wavelengths are extinguished more than longer wavelengths; that is, there is *reddening* of the spectrum of incident light upon transmission through a collection of sufficiently small spheres the optical properties of which (n and k) are not strongly dependent on wavelength over the region of interest.

5.2 THE ELECTROSTATICS APPROXIMATION

The absorption and scattering efficiencies of a small ($x \ll 1$, $|m|x \ll 1$) sphere may be written

$$Q_{abs} = 4x \operatorname{Im} \frac{\varepsilon_1 - \varepsilon_m}{\varepsilon_1 + 2\varepsilon_m}, \qquad Q_{sca} = \frac{8}{3} x^4 \left| \frac{\varepsilon_1 - \varepsilon_m}{\varepsilon_1 + 2\varepsilon_m} \right|^2, \qquad (5.12)$$

where ε_1 and ε_m are the permittivities of the sphere and the surrounding medium, respectively. The quantity $(\varepsilon_1 - \varepsilon_m)/(\varepsilon_1 + 2\varepsilon_m)$ appears in the problem of a sphere embedded in a uniform *static* electric field. This suggests a connection between electrostatics and scattering by particles small compared with the wavelength; in this section we examine the reasons for this connection.

Consider a homogeneous, isotropic sphere that is placed in an arbitrary medium in which there exists a uniform static electric field $\mathbf{E}_0 = E_0\hat{\mathbf{e}}_z$ (Fig. 5.3). If the permittivities of the sphere and medium are different, a charge will be induced on the surface of the sphere. Therefore, the initially uniform field will be distorted by the introduction of the sphere. The electric fields inside and outside the sphere, $\mathbf{E}_1$ and $\mathbf{E}_2$, respectively, are derivable from scalar potentials $\Phi_1(r, \theta)$ and $\Phi_2(r, \theta)$

$$\mathbf{E}_1 = -\nabla\Phi_1, \qquad \mathbf{E}_2 = -\nabla\Phi_2,$$

where

$$\nabla^2\Phi_1 = 0 \quad (r < a), \qquad \nabla^2\Phi_2 = 0 \quad (r > a).$$

Because of the symmetry of the problem, the potentials are independent of the azimuthal angle ϕ. At the boundary between sphere and medium the potentials must satisfy

$$\Phi_1 = \Phi_2, \qquad \varepsilon_1\frac{\partial\Phi_1}{\partial r} = \varepsilon_m\frac{\partial\Phi_2}{\partial r} \quad (r = a).$$

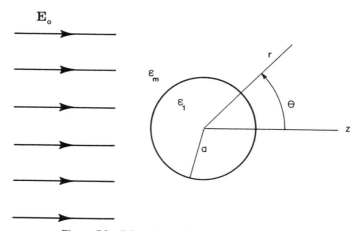

Figure 5.3 Sphere in a uniform static electric field.

In addition, we require that

$$\lim_{r \to \infty} \Phi_2 = -E_0 r \cos \theta = -E_0 z;$$

that is, at large distances from the sphere, the electric field is the unperturbed applied field. It is not difficult to show that the functions

$$\Phi_1 = -\frac{3\varepsilon_m}{\varepsilon_1 + 2\varepsilon_m} E_0 r \cos \theta,$$

$$\Phi_2 = -E_0 r \cos \theta + a^3 E_0 \frac{\varepsilon_1 - \varepsilon_m}{\varepsilon_1 + 2\varepsilon_m} \frac{\cos \theta}{r^2}, \tag{5.13}$$

satisfy the partial differential equations and boundary conditions above.

Consider now two point charges q and $-q$ which are separated by a distance d (Fig. 5.4). This configuration of charges is called a *dipole* with *dipole moment* $\mathbf{p} = p\mathbf{e}_z$, where $p = qd$. If the charges are embedded in a uniform unbounded medium with permittivity ε_m, the potential Φ of the dipole at any point P is

$$\Phi = \frac{q}{4\pi\varepsilon_m} \left(\frac{1}{r_+} - \frac{1}{r_-} \right),$$

$$r_+ = r \left(1 - \frac{\mathbf{r} \cdot \hat{\mathbf{e}}_z}{r^2} d + \frac{d^2}{4r^2} \right)^{1/2}, \qquad r_- = r \left(1 + \frac{\mathbf{r} \cdot \hat{\mathbf{e}}_z}{r^2} d + \frac{d^2}{4r^2} \right)^{1/2}.$$

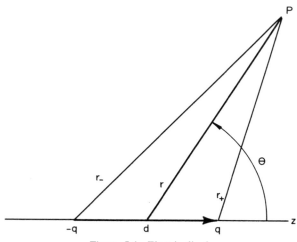

Figure 5.4 Electric dipole.

If we let d approach zero in such a way that the product qd remains constant, we obtain the potential of an *ideal dipole*

$$\Phi = \frac{\mathbf{p} \cdot \mathbf{r}}{4\pi\varepsilon_m r^3} = \frac{p\cos\theta}{4\pi\varepsilon_m r^2}. \tag{5.14}$$

Let us return now to the problem of a sphere in a uniform field. We note from (5.13) and (5.14) that the field outside the sphere is the superposition of the applied field and the field of an ideal dipole at the origin with dipole moment

$$\mathbf{p} = 4\pi\varepsilon_m a^3 \frac{\varepsilon_1 - \varepsilon_m}{\varepsilon_1 + 2\varepsilon_m} \mathbf{E}_0.$$

Thus, the applied field *induces* a dipole moment proportional to the field. The ease with which the sphere is polarized may be specified by the *polarizability* α defined by

$$\mathbf{p} = \varepsilon_m \alpha \mathbf{E}_0,$$

$$\alpha = 4\pi a^3 \frac{\varepsilon_1 - \varepsilon_m}{\varepsilon_1 + 2\varepsilon_m}. \tag{5.15}$$

Our analysis has been restricted to the response of a sphere to an applied uniform static electric field. But we are interested in scattering problems where the applied (incident) field is a plane wave that varies in space and time. We showed that a sphere in an electrostatic field is equivalent to an ideal dipole; therefore, let us assume that for purposes of calculations we may replace the sphere by an ideal dipole with dipole moment $\varepsilon_m \alpha \mathbf{E}_0$ even when the applied field is a plane wave. However, the permittivities in (5.15) are those appropriate to the frequency of the incident wave rather than the static field values.

The dipole moment $\mathbf{p} = \varepsilon_m \alpha E_0 e^{-i\omega t} \hat{\mathbf{e}}_x$ of an ideal dipole, located at $z = 0$ and illuminated by an x-polarized plane wave $E_0 \exp(ikz - i\omega t)\hat{\mathbf{e}}_x$, oscillates with the frequency of the applied field; therefore, the dipole radiates (i.e., scatters) an electric field $\mathbf{E}_s$ (Stratton, 1941, p. 453)

$$\mathbf{E}_s = \frac{e^{ikr}}{-ikr} \frac{ik^3}{4\pi\varepsilon_m} \hat{\mathbf{e}}_r \times (\hat{\mathbf{e}}_r \times \mathbf{p}), \qquad (kr \gg 1) \tag{5.16}$$

where we have omitted the time-dependent factor $e^{-i\omega t}$. After some manipulation (5.16) can be put in the form (3.21):

$$\mathbf{E}_s = \frac{e^{ik(r-z)}}{-ikr} X E, \qquad E = E_0 e^{ikz},$$

$$X = \frac{ik^3}{4\pi} \alpha \hat{\mathbf{e}}_r \times (\hat{\mathbf{e}}_r \times \hat{\mathbf{e}}_x). \tag{5.17}$$

Therefore, from (3.22) and (4.22) we have the scattering amplitudes

$$S_1 = \frac{-i\mathrm{k}^3\alpha}{4\pi}, \qquad S_2 = \frac{-i\mathrm{k}^3\alpha}{4\pi}\cos\theta,$$

which are equivalent to (5.4). The cross sections for extinction and scattering are obtained from (3.24) and (3.26):

$$C_{\mathrm{ext}} = \mathrm{k}\,\mathrm{Im}\{\alpha\} = \pi a^2 4x\,\mathrm{Im}\left\{\frac{\varepsilon_1 - \varepsilon_m}{\varepsilon_1 + 2\varepsilon_m}\right\}. \tag{5.18}$$

$$C_{\mathrm{sca}} = \frac{\mathrm{k}^4}{6\pi}|\alpha|^2 = \pi a^2 \frac{8}{3}x^4 \left|\frac{\varepsilon_1 - \varepsilon_m}{\varepsilon_1 + 2\varepsilon_m}\right|^2. \tag{5.19}$$

Equations (5.18) and (5.19) are similar to (5.12) with one exception: (5.18) is accurate only if scattering is small compared with absorption. In place of (5.18) we should therefore write

$$C_{\mathrm{abs}} = \mathrm{k}\,\mathrm{Im}\{\alpha\}.$$

Thus, replacing a sphere small compared with the wavelength by an ideal dipole has been justified—we obtain correct expressions for the scattering matrix elements and cross sections. However, let us briefly examine why this is so just to be sure that it is not a happy accident. At any instant the amplitude of the wave illuminating the sphere is $E_0\exp(i\mathrm{k}z)$; therefore, if $x = \mathrm{k}a \ll 1$, then $\exp(-i\mathrm{k}a) \simeq \exp(i\mathrm{k}a) \simeq 1$, and the field to which the sphere is exposed is approximately uniform over the region occupied by the sphere. Note also from (5.13) that the field inside the sphere is uniform in the electrostatic case. However, we would not expect the field in the sphere to be uniform when the external field is a plane wave unless $2\pi k_1 a/\lambda \ll 1$, where k_1 is the imaginary part of the particle's refractive index. The field changes over a characteristic time of order $\tau = 1/\omega$, where ω is the angular frequency of the incident field. The time τ^* required for a signal to propagate across the sphere is of order an_1/c, where n_1 is the real part of the particle's refractive index and c is the speed of light *in vacuo* [we have assumed that the group velocity coincides with the signal velocity and that the group velocity and phase velocity are approximately equal (see Stratton, 1941, pp. 333–340); these conditions will be satisfied at wavelengths not too close to strong absorption bands]. Thus, when the incident field changes, every point of the sphere will simultaneously get the message provided that $\tau^* \ll \tau$ or, equivalently, $2\pi n_1 a/\lambda \ll 1$. The two inequalities involving the real and imaginary parts of the refractive index may be combined into a single inequality: $|m|x \ll 1$.

Equations (5.12) were obtained from the exact theory in the limit $x \ll 1$ and $|m|x \ll 1$; these same equations can be obtained by treating the sphere as an ideal dipole with moment given by electrostatics theory. In the preceding paragraph, we gave a physical justification for this correspondence. However, the shape of the particle was not relevant in our considerations; for an arbitrary particle we need merely interpret a as a characteristic length. Therefore, it is now a straightforward, although possibly laborious task to calculate matrix elements and cross sections for other particles in the electrostatics approximation: we merely use electrostatics (potential theory) to calculate the polarizability of the particle. Thus, we have the means to obtain approximate solutions to a limited class of scattering problems which do not possess exact solutions.

Although magnetic particles are infrequently encountered, particularly at visible wavelengths, it is worth noting that our analysis would have to be modified somewhat if we wished to consider such particles. We have assumed that the secondary radiation is *electric dipole radiation*; but if μ_1 and μ_2 are appreciably different there will also be *magnetic dipole radiation*. The magnetic dipole moment can be calculated from magnetostatic theory and the resulting field radiated by the magnetic dipole added to that radiated by the electric dipole (see Stratton, 1941, p. 437, for a discussion of magnetic dipole radiation).

5.3 ELLIPSOID IN THE ELECTROSTATICS APPROXIMATION

The most general smooth particle—one without edges or corners—of regular shape is an ellipsoid with semiaxes $a > b > c$ (Fig. 5.5), the surface of which is specified by

$$\frac{x^2}{a^2} + \frac{y^2}{b^2} + \frac{z^2}{c^2} = 1.$$

The natural coordinates, albeit unfamiliar and not without their disagreeable features, for formulating the problem of determining the dipole moment of an ellipsoidal particle induced by a uniform electrostatic field are the *ellipsoidal coordinates* (ξ, η, ζ) defined by

$$\frac{x^2}{a^2 + \xi} + \frac{y^2}{b^2 + \xi} + \frac{z^2}{c^2 + \xi} = 1, \qquad -c^2 < \xi < \infty$$

$$\frac{x^2}{a^2 + \eta} + \frac{y^2}{b^2 + \eta} + \frac{z^2}{c^2 + \eta} = 1, \qquad -b^2 < \eta < -c^2$$

$$\frac{x^2}{a^2 + \zeta} + \frac{y^2}{b^2 + \zeta} + \frac{z^2}{c^2 + \zeta} = 1, \qquad -a^2 < \zeta < -b^2.$$

The surfaces $\xi =$ constant are confocal ellipsoids, and the particular ellipsoid $\xi = 0$ coincides with the boundary of the particle. The surfaces $\eta =$ constant

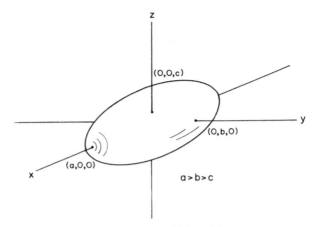

Figure 5.5 Ellipsoidal particle.

are hyperboloids of one sheet, and the surfaces ζ = constant are hyperboloids of two sheets. To any point (x, y, z) there corresponds one set of ellipsoidal coordinates (ξ, η, ζ); the converse, however, is not true. The coordinates (ξ, η, ζ) determine *eight* points symmetrically located in each of the octants into which space is partitioned by the *xyz* coordinate axes:

$$x^2 = \frac{(a^2 + \xi)(a^2 + \eta)(a^2 + \zeta)}{(b^2 - a^2)(c^2 - a^2)},$$

$$y^2 = \frac{(b^2 + \xi)(b^2 + \eta)(b^2 + \zeta)}{(a^2 - b^2)(c^2 - b^2)},$$

$$z^2 = \frac{(c^2 + \xi)(c^2 + \eta)(c^2 + \zeta)}{(a^2 - c^2)(b^2 - c^2)}.$$

This ambiguity may be removed by introducing the Weierstrassian elliptic function (Jones, 1964, p. 32). Fortunately, such a drastic step is not necessary in the problem at hand, a homogeneous ellipsoid in a uniform electrostatic field aligned along the z axis. In this instance the potential Φ has the symmetry properties

$$\Phi(x, y, z) = \Phi(-x, y, z) = \Phi(x, -y, z) = \Phi(-x, -y, z),$$

$$\Phi(x, y, -z) = \Phi(-x, y, -z) = \Phi(x, -y, -z) = \Phi(-x, -y, -z),$$

$$(5.20)$$

where x, y, z are positive. Thus, we need only consider the potential in two octants: one with positive z and one with negative z. The potential and its derivative with respect to z are also required to be continuous on the plane $z = 0$.

Let us consider the octant in which x, y, z are positive. We denote by Φ_1 the potential inside the particle; outside the particle the potential Φ_2 may be written as the superposition of the potential Φ_0 of the external field

$$\Phi_0 = -E_0 \left[\frac{(c^2 + \xi)(c^2 + \eta)(c^2 + \zeta)}{(a^2 - c^2)(b^2 - c^2)} \right]^{1/2}, \tag{5.21}$$

and the perturbing potential Φ_p caused by the particle. At sufficiently large distances from the particle the perturbing potential is negligible. We note that when $\xi \gg a^2$, then $\xi \simeq r^2$; therefore, we require that

$$\lim_{\xi \to \infty} \Phi_p = 0. \tag{5.22}$$

On the boundary of the particle the potentials are required to be continuous:

$$\Phi_1(0, \eta, \zeta) = \Phi_0(0, \eta, \zeta) + \Phi_p(0, \eta, \zeta). \tag{5.23}$$

Laplace's equation in ellipsoidal coordinates is

$$\nabla^2 \Phi = (\eta - \zeta) f(\xi) \frac{\partial}{\partial \xi} \left\{ f(\xi) \frac{\partial \Phi}{\partial \xi} \right\} + (\zeta - \xi) f(\eta) \frac{\partial}{\partial \eta} \left\{ f(\eta) \frac{\partial \Phi}{\partial \eta} \right\}$$

$$+ (\xi - \eta) f(\zeta) \frac{\partial}{\partial \zeta} \left\{ f(\zeta) \frac{\partial \Phi}{\partial \zeta} \right\} = 0, \tag{5.24}$$

where $f(q) = \{(q + a^2)(q + b^2)(q + c^2)\}^{1/2}$. At this point we could seek a complete set of solutions to (5.24) and expand the potentials in an infinite series of ellipsoidal harmonics. We can save ourselves an enormous amount of labor, however, by recognizing that, as in the case of scattering by a sphere, the form of these expansions is dictated by the form of the incident (external) field and the necessity of satisfying the boundary condition (5.23). Thus, because of (5.21), we postulate that the potentials Φ_1 and Φ_p are of the form

$$\Phi(\xi, \eta, \zeta) = F(\xi) \{(c^2 + \eta)(c^2 + \zeta)\}^{1/2},$$

where it follows from (5.24) that $F(\xi)$ satisfies the ordinary differential equation

$$f(\xi) \frac{d}{d\xi} \left\{ f(\xi) \frac{dF}{d\xi} \right\} - \left(\frac{a^2 + b^2}{4} + \frac{\xi}{2} \right) F(\xi) = 0. \tag{5.25}$$

The solution

$$F_1(\xi) = (c^2 + \xi)^{1/2}, \tag{5.26}$$

which can be verified by substitution in (5.25), follows from the fact that (5.21) satisfies Laplace's equation. A second linearly independent solution to (5.25) may be obtained by integration of (5.26) (see, e.g., Sokolnikoff and Redheffer, 1958, pp. 76–77):

$$F_2(\xi) = F_1(\xi) \int_\xi^\infty \frac{dq}{F_1^2(q) f(q)}, \tag{5.27}$$

with the property $\lim_{\xi \to \infty} F_2(\xi) = 0$. The function F_1 is not compatible with the requirement (5.22); therefore, the perturbing potential of the particle is

$$\Phi_p(\xi, \eta, \zeta) = C_2 F_2(\xi) \{ (c^2 + \eta)(c^2 + \zeta) \}^{1/2}, \tag{5.28}$$

where C_2 is a constant. If the potential inside the particle is to be finite at the origin, we must have

$$\Phi_1(\xi, \eta, \zeta) = C_1 F_1(\xi) \{ (c^2 + \eta)(c^2 + \zeta) \}^{1/2}, \tag{5.29}$$

where C_1 is a constant. Thus, the field inside the particle is uniform and parallel to the applied field. The boundary condition (5.23) yields one equation in the constants C_1 and C_2:

$$C_2 \int_0^\infty \frac{dq}{(c^2 + q) f(q)} - C_1 = \frac{E_0}{\{ (a^2 - c^2)(b^2 - c^2) \}^{1/2}},$$

and the requirement that the normal component of **D** be continuous at the boundary between particle and medium

$$\varepsilon_1 \frac{\partial \Phi_1}{\partial \xi} = \varepsilon_m \frac{\partial \Phi_0}{\partial \xi} + \varepsilon_m \frac{\partial \Phi_p}{\partial \xi} \qquad (\xi = 0),$$

yields a second equation

$$\varepsilon_m C_2 \left[\int_0^\infty \frac{dq}{(c^2 + q) f(q)} - \frac{2}{abc} \right] - \varepsilon_1 C_1 = \frac{\varepsilon_m E_0}{\{ (a^2 - c^2)(b^2 - c^2) \}^{1/2}}.$$

Therefore, the potentials inside and outside the particle are

$$\Phi_1 = \frac{\Phi_0}{1 + \dfrac{L_3(\varepsilon_1 - \varepsilon_m)}{\varepsilon_m}}, \tag{5.30}$$

$$\Phi_p = \Phi_0 \frac{\dfrac{abc}{2} \dfrac{\varepsilon_m - \varepsilon_1}{\varepsilon_m} \displaystyle\int_\xi^\infty \frac{dq}{(c^2 + q) f(q)}}{1 + \dfrac{L_3(\varepsilon_1 - \varepsilon_m)}{\varepsilon_m}}, \tag{5.31}$$

where

$$L_3 = \frac{abc}{2} \int_0^\infty \frac{dq}{(c^2 + q)f(q)}.$$

Although we considered only the octant with positive x, y, and z, it follows from the form of (5.30) and (5.31) that they are the potentials in the neighboring octant with z negative. Moreover, the eightfold degeneracy of the ellipsoidal coordinates implies that the conditions (5.20) are satisfied. Thus, (5.30) and (5.31) give the potential at all points in space; this fortunate result is a consequence of the fact that the particle has the same symmetry as the ellipsoidal coordinates. For particles with less symmetry we would have to attack the problem octant by octant.

At distances r from the origin which are much greater than the largest semiaxis a, the integral in (5.31) is approximately

$$\int_\xi^\infty \frac{dq}{(c^2 + q)f(q)} \simeq \int_\xi^\infty \frac{dq}{q^{5/2}} = \frac{2}{3}\xi^{-3/2} \qquad (\xi \simeq r^2 \gg a^2)$$

and therefore the potential Φ_p is given asymptotically by

$$\Phi_p \sim \frac{E_0\cos\theta}{r^2} \frac{\dfrac{abc}{3}\dfrac{\varepsilon_1 - \varepsilon_m}{\varepsilon_m}}{1 + \dfrac{L_3(\varepsilon_1 - \varepsilon_m)}{\varepsilon_m}}, \qquad (r \gg a),$$

which, from (5.14), we recognize as the potential of a dipole with moment

$$\mathbf{p} = 4\pi\varepsilon_m abc \frac{\varepsilon_1 - \varepsilon_m}{3\varepsilon_m + 3L_3(\varepsilon_1 - \varepsilon_m)}\mathbf{E}_0.$$

Therefore, the polarizability α_3 of an ellipsoid in a field parallel to one of its principal axes is

$$\alpha_3 = 4\pi abc \frac{\varepsilon_1 - \varepsilon_m}{3\varepsilon_m + 3L_3(\varepsilon_1 - \varepsilon_m)}. \qquad (5.32)$$

We chose the applied field to be parallel to the z axis; however, this axis has no special property to distinguish it from the other principal axes. Therefore, the polarizabilities α_1 and α_2 when the applied field is parallel to the x and y axes, respectively, are

$$\alpha_1 = 4\pi abc \frac{\varepsilon_1 - \varepsilon_m}{3\varepsilon_m + 3L_1(\varepsilon_1 - \varepsilon_m)},$$

$$\alpha_2 = 4\pi abc \frac{\varepsilon_1 - \varepsilon_m}{3\varepsilon_m + 3L_2(\varepsilon_1 - \varepsilon_m)},$$

where

$$L_1 = \frac{abc}{2} \int_0^\infty \frac{dq}{(a^2 + q)f(q)},$$

$$L_2 = \frac{abc}{2} \int_0^\infty \frac{dq}{(b^2 + q)f(q)}.$$

To check these results, we note that a sphere is a special ellipsoid with $a = b = c$; therefore,

$$L_1 = L_2 = L_3 = \frac{a^3}{2} \int_0^\infty \frac{dq}{(a^2 + q)^{5/2}} = \frac{1}{3},$$

and the polarizabilities reduce to that of a sphere (5.15) as required.

Only two of the three *geometrical factors* L_1, L_2, L_3 are independent because of the relation

$$L_1 + L_2 + L_3 = -abc \int_0^\infty \frac{d}{dq} \frac{1}{f(q)} dq = 1.$$

Moreover, they satisfy the inequalities $L_1 \leqslant L_2 \leqslant L_3$.

A special class of ellipsoids are the *spheroids*, which have two axes of equal length; therefore, only one of the geometrical factors is independent. The *prolate* (cigar-shaped) spheroids, for which $b = c$ and $L_2 = L_3$, are generated by rotating an ellipse about its *major* axis; the *oblate* (pancake-shaped) spheroids, for which $b = a$ and $L_1 = L_2$, are generated by rotating an ellipse about its *minor* axis. For spheroids, we have the following analytical expressions for L_1 as a function of the *eccentricity* e:

Prolate spheroid ($b = c$):

$$L_1 = \frac{1 - e^2}{e^2}\left(-1 + \frac{1}{2e}\ln\frac{1 + e}{1 - e} \right) \qquad e^2 = 1 - \frac{b^2}{a^2}, \qquad (5.33)$$

Oblate spheroid ($a = b$):

$$L_1 = \frac{g(e)}{2e^2}\left[\frac{\pi}{2} - \tan^{-1}g(e) \right] - \frac{g^2(e)}{2},$$

$$g(e) = \left(\frac{1 - e^2}{e^2} \right)^{1/2}, \qquad e^2 = 1 - \frac{c^2}{a^2}. \qquad (5.34)$$

The functions (5.33) and (5.34) are shown in Fig. 5.6. The shape of the oblate spheroid ranges from a *disk* ($e = 1$) to a *sphere* ($e = 0$); that of the prolate spheroid ranges from a *needle* ($e = 1$) to a sphere.

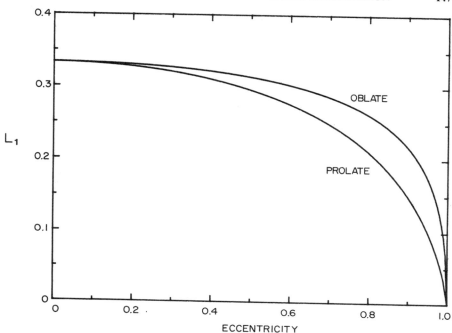

Figure 5.6 Geometrical factors for a spheroid.

The geometrical factors L_j are related to the *depolarization factors* $\bar{L}_j$ defined by

$$E_{1x} = E_{0x} - \bar{L}_1 P_{1x}, \qquad E_{1y} = E_{0y} - \bar{L}_2 P_{1y}, \qquad E_{1z} = E_{0z} - \bar{L}_3 P_{1z},$$

where $\mathbf{E}_1$ and $\mathbf{P}_1$ are the electric field and polarization induced in the particle by the applied field $\mathbf{E}_0$. The depolarization factors as defined above are meaningful quantities only for material particles (as opposed to *voids*, in which $\mathbf{P}_1 = 0$). The depolarization and geometrical factors are related by

$$\bar{L}_j = \frac{\varepsilon_1 - \varepsilon_m}{\varepsilon_1 - \varepsilon_0} \frac{L_j}{\varepsilon_m}.$$

Only when the particle is in free space ($\varepsilon_m = \varepsilon_0$) are its depolarization factors independent of composition; in this instance $\bar{L}_j = L_j/\varepsilon_0$. The induced field in an ellipsoidal particle is uniform but not necessarily parallel to an arbitrary applied field except in the special case of a sphere.

We do not restrict ourselves to material particles; ellipsoidal voids in an otherwise homogeneous medium scatter light in no way significantly different from that of "real" particles. Therefore, we shall avoid using the term depolarization factor, a concept that is meaningless for voids. Moreover, the word "*de*polarization" implies that the field inside the particle is *less* than the applied field; such is by no means always true. We shall encounter in later chapters examples where the internal field is *greater*, sometimes very much greater, than the applied field.

It might seem at first glance that arriving at the dipole moment $\mathbf{p}$ of an ellipsoidal particle via the asymptotic form of the potential Φ_p is a needlessly complicated procedure and that $\mathbf{p}$ is simply $v\mathbf{P}_1$, where v is the particle volume. However, this correspondence breaks down for a void, in which $\mathbf{P}_1 = 0$, but which nonetheless has a nonzero dipole moment. Because the medium is, in general, polarizable, $v\mathbf{P}_1$ is not equal to $\mathbf{p}$ even for a material particle except when it is in free space. In many applications of light scattering and absorption by small particles—in planetary atmospheres and interstellar space, for example—this condition is indeed satisfied. Laboratory experiments, however, are frequently carried out with particles suspended in some kind of medium such as water. It is for this reason that we have taken some care to ensure that the expressions for the polarizability of an ellipsoidal particle are completely general.

5.4   COATED ELLIPSOID

Although the necessary labor is increased, no new concepts are required to extend the results above for a homogeneous ellipsoid to a coated ellipsoid. We denote by ε_1 the permittivity of the inner or core ellipsoid with semiaxes a_1, b_1, c_1; ε_2 is the permittivity of the outer ellipsoid with semiaxes a_2, b_2, c_2. This coated ellipsoidal particle is in a medium with permittivity ε_m. As in the preceding section, we introduce ellipsoidal coordinates ξ, η, ζ:

$$\frac{x^2}{a_1^2 + \xi} + \frac{y^2}{b_1^2 + \xi} + \frac{z^2}{c_1^2 + \xi} = 1, \qquad -c_1^2 < \xi < \infty,$$

with similar expressions for η and ζ. Therefore, $\xi = 0$ is the equation of the surface of the inner ellipsoid and $\xi = t$ is that of the surface of the outer ellipsoid, where $a_1^2 + t = a_2^2$, $b_1^2 + t = b_2^2$, $c_1^2 + t = c_2^2$.

The potential of the applied field, which we take to be parallel to the z axis, is

$$\Phi_0 = -E_0 z = -E_0 F_1(\xi) G(\eta, \zeta),$$

$$F_1(\xi) = (c_1^2 + \xi)^{1/2}, \qquad G(\eta, \zeta) = \left[\frac{(c_1^2 + \eta)(c_1^2 + \zeta)}{(a_1^2 - c_1^2)(b_1^2 - c_1^2)} \right]^{1/2}.$$

The potentials Φ_1 and Φ_2 in the inner and outer ellipsoids, respectively, are

$$\Phi_1 = C_1 F_1(\xi)G(\eta, \zeta), \qquad\qquad -c_1^2 < \xi < 0,$$

$$\Phi_2 = [C_2 F_1(\xi) + C_3 F_2(\xi)]G(\eta, \zeta), \qquad 0 < \xi < t,$$

$$F_2(\xi) = F_1(\xi)\int_\xi^\infty \frac{dq}{(c_1^2 + q)f_1(q)},$$

$$f_1(q) = \left[(a_1^2 + q)(b_1^2 + q)(c_1^2 + q)\right]^{1/2}.$$

The potential Φ_3 in the surrounding medium is the sum of Φ_0 and the perturbing potential Φ_p of the particle:

$$\Phi_p = C_4 F_2(\xi)G(\eta, \zeta).$$

The requirement that Φ and $\varepsilon\partial\Phi/\partial\xi$ be continuous at boundaries gives us four linear equations in the unknown constants C_1, C_2, C_3, C_4, the solution to which yields the polarizability

$$\alpha_3 = \frac{v\big((\varepsilon_2 - \varepsilon_m)\big[\varepsilon_2 + (\varepsilon_1 - \varepsilon_2)(L_3^{(1)} - fL_3^{(2)})\big] + f\varepsilon_2(\varepsilon_1 - \varepsilon_2)\big)}{\big(\big[\varepsilon_2 + (\varepsilon_1 - \varepsilon_2)(L_3^{(1)} - fL_3^{(2)})\big]\big[\varepsilon_m + (\varepsilon_2 - \varepsilon_m)L_3^{(2)}\big] + fL_3^{(2)}\varepsilon_2(\varepsilon_1 - \varepsilon_2)\big)},$$

$$(5.35)$$

where $v = 4\pi a_2 b_2 c_2/3$ is the volume of the particle, $f = a_1 b_1 c_1/a_2 b_2 c_2$ is the fraction of the total particle volume occupied by the inner ellipsoid, and $L_3^{(1)}$ and $L_3^{(2)}$ are the geometrical factors for the inner and outer ellipsoids:

$$L_3^{(k)} = \frac{a_k b_k c_k}{2}\int_0^\infty \frac{dq}{(c_k^2 + q)f_k(q)} \qquad (k = 1, 2).$$

When $\varepsilon_1 = \varepsilon_2$, (5.35) is equivalent to (5.32), as required. Similar expressions for the polarizabilities are obtained when the field is applied along the x and y axes.

A special case is the *coated sphere* ($L_j^{(1)} = L_j^{(2)} = \frac{1}{3}$), for which $\alpha_1 = \alpha_2 = \alpha_3 = \alpha$:

$$\alpha = 4\pi a_2^3 \frac{(\varepsilon_2 - \varepsilon_m)(\varepsilon_1 + 2\varepsilon_2) + f(\varepsilon_1 - \varepsilon_2)(\varepsilon_m + 2\varepsilon_2)}{(\varepsilon_2 + 2\varepsilon_m)(\varepsilon_1 + 2\varepsilon_2) + f(2\varepsilon_2 - 2\varepsilon_m)(\varepsilon_1 - \varepsilon_2)}. \qquad (5.36)$$

We note that (5.36) implies that a homogeneous spherical particle will be *invisible* (i.e., $\alpha = 0$) if it is coated with a material such that the numerator in

(5.36) vanishes:

$$f\frac{\varepsilon_1 - \varepsilon_2}{\varepsilon_1 + 2\varepsilon_2} - \frac{\varepsilon_m - \varepsilon_2}{\varepsilon_m + 2\varepsilon_2} = 0.$$

5.5   THE POLARIZABILITY TENSOR

In the preceding sections the applied field was taken to be parallel to the *principal axes* of the ellipsoid. When the applied field $\mathbf{E}_0$ is arbitrarily directed, the induced dipole moment follows readily from superposition:

$$\mathbf{p} = \varepsilon_m(\alpha_1 E_{0x}\hat{\mathbf{e}}_x + \alpha_2 E_{0y}\hat{\mathbf{e}}_y + \alpha_3 E_{0z}\hat{\mathbf{e}}_z), \tag{5.37}$$

where E_{0x}, E_{0y}, E_{0z} are the components of $\mathbf{E}_0$ *relative to the principal axes of the ellipsoid.* In scattering problems, the coordinate axes are usually chosen to be fixed *relative to the incident beam.* Let $x'y'z'$ be such a coordinate system, where the direction of propagation is parallel to the z' axis. If the incident light is x'-polarized, we have from the optical theorem

$$C_{\text{abs},\,x'} = \frac{k\,\text{Im}\{p_{x'}\}}{\varepsilon_m E_{0x'}}. \tag{5.38}$$

To evaluate (5.38) we need the components of $\mathbf{p}$ relative to the primed axes. Equation (5.37) can be written in matrix form

$$\begin{pmatrix} p_x \\ p_y \\ p_z \end{pmatrix} = \varepsilon_m \begin{pmatrix} \alpha_1 & 0 & 0 \\ 0 & \alpha_2 & 0 \\ 0 & 0 & \alpha_3 \end{pmatrix} \begin{pmatrix} E_{0x} \\ E_{0y} \\ E_{0z} \end{pmatrix}. \tag{5.39}$$

In the interests of economy we shall write column vectors and matrices according to the following notational scheme:

$$[b] = \begin{pmatrix} b_x \\ b_y \\ b_z \end{pmatrix}, \qquad \overline{U} = \begin{pmatrix} u_{11} & u_{12} & u_{13} \\ u_{21} & u_{22} & u_{23} \\ u_{31} & u_{32} & u_{33} \end{pmatrix}.$$

With this notation, (5.39) is compactly written

$$[p] = \varepsilon_m \overline{\alpha}[E_0]. \tag{5.40}$$

The components of any vector $\mathbf{F}$ transform according to

$$[F] = \overline{A}[F'], \tag{5.41}$$

where $a_{11} = \hat{\mathbf{e}}_x \cdot \hat{\mathbf{e}}_{x'}$, $a_{12} = \hat{\mathbf{e}}_x \cdot \hat{\mathbf{e}}_{y'}$, and so on. Therefore, from (5.40) and the transformation (5.41), we have

$$[p'] = \varepsilon_m \bar{\alpha}'[E'_0],\tag{5.42}$$

$$\bar{\alpha}' = \bar{A}^T \bar{\alpha} \bar{A},\tag{5.43}$$

where the inverse of the matrix $\bar{A}$ is its transpose $\bar{A}^T$ because of the orthogonality of the coordinate axes. Thus, the polarizability of an ellipsoid is a *Cartesian tensor*; if its components are given relative to principal axes, then its components relative to rotated coordinate axes can be determined from (5.43). The absorption cross section for incident x'-polarized light follows in a straightforward manner:

$$C_{\text{abs}, x'} = k \operatorname{Im}\{\alpha_1 a_{11}^2 + \alpha_2 a_{21}^2 + \alpha_3 a_{31}^2\},$$

where $a_{11}^2 + a_{21}^2 + a_{31}^2 = 1$. Similarly, if the incident light is y'-polarized,

$$C_{\text{abs}, y'} = k \operatorname{Im}\{\alpha_1 a_{12}^2 + \alpha_2 a_{22}^2 + \alpha_3 a_{32}^2\},$$

where $a_{12}^2 + a_{22}^2 + a_{32}^2 = 1$.

If the vector scattering amplitude

$$\mathbf{X} = \frac{ik^3}{4\pi\varepsilon_m} \frac{\hat{\mathbf{e}}_r \times (\hat{\mathbf{e}}_r \times \mathbf{p})}{E_{0x'}}$$

for a dipole illuminated by x'-polarized light is substituted in (3.26), we obtain the scattering cross section

$$C_{\text{sca}, x'} = \frac{k^4}{6\pi}\left(|\alpha_1|^2 a_{11}^2 + |\alpha_2|^2 a_{21}^2 + |\alpha_3|^2 a_{31}^2\right),$$

where we have used $\bar{A}^T\bar{A} = \bar{A}\bar{A}^T = \bar{I}$, the identity matrix. A similar expression holds for the scattering cross section when the incident light is y'-polarized.

5.5.1   Randomly Oriented Ellipsoids

In most experiments and observations we are confronted with a collection of very many particles; unless special pains are taken to align the particles, or in the absence of a known alignment mechanism, we may reasonably assume that they are randomly oriented. Under these conditions the quantities of interest are the average cross sections $\langle C_{\text{abs}} \rangle$ and $\langle C_{\text{sca}} \rangle$, which are independent of the polarization of the incident light provided that the particles are not intrinsically optically active. Let $p(\hat{\Omega}) \, d\Omega$ be the probability that one of the axes fixed relative to a particle, the x axis, say, lies within a solid angle $d\Omega$ around the direction $\hat{\Omega}$. If the particles are randomly oriented, $p(\hat{\Omega}) = 1/4\pi$ and we have

$$\langle a_{11}^2 \rangle = \frac{1}{4\pi} \int_0^{2\pi} \int_0^{\pi} \cos^2\beta \sin\beta \, d\beta \, d\nu = \frac{1}{3},$$

where $a_{11} = \hat{\mathbf{e}}_x \cdot \hat{\mathbf{e}}_x' = \cos\beta$ and $\hat{\Omega}(\beta, \nu)$ is the direction of the x axis relative to the primed coordinate system. Similarly, we have $\langle a_{21}^2 \rangle = \langle a_{31}^2 \rangle = \frac{1}{3}$, and therefore

$$\langle C_{\text{abs}} \rangle = \text{k Im}\{\tfrac{1}{3}\alpha_1 + \tfrac{1}{3}\alpha_2 + \tfrac{1}{3}\alpha_3\}, \tag{5.44}$$

$$\langle C_{\text{sca}} \rangle = \frac{k^4}{6\pi}\left(\tfrac{1}{3}|\alpha_1|^2 + \tfrac{1}{3}|\alpha_2|^2 + \tfrac{1}{3}|\alpha_3|^2\right). \tag{5.45}$$

5.6 ANISOTROPIC SPHERE

We noted in the preceding section that the polarizability of an ellipsoid is anisotropic: the dipole moment induced by an applied uniform field is not, in general, parallel to that field. This anisotropy originates in the *shape* anisotropy of the ellipsoid. However, ellipsoids are not the only particles with an anisotropic polarizability; in fact, all the expressions above for cross sections are valid regardless of the origin of the anisotropy provided that there exists a coordinate system in which the polarizability tensor is diagonal.

Up to this point we have restricted consideration to materials for which the dielectric function is a scalar. However, except for amorphous materials and crystals with cubic symmetry, the dielectric function is a tensor; therefore, the constitutive relation connecting $\mathbf{D}$ and $\mathbf{E}$ is

$$\begin{pmatrix} D_x \\ D_y \\ D_z \end{pmatrix} = \begin{pmatrix} \varepsilon_{xx} & \varepsilon_{xy} & \varepsilon_{xz} \\ \varepsilon_{yx} & \varepsilon_{yy} & \varepsilon_{yz} \\ \varepsilon_{zx} & \varepsilon_{zy} & \varepsilon_{zz} \end{pmatrix} \begin{pmatrix} E_x \\ E_y \\ E_z \end{pmatrix}. \tag{5.46}$$

Let us consider a sphere composed of a material described by the constitutive relation (5.46). We assume that the principal axes of the real and imaginary parts of the permittivity tensor coincide; this condition is not necessarily satisfied except for crystals with at least orthorhombic symmetry (Born and Wolf, 1965, p. 708). If we take as coordinate axes the principal axes of the permittivity tensor, the constitutive relation (5.46) in the sphere is

$$D_{1x} = \varepsilon_{1,1}E_{1x}, \qquad D_{1y} = \varepsilon_{1,2}E_{1y}, \qquad D_{1z} = \varepsilon_{1,3}E_{1z}.$$

As before, the potential in the surrounding medium, which is taken to be isotropic, is the sum of the potentials of the applied field and the perturbing field, both of which satisfy Laplace's equation. In the sphere we have

$$\mathbf{E}_1 = -\nabla\Phi_1, \qquad \nabla \cdot \mathbf{D}_1 = 0.$$

If the applied field is parallel to one of the principal axes, the z axis, say, we might be tempted to guess, on the basis of previous experience, that the field in the sphere is uniform and parallel to the z axis:

$$\Phi_1 = C_1 z = C_1 r \cos \theta \qquad (r < a);$$

therefore, the divergence of $\mathbf{D}_1$ vanishes as required. We can show by direct substitution that

$$\Phi_2 = -E_0 r \cos \theta + C_2 \frac{\cos \theta}{r^2} \qquad (r > a)$$

is a solution to Laplace's equation and, moreover, that Φ and D_r, the radial component of $\mathbf{D}$, are continuous if

$$C_1 = -\frac{\varepsilon_m}{\varepsilon_{1,3} + 2\varepsilon_m} E_0, \qquad C_2 = a^3 \frac{\varepsilon_{1,3} - \varepsilon_m}{\varepsilon_{1,3} + 2\varepsilon_m} E_0.$$

Therefore, the polarizability when the field is applied along the z axis is

$$\alpha_3 = 4\pi a^3 \frac{\varepsilon_{1,3} - \varepsilon_m}{\varepsilon_{1,3} + 2\varepsilon_m}.$$

Similar expressions are obtained for the polarizabilities when the field is applied along the other two principal axes.

It is interesting to compare and contrast an isotropic ellipsoid and an anisotropic sphere; the polarizability of both particles is a tensor, the principal values of which are

$$\alpha_j^e = 4\pi abc \frac{\varepsilon_1 - \varepsilon_m}{3\varepsilon_m + 3L_j(\varepsilon_1 - \varepsilon_m)} \qquad \textit{isotropic ellipsoid}$$

$$\alpha_j^s = 4\pi a^3 \frac{\varepsilon_{1,j} - \varepsilon_m}{\varepsilon_{1,j} + 2\varepsilon_m} \qquad \textit{anisotropic sphere}$$

Although there are similarities between the two types of particle, they are not completely equivalent: given an anisotropic sphere in a particular medium, there does not exist, in general, an equal volume ellipsoid with the same polarizability. That this is so is evident from the fact that *six* parameters, the real and imaginary parts of the $\varepsilon_{1,j}$, determine the polarizability tensor of the sphere, whereas only *four* parameters, the real and imaginary parts of ε_1 together with two geometrical factors, determine the polarizability tensor of the ellipsoid. In the special case of a nonabsorbing sphere or when two principal values of the permittivity tensor are equal an isotropic ellipsoid can be found—on paper, at least—with the same polarizability tensor.

It is not difficult to generalize the results of this section to an anisotropic ellipsoid the axes of which coincide with the principal axes of its permittivity tensor. The principal values of the polarizability tensor of such a particle are

$$\alpha_j = 4\pi abc \frac{\varepsilon_{1,j} - \varepsilon_m}{3\varepsilon_m + 3L_j(\varepsilon_{1,j} - \varepsilon_m)}.$$

More general ellipsoidal particles in an anisotropic *medium*, where there is no restriction on the principal axes of either the real or imaginary parts of the permittivity tensors, have been treated by Jones (1945).

5.7   SCATTERING MATRIX

The proof is lengthy, but it follows from (5.16), (5.37), and (5.42) that the amplitude scattering matrix elements (3.12) for an anisotropic dipole are

$$S_1 = \frac{-ik^3}{4\pi}(\alpha_{11}\sin^2\phi - 2\alpha_{12}\sin\phi\cos\phi + \alpha_{22}\cos^2\phi),$$

$$S_2 = \frac{-ik^3}{4\pi}\Big[\cos\theta(\alpha_{11}\cos^2\phi + 2\alpha_{12}\sin\phi\cos\phi + \alpha_{22}\sin^2\phi)$$

$$- \sin\theta(\alpha_{13}\cos\phi + \alpha_{23}\sin\phi)\Big],$$

$$S_3 = \frac{-ik^3}{4\pi}\Big\{\cos\theta\big[\alpha_{11}\sin\phi\cos\phi + \alpha_{12}(\sin^2\phi - \cos^2\phi)$$

$$- \alpha_{22}\sin\phi\cos\phi\big] - \sin\theta(\alpha_{13}\sin\phi - \alpha_{23}\cos\phi)\Big\}, \qquad (5.47)$$

$$S_4 = \frac{-ik^3}{4\pi}\Big[\alpha_{11}\sin\phi\cos\phi + \alpha_{12}(\sin^2\phi - \cos^2\phi) - \alpha_{22}\sin\phi\cos\phi\Big],$$

where

$$\alpha_{ij} = \alpha_{ji} = \sum_{k=1}^{3} \alpha_k a_{ki} a_{kj}$$

are the components of the polarizability tensor in the coordinate system fixed relative to the incident beam [see (5.43)].

The scattering matrix elements S_{ij} corresponding to (5.47) can be obtained from (3.16). However, of possibly greater interest than the most general scattering matrix is that for a collection of identical, but randomly oriented, anisotropic dipoles; this scattering matrix is proportional to $\mathfrak{N}\langle S_{ij}\rangle$, where $\mathfrak{N}$ is the number of dipoles per unit volume and $\langle S_{ij}\rangle$ are the scattering matrix elements for a single anisotropic dipole averaged over all orientations. The $\langle S_{ij} \rangle$ are independent of the azimuthal angle ϕ; therefore, we can save ourselves a good bit of labor by choosing $\phi = 90°$ in (5.47) before computing the matrix elements (3.16) and performing the required averaging. A complete derivation of the average scattering matrix is less formidable than it might seem at first glance, but it is time consuming and best left to a long winter's evening. Most of the effort is bookkeeping; however, there are a few essential steps, which we outline in the following paragraphs.

The S_{ij} are of second degree in α_{kl} and hence of fourth degree in a_{mn}. Thus, we must calculate averages of the form $\langle a_{ij} a_{kl} a_{mn} a_{pq} \rangle$, many of which either vanish or are identical because of symmetry. If we assume that all orientations are equally probable, as in Section 5.5, then

$$\langle a_{11}^4 \rangle = \frac{1}{4\pi} \int_0^{2\pi} \int_0^{\pi} \cos^4\beta \sin \beta \, d\beta \, dv = \frac{1}{5}.$$

However, designating coordinate axes as x, y, z and so on, is arbitrary; therefore, the various averages must be invariant with respect to relabeling axes. Thus, it follows that $\langle a_{ij}^4 \rangle = \frac{1}{5}$ for all i and j. From the orthogonality of the transformation (5.41) we have

$$\sum_{k=1}^{3} a_{ki} a_{kj} = \sum_{k=1}^{3} a_{ik} a_{jk} = \delta_{ij}, \tag{5.48}$$

where δ_{ij}, the Kronecker delta, is 0 if $i \neq j$ and 1 otherwise. It follows from (5.48) that

$$\langle a_{11}^2 a_{21}^2 \rangle + \langle a_{11}^2 a_{31}^2 \rangle = \langle a_{11}^2 \rangle - \langle a_{11}^4 \rangle = \frac{2}{15}.$$

But again, by symmetry, $\langle a_{11}^2 a_{31}^2 \rangle = \langle a_{11}^2 a_{21}^2 \rangle = \langle a_{mj}^2 a_{nj}^2 \rangle = \langle a_{jm}^2 a_{jn}^2 \rangle = 1/15$, where j is not equal to both m and n. We also have $a_{11}a_{12} + a_{22}a_{21} + a_{31}a_{32} = 0$ from (5.48); therefore,

$$\langle a_{11}a_{12}a_{22}a_{21} \rangle + \langle a_{22}a_{21}a_{31}a_{32} \rangle = -\langle a_{22}^2 a_{21}^2 \rangle = -\frac{1}{15},$$

and by symmetry

$$\langle a_{11}a_{12}a_{22}a_{21} \rangle = \langle a_{22}a_{21}a_{31}a_{32} \rangle = -\frac{1}{30}.$$

Similarly, any average that can be obtained from $\langle a_{11}a_{12}a_{22}a_{21} \rangle = \langle (\hat{e}_x \cdot \hat{e}_{x'})(\hat{e}_x \cdot \hat{e}_{y'})(\hat{e}_y \cdot \hat{e}_{x'})(\hat{e}_y \cdot \hat{e}_{y'}) \rangle$ by relabeling axes is equal to $-\frac{1}{30}$. If we recall that $a_{12} = \hat{e}_x \cdot \hat{e}_{y'}$, $a_{13} = \hat{e}_x \cdot \hat{e}_{z'}$, and $\hat{e}_x = \cos \beta \hat{e}_{x'} + \sin \beta \cos v \hat{e}_{y'} + \sin \beta \sin v \hat{e}_{z'}$, then

$$\langle a_{12}^3 a_{13} \rangle = \frac{1}{4\pi} \int_0^{2\pi} \int_0^{\pi} \sin^4\beta \sin \beta \cos^3 v \sin v \, d\beta \, dv = 0, \tag{5.49}$$

and similarly for all equivalent averages. From (5.48) and symmetry we have $\langle a_{12}a_{13} \rangle = \langle a_{23}a_{22} \rangle = \langle a_{33}a_{32} \rangle = 0$, from which, together with (5.49), it follows that $\langle a_{12}^2 a_{23}a_{22} \rangle = 0$. The only remaining averages are those of the form $\langle a_{11}a_{13}a_{21}a_{22} \rangle$, which can readily be shown to vanish. Thus, in the expressions for the average matrix elements $\langle S_{ij} \rangle$, the only nonvanishing terms quadratic in the α_{kl} are the following:

$$\left\langle |\alpha_{11}|^2 \right\rangle = \left\langle |\alpha_{22}|^2 \right\rangle = \tfrac{1}{5}\left(|\alpha_1|^2 + |\alpha_2|^2 + |\alpha_3|^2 \right)$$

$$+ \tfrac{2}{15}\mathrm{Re}\{ \alpha_1^*\alpha_2 + \alpha_1^*\alpha_3 + \alpha_2^*\alpha_3 \},$$

$$\left\langle |\alpha_{12}|^2 \right\rangle = \left\langle |\alpha_{23}|^2 \right\rangle = \left\langle |\alpha_{13}|^2 \right\rangle$$

$$= \tfrac{1}{15}\left(|\alpha_1|^2 + |\alpha_2|^2 + |\alpha_3|^2 \right) - \tfrac{1}{15}\mathrm{Re}\{ \alpha_1^*\alpha_2 + \alpha_2^*\alpha_3 + \alpha_1^*\alpha_3 \},$$

$$\left\langle \alpha_{11}\alpha_{22}^* \right\rangle = \tfrac{1}{15}\left(|\alpha_1|^2 + |\alpha_2|^2 + |\alpha_3|^2 \right) + \tfrac{4}{15}\mathrm{Re}\{ \alpha_1^*\alpha_2 + \alpha_2^*\alpha_3 + \alpha_1^*\alpha_3 \}.$$

$$(5.50)$$

All the essential ingredients for calculating the average scattering matrix for a randomly oriented anisotropic dipole are now at hand. From the relations (5.50), after a good bit of algebra, we obtain

$$\begin{pmatrix} I_s \\ Q_s \\ U_s \\ V_s \end{pmatrix} = \frac{1}{k^2 r^2} \begin{pmatrix} \langle S_{11} \rangle & \langle S_{12} \rangle & 0 & 0 \\ \langle S_{12} \rangle & \langle S_{22} \rangle & 0 & 0 \\ 0 & 0 & \langle S_{33} \rangle & 0 \\ 0 & 0 & 0 & \langle S_{44} \rangle \end{pmatrix} \begin{pmatrix} I_i \\ Q_i \\ U_i \\ V_i \end{pmatrix},$$

$$\langle S_{11} \rangle = \frac{3k^2 \langle C_{\mathrm{sca}} \rangle}{8\pi} \frac{1}{2}\left(\frac{6 - M}{5} + \frac{2 + 3M}{5}\cos^2\theta \right),$$

$$\langle S_{12} \rangle = \frac{3k^2 \langle C_{\mathrm{sca}} \rangle}{8\pi} \frac{1}{2}(\cos^2\theta - 1)\frac{2 + 3M}{5},$$

$$\langle S_{22} \rangle = \frac{3k^2 \langle C_{\mathrm{sca}} \rangle}{8\pi} \frac{1}{2}(\cos^2\theta + 1)\frac{2 + 3M}{5},$$

$$\langle S_{33} \rangle = \frac{3k^2 \langle C_{\mathrm{sca}} \rangle}{8\pi} \frac{2 + 3M}{5}\cos\theta,$$

$$\langle S_{44} \rangle = \frac{3k^2 \langle C_{\mathrm{sca}} \rangle}{8\pi} M\cos\theta, \qquad (5.51)$$

where the average scattering cross section $\langle C_{\mathrm{sca}} \rangle$ was derived in Section 5.5;

the ratio

$$M = \frac{\mathrm{Re}\{\alpha_1^*\alpha_2 + \alpha_1^*\alpha_3 + \alpha_2^*\alpha_3\}}{|\alpha_1|^2 + |\alpha_2|^2 + |\alpha_3|^2} \tag{5.52}$$

satisfies the inequality $-\frac{1}{2} \leqslant M \leqslant 1$. For an isotropic dipole the α_j are all equal, $M = 1$, and (5.51) reduces to (5.5).

If the incident light is unpolarized, the Stokes parameters of the scattered light are $I_s = \langle S_{11} \rangle$, $Q_s = \langle S_{12} \rangle$, $U_s = V_s = 0$. Thus, the scattered light is partially polarized of degree P:

$$P = \frac{-\langle S_{12} \rangle}{\langle S_{11} \rangle} = \frac{1 - \cos^2\theta}{\dfrac{6 - M}{2 + 3M} + \cos^2\theta}. \tag{5.53}$$

P is positive for all scattering angles θ and all allowed values of M; therefore, the scattered light is partially polarized perpendicular to the scattering plane. The maximum degree of polarization occurs at $\theta = 90°$:

$$P(90°) = \frac{2 + 3M}{6 - M} \tag{5.54}$$

and lies between $\frac{1}{13}$ and 1, depending on the value of M. Unlike scattering by an isotropic sphere small compared with the wavelength, light scattered by a collection of randomly oriented anisotropic dipoles is not 100% polarized at 90°. This anisotropy effect is, in part, responsible for the lack of complete polarization of scattered skylight viewed normally to the direction of the sun: the molecules responsible for the scattering are not spherically symmetrical.

Chapter 6

Rayleigh–Gans Theory

If a particle has other than a regular geometrical shape then it is difficult, if not essentially impossible, to solve the scattering problem in its most general form. There are, however, frequently encountered situations, particularly in laboratory investigations of scattering, in which the particles are suspended in a medium with similar optical properties. If the particles, which are sometimes referred to as "soft" or "tenuous," are not too large (but they may be larger than in the Rayleigh theory discussed in the preceding chapter), it is possible to obtain relatively simple approximate expressions for the scattering matrix elements. Within the limits of this approximation, these expressions are valid for particles of arbitrary shape.

6.1 AMPLITUDE SCATTERING MATRIX ELEMENTS

If the amplitude scattering matrix elements (5.4) for a homogeneous, isotropic sphere of radius a are divided by the volume v, the resulting quotients approach finite limits as the sphere radius tends to zero:

$$s_1 = \lim_{a \to 0} \frac{S_1}{v} = -\frac{3i k^3}{4\pi} \frac{m^2 - 1}{m^2 + 2},$$

$$s_2 = \lim_{a \to 0} \frac{S_2}{v} = -\frac{3i k^3}{4\pi} \frac{m^2 - 1}{m^2 + 2} \cos \theta. \tag{6.1}$$

The quantities s_1 and s_2 may be interpreted as the scattering matrix elements per unit particle volume, and it is physically plausible that under certain conditions the matrix elements for an arbitrarily shaped particle may be approximated by a suitable integration of the s_j over the volume of the particle. This assumption is the basis of what is often called the Rayleigh–Gans theory. However, Kerker (1969, p. 414) has argued that Debye, not Gans, should share honors with Rayleigh. Rocard sometimes trails Rayleigh and Gans (Acquista, 1976). And in quantum-mechanical scattering, the analogous approximation is called the *Born approximation* (to be precise, the *first* Born approximation) (Merzbacher, 1970, p. 229); the name of Kirchhoff sometimes appears in tandem with that of Born (Saxon, 1955a). Buried in obscure journals there

undoubtedly rest papers the authors of which—or their heirs—could legitimately lay claim to having been first in print. Any day now we can expect
scholars to announce that the theory has been found scribbled in the margins
of one of Gauss's unpublished manuscripts; or in the notebooks of Leonardo;
or implicit in the writings of Aristotle; or painted in brilliant colors on the
walls of a French cave by Paleolithic men. Referring to the RGDRKBU
approximation—the U is reserved for as yet unknown claimants—seems a bit
unwieldly, although it assiduously avoids offense to anyone. Instead, we shall
content ourselves with the somewhat more prosaic Rayleigh–Gans theory.
What it lacks in historical accuracy it makes up for in brevity; moreover, it is
probably the term most familiar to the majority of readers.

The conditions for the validity of the Rayleigh–Gans approximation are

$$|m - 1| \ll 1, \tag{6.2}$$

$$kd|m - 1| \ll 1, \tag{6.3}$$

where d is a characteristic linear dimension of the particle and m is its complex
refractive index relative to that of the surrounding medium. It may be shown
rigorously from an integral equation formulation of the scattering problem
(Saxon, 1955a) that the Rayleigh–Gans approximation is obtained if the field
inside the particle is approximated by the incident field. Therefore, by analogy
with the problem of reflection and transmission by a homogeneous slab
(Section 2.8), we may interpret condition (6.2) as the requirement that the
incident wave is not appreciably "reflected" at the particle–medium interface;
condition (6.3) may be interpreted as the requirement that the incident wave
not undergo appreciable change of phase or amplitude after it enters the
particle. We emphasize, however, that this reasoning is heuristic.

Because of condition (6.2) it is customary (but not necessary) to write the
scattering matrix elements (6.1) as

$$s_1 = -\frac{ik^3}{2\pi}(m - 1), \qquad s_2 = -\frac{ik^3}{2\pi}(m - 1)\cos\theta, \tag{6.4}$$

where we have used

$$\frac{m^2 - 1}{m^2 + 2} = \frac{(m - 1)(m + 1)}{m^2 + 2} \simeq \frac{2}{3}(m - 1).$$

Consider an arbitrary particle illuminated by a plane wave propagating in
the z' direction (Fig. 6.1). The contribution of a volume element Δv located at
a point O' to the field scattered by the particle in a direction specified by the
unit vector $\hat{e}_r$ is

$$\begin{pmatrix} \Delta E_{\|s} \\ \Delta E_{\perp s} \end{pmatrix} = \frac{e^{ik(r'-z')}}{-ikr'}\Delta v \begin{pmatrix} s_2 & 0 \\ 0 & s_1 \end{pmatrix} \begin{pmatrix} E'_{\|i} \\ E'_{\perp i} \end{pmatrix}, \tag{6.5}$$

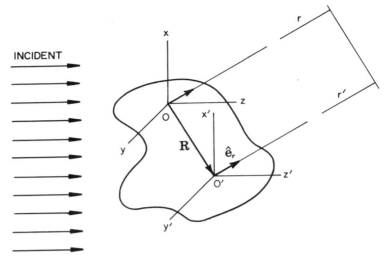

Figure 6.1 Coordinate systems for determining scattering by an arbitrary particle in the Rayleigh–Gans approximation.

where

$$E'_{\|i} = E'_{\|0}e^{ikz'}, \qquad E'_{\perp i} = E'_{\perp 0}e^{ikz'}, \tag{6.6}$$

and the amplitudes $E'_{\|0}$ and $E'_{\perp 0}$ are independent of position. A rectangular coordinate system (x', y', z') is centered at O'; let $-Z$ be the z' coordinate of the origin O of a reference coordinate system (x, y, z) the axes of which are parallel to those of the (x', y', z') system. The incident field at O is

$$E_{\|0} = E'_{\|i}(-Z) = E'_{\|0}e^{-ikZ},$$

$$E_{\perp 0} = E'_{\perp i}(-Z) = E'_{\perp 0}e^{-ikZ}. \tag{6.7}$$

It follows from (6.6) and (6.7) that $E'_{\|i} = E_{\|0}e^{ik(Z+z')}$ and $E'_{\perp i} = E_{\perp 0}e^{ik(Z+z')}$. If we use the relations $r' = r - \mathbf{R} \cdot \hat{\mathbf{e}}_r$, $Z = \mathbf{R} \cdot \hat{\mathbf{e}}_z$, then (6.5) can be written

$$\begin{pmatrix} \Delta E_{\|s} \\ \Delta E_{\perp s} \end{pmatrix} = \frac{e^{ik(r-z)}}{-ikr} \Delta v e^{i\delta} \begin{pmatrix} s_2 & 0 \\ 0 & s_1 \end{pmatrix} \begin{pmatrix} E_{\|i} \\ E_{\perp i} \end{pmatrix}, \tag{6.8}$$

where $\delta = k\mathbf{R} \cdot (\hat{\mathbf{e}}_z - \hat{\mathbf{e}}_r)$, $E_{\|i} = E_{\|0}e^{ikz}$, and $E_{\perp i} = E_{\perp 0}e^{ikz}$. Because the particle is much smaller than the distance to the point of observation, we may approximate the factor $1/kr'$ by $1/kr$. The total field $\mathbf{E}_s$ scattered in the direction $\hat{e}_r$ is obtained by integrating (6.8) over the particle volume v:

$$\begin{pmatrix} E_{\|s} \\ E_{\perp s} \end{pmatrix} = \frac{e^{ik(r-z)}}{-ikr} \begin{pmatrix} S_2 & 0 \\ 0 & S_1 \end{pmatrix} \begin{pmatrix} E_{\|i} \\ E_{\perp i} \end{pmatrix},$$

$$S_1 = -\frac{ik^3}{2\pi}(m-1)vf(\theta, \phi),$$

$$S_2 = -\frac{ik^3}{2\pi}(m-1)vf(\theta, \phi)\cos\theta. \tag{6.9}$$

The *form factor* $f(\theta, \phi)$ is

$$f(\theta, \phi) = \frac{1}{v}\int_v e^{i\delta}\, dv. \tag{6.10}$$

Implicit in the derivation of (6.9) is the assumption that the particle is homogeneous. This is not necessary, however; the particle may be composed of several distinct regions. The generalization of (6.9) to a heterogeneous particle is straightforward:

$$S_1 = -\frac{ik^3}{2\pi}\sum_j(m_j-1)v_jf_j(\theta, \phi),$$

$$S_2 = -\frac{ik^3}{2\pi}\sum_j(m_j-1)v_jf_j(\theta, \phi)\cos\theta,$$

where m_j is the relative refractive index of the jth region in the particle, v_j is its volume, and the associated form factor is

$$f_j = \frac{1}{v_j}\int_{v_j} e^{i\delta}\, dv.$$

In the forward direction ($\theta = 0°$) $\hat{e}_r = \hat{e}_z$ and, therefore, $f(0°) = 1$ for all particles; this, in turn, implies that $S_1(0°) = S_2(0°)$. Because our point of departure was the Rayleigh theory for an infinitesimal sphere, the Rayleigh–Gans theory shares some of its features. For example, the form of the 4×4 scattering matrix corresponding to (6.9) is the same as that for a Rayleigh sphere (5.5); the dependence of the individual scattering matrix elements on the scattering direction is, however, different in general.

The optical theorem yields the absorption cross section

$$C_{\text{abs}} = 2kv\,\text{Im}\{m\} \tag{6.11}$$

independent of the polarization of the incident light and the orientation of the particle. Equation (6.11) can be written in a more interesting form

$$C_{abs} = \alpha v,$$

where $\alpha = 4\pi k_1/\lambda$ is the absorption coefficient and k_1 is the imaginary part of the refractive index of the particle. The irradiance transmitted by a homogeneous slab of thickness h and absorption coefficient α is

$$I_t = I_i e^{-\alpha h},$$

where we have assumed that the reflectance is small (see Section 2.8). Therefore, the amount of energy W_{abs} absorbed by the slab is

$$W_{abs} = I_i(1 - e^{-\alpha h})A,$$

where A is the cross-sectional area. If we assume that $\alpha h \ll 1$, then

$$\frac{W_{abs}}{I_i} = \alpha V,$$

where V is the volume of the slab. Therefore, under conditions similar to (6.2) and (6.3), the absorption cross section W_{abs}/I_i of a slab has the same form as that of a tenuous particle. This strengthens the heuristic arguments at the beginning of this chapter which were made to give a physical basis to the conditions (6.2) and (6.3).

The scattering cross section, unlike the absorption cross section, depends on the state of polarization of the incident light unless S_1 and S_2 are independent of the azimuthal angle ϕ, which will be true for spherically symmetrical particles. Regardless of the shape of the particle, however, the degree of polarization of the scattered light is the same as that for a Rayleigh sphere; this follows from the fact that f merely multiplies the Rayleigh amplitude scattering matrix elements. The major difference between the Rayleigh and Rayleigh–Gans theories is the angular distribution of the scattered light.

6.2 HOMOGENEOUS SPHERE

The form factor for a particle of arbitrary shape can be calculated by numerical integration of (6.10). However, for certain regular geometrical shapes, it is possible to obtain analytical expressions for f. In this section we consider one such particle, a homogeneous sphere.

The vector $\hat{\mathbf{e}}_z - \hat{\mathbf{e}}_r$ is normal to planes $\mathbf{R} \cdot (\hat{\mathbf{e}}_z - \hat{\mathbf{e}}_r) = $ constant, over which the phase δ is constant; we can write the phase as

$$\delta = 2k \sin \frac{\theta}{2} R \cos(\mathbf{R}, \hat{\mathbf{e}}_z - \hat{\mathbf{e}}_r) = 2k\xi \sin \frac{\theta}{2},$$

where $|\xi| = |R \cos(\mathbf{R}, \hat{\mathbf{e}}_z - \hat{\mathbf{e}}_r)|$ is the distance from the origin to a plane of constant phase. Thus, the form factor can be expressed as an integral over the variable ξ:

$$f = \frac{1}{v} \int \exp\left(i2k\xi \sin\frac{\theta}{2}\right) A(\xi)\, d\xi, \qquad (6.12)$$

where $A(\xi)$ is the area of that portion of the plane $R \cos(\mathbf{R}, \hat{\mathbf{e}}_z - \hat{\mathbf{e}}_r) = \xi$ which lies within the boundaries of the particle. In general, the limits on ξ and the functional form of $A(\xi)$ depend on the direction $\hat{\mathbf{e}}_r$ of the scattered wave. However, for a sphere we have $A(\xi) = \pi(a^2 - \xi^2)$, $-a \le \xi \le a$, for all scattering directions, and it is not difficult to integrate (6.12):

$$f(\theta) = \frac{3}{u^3}(\sin u - u \cos u), \qquad u = 2x \sin\frac{\theta}{2}.$$

Note that f vanishes at those values of θ for which

$$\tan u - u = 0. \qquad (6.13)$$

The zeros u_n of (6.13) are to good approximation given by

$$u_n^2 = \left(n + \tfrac{1}{2}\right)^2 \pi^2 - 2, \qquad n = 1, 2, \ldots$$

from which, together with the inequalities $0 \le u \le 2x$, $u_1 < u_2 < u_3 \cdots$, it follows that f does not vanish for any angle θ unless $x > 2.25$.

6.3   FINITE CYLINDER

In Chapter 8 we shall derive the field scattered by an infinite cylinder of arbitrary radius and refractive index; we shall also consider scattering by a finite cylinder in the diffraction theory approximation. Although the finite cylinder scattering problem is not exactly soluble, we can obtain analytical expressions for the amplitude scattering matrix elements in the Rayleigh–Gans approximation.

Consider a cylinder of radius a and length $2L$, subject to the conditions (6.2) and (6.3), which is illuminated by a beam making an angle ζ with its axis (Fig. 6.2). The directions of the incident and scattered waves are specified by the unit vectors $\hat{\mathbf{e}}_i$ and $\hat{\mathbf{e}}_r$, respectively, where $\hat{\mathbf{e}}_i = \sin\zeta\hat{\mathbf{e}}_z - \cos\zeta\hat{\mathbf{e}}_x$, $\hat{\mathbf{e}}_r = \sin\theta\cos\phi\hat{\mathbf{e}}_x + \sin\theta\sin\phi\hat{\mathbf{e}}_y + \cos\theta\hat{\mathbf{e}}_z$; the position vector $\mathbf{R}$ of a point in the particle with cylindrical polar coordinates (ρ, ψ, x) is $x\hat{\mathbf{e}}_x + \rho\cos\psi\hat{\mathbf{e}}_y + \rho\sin\psi\hat{\mathbf{e}}_z$. Therefore, the form factor (6.10) is

$$f = \frac{1}{\pi a^2 2L} \int_{-L}^{L} e^{-ikAx}\, dx \int_0^a \rho\, d\rho \int_0^{2\pi} e^{-ik\rho(B\cos\psi + C\sin\psi)}\, d\psi,$$

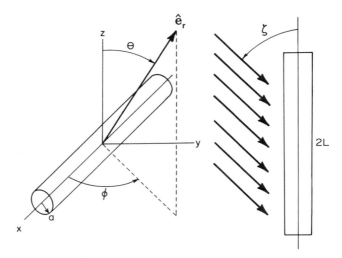

Figure 6.2 A finite cylinder illuminated obliquely.

where $A = \cos \zeta + \sin \theta \cos \phi$, $B = \sin \theta \sin \phi$, and $C = \cos \theta - \sin \zeta$. The integral over x is trivial:

$$\int_{-L}^{L} e^{-ikAx} dx = \frac{2 \sin(kAL)}{kA}.$$

If we write $B = M \cos \mu$, $C = M \sin \mu$, where $M = \sqrt{B^2 + C^2}$ and $\tan \mu = C/B$, then the integral over ψ is

$$\int_{0}^{2\pi} e^{-ik\rho M \cos(\psi - \mu)} \, d\psi = \int_{0}^{2\pi} e^{-ik\rho M \cos \psi} \, d\psi. \tag{6.14}$$

It follows from the integral representation of the Bessel function $J_0(z)$,

$$J_0(z) = \frac{1}{2\pi} \int_{0}^{2\pi} e^{-iz \cos \psi} \, d\psi,$$

where z is real, that the integral (6.14) is $2\pi J_0(k\rho M)$. The last integral,

$$\int_{0}^{a} \rho J_0(k\rho M) \, d\rho,$$

can be evaluated if we recall that $d(zJ_1)/dz = zJ_0$. Therefore, the form factor for a given angle of incidence ζ is

$$f(\theta, \phi; \zeta) = \frac{2 \sin(x \underline{\zeta} A)}{x \underline{\zeta} A} \frac{J_1(xM)}{xM}, \tag{6.15}$$

where $x = \mathrm{k}a$, and $\mathcal{L} = L/a$ is the ratio of cylinder length to diameter. If the incident light is normal to the cylinder axis ($\zeta = \pi/2$), then for scattering directions in a plane normal to the axis ($\phi = \pi/2$ or $3\pi/2$), (6.15) reduces to

$$f(\theta) = \frac{2J_1(u)}{u}, \qquad u = 2x\sin\frac{\theta}{2}.$$

NOTES AND COMMENTS

Chapter 7 of Kerker (1969) is a more thorough treatment of Rayleigh–Gans (RG) theory than we have given here. There is a good concise derivation of this theory in the appendix of a paper by Wyatt (1968).

Turner (1973) and McKellar (1976) applied RG theory to ensembles of randomly oriented particles of arbitrary shape; the former author included spheres with anisotropic optical constants. Optically active particles have been treated within the framework of the RG approximation by Bohren (1977).

Beginning with an integral equation, Acquista (1976) obtained an iterative solution to the problem of scattering by an arbitrary particle. The first iteration is just the RG expression. Agreement between approximate and exact theories of scattering by a sphere is considerably improved by a second iteration.

Chapter 7

Geometrical Optics

Concepts from geometrical optics—rays and all that—have occasionally intruded upon our discussions in preceding chapters, even in contexts where they were not strictly applicable. In spite of its limitations, geometrical optics is a simple and intuitively appealing approximate theory which need not be abandoned simply because the exact theory is at hand. In addition to its role in guiding intuition, geometrical optics can often provide quantitative answers to small-particle problems which are sufficiently accurate for many purposes, particularly when one soberly considers the accuracy with which many measurements can be made. In this chapter, therefore, we consider scattering and absorption by spheres within the framework of geometrical optics and compare these results with those of the exact theory. No book on light scattering by small particles would be complete without discussing the rainbow, the main features of which are quite adequately explained with simple geometrical optics; this then leads into atmospheric optical phenomena involving nonspherical particles for which there are no exact theories.

7.1 ABSORPTION AND SCATTERING CROSS SECTIONS

In this section we derive an approximate expression for the absorption cross section of a large weakly absorbing sphere. We assume that the incident plane wave can be subdivided into a large number of rays the behavior of which at interfaces is governed by the Fresnel equations and Snell's law (Section 2.7). A representative ray incident on the sphere at an angle Θ_i is shown in Fig. 7.1. At point 1 on the surface of the sphere the incident ray is divided into externally reflected and internally transmitted rays; these lie in the plane of incidence, which is determined by the normal to the sphere and the direction of the incident ray. If the polar coordinates of point 1 are (a, Θ_i, ϕ), the plane of incidence is the plane $\phi = $ constant. At point 2 the transmitted ray encounters another boundary and therefore is partially reflected and partially transmitted. In a like manner we can follow the path of the rays within the sphere, a path that does not deviate outside the plane of incidence. At each point where a ray encounters a boundary it is partially reflected internally and partially transmitted into the surrounding medium. On physical grounds we know that the absorption cross section cannot depend on the polarization of the incident

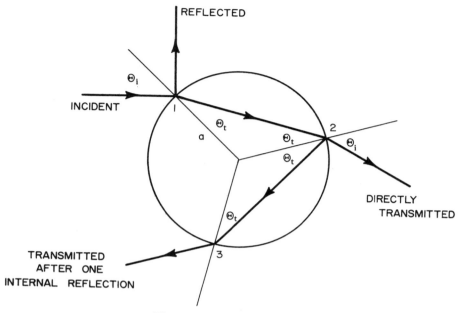

Figure 7.1 Ray-tracing diagram.

beam, which, for convenience, we may choose to be x-polarized. The components of the incident electric field parallel and perpendicular to the plane of incidence are $E_{\|i} = E_i\cos\phi$ and $E_{\perp i} = E_i\sin\phi$. We showed in Section 2.7 that these two components are reflected and transmitted independently of each other; therefore, we may consider each component in turn.

Let us first consider the component parallel to the plane of incidence. We assume that the imaginary part of the complex refractive index $m = n + ik$ of the sphere relative to the surrounding medium is small compared with the real part; therefore, the (real) angle of refraction Θ_t is to good approximation given by $\sin\Theta_t = \sin\Theta_i/m$. We may also ignore the imaginary part of m in the Fresnel equations (2.67)–(2.70). At point 1 the amplitude $E_{\|t}$ of the transmitted field is $\tilde{t}_\|(\Theta_i, n)E_{\|i}$, where the transmission coefficient $\tilde{t}_\|$ is given by (2.68). Therefore, the amount of energy that is transmitted through an element of area at 1 is

$$\frac{\mathrm{Re}\{N_1\}\cos\Theta_t|E_{\|t}|^2 a^2\sin\Theta_i\,d\Theta_i\,d\phi}{2Z_0},$$

where $Z_0 = \sqrt{\mu_0/\varepsilon_0}$ is the impedance of free space and N_1 is the refractive index of the sphere. The incident irradiance $I_{\|i}$ is $N|E_{\|i}|^2/2Z_0$, where N is the (real) refractive index of the surrounding medium. The expression above may be written more concisely as

$$T_{\parallel}(\Theta_i, n) I_{\parallel i}\cos \Theta_i a^2 \sin \Theta_i \, d\Theta_i \, d\phi,$$

where the *transmittance* is defined as

$$T_{\parallel}(\Theta_i, n) = \frac{n|\tilde{t}_{\parallel}(\Theta_i, n)|^2 \cos \Theta_t}{\cos \Theta_i}.$$

A fraction $1 - e^{-\alpha \xi}$ of the transmitted energy at 1 is absorbed as the ray traverses a path length $\xi = 2a\sqrt{n^2 - \sin^2 \Theta_i}\,/n$ between points 1 and 2, where α is the absorption coefficient of the sphere. The amplitude of the reflected field at 2 is

$$E_{\parallel i}\tilde{t}_{\parallel}(\Theta_i, n) e^{-\alpha \xi/2} \tilde{r}_{\parallel}\left(\Theta_t, \frac{1}{n}\right),$$

where $\tilde{r}_{\parallel}$ is the reflection coefficient; note that at 2 the angle of incidence is Θ_t and the relative refractive index is $1/n$. This reflected ray traverses the same path length ξ between 2 and 3; in so doing, an amount of energy

$$I_{\parallel i}T_{\parallel i}(\Theta_i, n) R_{\parallel}\left(\Theta_t, \frac{1}{n}\right) e^{-\alpha \xi} a^2 \cos \Theta_i \sin \Theta_i \, d\Theta_i \, d\phi (1 - e^{-\alpha \xi})$$

is absorbed in the sphere, where the reflectance $R_{\parallel}$ is $|\tilde{r}_{\parallel}|^2$. It is evident from the expression above that the amount of energy deposited by each successive internally reflected ray is $R_{\parallel}e^{-\alpha \xi}$ times that deposited by its predecessor.

If the incident field component is perpendicular to the plane of incidence, all the expressions for reflected, transmitted, and absorbed light are identical in form with those in the preceding paragraph; we need merely substitute $R_{\perp}$ and $T_{\perp}$ for $R_{\parallel}$ and $T_{\parallel}$.

The total energy W_{abs} absorbed in the sphere is obtained by summing the energy deposited by all internal rays for both components of polarization and a given incident ray and then integrating over all incident rays (i.e., all angles of incidence between 0 and $\pi/2$). The result is

$$W_{abs} = I_i 2\pi a^2 \int_0^{\pi/2} \left\{ T(\Theta_i, n) \sum_{j=0}^{\infty} \left[R(\Theta_t, 1/n) e^{-\alpha \xi} \right]^j \right.$$

$$\left. \times (1 - e^{-\alpha \xi}) \cos \Theta_i \sin \Theta_i \, d\Theta_i \right\}, \qquad (7.1)$$

where T and R are the transmittance and reflectance for unpolarized incident light

$$T = \tfrac{1}{2}(T_{\parallel} + T_{\perp}), \qquad R = \tfrac{1}{2}(R_{\parallel} + R_{\perp}),$$

and $I_i = I_{\|i} + I_{\perp i}$ is the total incident irradiance. The infinite series in (7.1) is readily summed:

$$\sum_{j=0}^{\infty} \left[R\left(\Theta_t, \frac{1}{n}\right) e^{-\alpha\xi} \right]^j = \frac{1}{1 - R(\Theta_t, 1/n) e^{-\alpha\xi}}.$$

Conservation of energy at boundaries requires that

$$R + T = 1,$$

and the reciprocal relations

$$R\left(\Theta_t, \frac{1}{n}\right) = R(\Theta_i, n), \qquad T\left(\Theta_t, \frac{1}{n}\right) = T(\Theta_i, n)$$

follow from the Fresnel formulas.

Up to this point we have only assumed that $k \ll n$; subject to this restriction and, of course, the assumption that geometrical optics combined with the Fresnel formulas is a good approximation, (7.1) is completely general. Let us further assume that the sphere is sufficiently weakly absorbing that $2a\alpha \ll 1$; with this assumption

$$1 - e^{-\alpha\xi} \simeq \alpha\xi, \qquad \frac{1}{1 - Re^{-\alpha\xi}} \simeq \frac{1}{T}$$

and (7.1) becomes

$$W_{abs} = \frac{4\pi a^3 \alpha}{n} I_i \int_0^{\pi/2} \cos \Theta_i \sqrt{n^2 - \sin^2 \Theta_i} \, \sin \Theta_i \, d\Theta_i,$$

which can be integrated to yield the absorption cross section $C_{abs} = W_{abs}/I_i$:

$$C_{abs} = \frac{4}{3} \pi a^3 \frac{\alpha}{n} \left[n^3 - (n^2 - 1)^{3/2} \right]. \tag{7.2}$$

Note that the absorption cross section of a weakly absorbing sphere, like that of a particle small compared with the wavelength (Chapter 5), is, in the geometrical optics limit, proportional to its volume. This proportionality does not hold, however, for indefinitely large particle radius. Energy is absorbed primarily in the outer layer of a highly absorbing sphere (i.e., $2a\alpha \gg 1$), and the interior of such a sphere plays no role in the absorption process. As the radius increases, therefore, the absorption cross section becomes proportional to *area* instead of *volume*.

Equation (7.2) was derived under the implicit assumption that the phase of the light could be ignored. In a like manner, we derived the transmittance of a slab in Section 2.8 by considering only irradiances and showed that the

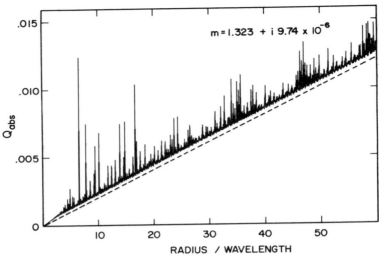

Figure 7.2 Absorption efficiency of a water droplet ($\lambda = 1.20 \ \mu$m); the dashed curve is the geometrical optics approximation (7.2).

resulting expression is valid under conditions for which interference effects are obliterated by sufficient departure from monochromaticity of the incident beam or planarity of the slab. C_{abs} in (7.2) should therefore be interpreted as the average cross section of a collection of particles with a distribution of radii sufficiently wide that interference effects may be neglected.

To assess the accuracy of this approximation, therefore, we should compare its predictions with exact calculations for a size distribution, although this is not really necessary. Accordingly, Fig. 7.2 shows the absorption efficiency of a single water droplet in air at a wavelength of 1.20 μm; the optical constants are taken from Irvine and Pollack (1968). If we ignore the extremely narrow peaks (ripple structure), which are smoothed by averaging over a distribution of radii (see Section 11.3, particularly Fig. 11.6), there is generally good agreement between the exact and approximate theories: the absorption efficiency increases linearly with radius in accordance with (7.2).

In collections of naturally occurring particles, such as water droplets in a cumulus cloud, there usually will be a considerable dispersion of radii. This will be true even in laboratory experiments unless special care is taken. For such applications, (7.2) is expected to be a good approximation provided, of course, that the particles satisfy the conditions under which it was derived: they must be large and weakly absorbing. For example, (7.2) has been incorporated in radiative transfer calculations for snow (Bohren and Barkstrom, 1974) and clouds (Twomey and Bohren, 1980).

In Fig. 7.3 we show the absorption efficiency for a water droplet at a wavelength (1.45 μm) where water is considerably more absorbing than at 1.20

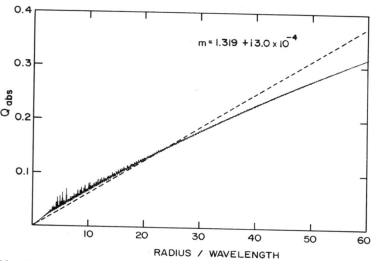

Figure 7.3 Absorption efficiency of a water droplet ($\lambda = 1.45 \ \mu$m); the dashed curve is the geometrical optics approximation (7.2).

μm. The agreement between exact and approximate theories begins to systematically worsen for radii such that $a\alpha > 0.1$; this gives us an approximate criterion for the range of validity of (7.2). As the particle radius increases, the absorption efficiency approaches a limiting value, which we shall discuss in the following paragraphs.

7.1.1 Asymptotic Absorption and Scattering Efficiencies

The total energy W_{sca} scattered by a large sphere may, to good approximation, be written as the sum of diffracted, reflected, and transmitted components:

$$W_{sca} = W_{diff} + W_{refl} + W_{tr}, \tag{7.3}$$

where the transmitted energy W_{tr} may be further subdivided into a directly transmitted component, a component transmitted after one internal reflection, and so on (see Fig. 7.1):

$$W_{tr} = \sum_{j=1}^{\infty} W_{tr,j}. \tag{7.4}$$

The externally reflected energy is given by

$$W_{refl} = I_i 2\pi a^2 \int_0^{\pi/2} R(\Theta_i)\cos\Theta_i \sin\Theta_i \, d\Theta_i,$$

and it is natural to define the *reflection cross section* C_{refl} as W_{refl}/I_i; the *reflection efficiency* is therefore

$$Q_{refl} = 2 \int_0^{\pi/2} R(\Theta_i) \cos \Theta_i \sin \Theta_i \, d\Theta_i. \tag{7.5}$$

All the light that enters a sufficiently large absorbing sphere ($a\alpha \gg 1$) will be absorbed; none of the unreflected light will be transmitted. We showed in Chapter 4 that the diffraction cross section of a large sphere is πa^2; thus, we might expect the limiting value of the scattering efficiency to be

$$\lim_{x \to \infty} Q_{sca} = 1 + Q_{refl}, \tag{7.6}$$

provided that $k \neq 0$. However, the asymptotic value (7.6) for the scattering efficiency is not universally accepted. Herman (1962) suggested that the scattering efficiency should approach the limit $1 + R(0°)$, where $R(0°) = |(m - 1)/(m + 1)|^2$ is the reflectance at normal incidence, on the grounds that a sphere becomes a planar surface as the size parameter x increases without limit. Deirmendjian (1969, pp. 34–41), however, argued that this conjecture is incorrect and that (7.6) is the proper limiting value. Nevertheless, Chýlek (1975) constructed a mathematical proof that $1 + R(0°)$ is the limiting value for Q_{sca}. Bohren and Herman (1979) subsequently concluded, on the basis of computations and physical arguments, that Chýlek's proof is in error, although they were not able to uncover the exact nature of this error; this was later done by Acquista et al. (1980).

To support the correctness of (7.6) we show in Fig. 7.4 Q_{sca} as a function of $1/x$, where $x = 2\pi a/\lambda$, computed from the exact theory for a sphere with refractive index $1.3 + i0.1$; for comparison, the limiting values $1 + R(0°)$ and $1 + Q_{refl}$ are also shown. It is clear from this figure that $1 + Q_{refl}$ is an increasingly better approximation to Q_{sca} as x increases. On physical grounds it is implausible that scalar diffraction theory and geometrical optics is a good approximation for a large range of the size parameter but then ceases to be so as x increases beyond some particular value; if the computed values are to reach the limit $1 + R(0°)$, however, this is what is required (see Fig. 7.4). At what value of x does (7.6) cease to be a good approximation, and what is the physical reason for such a value? We know of no such reason. Therefore, we conclude that $1 + R(0°)$ is not the asymptotic efficiency. A sphere is always a sphere and cannot be transformed into a slab by increasing its radius indefinitely, although at any point on its surface, a sphere can be considered *locally* planar to a degree of accuracy that increases with increasing radius. Nevertheless, even for an arbitrarily large sphere, the angle of incidence is a function of position and cannot be the same at all points.

We also showed in Chapter 4 that $\lim_{x \to \infty} Q_{ext} = 2$; this, together with (7.6), implies that

$$\lim_{x \to \infty} Q_{abs} = 1 - Q_{refl}. \tag{7.7}$$

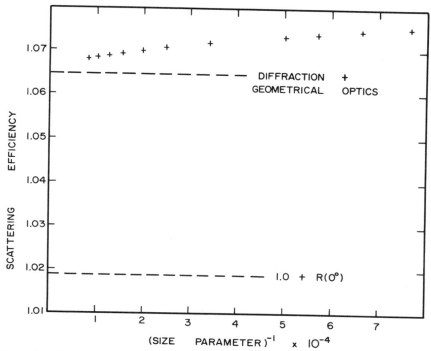

Figure 7.4 The crosses are computed from Mie theory. Scalar diffraction theory and geometrical optics predict the limiting value 1.067. From Bohren and Herman, 1979.

The physical interpretation of (7.7) is straightforward: all the geometrically incident light that is not externally reflected enters the sphere and is eventually absorbed provided, of course, that the absorptive part of the refractive index is not identically zero. As long as there is some absorption, no matter how small, all the transmitted light will be absorbed in the sphere for sufficiently large radius a.

7.1.2 Reflection and Transmission Efficiencies for a Nonabsorbing Sphere

It follows from (7.3) that the scattering efficiency of a large sphere can be written

$$Q_{sca} = Q_{diff} + Q_{refl} + Q_{tr},$$

where Q_{diff}, the diffraction efficiency, is unity and Q_{refl}, the (external) reflection efficiency, is given by (7.5). Furthermore, from (7.4), the transmission

Table 7.1 Reflection and Transmission Efficiencies for a Nonabsorbing Sphere with $m = 1.33$

Q_{refl}	$Q_{tr,1}$	$Q_{tr,2}$	$Q_{tr,3}$	$Q_{tr,4}$
0.06593	0.88451	0.04033	0.00607	0.00171

$$Q_{refl} + Q_{tr,1} + Q_{tr,2} + Q_{tr,3} + Q_{tr,4} = 0.99855$$

efficiency Q_{tr} can be written as a series

$$Q_{tr} = \sum_{j=1}^{\infty} Q_{tr,j},$$

where $Q_{tr,j}$ is the contribution to the total transmission efficiency from rays that have undergone $j - 1$ internal reflections (see Fig. 7.1).

The extinction efficiency of a *nonabsorbing* sphere is equal to its scattering efficiency; for such a sphere, provided that it is sufficiently large, it necessarily follows that $Q_{refl} + Q_{tr} = 1$ and

$$Q_{tr,j} = 2 \int_0^{\pi/2} T_j(\Theta_i) \cos \Theta_i \sin \Theta_i \, d\Theta_i,$$

$$T_j = \tfrac{1}{2} \left(R_{\parallel}^{j-1} T_{\parallel}^2 + R_{\perp}^{j-1} T_{\perp}^2 \right). \tag{7.8}$$

In Table 7.1 we give Q_{refl} and the first four terms in the series for Q_{tr}; the refractive index of the sphere is 1.33, which corresponds to a water droplet at visible wavelengths. One sometimes encounters the assertion that scattering by large transparent particles is the result of reflection at the particle–medium interface; indeed, reflection is sometimes used as a synonym for scattering. It should be clear from Table 7.1, however, that this is very much wide of the mark: only about 6.6% of the light scattered by a large water droplet—exclusive of that diffracted, which is confined to a narrow set of angles about the forward direction—is the result of reflection; most of the scattered light—over 88%—can be attributed to rays that are bent on a single traverse of the droplet, the directly transmitted rays.

7.2 ANGULAR DISTRIBUTION OF THE SCATTERED LIGHT: RAINBOW ANGLES

Externally reflected rays, directly transmitted rays, rays transmitted after undergoing one or more internal reflections, all can contribute to the light scattered into a unit solid angle about a particular direction. The incident rays are evenly spaced; but after reflection and transmission they can be concentrated in particular directions, much like the concentration or focusing of rays by a lens.

As in the preceding section, we may imagine that a ray incident on a sphere at an angle Θ_i between 0 and $\pi/2$ is decomposed into an infinite number of rays of varying strengths because of interaction with the sphere. For a given Θ_i the angle θ in which a particular type of ray is scattered can be determined from Fig. 7.1 by elementary trigonometry:

$$\theta = \pi - 2\Theta_i \qquad \qquad \textit{externally reflected}$$

$$\theta = 2\Theta_i - 2\Theta_t \qquad \qquad \textit{directly transmitted}$$

$$\theta = 2(n_r + 1)\Theta_t - 2\Theta_i \qquad \qquad \textit{transmitted after } n_r \textit{ internal} \qquad (7.9)$$
$$\textit{reflections}; \ n_r \textit{ even}$$

$$\theta = 2\Theta_i - 2(n_r + 1)\Theta_t + \pi \qquad \qquad \textit{transmitted after } n_r \textit{ internal}$$
$$\textit{reflections}; \ n_r \textit{ odd}$$

where $n_r = 1, 2, \ldots$ and $m \sin \Theta_t = \sin \Theta_i$; we have also assumed that $1 < m < 2$ and done some fiddling to ensure that θ lies between 0 and π.

The light incident on an element of area $a^2 \sin \Theta_i \, d\Theta_i \, d\phi$ of the sphere is scattered into a solid angle $\sin \theta \, d\theta \, d\phi$ about the direction (θ, ϕ). The contribution to this scattered light from rays of the kth type is of the form

$$F_k(\Theta_i, \phi) \cos \Theta_i \sin \Theta_i \, d\phi, \qquad (7.10)$$

where we have folded into F_k the sphere radius, the incident irradiance, and the various reflectances and transmittances appropriate to each type of ray. Because we are primarily interested in singularities in the scattering diagram predicted by geometrical optics rather than its precise mathematical form, we shall not need explicit expressions for F_k; suffice it to say that F_k is finite for all angles of incidence. It follows from (7.10) that the light scattered into a unit solid angle about (θ, ϕ)—the *intensity* in modern terminology—is the sum over terms of the form

$$F_k(\Theta_i, \phi) \frac{\cos \Theta_i \sin \Theta_i}{\sin \theta} \frac{d\Theta_i}{d\theta},$$

where $(d\Theta_i/d\theta)$ is obtained from (7.9); the factor $(\cos \Theta_i \sin \Theta_i)/\sin \theta$ is finite for all angles of incidence. For externally reflected rays $d\Theta_i/d\theta$ is merely a constant; but for the various transmitted rays we have

$$\frac{d\Theta_i}{d\theta} = \frac{\pm 1}{2 - 2(n_r + 1)\dfrac{d\Theta_t}{d\Theta_i}}, \qquad (7.11)$$

$$\frac{d\Theta_t}{d\Theta_i} = \frac{\cos \Theta_i}{\sqrt{m^2 - \sin^2 \Theta_i}}, \qquad (7.12)$$

where $n_r = 0, 1, 2, \ldots$. Therefore, according to geometrical optics, the scattered intensity is infinite at those scattering angles where one of the conditions

$$\frac{d\Theta_t}{d\Theta_i} = \frac{1}{n_r + 1} \qquad (n_r = 0, 1, 2, \ldots) \tag{7.13}$$

is satisfied. It is clear that (7.13) is not satisfied for the directly transmitted ray $(n_r = 0)$, but can be satisfied for the various internally reflected rays. Equations (7.12) and (7.13) can be combined to give

$$\cos \Theta_i = \sqrt{\frac{m^2 - 1}{n_r(n_r + 2)}} .$$

The scattering angles at which the intensity is infinite according to geometrical optics are called *rainbow angles*: the *primary* rainbow corresponds to rays that have undergone a single internal reflection; the *secondary* rainbow to rays that have undergone two internal reflections; the *tertiary* rainbow to rays that have undergone three internal reflections; and so on.

Although the scattered intensity can be quite large, it cannot be infinite at any angle; consequently, geometrical optics is strictly incorrect. This is similar to the result from geometrical optics that the intensity at the focal point of a lens, through which passes a finite amount of light, is infinite. Such singularities in the geometrical optics light field are called *caustics*. Rainbows are therefore caustics.

Although geometrical optics does not correctly give the magnitude of strong intensity maxima in the scattering diagram of large transparent spheres, it can give their positions to good approximation. This is readily verified by observations on rainbows formed by water droplets illuminated by the sun. The condition for the primary rainbow is

$$\cos \Theta_i = \sqrt{\frac{m^2 - 1}{3}} \tag{7.14}$$

and that for the secondary rainbow is

$$\cos \Theta_i = \sqrt{\frac{m^2 - 1}{8}} ; \tag{7.15}$$

the corresponding scattering angles are determined from (7.9). Note that there cannot be a primary rainbow angle if m is greater than 2.

If we take $m = 1.333$ as the average refractive index of water over the visible spectrum, we obtain

$$\theta = 137.9° \qquad \textit{primary rainbow}$$

$$\theta = 129.1° \qquad \textit{secondary rainbow}$$

Rainbows of order higher than the second are not observed in the atmosphere;

they fade into the background illumination. In the laboratory, however, it is possible to observe higher-order rainbows and, moreover, to use liquids other than water. The highest-order rainbow ever observed has so far escaped the attention of the compilers of *Guinness' Book of World Records*, but in a paper highly recommended to rainbow aficionados, Walker (1976) has described observations of, among other things, a seventeenth-order corn syrup bow.

Rainbows may be seen during showers when the sun is behind the observer; the direction of the sunlight determines the forward direction (or line of sight). The angular positions of the primary and secondary rainbows relative to the observer's line of sight are $180° - 137.9° = 42.1°$ and $180° - 129.1° = 50.9°$, respectively. The fraction of the total rainbow that can be seen depends on the solar elevation. When the sun is greater than about 51° above the horizon, no rainbow can be seen even though conditions are otherwise favorable. On the other hand, the complete rainbow—one that forms a complete circle—may be seen from an airplane.

Were it not for dispersion—the refractive index depends upon wavelength— the aesthetic appeal of rainbows would be greatly diminished. Indeed, the word "rainbow" used in everyday speech evokes images of a profusion of colors—the colors of the rainbow—rather than just an intensely bright arc in the sky. If we take $m = 1.343$ as the refractive index of "violet" light ($\lambda \simeq 0.4$ μm) and $m = 1.331$ as the refractive index of "red" light ($\lambda \simeq 0.65$ μm) (Irvine and Pollack, 1968), then the angular widths of the primary and secondary rainbows are about 1.7° and 3.1°, respectively.

Rainbow angles correspond to those angles of incidence at which the denominator of (7.11) vanishes; because $d\theta/d\Theta_i = 1/(d\Theta_i/d\theta)$, this in turn implies that for given n_r the rainbow angles correspond to *extrema* of the function $\theta(\Theta_i)$. The nature of these extrema are determined by the sign of $d^2\theta/d\Theta_i^2$: minima for odd n_r, maxima for even n_r. Thus, the primary rainbow angle 137.9° corresponds to a *minimum* of θ; the secondary rainbow angle 129.1° corresponds to a *maximum* of θ. Consequently, there is a dark band, *Alexander's dark band*, about 9° wide between the primary and secondary rainbows. Note, however, that both rainbow angles are angles of *minimum deviation*: they correspond to rays that have suffered the least *total* deviation.

Geometrical optics successfully explains the major features of the rainbow: the angular positions of the primary and secondary bows, their widths, color separation, and Alexander's dark band. As might be expected, it cannot explain all the observed features. According to geometrical optics, all rainbows are equal under conditions of similar illumination: the size distribution of droplets is irrelevant. But it is a matter of common experience that some rainbows are more vivid than others. *Supernumerary bows* (i.e., ones that should not exist according to geometrical optics) are sometimes observed below the primary bow; these require invoking interference arguments for their explanation. All this is beyond the scope of geometrical optics. For more details about rainbows we direct the reader to the list of references at the end of this chapter.

7.3   SCATTERING BY PRISMS: ICE CRYSTAL HALOES

It might naively be supposed that a collection of nonspherical particles, if randomly oriented, is equivalent to a suitably chosen collection of spheres. Evidence to the contrary is freely available—more frequently than is commonly thought—to all who would look toward the sun through thin veils of high cirrus clouds and observe the various halo phenomena that owe their existence to scattering by ice crystals. By simple geometrical arguments, similar to those in the preceding section, we can acquire an understanding of many of the features of ice crystal haloes and their associated arcs.

Consider a transparent triangular prism with refractive index m and vertex angle Δ, which is illuminated by a ray parallel to a right section of the prism; the angle between the ray and the normal to the lateral face is Θ_i (Fig. 7.5). After two refractions, the incident ray dutifully emerges from the opposite face (provided that Θ_i is sufficiently large), having been deviated through a total angle θ relative to its original direction:

$$\theta = \Theta_i - \Theta_t + \Theta'_t - \Theta'_i,$$

where $\sin \Theta_i = m \sin \Theta_t$, $\sin \Theta'_t = m \sin \Theta'_i$, and $\Theta'_i = \Delta - \Theta_t$. The deviation is an extremum for the angle of incidence where

$$\frac{d\theta}{d\Theta_i} = 1 - \frac{\cos \Theta'_i \cos \Theta_i}{\cos \Theta'_t \cos \Theta_t} = 0,$$

and by examining the sign of $d^2\theta/d\Theta_i^2$ we can show that the extremum is a minimum. The path of the minimally deviated ray is such that $\Theta_i = \Theta'_t$ and

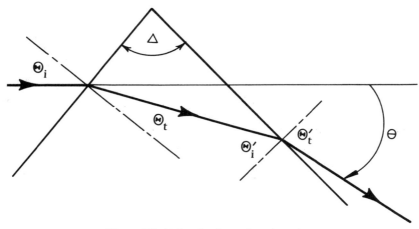

Figure 7.5   Refraction by a triangular prism.

$\Theta_t = \Theta_i'$; therefore, the angle of minimum deviation θ_m is

$$\theta_m = 2\sin^{-1}\left(m\sin\frac{\Delta}{2}\right) - \Delta. \qquad (7.16)$$

By analogy with the primary rainbow angle—an angle of minimum deviation—we expect maxima (caustics) in the scattering diagrams of prismatic particles at angles given by (7.16). The crystallographic form of ice is hexagonal; therefore, hexagonal ice crystals occasionally inhabit the atmosphere. And if such crystals are interposed between the sun and an informed observer, haloes and a myriad of other optical displays may be seen.

There are two haloes that may be attributed to minimum angles of deviation associated with ice crystals: the 22° halo and the 46° halo, the former being the most common. But both are much more common than is generally realized; their frequency is a function of the state of awareness of the observer. To test this, one can ask a class of undergraduates—or an audience of Ph.D. physicists for that matter—if anyone has ever seen a halo or a sun dog. The response is likely to be feeble. But within a few days after explaining these phenomena, you are usually deluged with phone calls or confronted by breathless students all aglow from having seen what had previously been a "rare" phenomenon.

The angular positions of the two haloes can be determined from (7.16); the smaller halo is associated with a vertex angle of 60° and the larger with 90° (Fig. 7.6). Ice is slightly birefringent (Hobbs, 1974, p. 202), but we can ignore this and take $m = 1.318$ as the refractive index of "blue" light and $m = 1.308$ as that of "red" light. The corresponding angles of minimum deviation

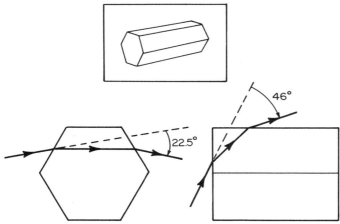

Figure 7.6 Refraction by a hexagonal ice crystal showing the rays associated with the 22° and 46° haloes.

calculated from (7.16) are

$$\theta_m(\text{blue}) = 22.4°,$$
$$\theta_m(\text{red}) = 21.7°, \qquad \Delta = 60°$$

$$\theta_m(\text{blue}) = 47.5°,$$
$$\theta_m(\text{red}) = 45.3°, \qquad \Delta = 90°.$$

Geometrical optics therefore correctly predicts the approximate angular positions of the two most commonly observed ice crystal haloes. Note that red light is deviated least, so the haloes may exhibit color separation with red appearing at their inner edges.

Complete haloes are traditionally attributed to scattering by randomly oriented ice crystals (but, see Fraser, 1979, for a contrary viewpoint); sun dogs or mock suns—brights spots on one or both sides of the sun, at about 22° when the sun is low—are the result of scattering by oriented crystals. This, however, by no means exhausts the list of atmospheric optical phenomena associated with ice crystals; there are many more, some of which are quite rare. But these are beyond the scope of this section, which is meant to be only a simple application of geometrical optics to light scattering by nonspherical particles. Suggestions for further reading are given below.

NOTES AND COMMENTS

Justification for dividing the light scattered by large particles into diffracted, reflected, and transmitted components is provided by the *localization principle* (van de Hulst, 1957, pp. 208–214) whereby the terms in the Mie series are associated with each of these components.

The quality of rainbow colors is not uniform across the bow; the reasons for this are given by Fraser (1972).

Neuberger (1951, p. 174) reports that halo phenomena were observed an average of 74 days a year over a 16-year period in State College, Pennsylvania. They are just as common on the other side of the Atlantic: near Bristol, England, Brain (1972) observed halo phenomena on 80 occasions in 66 days.

Haloes and other ice crystal phenomena are discussed by Minnaert (1954), Tricker (1970, 1979), and Greenler (1980). The book by the latter author is particularly recommended for its clarity and its superb color photographs.

The *Journal of the Optical Society of America* for August 1979 is almost entirely devoted to meteorological optics; it contains several papers on rainbows, the glory, and ice crystal phenomena.

Chapter 8

A Potpourri of Particles

Although theoretical analysis of absorption and scattering by a homogeneous isotropic sphere is complicated, it is nonetheless manageable: one follows a straightforward route guided by well-known techniques in mathematical physics. Such techniques can be extended to particles of regular shape without great difficulty. The route can become more tortuous, however, if we inquire into scattering by inhomogeneous particles, or ones with anisotropic optical properties, or irregular particles. This is particularly true if the particles are neither very large nor very small compared with the wavelength.

In this chapter we consider theories of scattering by particles that are either inhomogeneous, anisotropic, or nonspherical. No attempt will be made to be comprehensive: our choice of examples is guided solely by personal taste. First we consider a special example of inhomogeneity, a layered sphere. Then we briefly discuss anisotropic spheres, including an exactly soluble problem. Isotropic optically active particles, ones with mirror asymmetry, are then considered. Cylindrical particles are not uncommon in nature—spider webs, viruses, various fibers—and we therefore devote considerable space to scattering by a right circular cylinder.

A discussion of some theoretical approaches to scattering by randomly inhomogeneous particles is followed in the final section by an outline of recent progress in constructing solutions to problems of scattering by nonspherical particles, including those of arbitrary shape.

8.1 COATED SPHERE

The field scattered by any spherically symmetrical particle composed of materials described by the constitutive relations (2.7)–(2.9) has the same form as that scattered by the homogeneous sphere considered in Chapter 4. However, the functional form of the coefficients a_n and b_n depends on the radial variation of ε and μ. In this section we consider the problem of scattering by a homogeneous sphere coated with a homogeneous layer of uniform thickness, the solution to which was first obtained by Aden and Kerker (1951). This is one of the simplest examples of a particle with a spatially variable refractive index, and it can readily be generalized to a multilayered sphere.

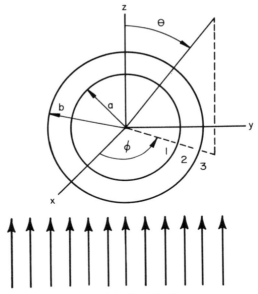

Figure 8.1   Coated sphere.

Suppose that the electromagnetic wave (4.37), (4.38) is incident on a coated sphere with inner radius a and outer radius b(Fig. 8.1). The electromagnetic field $(\mathbf{E}_1, \mathbf{H}_1)$ in the region $0 \leq r \leq a$ is given by (4.40) and the scattered field $(\mathbf{E}_s, \mathbf{H}_s)$ by (4.45). Because of the requirement of finiteness at the origin, the radial part of the functions (4.15) and (4.16), which generate the vector harmonics in the expansion of $(\mathbf{E}_1, \mathbf{H}_1)$ is constrained to be j_n. However, in the region $a \leq r \leq b$ $(a \neq 0)$, both spherical Bessel functions j_n and y_n are finite; as a consequence, the expansion of the field $(\mathbf{E}_2, \mathbf{H}_2)$ in this region is

$$\mathbf{E}_2 = \sum_{n=1}^{\infty} E_n \left[f_n \mathbf{M}_{o1n}^{(1)} - i g_n \mathbf{N}_{e1n}^{(1)} + v_n \mathbf{M}_{o1n}^{(2)} - i w_n \mathbf{N}_{e1n}^{(2)} \right],$$

$$\mathbf{H}_2 = -\frac{k_2}{\omega \mu_2} \sum_{n=1}^{\infty} E_n \left[g_n \mathbf{M}_{e1n}^{(1)} + i f_n \mathbf{N}_{o1n}^{(1)} + w_n \mathbf{M}_{e1n}^{(2)} + i v_n \mathbf{N}_{o1n}^{(2)} \right],$$

where the vector harmonics $\mathbf{M}_{e1n}^{(2)}$, and so on, are generated by functions of the form (4.15) and (4.16) with radial dependence $y_n(k_2 r)$. The boundary conditions

$$(\mathbf{E}_2 - \mathbf{E}_1) \times \hat{\mathbf{e}}_r = 0, \qquad (\mathbf{H}_2 - \mathbf{H}_1) \times \hat{\mathbf{e}}_r = 0, \qquad r = a$$

$$(\mathbf{E}_s + \mathbf{E}_i - \mathbf{E}_2) \times \hat{\mathbf{e}}_r = 0, \qquad (\mathbf{H}_s + \mathbf{H}_i - \mathbf{H}_2) \times \hat{\mathbf{e}}_r = 0, \qquad r = b$$

yield eight equations in the coefficients a_n, b_n, c_n, d_n, f_n, g_n, v_n, w_n:

$$f_n m_1 \psi_n(m_2 x) - v_n m_1 \chi_n(m_2 x) - c_n m_2 \psi_n(m_1 x) = 0,$$

$$w_n m_1 \chi_n'(m_2 x) - g_n m_1 \psi_n'(m_2 x) + d_n m_2 \psi_n'(m_1 x) = 0,$$

$$v_n \mu_1 \chi_n'(m_2 x) - f_n \mu_1 \psi_n'(m_2 x) + c_n \mu_2 \psi_n'(m_1 x) = 0,$$

$$g_n \mu_1 \psi_n(m_2 x) - w_n \mu_1 \chi_n(m_2 x) - d_n \mu_2 \psi_n(m_1 x) = 0,$$

$$m_2 \psi_n'(y) - a_n m_2 \xi_n'(y) - g_n \psi_n'(m_2 y) + w_n \chi_n'(m_2 y) = 0, \quad (8.1)$$

$$m_2 b_n \xi_n(y) - m_2 \psi_n(y) + f_n \psi_n(m_2 y) - v_n \chi_n(m_2 y) = 0,$$

$$\mu_2 \psi_n(y) - a_n \mu_2 \xi_n(y) - g_n \mu \psi_n(m_2 y) + w_n \mu \chi_n(m_2 y) = 0,$$

$$b_n \mu_2 \xi_n'(y) - \mu_2 \psi_n'(y) + f_n \mu \psi_n'(m_2 y) - v_n \mu \chi_n'(m_2 y) = 0,$$

where m_1 and m_2 are the refractive indices of the core and coating relative to the surrounding medium; μ, μ_1, μ_2 are the permeabilities of the surrounding medium, core, and coating; and $x = ka$, $y = kb$. The Riccati–Bessel function $\chi_n(z)$ is $-zy_n(z)$. Let us assume for simplicity that $\mu = \mu_1 = \mu_2$ and solve the set of equations (8.1) for the scattering coefficients a_n and b_n:

$$a_n = \frac{\psi_n(y)[\psi_n'(m_2 y) - A_n \chi_n'(m_2 y)] - m_2 \psi_n'(y)[\psi_n(m_2 y) - A_n \chi_n(m_2 y)]}{\xi_n(y)[\psi_n'(m_2 y) - A_n \chi_n'(m_2 y)] - m_2 \xi_n'(y)[\psi_n(m_2 y) - A_n \chi_n(m_2 y)]},$$

$$b_n = \frac{m_2 \psi_n(y)[\psi_n'(m_2 y) - B_n \chi_n'(m_2 y)] - \psi_n'(y)[\psi_n(m_2 y) - B_n \chi_n(m_2 y)]}{m_2 \xi_n(y)[\psi_n'(m_2 y) - B_n \chi_n'(m_2 y)] - \xi_n'(y)[\psi_n(m_2 y) - B_n \chi_n(m_2 y)]},$$

$$A_n = \frac{m_2 \psi_n(m_2 x) \psi_n'(m_1 x) - m_1 \psi_n'(m_2 x) \psi_n(m_1 x)}{m_2 \chi_n(m_2 x) \psi_n'(m_1 x) - m_1 \chi_n'(m_2 x) \psi_n(m_1 x)},$$

$$B_n = \frac{m_2 \psi_n(m_1 x) \psi_n'(m_2 x) - m_1 \psi_n(m_2 x) \psi_n'(m_1 x)}{m_2 \chi_n'(m_2 x) \psi_n(m_1 x) - m_1 \psi_n'(m_1 x) \chi_n(m_2 x)}. \quad (8.2)$$

A program for computing a_n and b_n is given in Appendix B.

If $m_1 = m_2$, then $A_n = B_n = 0$ and the coefficients (8.2) reduce to those for a homogeneous sphere. We also have $\lim_{a \to 0} A_n = \lim_{a \to 0} B_n = 0$; therefore, in the limit of zero core radius the coefficients (8.2) reduce to those for a homogeneous sphere of radius b and relative refractive index m_2, as required. When $m_2 = 1$, the coefficients reduce to those for a sphere of radius a and relative refractive index m_1; this gives us yet another check on the correctness of our solution.

8.2 ANISOTROPIC SPHERE

We have discussed *intrinsically* anisotropic particles—ones with anisotropy originating in their optical constants rather than their shape—in previous chapters. In Section 5.6 we gave the solution to the problem of scattering by an anisotropic sphere in the Rayleigh approximation. From the results of that section and Section 5.5 it follows that the average cross section $\langle C \rangle$ (scattering or absorption) of a collection of randomly oriented, *sufficiently small*, anisotropic spheres is

$$\langle C \rangle = \tfrac{1}{3}C_1 + \tfrac{1}{3}C_2 + \tfrac{1}{3}C_3, \qquad (8.3)$$

where C_j is the cross section for an isotropic sphere with dielectric function ϵ_j, one of the three distinct (in general) principal values of the dielectric function tensor. Equation (8.3) has been used not only for spheres small compared with the wavelength but for larger spheres as well. Perhaps (8.3) is valid without qualification (although we believe otherwise); nevertheless, to our knowledge this has never been demonstrated. The reason for this is that no exact solution to the problem of scattering by an anisotropic sphere of arbitrary radius has been published. Therefore, it is difficult to determine by computations the limits of validity—if any—of (8.3). In the absence of an exact theory, we are forced to fall back on physical reasoning to make an educated guess about the conditions under which (8.3) might be expected to fail.

Let us consider for simplicity a sphere composed of a uniaxial material (see Section 9.3). We denote by $k_\parallel$ and $k_\perp$ the wave numbers corresponding to the two principal values of the dielectric function tensor. It is reasonable to assert on physical grounds that anisotropy is only a perturbation if

$$|(k_\parallel - k_\perp)a| \ll 1,$$

where a is the radius of the sphere; that is, the two kinds of plane waves that can propagate in a uniaxial material undergo the same change in phase and amplitude over a distance comparable with the size of the particle. A corollary of this is that the effect of anisotropy becomes appreciable when

$$|(k_\parallel - k_\perp)a| > 1,$$

and if this condition holds, one cannot reasonably expect to obtain the solution to the anisotropic sphere problem by patching together solutions for an isotropic sphere. We emphasize, however, that this criterion is only our best guess in the absence of anything better.

The reason for the intractability of the anisotropic sphere scattering problem is the fundamental mismatch between the symmetry of the optical constants and the shape of the particle. For example, the vector wave equation for a uniaxial material is separable in cylindrical coordinates; that is, the solutions to the field equations are cylindrical waves. But the bounding surface of the particle is a sphere, and troubles arise when we try to satisfy the boundary conditions. Thus, techniques for extending the theory of scattering by an anisotropic sphere beyond the Rayleigh limit are not likely to be based on separation of variables.

A special anisotropic particle scattering problem has been treated by Roth and Dignam (1973), who considered an isotropic sphere coated with a uniform film with constitutive relations

$$D_r = \varepsilon_n E_r, \qquad D_\theta = \varepsilon_t E_\theta, \qquad D_\phi = \varepsilon_t E_\phi, \tag{8.4}$$

where the permittivities ε_n and ε_t are independent of position and (E_r, E_θ, E_ϕ) are the field components relative to a spherical polar coordinate system. The constitutive relations (8.4) may be interpreted as applying to an oriented layer, or one that is "locally uniaxial." The solution to this problem is exact, but involves Bessel functions of complex *order*. Computation of ordinary Bessel functions is difficult enough, but mere contemplation of complex-order Bessel functions is sufficient to make strong men weep. It is perhaps for these apparent computational difficulties that the ramifications of the solution of Roth and Dignam have not been fully explored.

8.3   OPTICALLY ACTIVE PARTICLES

Almost all the particles we have considered have been composed of linear isotropic media described by the constitutive relations (2.7)–(2.9), which are not universally valid. Unfortunately, as soon as we depart from these relations we are usually confronted with problems that are not exactly soluble. However, exact solutions are obtainable for particles of regular geometrical shape composed of isotropic, *optically active* media. Such media are ones in which plane harmonic waves can propagate without change in polarization, but only if they are circularly polarized with either handedness; the complex refractive indices for left-circularly and right-circularly polarized waves are different. A simple conceptual model of an isotropic, optically active medium is a random array of screws, which is invariant under rotation through any angle, but under reflection the handedness of the screws changes. An example of an isotropic, optically active particle is a sugar-water drop.

The constitutive relations

$$\mathbf{D} = \varepsilon\mathbf{E} + \gamma\varepsilon\nabla \times \mathbf{E}, \qquad \mathbf{B} = \mu\mathbf{H} + \beta\mu\nabla \times \mathbf{H}, \tag{8.5}$$

where ε, μ, γ, and β are scalar phenomenological coefficients, are sufficient for a macroscopic description of optical activity. That is, plane homogeneous electromagnetic waves can propagate in media described by (8.5) only if they are circularly polarized. If we take γ and β to be equal, the phenomenological coefficients are related to the complex refractive indices N_L and N_R by

$$\beta = \frac{1}{2}\left(\frac{1}{k_R} - \frac{1}{k_L}\right), \qquad \omega\sqrt{\varepsilon\mu} = \frac{1}{\frac{1}{2}(1/k_R + 1/k_L)},$$

where the wave numbers k_L and k_R are

$$k_L = \frac{2\pi}{\lambda} N_L, \qquad k_R = \frac{2\pi}{\lambda} N_R.$$

If we assume harmonic time dependence $e^{-i\omega t}$, the Maxwell equations (2.1)–(2.4) may be written

$$\nabla \cdot \left(\mathbf{D} + \frac{i}{\omega} \mathbf{J}_F \right) = 0, \qquad \nabla \cdot \mathbf{B} = 0, \tag{8.6}$$

$$\nabla \times \mathbf{E} - i\omega \mathbf{B} = 0, \qquad \nabla \times \mathbf{H} + i\omega \left(\mathbf{D} + \frac{i}{\omega} \mathbf{J}_F \right) = 0. \tag{8.7}$$

Note that $\mathbf{D}$ and $\mathbf{J}_F$ do not appear separately in (8.6) and (8.7) but only in the combination $\mathbf{D} + i\mathbf{J}_F/\omega$, which we may interpret as the total electric displacement and assume that $\mathbf{D}$ in (8.5) is this quantity. In fact, we could have done this in previous chapters but refrained from doing so because the notion of conductivity is well established. However, it is not possible to determine from macroscopic experiments of the type discussed in this book if the imaginary part of the refractive index originates from "free" or "bound" charge currents. Thus, we need not make separate assumptions about the relations between $\mathbf{D}$ and $\mathbf{E}$ and between $\mathbf{J}_F$ and $\mathbf{E}$.

The constitutive relations (8.5) and the field equations (8.6), (8.7) can be written compactly in matrix form:

$$\nabla^2 \begin{pmatrix} \mathbf{E} \\ \mathbf{H} \end{pmatrix} + \mathsf{K}^2 \begin{pmatrix} \mathbf{E} \\ \mathbf{H} \end{pmatrix} = 0,$$

$$\nabla \times \begin{pmatrix} \mathbf{E} \\ \mathbf{H} \end{pmatrix} = \mathsf{K} \begin{pmatrix} \mathbf{E} \\ \mathbf{H} \end{pmatrix},$$

$$\nabla \cdot \begin{pmatrix} \mathbf{E} \\ \mathbf{H} \end{pmatrix} = 0,$$

$$\mathsf{K} = \frac{i\omega}{1 - \beta^2 \varepsilon\mu\omega^2} \begin{pmatrix} -i\beta\varepsilon\mu\omega & \mu \\ -\varepsilon & -i\beta\varepsilon\mu\omega \end{pmatrix}. \tag{8.8}$$

We often refer to the electromagnetic field and then go on to treat the electric and magnetic fields as separate entities, a slight inconsistency. However, we have shown that the field equations can be written in such a way that the electromagnetic field, the column vector with elements $\mathbf{E}$ and $\mathbf{H}$, is treated as a single entity. Thus, (8.8) has an aesthetic appeal, which transcends its immediate usefulness to the problem at hand.

A linear transformation of the electromagnetic field

$$\begin{pmatrix} \mathbf{E} \\ \mathbf{H} \end{pmatrix} = \mathsf{A} \begin{pmatrix} \mathbf{Q}_L \\ \mathbf{Q}_R \end{pmatrix} \tag{8.9}$$

diagonalizes K:

$$\Lambda = \mathsf{A}^{-1} \mathsf{K} \mathsf{A},$$

where

$$\Lambda = \begin{pmatrix} k_L & 0 \\ 0 & -k_R \end{pmatrix}, \qquad A = \begin{pmatrix} 1 & a_R \\ a_L & 1 \end{pmatrix},$$

$$a_R = -i\sqrt{\mu/\varepsilon}, \qquad a_L = -i\sqrt{\varepsilon/\mu}.$$

The transformed fields $\mathbf{Q}_L$ and $\mathbf{Q}_R$ independently satisfy equations of the form

$$\nabla^2 \mathbf{Q} + k^2 \mathbf{Q} = 0, \tag{8.10}$$

$$\nabla \times \mathbf{Q} = k\mathbf{Q}, \tag{8.11}$$

$$\nabla \cdot \mathbf{Q} = 0, \tag{8.12}$$

where $k = k_L$ for $\mathbf{Q} = \mathbf{Q}_L$ and $k = -k_R$ for $\mathbf{Q} = \mathbf{Q}_R$. Therefore, the most general electromagnetic wave in an optically active medium is a superposition of waves of left-handed and right-handed types.

At the beginning of this section we stated, without proof, that only circularly polarized plane waves can propagate in media described by the constitutive relations (8.5). It is now a relatively easy matter to show that this is so. Let us consider plane homogeneous waves $\exp(i\mathbf{k} \cdot \mathbf{x})$ propagating in the $\hat{\mathbf{e}}$ direction, where $\mathbf{k} = k\hat{\mathbf{e}}$. From (8.11) we have $i\hat{\mathbf{e}} \times \mathbf{Q} = \pm\mathbf{Q}$, and the transversality condition (8.12) implies that $\hat{\mathbf{e}} \cdot \mathbf{Q} = 0$. Therefore, we can write $\mathbf{Q} = Q_{\parallel}\hat{\mathbf{e}}_{\parallel} + Q_{\perp}\hat{\mathbf{e}}_{\perp}$, where $\hat{\mathbf{e}} \times \hat{\mathbf{e}}_{\perp} = \hat{\mathbf{e}}_{\parallel}$ and $\hat{\mathbf{e}} \times \hat{\mathbf{e}}_{\parallel} = -\hat{\mathbf{e}}_{\perp}$, which yields

$$\mathbf{Q} = Q_{\parallel}(\hat{\mathbf{e}}_{\parallel} \pm i\hat{\mathbf{e}}_{\perp}),$$

where the plus sign holds when $\mathbf{Q} = \mathbf{Q}_R$ and the minus when $\mathbf{Q} = \mathbf{Q}_L$. We showed in Section 2.11 that $\hat{\mathbf{e}}_{\parallel} + i\hat{\mathbf{e}}_{\perp}$ represents a right-circularly polarized wave and $\hat{\mathbf{e}}_{\parallel} - i\hat{\mathbf{e}}_{\perp}$ a left-circularly polarized wave. Thus, it is clear that plane wave solutions to the field equations (8.10)–(8.12) are necessarily circularly polarized. In general, therefore, it is natural to refer to $\mathbf{Q}_L$ as a wave of left-handed type and $\mathbf{Q}_R$ as a wave of right-handed type.

Consider now the field scattered by an isotropic, optically active sphere of radius a, which is embedded in a nonactive medium with wave number k and illuminated by an x-polarized wave. Most of the groundwork for the solution to this problem has been laid in Chapter 4, where the expansions (4.37) and (4.38) of the incident electric and magnetic fields are given. Equation (8.11) requires that the expansion functions for $\mathbf{Q}$ be of the form $\mathbf{M} \pm \mathbf{N}$; therefore, the vector spherical harmonics expansions of the fields inside the sphere are

$$\mathbf{Q}_L = \sum_{n=1}^{\infty} E_n \big\{ f_{on} \big[\mathbf{M}_{o1n}^{(1)}(k_L) + \mathbf{N}_{o1n}^{(1)}(k_L) \big]$$

$$+ f_{en} \big[\mathbf{M}_{e1n}^{(1)}(k_L) + \mathbf{N}_{e1n}^{(1)}(k_L) \big] \big\},$$

$$\mathbf{Q}_R = \sum_{n=1}^{\infty} E_n \big\{ g_{on} \big[\mathbf{M}_{o1n}^{(1)}(k_R) - \mathbf{N}_{o1n}^{(1)}(k_R) \big]$$

$$+ g_{en} \big[\mathbf{M}_{e1n}^{(1)}(k_R) - \mathbf{N}_{e1n}^{(1)}(k_R) \big] \big\}, \tag{8.13}$$

where $E_n = E_0 i^n (2n + 1)/n(n + 1)$ and k_R or k_L in the argument of the vector harmonics indicates that $\rho = k_R r$ or $\rho = k_L r$ are the arguments of the spherical Bessel function $j_n(\rho)$ in the generating functions for these harmonics. The expansions of the scattered field are

$$\mathbf{E}_s = \sum_{n=1}^{\infty} E_n \left[i a_n \mathbf{N}_{e1n}^{(3)} - b_n \mathbf{M}_{o1n}^{(3)} + c_n \mathbf{M}_{e1n}^{(3)} - i d_n \mathbf{N}_{o1n}^{(3)} \right],$$

$$\mathbf{H}_s = \frac{k}{\omega\mu} \sum_{n=1}^{\infty} E_n \left[a_n \mathbf{M}_{e1n}^{(3)} + i b_n \mathbf{N}_{o1n}^{(3)} - i c_n \mathbf{N}_{e1n}^{(3)} - d_n \mathbf{M}_{o1n}^{(3)} \right].$$

The electromagnetic field $(\mathbf{E}_1, \mathbf{H}_1)$ inside the sphere is obtained from (8.13) and the transformation (8.9). We saw in Chapter 4 that, for given n, there are four unknown coefficients in the expansions for the fields when the sphere is nonactive; optical activity doubles the number of coefficients, which are determined by applying the conditions (4.39) at the boundary between sphere and surrounding medium and solving the resulting system of eight linear equations. We are interested primarily in the coefficients of the scattered field:

$$a_n = \frac{V_n(R)A_n(L) + V_n(L)A_n(R)}{W_n(L)V_n(R) + V_n(L)W_n(R)},$$

$$b_n = \frac{W_n(L)B_n(R) + W_n(R)B_n(L)}{W_n(L)V_n(R) + V_n(L)W_n(R)},$$

$$c_n = i\frac{W_n(R)A_n(L) - W_n(L)A_n(R)}{W_n(L)V_n(R) + V_n(L)W_n(R)} = -d_n,$$

$$W_n(J) = m\psi_n(m_J x)\xi_n'(x) - \xi_n(x)\psi_n'(m_J x),$$

$$V_n(J) = \psi_n(m_J x)\xi_n'(x) - m\xi_n(x)\psi_n'(m_J x),$$

$$A_n(J) = m\psi_n(m_J x)\psi_n'(x) - \psi_n(x)\psi_n'(m_J x),$$

$$B_n(J) = \psi_n(m_J x)\psi_n'(x) - m\psi_n(x)\psi_n'(m_J x).$$

J is L or R. The relative refractive indices m_L, m_R and the mean refractive index m are defined as follows:

$$m_L = \frac{N_L}{N}, \qquad m_R = \frac{N_R}{N}, \qquad \frac{1}{m} = \frac{1}{2}\left(\frac{1}{m_R} + \frac{1}{m_L}\right)\frac{\mu_1}{\mu},$$

where N is the refractive index and μ the permeability of the surrounding medium. The difference Δm between the refractive indices m_R and m_L is usually small; therefore, m is $(m_R + m_L)/2$ to terms of order $(\Delta m)^2$ (we have also assumed that $\mu_1 = \mu$). If there is no optical activity $(m_L = m_R)$, then a_n and b_n reduce to (4.53) and c_n vanishes identically.

8.3.1   Matrix Elements and Cross Sections

In Section 4.4 we showed that the off-diagonal elements of the amplitude scattering matrix (3.12) are zero for a nonactive sphere. If the sphere is optically active, however, the matrix elements are

$$S_1 = \sum_n \frac{2n + 1}{n(n + 1)} (a_n \pi_n + b_n \tau_n),$$

$$S_2 = \sum_n \frac{2n + 1}{n(n + 1)} (a_n \tau_n + b_n \pi_n),$$

$$S_3 = \sum_n \frac{2n + 1}{n(n + 1)} c_n (\pi_n + \tau_n) = -S_4.$$

In problems involving optically active particles it is usually more convenient to use the amplitude scattering matrix in the circular polarization representation. The transformation from linearly to circularly polarized electric field components is

$$\begin{pmatrix} E_L \\ E_R \end{pmatrix} = \frac{1}{\sqrt{2}} \begin{pmatrix} 1 & i \\ 1 & -i \end{pmatrix} \begin{pmatrix} E_\parallel \\ E_\perp \end{pmatrix}, \tag{8.14}$$

and the inverse transformation is

$$\begin{pmatrix} E_\parallel \\ E_\perp \end{pmatrix} = \frac{1}{\sqrt{2}} \begin{pmatrix} 1 & 1 \\ -i & i \end{pmatrix} \begin{pmatrix} E_L \\ E_R \end{pmatrix}. \tag{8.15}$$

If the fields in (3.12) are transformed according to (8.14) and (8.15), the relation between incident and scattered fields becomes

$$\begin{pmatrix} E_{Ls} \\ E_{Rs} \end{pmatrix} = \frac{e^{ik(r-z)}}{-ikr} \begin{pmatrix} S_{2c} & S_{3c} \\ S_{4c} & S_{1c} \end{pmatrix} \begin{pmatrix} E_{Li} \\ E_{Ri} \end{pmatrix},$$

$$S_{1c} = \tfrac{1}{2}(S_2 + S_1 - iS_4 + iS_3),$$

$$S_{2c} = \tfrac{1}{2}(S_2 + S_1 + iS_4 - iS_3),$$

$$S_{3c} = \tfrac{1}{2}(S_2 - S_1 + iS_4 + iS_3),$$

$$S_{4c} = \tfrac{1}{2}(S_2 - S_1 - iS_4 - iS_3).$$

This relation is not restricted to a specific particle; for an optically active sphere, however, two of the matrix elements are equal: $S_{3c} = S_{4c}$.

The (4×4) scattering matrix elements (3.16) for an optically active sphere satisfy the following six relations:

$$S_{31} = -S_{13}, \quad S_{32} = -S_{23}, \quad S_{43} = -S_{34},$$

$$S_{41} = S_{14}, \quad S_{42} = S_{24}, \quad S_{21} = S_{12}. \tag{8.16}$$

The cross sections for extinction and scattering by an optically active particle are different for incident left-circularly and right-circularly polarized light. For an optically active sphere, the cross sections can be obtained in a manner similar to that for a nonactive sphere (Section 4.4). Therefore, we give only the results and omit the details:

$$C_{sca, L} = \frac{2\pi}{k^2} \sum_{n=1}^{\infty} (2n + 1)\left[|a_n|^2 + |b_n|^2 + 2|c_n|^2\right.$$

$$\left. - 2 \, \text{Im}\{(a_n + b_n)c_n^*\}\right],$$

$$C_{sca, R} = \frac{2\pi}{k^2} \sum_{n=1}^{\infty} (2n + 1)\left[|a_n|^2 + |b_n|^2 + 2|c_n|^2\right.$$

$$\left. + 2 \, \text{Im}\{(a_n + b_n)c_n^*\}\right],$$

$$C_{ext, L} = \frac{4\pi}{k^2} \, \text{Re}\{S_L\}$$

$$= \frac{2\pi}{k^2} \sum_{n=1}^{\infty} (2n + 1)\text{Re}\{a_n + b_n - 2ic_n\},$$

$$C_{ext, R} = \frac{4\pi}{k^2} \, \text{Re}\{S_R\}$$

$$= \frac{2\pi}{k^2} \sum_{n=1}^{\infty} (2n + 1)\text{Re}\{a_n + b_n + 2ic_n\}, \tag{8.17}$$

where $S_L = S_{2c}(0°)$ and $S_R = S_{1c}(0°)$ are the amplitude scattering matrix elements in the forward direction.

8.3.2 Circular Dichroism and Optical Rotation

We have shown that only circularly polarized waves may propagate in optically active media without change in their state of polarization. However, the polarization of a wave that is *linearly* polarized at a particular point, say $z = 0$, changes continuously as it propagates in the z direction. At $z = h$ the wave will be *elliptically* polarized with ellipticity $|\Theta_T|$, and the azimuth of the vibration ellipse will have rotated through an angle Φ_T relative to the direction of polarization at $z = 0$. These quantities are related to the complex refractive indices $N_L = n_L + ik_L$ and $N_R = n_R + ik_R$ for left-circularly and right-circularly polarized waves by

$$\Phi_T + i\Theta_T = \frac{\pi}{\lambda}(N_L - N_R)h,$$

provided that $|2\pi(k_L - k_R)h/\lambda| \ll 1$. If Φ_T is positive the vibration ellipse rotates in the clockwise sense.

A medium is said to be *circularly dichroic*—it absorbs differently according to the state of circular polarization of the light—if $k_L - k_R \neq 0$; it is *circularly birefringent*, which is manifested by *optical rotation*, if $n_L - n_R \neq 0$. Optical rotation and circular dichroism are not independent phenomena, but are connected by Kramers–Kronig relations:

$$\phi(\omega) = \frac{2\omega^2}{\pi} P \int_0^\infty \frac{\theta(\Omega)}{\Omega(\Omega^2 - \omega^2)} d\Omega,$$

$$\theta(\omega) = \frac{-2\omega}{\pi} P \int_0^\infty \frac{\phi(\Omega)}{\Omega^2 - \omega^2} d\Omega,$$

(8.18)

where $\phi = \Phi_T/h$ and $\theta = \Theta_T/h$ are the rotation and change of ellipticity per unit path length. A derivation of (8.18) in the same spirit as the derivations of dispersion relations in Section 2.3 has been given by Emeis et al. (1967). Note that (8.18) do not follow by naively applying the dispersion relations (2.49) and (2.50) to N_L and N_R separately and then subtracting the resulting expressions; the reasons for this have been discussed by Smith (1976).

Circular dichroism and optical rotation for *particulate* media may be operationally defined in terms of the Stokes parameters (2.80), which in the circular polarization representation are written

$$I = E_L E_L^* + E_R E_R^*, \qquad Q = E_L^* E_R + E_R^* E_L,$$

$$U = i(E_L^* E_R - E_L E_R^*), \qquad V = E_R E_R^* - E_L E_L^*. \qquad (8.19)$$

The azimuth γ and ellipticity $|\tan \eta|$ of the vibration ellipse for an arbitrary beam can be determined from the Stokes parameters by (2.82).

If optical rotation Φ_T for a collection of particles is defined as the change in azimuth of a horizontally polarized incident beam ($\gamma_i = 0$) after it is transmitted through this medium, then

$$\Phi_T = \gamma_t - \gamma_i = \tfrac{1}{2} \tan^{-1} \frac{U_t}{Q_t}, \qquad (8.20)$$

where subscripts i and t denote incident and transmitted beams. Similarly, if circular dichroism Θ_T for the particulate medium is defined as the change in ellipticity of a horizontally polarized beam ($\eta_i = 0$) after it is transmitted through the medium, then

$$\Theta_T = \tan \eta_t; \qquad \eta_t = \tfrac{1}{2} \tan^{-1} \frac{V_t}{\sqrt{Q_t^2 + U_t^2}}. \qquad (8.21)$$

Note that these definitions of optical rotation and circular dichroism for a particulate medium depend on the choice of the horizontal direction unless the medium is invariant with respect to arbitrary rotation about an axis parallel to the incident beam.

Equations (8.20) and (8.21) are completely general; no assumptions have been made about the nature of the particles. Let us now consider a more specific example: $\mathfrak{N}$ *identical* particles per unit volume of a slab of thickness h; the medium surrounding the particles is nonabsorbing and nonactive. If the amplitude scattering matrix in the circular polarization representation is diagonal in the forward direction $[S_{3c}(0°) = S_{4c}(0°) = 0]$, then by following a line of reasoning similar to that which led to (3.39) we obtain the left-handed and right-handed components of the transmitted electric field:

$$E_L = \frac{1}{\sqrt{2}} E \left(1 - \frac{2\pi}{k^2} \mathfrak{N} S_L h \right),$$

$$ \qquad (8.22)$$

$$E_R = \frac{1}{\sqrt{2}} E \left(1 - \frac{2\pi}{k^2} \mathfrak{N} S_R h \right),$$

where $E = E_0 \exp(ikz)$ is the incident electric field. We have also assumed in the derivation of (8.22) that $|2\pi \, \mathfrak{N} S h / k^2| \ll 1$. If the Stokes parameters (8.19) corresponding to (8.22) are inserted into (8.20) and (8.21), we obtain

$$\phi + i\theta = \frac{\pi}{k^2} \mathfrak{N} i (S_L - S_R). \qquad (8.23)$$

Thus, the difference between the diagonal elements of the forward amplitude scattering matrix in the circular polarization representation has a simple physical interpretation. Although we considered identical particles for convenience, (8.23) is readily generalized to a suspension of nonidentical particles: the optical rotation and circular dichroism of such a suspension is merely the sum of the contributions from the individual components. We may also relax the requirement that $|2\pi \mathfrak{N} Sh/k^2| \ll 1$ provided that multiple scattering is negligible; see the discussion following (3.46).

If the particles are homogeneous spheres, then from the results of the preceding section we have

$$i(S_L - S_R) = 2S_3(0°) = 2 \sum_{n=1}^{\infty} (2n+1)c_n. \qquad (8.24)$$

It also follows from (8.17) and (8.23) that the circular dichroism of a suspension of spheres is proportional to the difference between the extinction cross sections for left-circularly and right-circularly polarized light:

$$\theta = \tfrac{1}{4}\mathfrak{N}(C_{ext,\,L} - C_{ext,\,R}) = \theta_{sca} + \theta_{abs},$$

$$\theta_{sca} = \tfrac{1}{4}\mathfrak{N}(C_{sca,\,L} - C_{sca,\,R}), \qquad \theta_{abs} = \tfrac{1}{4}\mathfrak{N}(C_{abs,\,L} - C_{abs,\,R}). \qquad (8.25)$$

However, (8.25) is not restricted to spheres but holds for particles of arbitrary shape. Thus, circular dichroism in particulate media includes a component that is the result of differential scattering, in contrast with circular dichroism in homogeneous media, which arises solely from differential absorption of left-circularly and right-circularly polarized light.

If the spheres are sufficiently small ($x \ll 1$, $|mx| \ll 1$), the series (8.24) can be truncated after the first term; the leading coefficient correct to terms of order x^3 is

$$c_1 = \frac{x^3}{3} \frac{m_L - m_R}{m^2 + 2},$$

where we have used the series expansions (5.3) of the various Bessel functions and their derivatives. Therefore, (8.23) in the small particle limit is

$$\phi + i\theta = f\frac{\pi}{\lambda}(N_L - N_R)\frac{3}{m^2 + 2},$$

where $f = 4\pi a^3 \mathfrak{N}/3$ is the fraction of the suspension volume occupied by the particles. The quantity $\pi(N_L - N_R)/\lambda$ is the intrinsic optical rotation and circular dichroism of the spheres; $3/(m^2 + 2)$ may be interpreted as the effect of the surrounding medium—a "solvent" correction.

If m, the average relative refractive index, varies only slightly over the frequency region of interest, then the circular dichroism (CD) spectrum $\theta(\omega)$ and the optical rotatory dispersion (ORD) spectrum $\phi(\omega)$ of spheres small compared with the wavelength are essentially the same as those of the homogeneous parent material. But this is not necessarily true for spheres, or particles of any shape, comparable with or larger than the wavelength. Indeed, CD (or ORD) spectra for the same material in the homogeneous and particulate states may bear little resemblance to each other. This has important implications for interpreting CD and ORD spectra of biological particles.

8.4 INFINITE RIGHT CIRCULAR CYLINDER

There are many naturally occurring particles, such as some viruses and asbestos fibers, which are best represented as cylinders long compared with their diameter. Therefore, in this section we shall construct the exact solution to the problem of absorption and scattering by an infinitely long right circular cylinder and examine some of the properties of this solution.

As in the problem of scattering by a sphere (Chapter 4), our starting point is the scalar wave equation $\nabla^2\psi + k^2\psi = 0$, which in cylindrical polar coordinates r, ϕ, z(Fig. 8.2) is

$$\frac{1}{r}\frac{\partial}{\partial r}\left(r\frac{\partial\psi}{\partial r}\right) + \frac{1}{r^2}\frac{\partial^2\psi}{\partial\phi^2} + \frac{\partial^2\psi}{\partial z^2} + k^2\psi = 0. \qquad (8.26)$$

Separable solutions to (8.26) that are single-valued functions of ϕ are of the form

$$\psi_n(r, \phi, z) = Z_n(\rho)e^{in\phi}e^{ihz} \qquad (n = 0, \pm 1, \ldots), \qquad (8.27)$$

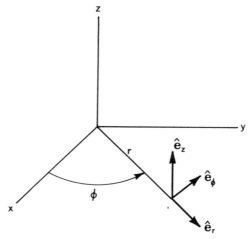

Figure 8.2 Cylindrical polar coordinate system. The z axis lies along the axis of the infinite cylinder.

where $\rho = r\sqrt{k^2 - h^2}$ and Z_n is a solution to the Bessel equation

$$\rho \frac{d}{d\rho}\left(\rho \frac{d}{d\rho}Z_n\right) + (\rho^2 - n^2)Z_n = 0. \tag{8.28}$$

The linearly independent solutions to (8.28) are the Bessel functions of first and second kind, J_n and Y_n, of integral order n. In general, the separation constant h is unrestricted, although in the problems with which we shall deal, h is dictated by the form of the incident field and the necessity of satisfying the conditions (3.7) at the boundary between the cylinder and the surrounding medium.

The vector cylindrical harmonics generated by (8.27) are

$$\mathbf{M}_n = \nabla \times (\hat{\mathbf{e}}_z \psi_n), \qquad \mathbf{N}_n = \frac{\nabla \times \mathbf{M}_n}{k},$$

where we have taken as pilot vector the unit vector $\hat{\mathbf{e}}_z$ parallel to the cylinder axis (Fig. 8.3). In component form these vector harmonics are

$$\mathbf{M}_n = \sqrt{k^2 - h^2}\left(in\frac{Z_n(\rho)}{\rho}\hat{\mathbf{e}}_r - Z_n'(\rho)\hat{\mathbf{e}}_\phi\right)e^{i(n\phi + hz)},$$

$$\mathbf{N}_n = \frac{\sqrt{k^2 - h^2}}{k}\left(ihZ_n'(\rho)\hat{\mathbf{e}}_r - hn\frac{Z_n(\rho)}{\rho}\hat{\mathbf{e}}_\phi\right.$$

$$\left. + \sqrt{k^2 - h^2}\,Z_n(\rho)\hat{\mathbf{e}}_z\right)e^{i(n\phi + hz)}.$$

The vector harmonics are orthogonal in the sense that

$$\int_0^{2\pi}\mathbf{M}_n \cdot \mathbf{M}_m^* \, d\phi = \int_0^{2\pi}\mathbf{N}_n \cdot \mathbf{N}_m^* \, d\phi = \int_0^{2\pi}\mathbf{M}_n \cdot \mathbf{N}_m^* \, d\phi = 0 \qquad (n \neq m).$$

Let us now consider an infinite right circular cylinder of radius a, which is illuminated by a plane homogeneous wave $\mathbf{E}_i = \mathbf{E}_0 e^{ik\hat{\mathbf{e}}_i \cdot \mathbf{x}}$ propagating in the direction $\hat{\mathbf{e}}_i = -\sin\zeta\hat{\mathbf{e}}_x - \cos\zeta\hat{\mathbf{e}}_z$, where ζ is the angle between the incident wave and the cylinder axis (Fig. 8.3). There are two possible orthogonal polarization states of the incident wave: electric field polarized *parallel* to the xz plane; and electric field polarized *perpendicular* to the xz plane. We shall consider each of these polarizations in turn.

Case I. Incident Electric Field Parallel to the xz Plane. The first step is to expand the incident electric field

$$\mathbf{E}_i = E_0(\sin\zeta\hat{\mathbf{e}}_z - \cos\zeta\hat{\mathbf{e}}_x)e^{-ik(r\sin\zeta\cos\phi + z\cos\zeta)}$$

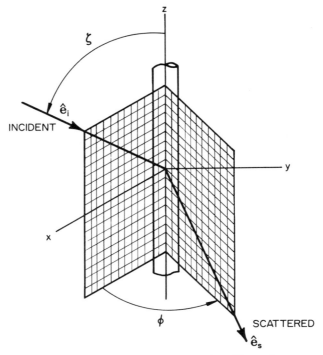

Figure 8.3 Infinite cylinder obliquely illuminated by a plane wave.

in vector cylindrical harmonics. In order for the expansion to be finite at $r = 0$ we must exclude the Bessel function Y_n as the radially dependent part of the generating function; it is also clear from the form of $\mathbf{E}_i$ that h in (8.27) must be $-k\cos\zeta$. Thus, the expansion of $\mathbf{E}_i$ is

$$\mathbf{E}_i = \sum_{n=-\infty}^{\infty} \left[A_n \mathbf{M}_n^{(1)} + B_n \mathbf{N}_n^{(1)} \right],$$

where the vector harmonics are generated by $J_n(kr\sin\zeta)e^{in\phi}e^{-ikz\cos\zeta}$. To determine the coefficients A_n and B_n we use the orthogonality of the vector harmonics, which requires that we evaluate the integrals

$$\mathcal{G}_n^{(1)} = \int_0^{2\pi} e^{-i(n\phi + \rho\cos\phi)}\, d\phi,$$

$$\mathcal{G}_n^{(2)} = \int_0^{2\pi} e^{-i(n\phi + \rho\cos\phi)}\cos\phi\, d\phi,$$

$$\mathcal{G}_n^{(3)} = \int_0^{2\pi} e^{-i(n\phi + \rho\cos\phi)}\sin\phi\, d\phi,$$

where $\rho = kr \sin \zeta$. From the integral representation of the Bessel function $J_n(\rho)$, which is real for real ρ,

$$J_n(\rho) = \frac{i^{-n}}{2\pi} \int_0^{2\pi} e^{i(n\phi + \rho \cos \phi)} \, d\phi,$$

it immediately follows that $\mathcal{G}_n^{(1)} = 2\pi(-i)^n J_n(\rho)$; $\mathcal{G}_n^{(2)} = 2\pi i(-i)^n J_n'(\rho)$ is obtained by differentiating $\mathcal{G}_n^{(1)}$ with respect to ρ. The third integral may be written

$$2i\mathcal{G}_n^{(3)} = \mathcal{G}_{n-1}^{(1)} - \mathcal{G}_{n+1}^{(1)},$$

and if we use the identity

$$\frac{2nZ_n}{\rho} = Z_{n-1} + Z_{n+1},$$

it follows that $\mathcal{G}_n^{(3)} = 2\pi(-i)^n J_n(\rho)n/\rho$. All that is required now is a good bit of patience to show that

$$A_n = 0, \qquad B_n = \frac{E_0(-i)^n}{k \sin \zeta};$$

therefore, the expansion of the incident electromagnetic field is

$$\mathbf{E}_i = \sum_{n=-\infty}^{\infty} E_n \mathbf{N}_n^{(1)}, \qquad \mathbf{H}_i = \frac{-ik}{\omega\mu} \sum_{n=-\infty}^{\infty} E_n \mathbf{M}_n^{(1)},$$

where $E_n = E_0(-i)^n/k \sin \zeta$.

In order to satisfy the continuity conditions (3.7) for all values of z on the boundary of the cylinder, the separation constant h in the wave functions that generate the vector harmonics of the field inside the cylinder must also be $-k \cos \zeta$; finiteness at the origin requires that J_n is the appropriate Bessel function. Thus the generating functions for the internal field $(\mathbf{E}_1, \mathbf{H}_1)$ are $J_n(kr\sqrt{m^2 - \cos^2\zeta})e^{in\phi}e^{-ikz \cos\zeta}$, where m is the refractive index of the cylinder relative to that of the surrounding medium. The corresponding expansions are

$$\mathbf{E}_1 = \sum_{n=-\infty}^{\infty} E_n \left[g_n \mathbf{M}_n^{(1)} + f_n \mathbf{N}_n^{(1)} \right],$$

$$\mathbf{H}_1 = \frac{-ik_1}{\omega\mu_1} \sum_{n=-\infty}^{\infty} E_n \left[g_n \mathbf{N}_n^{(1)} + f_n \mathbf{M}_n^{(1)} \right].$$

The Hankel functions $H_n^{(1)} = J_n + iY_n$ and $H_n^{(2)} = J_n - iY_n$ are also linearly independent solutions to (8.28); they are given asymptotically by (4.41):

$$H_n^{(1)}(\rho) \sim \sqrt{\frac{2}{\pi\rho}} \, e^{i\rho}(-i)^n e^{-i\pi/4},$$

$$|\rho| \gg n^2.$$

$$H_n^{(2)}(\rho) \sim \sqrt{\frac{2}{\pi\rho}} \, e^{-i\rho} i^n e^{i\pi/4},$$

Therefore, if the scattered field $(\mathbf{E}_s, \mathbf{H}_s)$ is to be an outgoing wave at large distances from the cylinder, the generating functions in the expansions

$$\mathbf{E}_s = -\sum_{n=-\infty}^{\infty} E_n \left[b_{nI} \mathbf{N}_n^{(3)} + i a_{nI} \mathbf{M}_n^{(3)} \right],$$

$$\mathbf{H}_s = \frac{ik}{\omega\mu} \sum_{n=-\infty}^{\infty} E_n \left[b_{nI} \mathbf{M}_n^{(3)} + i a_{nI} \mathbf{N}_n^{(3)} \right],$$

must be $H_n^{(1)}(kr \sin \zeta) e^{in\phi} e^{-ikz \cos \zeta}$.

If we apply the conditions (3.7) at $r = a$, we obtain four equations in the four expansion coefficients, which can be solved for the coefficients a_{nI}, b_{nI} of the scattered field:

$$a_{nI} = \frac{C_n V_n - B_n D_n}{W_n V_n + i D_n^2}, \qquad b_{nI} = \frac{W_n B_n + i D_n C_n}{W_n V_n + i D_n^2},$$

$$D_n = n \cos \zeta \, \eta J_n(\eta) H_n^{(1)}(\xi) \left(\frac{\xi^2}{\eta^2} - 1 \right),$$

$$B_n = \xi \left[m^2 \xi J_n'(\eta) J_n(\xi) - \eta J_n(\eta) J_n'(\xi) \right],$$

$$C_n = n \cos \zeta \, \eta J_n(\eta) J_n(\xi) \left(\frac{\xi^2}{\eta^2} - 1 \right), \tag{8.29}$$

$$V_n = \xi \left[m^2 \xi J_n'(\eta) H_n^{(1)}(\xi) - \eta J_n(\eta) H_n^{(1)\prime}(\xi) \right],$$

$$W_n = i\xi \left[\eta J_n(\eta) H_n^{(1)\prime}(\xi) - \xi J_n'(\eta) H_n^{(1)}(\xi) \right],$$

where $\xi = x \sin \zeta$, $\eta = x\sqrt{m^2 - \cos^2 \zeta}$, $x = ka$, and we have taken $\mu = \mu_1$. It follows from the relations $J_{-n} = (-1)^n J_n$ and $Y_{-n} = (-1)^n Y_n$ that

$$a_{-nI} = -a_{nI}, \qquad b_{-nI} = b_{nI}, \qquad a_{0I} = 0.$$

When the incident light is normal to the cylinder axis ($\zeta = 90°$), a_{nI} vanishes and

$$b_{nI}(\zeta = 90°) = b_n = \frac{J_n(mx)J_n'(x) - mJ_n'(mx)J_n(x)}{J_n(mx)H_n^{(1)'}(x) - mJ_n'(mx)H_n^{(1)}(x)}. \quad (8.30)$$

Case II.   Incident Electric Field Perpendicular to the xz Plane.   The expansion of the incident electric field $\mathbf{E}_i = E_0\hat{\mathbf{e}}_y e^{-ik(r\sin\zeta\cos\phi + z\cos\zeta)}$ is

$$\mathbf{E}_i = -i\sum_{n=-\infty}^{\infty} E_n \mathbf{M}_n^{(1)},$$

the curl of which gives the incident magnetic field. The coefficients of the scattered field

$$\mathbf{E}_s = \sum_{n=-\infty}^{\infty} E_n\left[ia_{nII}\mathbf{M}_n^{(3)} + b_{nII}\mathbf{N}_n^{(3)}\right],$$

can be written in the form

$$a_{nII} = -\frac{A_n V_n - iC_n D_n}{W_n V_n + iD_n^2}, \qquad b_{nII} = -i\frac{C_n W_n + A_n D_n}{W_n V_n + iD_n^2}, \quad (8.31)$$

where D_n, C_n, and so on, were defined in the preceding section and

$$A_n = i\xi\left[\xi J_n'(\eta) J_n(\xi) - \eta J_n(\eta) J_n'(\xi)\right].$$

It follows from the properties of the Bessel functions that

$$a_{-nII} = a_{nII}, \qquad b_{-nII} = -b_{nII}, \qquad b_{0II} = 0.$$

Although it is not obvious, it is not too difficult to show that

$$a_{nI} = -b_{nII}.$$

This was pointed out by Kerker et al. (1966). When the incident light is normal to the cylinder axis, b_{nII} vanishes and

$$a_{nII}(\zeta = 90°) = a_n = \frac{mJ_n'(x)J_n(mx) - J_n(x)J_n'(mx)}{mJ_n(mx)H_n^{(1)'}(x) - J_n'(mx)H_n^{(1)}(x)}. \quad (8.32)$$

8.4.1   Asymptotic Scattered Field

At large distances from the cylinder $(kr\sin\zeta \gg 1)$, the scattered field (Case I) is given asymptotically by

$$\mathbf{E}_s \sim -E_0 e^{-i\pi/4}\sqrt{\frac{2}{\pi kr\sin\zeta}}\, e^{ik(r\sin\zeta - z\cos\zeta)}$$

$$\times \sum_n (-1)^n e^{in\phi}\left[a_{nI}\hat{\mathbf{e}}_\phi + b_{nI}(\cos\zeta\hat{\mathbf{e}}_r + \sin\zeta\hat{\mathbf{e}}_z)\right]. \quad (8.33)$$

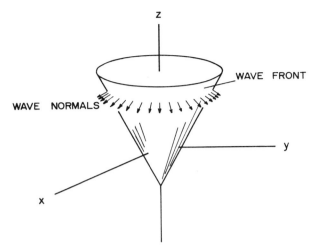

Figure 8.4 Wave front and wave normals of light scattered by an infinite cylinder.

When the incident electric field is polarized perpendicular to the xz plane (Case II), the asymptotic scattered field is given by (8.33) with a_{nI} replaced by $-a_{nII}$ and b_{nI} replaced by $-b_{nII}$.

The surfaces of constant phase, or *wave fronts*, of the scattered wave (8.33), the points on which satisfy

$$f(x, y, z) = r \sin \zeta - z \cos \zeta = C,$$

are *cones* of half-angle ζ and apexes at $z = -C/\cos \zeta$ (Fig. 8.4). Thus, we may visualize the propagation of the scattered wave as a cone that is sliding down the cylinder. At any point on the cone, the direction of propagation, or *wave normal* $\hat{\mathbf{e}}_s$, is

$$\hat{\mathbf{e}}_s = \nabla f = \sin \zeta \hat{\mathbf{e}}_r - \cos \zeta \hat{\mathbf{e}}_z.$$

The Poynting vector, therefore, is in the direction $\hat{\mathbf{e}}_s$. When the incident beam is normal to the cylinder axis ($\zeta = 90°$), the cone reduces to a cylinder.

A convincing and aesthetically pleasing demonstration that the light scattered by a long cylinder is a conical wave can be made by illuminating a fiber with a narrow laser beam. If a screen perpendicular to the incident beam is placed at some distance from the fiber, the resulting patterns formed on the screen will be *conic sections*. When the incident light is normal to the fiber axis, the pattern is a straight line. As ζ is decreased, a succession of *hyperbolas* are traced out, and at $\zeta = 45°$ the pattern is a *parabola*. For angles of incidence less than $45°$, *ellipses* appear on the screen, the eccentricities of which decrease with decreasing ζ; as ζ approaches $0°$ the pattern approaches a circle. In Fig. 8.5 we show a series of photographs, taken with an oscilloscope camera, of the

Figure 8.5   Conic sections formed by scattering of a laser beam by a thin fiber.

patterns formed on a screen when a laser beam is scattered by a thin fiber. In addition to showing the various conic sections that result from varying the angle ζ, this figure also shows some of the maxima and minima of the scattering diagram.

8.4.2   Amplitude Scattering Matrix

In Chapter 3 we derived a general expression for the amplitude scattering matrix for an arbitrary particle. An unstated assumption underlying that derivation is that the particle is confined within a bounded region, a condition that is not satisfied by an infinite cylinder. Nevertheless, we can express the field scattered by such a cylinder in a concise form by resolving the incident and scattered fields into components parallel and perpendicular to planes determined by the cylinder axis ($\hat{\mathbf{e}}_z$) and the appropriate wave normals (see Fig. 8.3). That is, we write the incident field

$$\mathbf{E}_i = \left(E_{\|i} \hat{\mathbf{e}}_{\|i} + E_{\perp i} \hat{\mathbf{e}}_{\perp i} \right) e^{i \mathbf{k} \cdot \mathbf{x}},$$

$$\hat{\mathbf{e}}_{\|i} = \sin \zeta \hat{\mathbf{e}}_z - \cos \zeta \hat{\mathbf{e}}_x, \qquad \hat{\mathbf{e}}_{\perp i} = -\hat{\mathbf{e}}_y, \qquad \hat{\mathbf{e}}_{\perp i} \times \hat{\mathbf{e}}_{\|i} = \hat{\mathbf{e}}_i.$$

The scattered field is the sum of components parallel and perpendicular to the plane ϕ = constant:

$$\mathbf{E}_s = E_{\|s}\hat{\mathbf{e}}_{\|s} + E_{\perp s}\hat{\mathbf{e}}_{\perp s},$$

$$\hat{\mathbf{e}}_{\|s} = \cos\zeta\hat{\mathbf{e}}_r + \sin\zeta\hat{\mathbf{e}}_z, \qquad \hat{\mathbf{e}}_{\perp s} = \hat{\mathbf{e}}_\phi, \qquad \hat{\mathbf{e}}_{\perp s} \times \hat{\mathbf{e}}_{\|s} = \hat{\mathbf{e}}_s.$$

We may now write the relation between incident and scattered fields in matrix form:

$$\begin{pmatrix} E_{\|s} \\ E_{\perp s} \end{pmatrix} = e^{i3\pi/4}\sqrt{\frac{2}{\pi kr\sin\zeta}}\, e^{ik(r\sin\zeta - z\cos\zeta)}\begin{pmatrix} T_1 & T_4 \\ T_3 & T_2 \end{pmatrix}\begin{pmatrix} E_{\|i} \\ E_{\perp i} \end{pmatrix},$$

$$T_1 = \sum_{-\infty}^{\infty} b_{n\mathrm{I}}e^{-in\Theta} = b_{0\mathrm{I}} + 2\sum_{n=1}^{\infty} b_{n\mathrm{I}}\cos(n\Theta),$$

$$T_2 = \sum_{-\infty}^{\infty} a_{n\mathrm{II}}e^{-in\Theta} = a_{0\mathrm{II}} + 2\sum_{n=1}^{\infty} a_{n\mathrm{II}}\cos(n\Theta), \tag{8.34}$$

$$T_3 = \sum_{-\infty}^{\infty} a_{n\mathrm{I}}e^{-in\Theta} = -2i\sum_{n=1}^{\infty} a_{n\mathrm{I}}\sin(n\Theta),$$

$$T_4 = \sum_{-\infty}^{\infty} b_{n\mathrm{II}}e^{-in\Theta} = -2i\sum_{n=1}^{\infty} b_{n\mathrm{II}}\sin(n\Theta) = -T_3,$$

where we have transformed the angle variable ϕ to $\Theta = \pi - \phi$.

Unlike the amplitude scattering matrix for a sphere (Chapter 4), the off-diagonal elements of the amplitude scattering matrix for a cylinder are, in general, nonzero. Thus, if the incident light is polarized *parallel* (perpendicular) to the xz plane, the scattered light has a component *perpendicular* (parallel) to the plane determined by the cylinder axis and the scattering direction $\hat{\mathbf{e}}_s$. However, symmetry requires that T_3 and T_4 vanish when $\hat{\mathbf{e}}_s$ lies in the forward scattering plane ($\Theta = 0°$) and the backward scattering plane ($\Theta = 180°$). When the incident light is normal to the cylinder axis, T_3 and T_4 are identically zero for all Θ.

8.4.3   Cross Sections

Although we have repeatedly referred to an "infinite" cylinder, it is clear that no such cylinder exists except as an idealization. So what we really have in mind is a cylinder long compared with its diameter. Later in this section we shall try to acquire some insight into how long a cylinder must be before it is effectively infinite by considering scattering in the diffraction theory approximation.

The scattering and absorption cross sections of an infinite cylinder are, of course, infinite; however, the light scattered and absorbed per unit length of such a cylinder is finite. If we ignore end effects the ratio $W_s/L(W_a/L)$ for a finite cylinder of length L can be approximated by the amount of light scattered (absorbed) per unit length of an infinite cylinder; this approximation will be increasingly better the greater the ratio of cylinder length to diameter.

In a manner similar to that in Section 3.4, where we considered cross sections of finite particles, we can calculate cross sections *per unit length* of an infinite cylinder by constructing an imaginary closed concentric surface A of length L and radius R (Fig. 8.6). The rate W_a at which energy is absorbed within this surface is

$$W_a = -\int_A \mathbf{S} \cdot \hat{\mathbf{n}}\, dA = W_{\text{ext}} - W_s,$$

where $\mathbf{S}$ is the Poynting vector. There is no net contribution to W_a from the ends of A; therefore,

$$W_s = RL\int_0^{2\pi} (\mathbf{S}_s)_r\, d\phi, \qquad W_{\text{ext}} = RL\int_0^{2\pi} (\mathbf{S}_{\text{ext}})_r\, d\phi, \qquad (8.35)$$

where $(\mathbf{S}_s)_r$ and $(\mathbf{S}_{\text{ext}})_r$ are the radial components of

$$\mathbf{S}_s = \tfrac{1}{2}\,\text{Re}\{\mathbf{E}_s \times \mathbf{H}_s^*\}, \qquad \mathbf{S}_{\text{ext}} = \tfrac{1}{2}\,\text{Re}\{\mathbf{E}_i \times \mathbf{H}_s^* + \mathbf{E}_s \times \mathbf{H}_i^*\}.$$

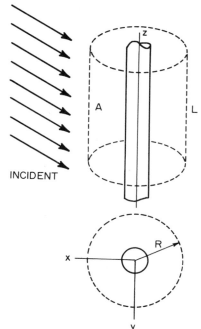

INCIDENT

Figure 8.6 Surface of integration (dashed lines).

Let us consider the incident electric field parallel to the xz plane (Case I). If the series expansions of the incident and scattered fields are substituted into the expressions for the vectors $\mathbf{S}_s$ and $\mathbf{S}_{\text{ext}}$, then after performing the required integrations in (8.35) we obtain the efficiencies $Q_{\text{sca},\text{I}}$ and $Q_{\text{ext},\text{I}}$ for scattering and extinction:

$$Q_{\text{sca},\text{I}} = \frac{W_{s,\text{I}}}{2aLI_i} = \frac{2}{x}\left[|b_{0\text{I}}|^2 + 2\sum_{n=1}^{\infty}\left(|b_{n\text{I}}|^2 + |a_{n\text{I}}|^2\right)\right],$$

$$Q_{\text{ext},\text{I}} = \frac{W_{\text{ext},\text{I}}}{2aLI_i} = \frac{2}{x}\,\text{Re}\left\{b_{0\text{I}} + 2\sum_{n=1}^{\infty}b_{n\text{I}}\right\}. \tag{8.36}$$

Note that

$$Q_{\text{ext},\text{I}} = \frac{2}{x}\,\text{Re}\{T_1(\Theta = 0^\circ)\},$$

which is the *optical theorem* for an infinite cylinder (see Section 3.4).

In a similar manner we obtain the scattering and extinction efficiencies when the incident electric field is polarized perpendicular to the xz plane (Case II):

$$Q_{\text{sca},\text{II}} = \frac{2}{x}\left[|a_{0\text{II}}|^2 + 2\sum_{n=1}^{\infty}\left(|a_{n\text{II}}|^2 + |b_{n\text{II}}|^2\right)\right],$$

$$Q_{\text{ext},\text{II}} = \frac{2}{x}\,\text{Re}\left\{a_{0\text{II}} + 2\sum_{n=1}^{\infty}a_{n\text{II}}\right\} = \frac{2}{x}\,\text{Re}\{T_2(\Theta = 0^\circ)\}. \tag{8.37}$$

If the incident light is *unpolarized*, the efficiencies are

$$Q_{\text{sca}} = \tfrac{1}{2}(Q_{\text{sca},\text{I}} + Q_{\text{sca},\text{II}}), \qquad Q_{\text{ext}} = \tfrac{1}{2}(Q_{\text{ext},\text{I}} + Q_{\text{ext},\text{II}}).$$

8.4.4 Normally Incident Light

If the incident light is normal to the cylinder axis the scattering coefficients have their simplest form. However, the coefficients (8.30) and (8.32) are not in the form most suitable for computations. If we introduce the logarithmic derivative

$$D_n(\rho) = \frac{J_n'(\rho)}{J_n(\rho)}$$

and use the recurrence relation

$$Z_n'(x) = Z_{n-1}(x) - \frac{n}{x}Z_n(x),$$

where Z_n is any Bessel function, the scattering coefficients may be written

$$a_n = \frac{[D_n(mx)/m + n/x]J_n(x) - J_{n-1}(x)}{[D_n(mx)/m + n/x]H_n^{(1)}(x) - H_{n-1}^{(1)}(x)},$$

(8.38)

$$b_n = \frac{[mD_n(mx) + n/x]J_n(x) - J_{n-1}(x)}{[mD_n(mx) + n/x]H_n^{(1)}(x) - H_{n-1}^{(1)}(x)}.$$

The logarithmic derivative satisfies the recurrence relation

$$D_{n-1}(z) = \frac{n-1}{z} - \frac{1}{(n/z) + D_n(z)}.$$

(8.39)

A computer program for calculating the scattering coefficients (8.38) and the corresponding cross sections and scattering matrix elements is described in Appendix C; all the examples in this section were obtained with this program.

The amplitude scattering matrix for a normally illuminated cylinder is diagonal:

$$\begin{pmatrix} E_{\parallel s} \\ E_{\perp s} \end{pmatrix} = e^{i3\pi/4}\sqrt{\frac{2}{\pi kr}}\, e^{ikr} \begin{pmatrix} T_1 & 0 \\ 0 & T_2 \end{pmatrix} \begin{pmatrix} E_{\parallel i} \\ E_{\perp i} \end{pmatrix}.$$

(8.40)

Note that the surfaces of constant phase of the scattered wave are cylinders. Thus, the light scattered by a cylinder long compared with its diameter will not be seen by a suitably collimated detector unless it "looks" in a direction perpendicular to the cylinder axis. Many examples of this are commonly encountered, most of which probably go unnoticed. For example, randomly oriented scratches on a car windshield form a circular pattern when illuminated by the headlights of an oncoming car; dust on the windshield, on the other hand, forms no such obvious pattern. The scratches are sufficiently long compared with their lateral dimensions that they may be considered to be infinite cylinders; thus, only those scratches perpendicular to a line from the eye to a scratch will scatter light into the eye, and such scratches lie on circles centered on the line of sight. If a point source is viewed through a fibrous material, one may also see circular patterns: Christmas tree lights embedded in wisps of "angel's hair" are seen to be surrounded by haloes of scattered light; steel wool will yield the same effect.

The scattering matrix corresponding to (8.40)

$$
\begin{pmatrix} I_s \\ Q_s \\ U_s \\ V_s \end{pmatrix} = \frac{2}{\pi k r} \begin{pmatrix} T_{11} & T_{12} & 0 & 0 \\ T_{12} & T_{11} & 0 & 0 \\ 0 & 0 & T_{33} & T_{34} \\ 0 & 0 & -T_{34} & T_{33} \end{pmatrix} \begin{pmatrix} I_i \\ Q_i \\ U_i \\ V_i \end{pmatrix}
$$

$$ T_{11} = \tfrac{1}{2}(|T_1|^2 + |T_2|^2), \qquad T_{12} = \tfrac{1}{2}(|T_1|^2 - |T_2|^2), $$

$$ T_{33} = \mathrm{Re}\{T_1 T_2^*\}, \qquad\qquad T_{34} = \mathrm{Im}\{T_1 T_2^*\}, $$

has the same form as that for a sphere (4.77). However, there are appreciable differences between scattering by a sphere and by a normally illuminated cylinder. For example, the ratio $P = T_{12}/T_{11}$, which is analogous to (4.78), does not necessarily vanish in the forward (or backward) direction. Thus, if the incident light is unpolarized, the forward scattered light will, in general, be partially polarized. Scattering and extinction cross sections for a nonactive sphere are independent of the state of polarization of the incident light; however, these quantities for a cylinder may be quite different for different polarizations. This is illustrated in Fig. 8.7, where the scattering cross sections

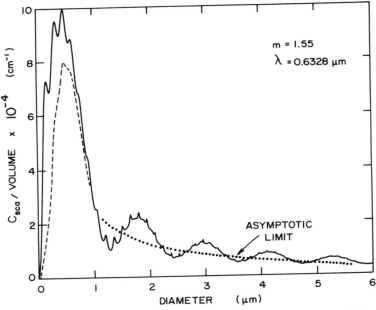

Figure 8.7 Scattering cross section per unit particle volume for normally incident light polarized parallel (——) and perpendicular (------) to the axis of an infinite cylinder in air.

per unit particle volume for incident light polarized parallel and perpendicular to the cylinder axis are plotted as functions of diameter. The cylinder is in air and is taken to be nonabsorbing with a refractive index of 1.55; this corresponds approximately to many silicates at visible wavelengths. The wavelength 0.6328 μm is that of the He–Ne laser. In Section 4.4 we showed that the extinction cross section of an object large compared with the wavelength is twice its geometrical cross section G. Therefore, the scattering cross section per unit volume of a nonabsorbing cylinder asymptotically approaches the limiting value

$$\frac{C_{sca}}{v} \sim \frac{2G}{v} = \frac{4}{\pi a} \qquad (8.41)$$

independently of the state of polarization of the incident light; the limit (8.41) is also plotted in Fig. 8.7. This figure deserves careful study, for there is much to be learned from it. If the size parameter is greater than about 5, there is essentially no difference between scattering of incident light polarized parallel or perpendicular to the cylinder axis. However, for smaller size parameters, there can be appreciable difference; the maximum occurs for $x \simeq 1$. Note also that the scattering cross section per unit particle volume has its greatest value for $x \simeq 2.5$, which corresponds to a particle diameter of about 0.5 μm; thus,

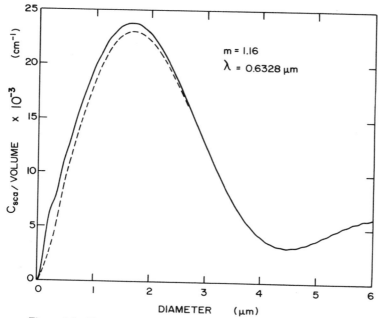

Figure 8.8 The same cylinder as in Fig. 8.7 but surrounded by water.

cylindrical particles of this size are the most effective at scattering light from the beam. The consequences of changing the surrounding medium from air to water are striking; this is shown in Fig. 8.8. Regardless of the particle diameter, the scattering cross section is only weakly dependent upon the polarization state of the incident light. The shapes of the two sets of curves are markedly different: when the cylinder is in water the ripples disappear and the diameter at which C_{sca}/v is a maximum shifts from 0.5 μm to about 1.1 μm.

8.4.5   Small-Particle Limit

If x and $|m|x$ are sufficiently small, then the scattering coefficients a_0 and b_0 are approximately

$$a_0 \simeq \frac{-i\pi x^4(m^2-1)}{32}, \qquad b_0 \simeq \frac{-i\pi x^2(m^2-1)}{4}. \qquad (8.42)$$

To obtain (8.42) we substituted the following expressions for the Bessel functions and their derivatives

$$J_0(z) \simeq 1 - \frac{z^2}{4}, \qquad J_0'(z) \simeq -\frac{z}{2} + \frac{z^3}{16},$$

$$\qquad\qquad\qquad\qquad\qquad\qquad\qquad |z| \ll 1$$

$$Y_0(z) \simeq \frac{2}{\pi}\log\!\left(\frac{z}{2}\right), \qquad Y_0'(z) \simeq \frac{2}{\pi z},$$

in (8.30) and (8.32) and retained the terms of smallest degree in x. Similarly, if we use

$$J_1(z) \simeq \frac{z}{2} - \frac{z^3}{16}, \qquad J_1'(z) \simeq \frac{1}{2} - \frac{3z^2}{16},$$

$$\qquad\qquad\qquad\qquad\qquad\qquad\qquad |z| \ll 1$$

$$Y_1(z) \simeq -\frac{2}{\pi z}, \qquad Y_1'(z) \simeq \frac{2}{\pi z^2},$$

we obtain the following approximations for a_1 and b_1:

$$a_1 \simeq \frac{-i\pi x^2}{4}\frac{m^2-1}{m^2+1}, \qquad b_1 \simeq \frac{-i\pi x^4(m^2-1)}{32}.$$

The amplitude scattering matrix elements correct to terms of order x^2 are

$$T_1 = b_0, \qquad T_2 = 2a_1\cos\Theta.$$

Thus, the degree of polarization of scattered light, given unpolarized incident light, is

$$P = \frac{|m^2 + 1|^2 - 4\cos^2\Theta}{|m^2 + 1|^2 + 4\cos^2\Theta}.$$

At $\Theta = 90°$, therefore, the scattered light is 100% polarized along the cylinder axis.

8.4.6 Anisotropic Cylinder

Scattering problems in which the particle is composed of an anisotropic material are generally intractable. One of the few exceptions to this generalization is a normally illuminated cylinder composed of a uniaxial material, where the cylinder axis coincides with the optic axis. That is, if the constitutive relation connecting $\mathbf{D}$ and $\mathbf{E}$ is

$$\begin{pmatrix} D_x \\ D_y \\ D_z \end{pmatrix} = \begin{pmatrix} \varepsilon_\perp & 0 & 0 \\ 0 & \varepsilon_\perp & 0 \\ 0 & 0 & \varepsilon_\| \end{pmatrix} \begin{pmatrix} E_x \\ E_y \\ E_z \end{pmatrix},$$

where the z axis is parallel to the cylinder axis, then the scattering problem has an exact solution. It is not difficult to show that the scattering coefficients a_n and b_n are of the form (8.32) and (8.30):

$$a_n = a_n(x, m_\perp), \qquad b_n = b_n(x, m_\|),$$

where $m_\perp$ and $m_\|$ are the complex refractive indices corresponding to the principal values $\varepsilon_\perp$ and $\varepsilon_\|$ of the permittivity tensor. Thus, if the incident light is polarized parallel to the cylinder axis, the cylinder scatters and absorbs light as if it were isotropic with permittivity $\varepsilon_\|$; on the other hand, incident light polarized perpendicular to the cylinder axis is oblivious to the permittivity along the axis and responds only to $\varepsilon_\perp$.

8.4.7 Diffraction Theory

Scattering by a cylinder of *finite* length cannot be treated exactly by constructing separable solutions to the scalar wave equation and expanding the various fields in the corresponding vector harmonics, a method that has to this point served us quite well. A finite cylinder has an edge, which is the bane of this scattering problem. However, by considering the diffraction theory approximation, we can acquire some insight into scattering by a finite cylinder. According to this theory, which was discussed in Section 4.4, an opaque cylinder, an opaque rectangular obstacle, and a rectangular aperture in an opaque screen, all with the same projected cross sectional area, scatter light in the same

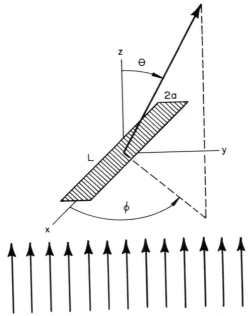

Figure 8.9   An opaque rectangular obstacle illuminated by a plane wave.

manner independently of the polarization of the incident light. Let us therefore consider scattering (diffraction) by the opaque obstacle shown in Fig. 8.9. The amplitude of the scattered wave is, from (4.70), given by

$$S(\theta, \phi) = \frac{k^2(1 + \cos \theta)}{4\pi} \int_{-a}^{a} \int_{-L/2}^{L/2} e^{-ik \sin \theta(\xi \cos \phi + \eta \sin \phi)} d\xi \, d\eta,$$

which can be integrated readily to yield

$$S(\theta, \phi) = \frac{(1 + \cos \theta)}{\pi} \frac{x \sin(x \sin \theta \sin \phi)}{x \sin \theta \sin \phi} \frac{xR \sin(xR \sin \theta \cos \phi)}{xR \sin \theta \cos \phi},$$

where R is the ratio of length to diameter ($L/2a$) and $x = ka$. The diffraction cross section is equal to the geometrical cross section G:

$$\int_0^{2\pi} \int_0^{\pi} \frac{|S(\theta, \phi)|^2}{k^2} \sin \theta \, d\theta \, d\phi = 2aL = 4a^2R.$$

The phase function $p(\theta, \phi) = |S(\theta, \phi)|^2/4x^2R$, which is the fraction of the total scattered light that is scattered into a unit solid angle about a given direction (θ, ϕ), has its greatest value in the forward direction:

$$p(0°) = \frac{x^2 R}{\pi^2} = \frac{k^2 aL}{2\pi^2}.$$

Therefore, the greater the dimensions of the particle (radius or length), the more the scattered light is concentrated about the forward direction.

It is instructive to consider how p varies with scattering angle θ for the two azimuthal angles $0°$ and $90°$. For scattering directions in a plane *perpendicular* to the cylinder axis the phase function $p(\theta, 90°)$ is $p_e(\theta, 90°)\sin^2(x \sin \theta)$, where the *envelope*

$$p_e(\theta, 90°) = \left(\frac{1 + \cos \theta}{2} \right)^2 \frac{p(0°)}{x^2 \sin^2 \theta}$$

is modulated by the rapidly oscillating function $\sin^2(x \sin \theta)$. Similarly, for directions in a plane *parallel* to the cylinder axis, we have $p(\theta, 0°) = p_e(\theta, 0°)\sin^2(xR \sin \theta)$, where

$$p_e(\theta, 0°) = \left(\frac{1 + \cos \theta}{2} \right)^2 \frac{p(0°)}{x^2 R^2 \sin^2 \theta}.$$

The ratio of the envelopes is

$$\frac{p_e(\theta, 90°)}{p_e(\theta, 0°)} = R^2. \tag{8.43}$$

Equation (8.43) provides us with an approximate criterion, subject to the limitations of diffraction theory, for when a finite cylinder may be regarded as effectively infinite: if $R > 10$, say, there will be comparatively little light scattered in directions other than those in a plane perpendicular to the cylinder axis. The greater is R, the more the scattered light is concentrated in this plane; in the limit of indefinitely large R, no light is scattered in directions other than in this plane. We may show this as follows. The phase function may be written in the form $p(\theta, \phi) = G(\theta, \phi)F(\theta, \phi)$, where

$$G(\theta, \phi) = \left[\frac{1 + \cos \theta}{\pi} \frac{x \sin(x \sin \theta \sin \phi)}{x \sin \theta \sin \phi} \right]^2,$$

$$F(\theta, \phi) = \frac{R}{4} \left[\frac{\sin(xR \sin \theta \cos \phi)}{xR \sin \theta \cos \phi} \right]^2.$$

For all azimuthal angles except $90°$ and $270°$

$$\lim_{R \to \infty} F(\theta, \phi) = 0 \qquad (\theta \neq 0).$$

However, the integrals

$$\int_0^\pi F(\theta, \phi)\, d\phi, \qquad \int_\pi^{2\pi} F(\theta, \phi)\, d\phi,$$

are nonzero for all R and have the limiting value $\pi/4x\sin\theta$ as R approaches infinity. Therefore,

$$\lim_{R\to\infty} F(\theta, \phi) = \frac{\pi}{4x\sin\theta}\left[\delta\left(\phi - \frac{\pi}{2}\right) + \delta\left(\phi - \frac{3\pi}{2}\right)\right],$$

where δ is the Dirac delta function. If we denote the phase function for indefinitely large R by $\bar{p}$, then

$$\int_0^{2\pi}\int_0^\pi \bar{p}(\theta, \phi)\sin\theta\, d\theta\, d\phi = \frac{\pi}{2x}\int_0^\pi\left[\frac{1 + \cos\theta}{\pi}\frac{x\sin(x\sin\theta)}{x\sin\theta}\right]^2 d\theta = 1.$$

We need consider only scattering directions in the plane $\phi = \pi/2$ (or $\phi = 3\pi/2$) because $\bar{p}$ vanishes outside this plane; we also have $\Theta = \theta$ when $\phi = \pi/2$ and $\Theta = -\theta$ when $\phi = 3\pi/2$, where $\Theta = 0$ is the forward direction. Thus, we may take the phase function for scattering by an infinite cylinder in the diffraction theory approximation to be

$$p(\Theta) = \frac{\pi}{4x}\left[\frac{1 + \cos\Theta}{\pi}\frac{x\sin(x\sin\Theta)}{x\sin\Theta}\right]^2, \qquad (8.44)$$

which is normalized:

$$\int_{-\pi}^\pi p(\Theta)\, d\Theta = 1.$$

The phase function (8.44) vanishes in the backward direction ($\Theta = \pi$) and at those angles for which $\sin\Theta = n\pi/x$, where n is an integer. This gives us a means for estimating the diameter of cylinders sufficiently large that diffraction theory is a good approximation: because $|\sin\Theta| \leqslant 1$, the diameter d is

$$d \simeq \lambda n_{\min}, \qquad (8.45)$$

where $n_{\min}$ is the number of minima in the phase function (8.44) between 0 and 90°. Diffraction theory should be at its best for a large opaque cylinder. An opaque particle is one that does not reflect incident light and that absorbs all transmitted light. This is an idealization: no such cylinder exists (i.e., there are no values of m and x such that these criteria are strictly satisfied). But we can conjure up an approximately opaque cylinder by taking the real part of its refractive index to be that of the surrounding medium; therefore, the reflectance at normal incidence is, from (2.58), $k^2/(4 + k^2)$. Thus, we want k to be as small as possible subject to the constraint that $kx > 1$. In Fig. 8.10 we show the phase function calculated from the exact theory for a cylinder with $x = 20$

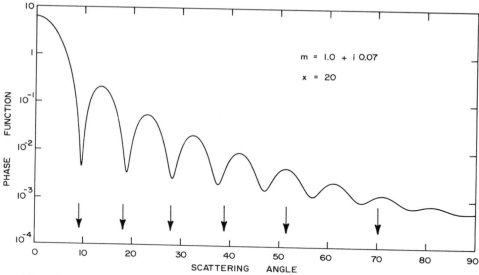

Figure 8.10 Phase function for scattering of unpolarized light by an infinite cylinder. The arrows indicate minima according to diffraction theory.

and $m = 1.0 + i0.07$; the incident light is unpolarized. Also shown are the minima in the diffraction theory phase function (8.44). The positions of the first few minima of the exact phase function are predicted quite well by diffraction theory. As the scattering angle increases, however, diffraction theory becomes an increasingly poorer approximation.

8.5 INHOMOGENEOUS PARTICLES: AVERAGE DIELECTRIC FUNCTION

Although inhomogeneous particles exist—in planetary atmospheres, for example, where particles of different composition are continually coagulating—little attention has been given to absorption and scattering by such particles except those that are regularly inhomogeneous—coated spheres, for example (Section 8.1). But we would not be tempted to recast the problem of scattering by a coated sphere in the form of scattering by an equivalent homogeneous sphere with an average, or effective, dielectric function obtained by combining somehow the dielectric functions of core and coating. For although our prescription might give good results for one size at a particular wavelength, it would not do so for other sizes and wavelengths; we would be faced with the task of continuously modifying the prescription to force the scattering and absorbing properties of the equivalent homogeneous particle into congruence with those of the inhomogeneous particle. Clearly, therefore, some kind of statistical irregularity on an appropriate scale is a necessary concomitant of a well-defined average dielectric function of an inhomogeneous medium.

The notion of homogeneity is not absolute: all substances are inhomogeneous upon sufficiently close inspection. Thus, the description of the interaction of an electromagnetic wave with *any* medium by means of a spatially uniform dielectric function is ultimately statistical, and its validity requires that the constituents—whatever their nature—be small compared with the wavelength. It is for this reason that the optical properties of media usually considered to be homogeneous—pure liquids, for example—are adequately described to first approximation by a dielectric function. There is no sharp distinction between such molecular media and those composed of small particles each of which contains sufficiently many molecules that they can be individually assigned a bulk dielectric function: we may consider the particles to be giant "molecules" with polarizabilities determined by their composition and shape.

It is not easy, however, to determine the average dielectric function of an inhomogeneous medium given the properties of its constituents: the interactions among them lead to problems which are insoluble except by approximate methods. Each type of approximation leads to a different dielectric function. As a consequence, there is a bewildering array of choices in the scientific literature, all of which are at least superficially different; moreover, there are apparently conflicting claims about the relative merits and ranges of validity of the various theories. Perhaps the Maxwell Garnett theory has the greatest following. In the following paragraphs we generalize the Maxwell Garnett theory, explore some of its properties, and discuss some experimental data on scattering by inhomogeneous particles.

We take as our model of an inhomogeneous medium a two-component mixture composed of *inclusions* embedded in an otherwise homogeneous *matrix*, where ϵ and ϵ_m are their respective dielectric functions. The inclusions are identical in composition but may be different in volume, shape, and orientation; we shall restrict ourselves, however, to ellipsoidal inclusions. The average electric field $\langle \mathbf{E} \rangle$ over a volume V surrounding the point x is defined as

$$\langle \mathbf{E}(\mathbf{x}) \rangle = \frac{1}{V} \int_V \mathbf{E}(\mathbf{x} + \boldsymbol{\xi})\, d\boldsymbol{\xi},$$

where V contains many inclusions but is otherwise arbitrary. Note that $\mathbf{E}$ itself has been obtained from a microscopic field by suitably averaging over many molecules (see, e.g., de Groot, 1969); thus, $\langle \mathbf{E} \rangle$ is an even less finely grained macroscopic field than $\mathbf{E}$. V is composed of the matrix volume and the volume of all the inclusions; therefore, we may write

$$\langle \mathbf{E}(\mathbf{x}) \rangle = (1 - f) \langle \mathbf{E}_m(\mathbf{x}) \rangle + f \sum_k w_k \langle \mathbf{E}_k(\mathbf{x}) \rangle,$$

$$\langle \mathbf{E}_m(\mathbf{x}) \rangle = \frac{1}{V_m} \int_{V_m} \mathbf{E}(\mathbf{x} + \boldsymbol{\xi})\, d\boldsymbol{\xi}, \qquad \langle \mathbf{E}_k(\mathbf{x}) \rangle = \frac{1}{v_k} \int_{v_k} \mathbf{E}(\mathbf{x} + \boldsymbol{\xi})\, d\boldsymbol{\xi},$$

where V_m is the matrix volume, v_k is the volume of the kth inclusion, f is the volume fraction of inclusions, w_k is f_k/f, and f_k is v_k/V. Similarly, the average polarization is given by

$$\langle \mathbf{P}(\mathbf{x}) \rangle = (1 - f)\langle \mathbf{P}_m(\mathbf{x}) \rangle + f\sum_k w_k \langle \mathbf{P}_k(\mathbf{x}) \rangle.$$

If we assume that constitutive relations of the form (2.23) are valid for the matrix and inclusions, it follows that

$$\langle \mathbf{P}_m(\mathbf{x}) \rangle = \varepsilon_0 \chi_m \langle \mathbf{E}_m(\mathbf{x}) \rangle, \qquad \langle \mathbf{P}_k(\mathbf{x}) \rangle = \varepsilon_0 \chi \langle \mathbf{E}_k(\mathbf{x}) \rangle,$$

where $\chi_m = \epsilon_m - 1$ is the susceptibility of the matrix and $\chi = \epsilon - 1$ is that of the inclusions. The average susceptibility tensor χ_{av} of the composite medium is *defined* by

$$\langle \mathbf{P}(\mathbf{x}) \rangle = \epsilon_0 \chi_{av} \cdot \langle \mathbf{E}(\mathbf{x}) \rangle,$$

where χ_{av} is independent of position if the medium is statistically homogeneous. The preceding equations can be combined to yield

$$(1 - f)(\epsilon_{av} - \epsilon_m \mathbf{1}) \cdot \langle \mathbf{E}_m(\mathbf{x}) \rangle + f\sum_k w_k(\epsilon_{av} - \epsilon \mathbf{1}) \cdot \langle \mathbf{E}_k(\mathbf{x}) \rangle = 0, \quad (8.46)$$

where ϵ_{av} is the average dielectric tensor and $\mathbf{1}$ is the unit tensor. Clearly, if ϵ_{av} is to be independent of position, then $\langle \mathbf{E}_m \rangle$ and $\langle \mathbf{E}_k \rangle$ must be linearly related.

Consider an *isolated* ellipsoid in a *uniform* field $\mathbf{E}_m$; the uniform field $\mathbf{E}_k$ in the ellipsoid is given by $\mathbf{E}_k = \lambda_k \cdot \mathbf{E}_m$, where the principal components of the tensor λ_k are (see Section 5.3)

$$\lambda_j = \frac{\epsilon_m}{\epsilon_m + L_j(\epsilon - \epsilon_m)} \qquad (j = 1, 2, 3).$$

With the *assumption*—and this is our major assumption—that the average fields are similarly related, that is, $\langle \mathbf{E}_k \rangle = \lambda_k \cdot \langle \mathbf{E}_m \rangle$, (8.46) becomes

$$(1 - f)(\epsilon_{av} - \epsilon_m \mathbf{1}) + f(\epsilon_{av} - \epsilon \mathbf{1}) \cdot \sum_k w_k \lambda_k = 0. \qquad (8.47)$$

So our task reduces to that of determining the sum in (8.47); λ_k depends on the shape and orientation of the kth ellipsoid and w_k is the ratio of its volume to that of all ellipsoids. It is convenient to approximate the sum in (8.47) by an integral

$$\sum_k w_k \lambda_k \simeq \int_k w(k) \lambda(k)\, dk,$$

where the continuous variable k in the integral represents all the variables that specify an ellipsoid: shape, volume, and orientation. We now make several assumptions: there is no correlation between the volume of an inclusion and its shape or orientation; there is no correlation between shape and orientation; all orientations are equally probable. It follows from these assumptions that

$$\sum_k w_k \lambda_k \simeq \beta \mathbf{1},$$

$$\beta = \int \int \mathscr{P}(L_1, L_2) \frac{\lambda_1 + \lambda_2 + \lambda_3}{3} dL_1 \, dL_2, \tag{8.48}$$

where $\mathscr{P}(L_1, L_2)$ is the shape probability distribution function (see Section 12.2); the average (scalar) dielectric function is therefore

$$\epsilon_{av} = \frac{(1 - f)\epsilon_m + f\beta\epsilon}{1 - f + f\beta}. \tag{8.49}$$

Note that (8.49), which is a generalization of the Maxwell Garnett dielectric function (8.50), is not invariant with respect to interchanging the roles of matrix and inclusions: if we make the substitutions $\epsilon \to \epsilon_m$, $\epsilon_m \to \epsilon$, and $f \to (1 - f)$, then ϵ_{av} is not, in general, unchanged. If, therefore, a two-component mixture is to be described by (8.49), a choice must be made as to which component is the matrix and which the inclusions (there may be physical reasons to guide this choice). The limiting values of ϵ_{av} are independent of β:

$$\lim_{f \to 1} \epsilon_{av} = \epsilon, \qquad \lim_{f \to 0} \epsilon_{av} = \epsilon_m.$$

It is not difficult to extend (8.49) or (8.47) to multicomponent mixtures. If we make the same assumptions for each inclusion that were made preceding (8.48), then the average dielectric function is

$$\epsilon_{av} = \frac{(1 - f)\epsilon_m + \sum f_j \beta_j \epsilon_j}{1 - f + \sum f_j \beta_j},$$

where the volume fraction of the jth inclusion with dielectric function ϵ_j is f_j, $f = \sum f_j$, and β_j has the same form as (8.48) with ϵ replaced by ϵ_j.

It is instructive to expand (8.49) in powers of $\delta = \Delta/\epsilon_m$, where $\Delta = \epsilon - \epsilon_m$:

$$\epsilon_{av} = \epsilon_m \left[1 + f\delta - \frac{f(1 - f)}{3}\delta^2 + \frac{f(1 - f)(3\langle L^2 \rangle - f)}{9}\delta^3 + \cdots \right],$$

$$\langle L^2 \rangle = \int \int \mathscr{P}(L_1, L_2)(L_1^2 + L_2^2 + L_3^2) \, dL_1 \, dL_2.$$

To terms of order δ^2 the average dielectric function is independent of the shape of the inclusions. Also, to terms of order δ (8.49) is symmetric in matrix and inclusions. Bohren and Battan (1980) similarly expanded various average dielectric functions—Maxwell Garnett, Bruggeman (often called effective medium), Debye—and showed that they all agree to at least terms of order δ. Therefore, if we are dealing with mixtures the components of which are similar ($|\delta| \ll 1$), we need not worry unduly about which component to designate as the inclusions nor their shape. Neither need we agonize over which among the various competing expressions for the average dielectric function is best: any of them will do. A more stringent test of their relative merits would be how well they do for mixtures of very dissimilar substances.

If all the inclusions are spherical, $\beta = 3\epsilon_m/(\epsilon + 2\epsilon_m)$ and (8.49) reduces to

$$\epsilon_{av} = \epsilon_m \left[1 + \frac{3f\left(\dfrac{\epsilon - \epsilon_m}{\epsilon + 2\epsilon_m} \right)}{1 - f\left(\dfrac{\epsilon - \epsilon_m}{\epsilon + 2\epsilon_m} \right)} \right]. \tag{8.50}$$

The average dielectric function (8.50) was first derived by Maxwell Garnett (1904); subsequently, it has been rederived under various sets of assumptions (see, for example, Genzel and Martin, 1972, 1973; Barker, 1973; Bohren and Wickramasinghe, 1977).

The following expression for an average dielectric function was first obtained by Bruggeman (1935):

$$f\frac{\epsilon - \epsilon_{av}}{\epsilon + 2\epsilon_{av}} + (1 - f)\frac{\epsilon_m - \epsilon_{av}}{\epsilon_m + 2\epsilon_{av}} = 0. \tag{8.51}$$

Both the Maxwell Garnett and Bruggeman dielectric functions have been shown by Stroud (1975) to follow from the same integral equation: either one or the other is obtained depending on the approximations that are made. So we may consider (8.50) and (8.51) to be related: they issue from the same parent equation. Note that the Bruggeman dielectric function applies to a two-component mixture in which there are no distinguishable inclusions embedded in a definite matrix: both components are treated symmetrically. It might be more correct to say that it applies to a completely randomly inhomogeneous medium; it does not strictly apply to a particulate medium because there is no way to decide which component is the particles and which the surrounding medium. This difference between the Maxwell Garnett and Bruggeman theories shows up in comparisons with experimental data. Abeles and Gittleman (1976) measured transmission by the composite metal–insulator system Ag–SiO_2 and the composite semiconductor–insulator systems Si–SiC and Ge–Al_2O_3 and compared their results with calculations based on the two average dielectric functions. They concluded that only the Maxwell Garnett theory was compatible with their observations. In particular, the Ag–SiO$_2$ transmission spectrum showed a band which the Bruggeman theory completely failed to predict. Although this band was not explicitly identified as a manifestation of surface mode excitation in silver particles, it is likely that this is so: we shall show in Section 12.4 that small silver particles can have strong shape-dependent absorption bands at visible and near-infrared wavelengths (see Fig. 12.18). As these bands depend very much on the particulate nature of the composite medium, it comes as no great surprise that the Bruggeman theory is inadequate in this instance. Landauer (1952), however, found good agreement between conductivities calculated from the Bruggeman theory and measured values, even for mixtures with components having greatly different conductivities. So it is certainly not true that the Maxwell Garnett theory is necessarily superior to the Bruggeman theory in every instance: both theories have their successes. Part of the ambiguity surely arises from the difference between a randomly inhomogeneous medium and one in which there are distinguishable inclusions embedded in a definite matrix.

Bohren and Battan (1980) tested the applicability of (8.50) and other average dielectric functions to inhomogeneous particles by calculating refractive indices of ice–water mixtures at a wavelength of 5.05 cm—the microwave dielectric functions of ice and liquid water are vastly different (Section 10.3). Radar backscattering calculations for ice spheres coated with ice–water mixtures were compared with measured cross sections. The evidence favoring the Maxwell Garnett theory over the Bruggeman theory was not compelling: both agreed reasonably well with measurements. Much more data needs to be gathered and analyzed, however, before the problem of scattering and absorption by inhomogeneous particles can be considered to have been "solved" (if, indeed, a general solution is possible). In the interim, we would be inclined to use (8.49) or (8.50) in small-particle calculations.

Measurements are sometimes made on inhomogeneous samples for which optical constants are inferred under the assumption of homogeneity and then used in all kinds of small-particle scattering and absorption calculations. For example, measured reflectances of inhomogeneous samples can be inverted to obtain optical constants from expressions which strictly apply only to homogeneous media (see Section 2.7). Now one cannot deny that this might give some sort of average optical constants. But they are certain to be applicable only to the exact experiment from which they were extracted. It would not generally be correct, for example, to use them in Mie calculations of scattering from a more or less spherical aggregation of inhomogeneous material.

Another kind of effective or average optical constants involves mixtures of different particles such as atmospheric aerosols or soils. Effective optical constants for compacted samples of these mixtures might be inferred from reflectance and transmittance measurements as if the samples were homogeneous. But scattering or extinction calculations based on these optical constants would not necessarily be correct.

An example of practical importance in atmospheric physics is the inference of effective optical constants for atmospheric aerosols composed of various kinds of particles and the subsequent use of these optical constants in other ways. One might infer effective n and k from measurements—made either in the laboratory or remotely by, for example, using bistatic lidar—of angular scattering; fitting the experimental data with Mie theory would give "effective" optical constants. But how *effectual* would they be? Would they have more than a limited applicability? Would they be more than merely consistent with an experiment of limited scope? It is by no means certain that they would lead to correct calculations of extinction; or backscattering; or absorption. We shall return to these questions in Section 14.2.

8.6 A SURVEY OF NONSPHERICAL PARTICLES, REGULAR AND IRREGULAR

Because homogeneous, spherical particles are the exception in nature rather than the rule, various exact and approximate methods have been devised for determining absorption and scattering by nonspherical particles, both those with regular shape (i.e., with boundaries specified by simple, smooth functions) and with irregular shape; some of these have already been discussed. Rayleigh theory and geometrical optics combined with diffraction theory, which are conceptually simple, apply to particles small and large compared with the wavelength, respectively. It is between these two limits where the major difficulties lie and to which most recent theoretical effort has been directed. Although it might be thought that all scattering problems are tractable by simple numerical methods, computational time can be excessive. Therefore, accurate and efficient methods for computing scattering by nonspherical particles are actively being developed.

The following is only a brief discussion of a few selected methods, their physical basis, advantages, and disadvantages. Further details may be found in the references cited. Yeh and Mei (1980) have given a succinct overview of some of these methods.

8.6.1 Separation of Variables

The classical method of solving scattering problems, separation of variables, has been applied previously in this book to a homogeneous sphere, a coated sphere (a simple example of an inhomogeneous particle), and an infinite right circular cylinder. It is applicable to particles with boundaries coinciding with coordinate surfaces of coordinate systems in which the wave equation is separable. By this method Asano and Yamamoto (1975) obtained an exact solution to the problem of scattering by an arbitrary spheroid (prolate or oblate) and numerical results have been obtained for spheroids of various shape, orientation, and refractive index (Asano, 1979; Asano and Sato, 1980).

Although this solution is exact it shares the drawbacks of other solutions for nonspherical particles: computations can be complicated and lengthy, particularly for larger particles and for averages over all orientations, not to mention size and shape distributions. Because of this, perhaps, the spheroid solution has been slow to be adopted by other workers. Nevertheless, the results of Asano and Sato (1980) give some of the best insights available into the systematic effects of nonsphericity in light scattering. Several examples from the work of Asano and his collaborators are included in Chapters 11 and 13.

8.6.2 Point Matching

In the point matching method (Oguchi, 1973; Bates, 1975) the fields inside and outside a particle are expanded in vector spherical harmonics and the resulting series truncated; the tangential field components are required to be continuous at a finite number of points on the particle boundary. Although easy to describe and to understand, the practical usefulness of this method is limited to nearly spherical particles; large demands on computer time and uncertain convergence are also drawbacks (Yeh and Mei, 1980).

8.6.3 Perturbation Methods

A nonspherical particle may be looked upon as a sphere the boundary of which is distorted, or "perturbed," by different amounts at different points. The field scattered by such a particle is, formally at least, given by an infinite series in a perturbation parameter the first term of which is the Mie solution. Perturbation solutions to the problem of scattering by nonspherical particles have been advanced by Yeh (1964) and by Erma (1968ab, 1969). The former author developed his theory only to first order in the perturbation parameter. The latter author claims an exact solution valid for all particles for which the series converge, without, however, specifying the conditions for convergence or giving numerical results so that the practicality of his method may be assessed. An infinite series solution that is formally exact is of little practical use unless it can be truncated after a reasonable number of terms without appreciable error. It seems likely that the practical usefulness of both these perturbation methods is limited to nearly spherical particles. This is a feature common to all perturbation methods: the further the system of interest deviates from the unperturbed system, the more terms are required and the more cumbersome and time consuming the computations become.

8.6.4 Purcell–Pennypacker Method

A solution to the problem of scattering by an arbitrary particle could be obtained in principle by computing the dipole moment induced in each of its constituent atoms by the incident field and the combined fields of all the other atoms. But even a small particle (say $\sim 1~\mu$m) contains more than 10^{10} atoms, which means that a comparable number of equations would have to be solved iteratively, clearly a formidable task. Purcell and Pennypacker (1973) adopted the basic methodology of this approach while reducing it to more manageable proportions.

A particle is subdivided into a small number of identical elements, perhaps 100 or more, each of which contains many atoms but is still sufficiently small to be represented as a dipole oscillator. These elements are arranged on a cubic lattice and their polarizability is such that when inserted into the Clausius–Mossotti relation the bulk dielectric function of the particle material is obtained. The vector amplitude of the field scattered by each dipole oscillator, driven by the incident field and that of all the other oscillators, is determined iteratively. The total scattered field, from which cross sections and scattering diagrams can be calculated, is the sum of all these dipolar fields.

Purcell and Pennypacker calculated scattering and absorption by various rectangular particles with different refractive indices for size parameters up to about two. More recently, Kattawar and Humphreys (1980) used the Purcell–Pennypacker method to investigate scattering by two spheres as a function of separation. Arbitrary shapes can be treated by this method but excessive computation time appears to preclude its use for large size parameters.

8.6.5   T-Matrix Method

A promising method based on an integral equation formulation of the problem of scattering by an arbitrary particle has come into prominence in recent years. It was developed by Waterman, first for a perfect conductor (1965), later for a particle with less restricted optical properties (1971). More recently it has been applied to various scattering problems under the name Extended Boundary Condition Method, although we shall follow Waterman's preference for the designation T-matrix method. Barber and Yeh (1975) have given an alternative derivation of this method.

In Chapter 4 a plane wave incident on a sphere was expanded in an infinite series of vector spherical harmonics as were the scattered and internal fields. Such expansions, however, are possible for arbitrary particles and incident fields. It is the scattered field that is of primary interest because from it various observable quantities can be obtained. Linearity of the Maxwell equations and the boundary conditions (3.7) implies that the coefficients of the scattered field are linearly related to those of the incident field. The linear transformation connecting these two sets of coefficients is called the T (for transition) matrix. If the particle is spherical, then the T matrix is diagonal.

Explicit expressions for the T matrix can be obtained by casting the scattering problem in integral rather than differential form. Details are given in the references cited above. We also recommend an expository article by Ström (1975).

Scattering by homogeneous spheroids and finite cylinders has been investigated by Barber and Yeh (1975) using the T-matrix method. Single coated prolate spheroids (Wang and Barber, 1979) as well as polydispersions of such particles (Wang et al., 1979) have also been treated. Because numerical results for particles of arbitrary shape and large volume can be obtained efficiently (Yeh and Mei, 1980) this method is actively being used at present.

8.6.6 The Need for a Statistical Approach

Many calculations for nonspherical particles have been, and are being, done with the methods outlined above. Although they may be quite suitable for single particles, ensembles of particles distributed in size, shape, and orientation are often of more interest. Of course, the properties of such ensembles follow from those of their individual members—but not trivially. With increasing dispersion, calculations can quickly escalate to unmanageable proportions, although heroic efforts have been made for randomly oriented, homogeneous (Asano and Sato, 1980) and inhomogeneous (Wang et al., 1979) spheroids.

We may imagine that an irregular particle is generated from a spheroid by chipping it at random. A spheroid requires four parameters for its characterization: two for size and shape, two for orientation. Each time a chip is made additional parameters are required. Even if an exact solution to the problem of scattering by such a chipped spheroid were obtainable, would it be of great value, particularly if we were interested only in the average properties of an ensemble of similar particles?

The effect of averaging over one or more particle parameters—size, shape, orientation—is to efface details: extinction fine structure, particularly ripple structure, to a lesser extent interference structure (Chapter 11); and undulations in scattering diagrams. If the details disappear upon averaging over an ensemble perhaps the best strategy in this instance would be to avoid the details of individual-particle scattering altogether and reformulate the problem statistically.

Further progress in the theory of light scattering by irregular particles may come from a statistical formulation of this problem. A small step in this direction is taken in Chapter 12, where extinction spectra of small irregular particles are approximated by averages over Rayleigh ellipsoids. The modest success of this approach suggests that it might profitably be extended to larger particles.

NOTES AND COMMENTS

The solution to the problem of scattering by a sphere described by the constitutive relations (8.5) was obtained by Bohren (1974). This was then extended to optically active spherical shells (Bohren, 1975) and cylinders (Bohren, 1978).

A good discussion of the dielectric properties of a mixture, independent of a specific model, has been given by Landau and Lifshitz (1960); we also recommend a paper by Niklasson et al. (1981) for its discussion of several aspects of the mixture problem.

An approach somewhat similar to that in Section 8.5 was taken by O'Neill and Ignatiev (1978), who obtained an expression for the average dielectric function of a mixture containing spheroidal inclusions with ratios of semiminor to semimajor axes given by a probability distribution function.

Further work on the theory of backscattering by inhomogeneous particles and its comparison with measurements has been done by Bohren and Battan (1981).

Banderman and Kemp (1973) have treated scattering by arbitrarily shaped particles in a manner similar to that of Purcell and Pennypacker (1973).

Some of the most recent work on scattering by irregularly shaped particles is contained in a collection of papers edited by Schuerman (1980).

Part 2
Optical Properties of Bulk Matter

Chapter 9

Classical Theories
of Optical Constants

A new silver coin or a polished aluminum cooking pan is shiny but opaque to transmitted light; pure water or a piece of window glass is transparent but weakly reflecting; the light transmitted by ruby gemstones is red. Such commonly observed phenomena are consequences of the many different ways in which objects reflect and transmit visible light; but these properties are determined by the more fundamental optical constants of the materials of which the objects are composed. Relations between optical constants and reflection and transmission at plane interfaces were derived in Chapter 2. In this chapter we show how the variation of optical constants can be understood by appealing to simple models of the microscopic structure of matter.

There are two sets of quantities that are often used to describe optical properties: the real and imaginary parts of the *complex refractive index* $N = n + ik$ and the real and imaginary parts of the *complex dielectric function* (or relative permittivity) $\epsilon = \epsilon' + i\epsilon''$. These two sets of quantities are not independent; either may be thought of as describing the intrinsic optical properties of matter. The relations between the two are, from (2.47) and (2.48),

$$\epsilon' = \frac{\varepsilon'}{\varepsilon_0} = n^2 - k^2,$$

$$\epsilon'' = \frac{\varepsilon''}{\varepsilon_0} = 2nk, \tag{9.1}$$

$$n = \sqrt{\frac{\sqrt{\epsilon'^2 + \epsilon''^2} + \epsilon'}{2}},$$

$$k = \sqrt{\frac{\sqrt{\epsilon'^2 + \epsilon''^2} - \epsilon'}{2}}, \tag{9.2}$$

where we have assumed that the material is nonmagnetic ($\mu = \mu_0$).

In Part 2 we shall usually give both sets of optical constants, (n, k) and (ϵ', ϵ''), side by side. For some purposes one set is preferable, and for other purposes the other set is preferred. For example, most people have a better intuitive understanding of n and k because they are related to the phase velocity and attenuation of plane waves in matter. In considerations of wave propagation, therefore, n and k are preferred. However, in considering the microscopic mechanisms that are responsible for optical effects, ϵ' and ϵ'' are more appropriate. Reflection and transmission by slabs and plane interfaces are described more simply with n and k, while equations for absorption and scattering by particles small compared with the wavelength are more simply written using ϵ' and ϵ''. Therefore, at the slight risk of undue repetition, we shall display both sets of optical constants.

9.1   THE LORENTZ MODEL

Near the beginning of this century H. A. Lorentz developed a classical theory of optical properties in which the electrons and ions of matter were treated as simple harmonic oscillators (i.e., "springs") subject to the driving force of applied electromagnetic fields. The results obtained therefrom are formally identical to those of quantum-mechanical treatments, although various quantities are interpreted differently in the classical and quantum-mechanical theories. As most of us are more comfortable thinking in classical terms, it is fortunate that we may do so without doing violence to the correct results. This undoubtedly explains why the Lorentz model remains so useful, not only in guiding our intuition, but also in quantitatively analyzing experimental data.

Following Lorentz, we take as our microscopic model of polarizable matter a collection of identical, independent, isotropic harmonic oscillators (Fig. 9.1). We shall later generalize to more than one kind of oscillator and to anisotropic oscillators. An oscillator with mass m and charge e is acted upon by: a linear restoring force $K\mathbf{x}$, where K is the spring constant (stiffness) and $\mathbf{x}$ is the displacement from equilibrium; a damping force $b\dot{\mathbf{x}}$, where b is the damping constant; and a driving force produced by the *local* electric field $\mathbf{E}_{local}$ (magnetic forces may usually be neglected compared with electrical forces). We neglect radiation reaction. The equation of motion of such an oscillator is

$$m\ddot{\mathbf{x}} + b\dot{\mathbf{x}} + K\mathbf{x} = e\mathbf{E}_{local}. \tag{9.3}$$

The *local field* $\mathbf{E}_{local}$ "seen" by a single oscillator and the *macroscopic field* $\mathbf{E}$, which is an average over a region containing many oscillators, are different in general. However, we shall ignore this difference because it does not affect our simple model of optical constants, and proper treatment of the local field would be a fruitless digression at this point. A good elementary discussion of the local field is given in Kittel (1971, pp. 454–458).

The electric field is taken to be time harmonic with frequency ω. As in previous chapters, we shall deal with the complex representations of the real

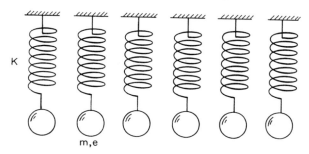

Figure 9.1 The Lorentz model of matter.

time-harmonic quantities (see Section 2.2). The solution to (9.3) is composed of a transient part, which dies away because of damping, and an oscillatory part with the same frequency as the driving field. We shall be interested only in the oscillatory part

$$\mathbf{x} = \frac{(e/m)\mathbf{E}}{\omega_0^2 - \omega^2 - i\gamma\omega},$$ (9.4)

where $\omega_0^2 = K/m$ and $\gamma = b/m$. If $\gamma \neq 0$, the proportionality factor between $\mathbf{x}$ and $\mathbf{E}$ is complex; therefore, the displacement and field are not, in general, in phase. In order to discuss the consequences of this phase difference, we write the displacement as $Ae^{i\Theta}(e\mathbf{E}/m)$, where the amplitude A and phase angle Θ are

$$A = \frac{1}{\left[\left(\omega_0^2 - \omega^2 \right)^2 + \gamma^2\omega^2 \right]^{1/2}},$$

$$\Theta = \tan^{-1} \frac{\gamma\omega}{\omega_0^2 - \omega^2}.$$ (9.5)

These are shown as functions of frequency in Fig. 9.2a. Note that the amplitude is a maximum at $\omega \simeq \omega_0$, where the height of this maximum is inversely proportional to γ, and the width at half-maximum is proportional to γ (provided that $\gamma \ll \omega_0$). At low frequencies ($\omega \ll \omega_0$) the oscillator responds in phase with the driving force ($\Theta \simeq 0°$), whereas at high frequencies ($\omega \gg \omega_0$) the two are 180° out of phase; the 180° phase change occurs in the vicinity of the resonant frequency ω_0. The behavior of the phase will be helpful in understanding the speed of light in matter.

Given the response of a single oscillator to a time-harmonic electric field, the optical constants appropriate to a collection of such oscillators readily follow. The induced dipole moment $\mathbf{p}$ of an oscillator is $e\mathbf{x}$. If $\mathfrak{N}$ is the number of oscillators per unit volume, the polarization $\mathbf{P}$ (dipole moment per unit

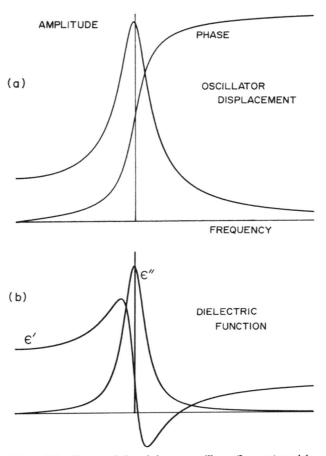

Figure 9.2   Characteristics of the one-oscillator (Lorentz) model.

volume) is $\mathfrak{N}\mathbf{p} = \mathfrak{N}e\mathbf{x}$, and from (9.4) we have

$$\mathbf{P} = \frac{\omega_p^2}{\omega_0^2 - \omega^2 - i\gamma\omega}\varepsilon_0\mathbf{E}, \tag{9.6}$$

where the *plasma frequency* is defined by $\omega_p^2 = \mathfrak{N}e^2/m\varepsilon_0$. Equation (9.6) is a particular example of the constitutive relation (2.9). Therefore, the dielectric function for our system of simple harmonic oscillators is

$$\epsilon = 1 + \chi = 1 + \frac{\omega_p^2}{\omega_0^2 - \omega^2 - i\gamma\omega}, \tag{9.7}$$

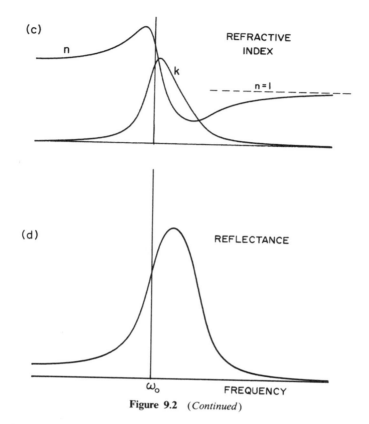

Figure 9.2 (*Continued*)

with real and imaginary parts

$$\epsilon' = 1 + \chi' = 1 + \frac{\omega_p^2 \left(\omega_0^2 - \omega^2 \right)}{\left(\omega_0^2 - \omega^2 \right)^2 + \gamma^2 \omega^2}, \tag{9.8}$$

$$\epsilon'' = \chi'' = \frac{\omega_p^2 \gamma \omega}{\left(\omega_0^2 - \omega^2 \right)^2 + \gamma^2 \omega^2}. \tag{9.9}$$

The proof is lengthy, although straightforward, and will be omitted here, but it can be shown by direct substitution and integration that χ' and χ'' satisfy the Kramers–Kronig relations (2.36) and (2.37).

The frequency dependence of the real and imaginary parts of the dielectric function (9.7) are shown schematically in Fig. 9.2b; following this are the corresponding real and imaginary parts of the refractive index (Fig. 9.2c). The reflectance at normal incidence (2.58) is also of interest and is shown in Fig. 9.2d. Many features of the optical properties of materials are illustrated by these characteristics of the simple oscillator model because such an oscillator can describe many different kinds of optical excitations. Several examples of actual data will be compared with the results of the oscillator model later in the chapter. First, however, there are some general oscillator characteristics that should be noted:

The region of high absorption at frequencies around ω_0 gives rise to an associated region of high reflectance provided, of course, that the oscillator parameters are such that $k \gg 1$ in this region. The high reflectance allows little light to get past the bounding surface of the material, and that which does is rapidly attenuated.

On both sides of the resonance region n increases with increasing frequency, which is called *normal dispersion*. Only in the immediate vicinity of the resonance frequency does n decrease with frequency, so-called *anomalous dispersion*. Such a reversal of dispersion, if it occurred in transparent regions, would provide a much-needed material for designing color-corrected lenses. Unfortunately, anomalous dispersion occurs only in regions of high absorption where no appreciable light is transmitted.

The maximum value of ϵ'' occurs approximately at ω_0 provided that $\gamma \ll \omega_0$. For frequencies in the neighborhood of ω_0 the dielectric functions (9.8) and (9.9) may be approximated by

$$\epsilon' \simeq 1 + \frac{\omega_p^2 (\omega_0 - \omega)/2\omega_0}{(\omega_0 - \omega)^2 + (\gamma/2)^2}, \qquad (9.10)$$

$$\epsilon'' \simeq \frac{\gamma \omega_p^2 / 4\omega_0}{(\omega_0 - \omega)^2 + (\gamma/2)^2}. \qquad (9.11)$$

It is obvious from (9.11) that the maximum value of ϵ'' is approximately $\omega_p^2/\gamma\omega_0$ and the width of the bell-shaped curve $\epsilon''(\omega)$ is γ (i.e., ϵ'' falls to one-half its maximum value when $\omega_0 - \omega = \pm\gamma/2$). If we set the derivative of (9.10) with respect to the variable $\omega_0 - \omega$ equal to zero, it follows that the extreme values of ϵ', $\epsilon'_{max} = 1 + \epsilon''_{max}/2$ and $\epsilon'_{min} = 1 - \epsilon''_{max}/2$, occur at $\omega = \omega_0 - \gamma/2$ and $\omega = \omega_0 + \gamma/2$, respectively. These properties of a narrow Lorentzian line are very helpful in quickly visualizing the shape of $\epsilon(\omega)$ from the parameters of (9.7).

9.1.1 Comparison of Classical and Quantum-Mechanical Concepts

The models of optical properties in this book are strictly classical. However, modern theoretical work aimed at understanding in detail the microphysics of optical properties is mostly quantum mechanical. Therefore, in this section we briefly discuss a few relevant quantum-mechanical concepts and also show that there is an analogy between the classical and quantum-mechanical descriptions of optical properties.

Most readers are no doubt comfortable with the alternative descriptions of light as either a stream of *photons*—quanta of electromagnetic energy—or electromagnetic *waves*. The relation between the two is $\mathscr{E} = \hbar\omega$, where ω is the angular frequency of the electromagnetic wave, $\mathscr{E}$ is the energy of the associated photon, and $\hbar$ is Planck's constant divided by 2π. In discussing the interaction of photons with matter, we must acknowledge that they do so in a discrete manner: a photon absorbed in matter gives up its energy and momentum to produce quantum *excitations*. Each excitation has its own quantized energy and momentum. A large part of the research effort in solid-state physics is involved with investigating the spectrum of such elementary excitations.

One important example is the *phonon*, a quantum of lattice vibration. An incident photon may give up its energy in a solid and create a phonon in the process. A less likely outcome is the production of two or more phonons within the bounds imposed by energy and momentum conservation. It is also possible for a photon to interact in a solid by exciting a phonon of lesser energy, the energy difference being carried off by another photon; such processes, called *inelastic*, include Raman and Brillouin scattering and are beyond the scope of this book. The following is a partial list of quantum excitations; we shall discuss some of these later.

Phonon   quantized lattice wave

Plasmon   quantized plasma (charge density) wave

Magnon   quantized magnetic spin wave

Exciton   quantum of electronic excitation consisting of an electron–hole pair

Polariton   quantum of coupled electromagnetic wave and another excitation such as a phonon or plasmon

The quantum-mechanical expression for the dielectric function may be written in the form (see, e.g., Ziman, 1972, Chap. 8, for an elementary discussion of the quantum theory of optical properties)

$$\epsilon(\omega) = 1 + \sum_j \frac{(\mathfrak{N}e^2/m\varepsilon_0)f_{ij}}{\omega_{ij}^2 - \omega^2 - i\gamma_j\omega}. \tag{9.12}$$

Note that if $j = 1$, (9.12) is *formally* identical with the classical expression (9.7); the classical multiple oscillator model, which will be discussed in Section 9.2, is even more closely analogous to (9.12). However, the *interpretations* of the terms in the quantum and classical expressions are quite different. Classically, ω_0 is the resonance frequency of the simple harmonic oscillator; quantum mechanically ω_{ij} is the energy difference (divided by $\hbar$) between the initial or ground state i and excited state j. Classically, γ is a damping factor such as that caused by drag on an object moving in a viscous fluid; quantum mechanically, γ_j

relates to the probabilities of transition to all other quantum states. The *oscillator strengths* f_{ij} represent the probability of an excitation from state i to state j; they are calculated through the matrix elements of the dipole moment operator. Despite considerable underlying conceptual differences, the formal similarities between the classical and quantum-mechanical theories allow us to think classically and not be far wrong.

9.1.2   High- and Low-Frequency Limiting Behavior

For frequencies much *greater* than the resonance frequency, the real and imaginary parts of the dielectric function (9.7) are approximately

$$\epsilon' \simeq 1 - \frac{\omega_p^2}{\omega^2},$$
$$\epsilon'' \simeq \frac{\gamma \omega_p^2}{\omega^3}. \qquad \omega \gg \omega_0 \qquad (9.13)$$

The real part of the dielectric function approaches unity from below, while the imaginary part tends to zero as the inverse third power of the frequency. The corresponding components of the refractive index are

$$n \simeq \sqrt{\epsilon'} \simeq 1 - \frac{\omega_p^2}{2\omega^2},$$
$$k \simeq \frac{\epsilon''}{2} \simeq \frac{\gamma \omega_p^2}{2\omega^3}, \qquad \omega \gg \omega_0 \qquad (9.14)$$

and the reflectance at normal incidence is, from (9.14) and (2.58),

$$R \simeq \left(\frac{\omega_p}{2\omega} \right)^4 \qquad \omega \gg \omega_0. \qquad (9.15)$$

The $1/\omega^4$ high frequency limit for R can be useful in determining optical constants from Kramers–Kronig analysis of reflectance data (see Section 2.7). Reflectances at frequencies higher than the greatest far-ultraviolet frequency for which measurements are made can be calculated from (9.15) and used to complete the Kramers–Kronig integral to infinite frequency.

For frequencies much *less* than the resonance frequency, the real and imaginary parts of (9.7) are

$$\epsilon' \simeq 1 + \frac{\omega_p^2}{\omega_0^2},$$
$$\epsilon'' \simeq \frac{\gamma \omega_p^2 \omega}{\omega_0^4} \qquad \omega \ll \omega_0. \qquad (9.16)$$

The imaginary part tends to zero as ω, and the real part approaches a constant that depends on the number density of oscillators and their mass.

Notions of "high"- and "low"-frequency limiting behavior depend on one's point of view, and the notation reflects this: what is low frequency to an ultraviolet spectroscopist may be high frequency to an infrared spectroscopist. For insulating solids the value of ϵ' in the near infrared is often denoted as ϵ_0 by ultraviolet spectroscopists; it refers to frequencies low compared with certain oscillators—electrons in this example—which may, however, be high compared with lattice vibrational frequencies. Consequently, this same limiting value is denoted as ϵ_∞ by infrared workers.

The low-frequency limit of ϵ'' (9.16) correctly describes the far-infrared ($1/\lambda$ less than about 100 cm^{-1}) behavior of many crystalline solids because their strong vibrational absorption bands are at higher frequencies. This limiting value for the bulk absorption, coupled with the absorption efficiency in the Rayleigh limit (Section 5.1), gives an ω^2 dependence for absorption by small particles; this is expected to be valid for many particles at far-infrared wavelengths.

9.1.3 Index of Refraction Less Than 1

When encountered for the first time, a refractive index n less than 1, such as at frequencies greater than ω_0 in Fig. 9.2c, often causes great anguish—sometimes followed by heated debate. For one of the consequences of special relativity is that speeds greater than c, the free-space speed of light, are not attainable; yet the phase velocity of a plane harmonic wave is c/n (Section 2.6). In fact, this was used as a stick—to no effect—with which to beat the theory of relativity during its infancy. In elementary texts this problem is sometimes disposed of by invoking the *group velocity* $c/[d(\omega n)/d\omega]$ and stating that it is this velocity which cannot exceed c. This may be so in regions of normal dispersion, but in regions of anomalous dispersion the group velocity can exceed c and, indeed, can be negative. So invoking the group velocity is only likely to compound the anguish.

This apparent clash with special relativity arises from the mistaken assumption that all quantities with dimensions of velocity are constrained to be less than or equal to c. Special relativity only places an upper bound on the velocity of *material bodies*, and we can extend this to mean *signals* as well. The phase velocity of a plane harmonic wave is the velocity of neither a material body nor a signal; an observer equipped with a detector and situated in the medium does not measure c/n as the speed at which the troughs and crests of the wave go by him. The plane wave originates from a source which has been turned on for a sufficient period of time to allow the wave to be established throughout the medium. The only way to measure the velocity of a signal is to turn off the source, note the distance between source and detector, turn on the source, and wait for a response. The distance to the source divided by the time elapsed between turning on the source and noting the first response of the detector cannot be greater than c. To show this we must consider the propagation of a pulse, which can be decomposed into its Fourier components (Section 2.3), each of which propagates with a different phase velocity—some greater than and some less than c. As a consequence of dispersion—n depends on frequency—the pulse changes its shape in a complicated way as it progresses through the medium. The proof is difficult, but Brillouin and Sommerfeld showed that in a medium described by a realistic dielectric function, no signal can be propagated faster than c; for details we refer the reader to the book by Brillouin (1960), in which English translations of the original papers on this subject can be found. Subsequently, Kramers showed that a necessary and sufficient condition for the signal velocity in a medium to be less than c is that the real and imaginary parts of its refractive index satisfy the dispersion relations (2.49), (2.50); therefore, the proof of Brillouin and Sommerfeld, who used such a refractive index in their analysis, is a special case of a more general result.

The proof that the signal velocity cannot be greater than c is complicated; but it is possible to obtain some insight into refractive indices less than 1—or greater than 1 for that matter—by simple reasoning based on the oscillator model discussed above. To do so, we adopt a microscopic approach according to which the oscillators in a sample of matter are driven by an incident plane wave and reradiate (scatter) waves in all directions. In directions other than that of the incident wave, there is almost complete destructive interference of the scattered waves. In the direction of propagation, however, the incident and scattered waves combine to produce a transmitted wave with a phase either retarded or advanced relative to that of the incident wave depending on the phase difference between an oscillator and the incident wave; as shown in Fig. 9.2a, this relative phase can vary from $0°$ to $180°$.

Consider the waves scattered by isotropic dipole oscillators in the thin slab of matter shown in Fig. 9.3; only part a is of concern at the moment. These waves add vectorially at point P to produce the resultant forward-scattered wave s; the important point, which is by no means obvious yet, is that this resultant scattered wave is phase shifted $90°$ relative to the incident wave (in addition to the phase shift between the oscillators and the incident wave). The background necessary to show this has been presented in Chapter 3; Fig. 3.8 is similar to Fig. 9.3 except that in the former the scatterers were arbitrary particles. The transmitted field t at P is the vector sum of the incident field and the fields scattered by all the oscillators. If we assume that the direction of vibration of the incident wave is not rotated as it propagates through the slab, the transmitted field is given by (3.39):

$$\mathbf{E}_t = E_0 e^{ikd} \left[1 - \frac{2\pi \mathfrak{N} h}{k^2} (\hat{\mathbf{e}}_x \cdot \mathbf{X})_{\theta=0} \right] \hat{\mathbf{e}}_x, \tag{9.17}$$

where $\mathfrak{N}$ is the number of oscillators per unit volume. This result has often been obtained using scalar diffraction theory. Strictly speaking, (9.17) only applies to a dilute collection of oscillators, but interactions among them will

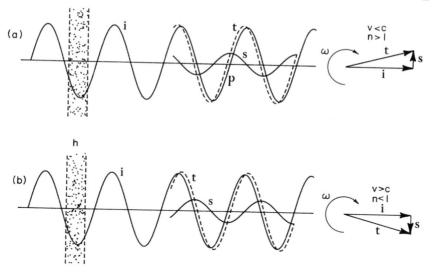

Figure 9.3 The wave transmitted (t) by a slab of dipole oscillators is the sum of incident (i) and scattered (s) waves.

not affect the qualitative aspects of our analysis, which is intended to be descriptive rather than completely rigorous. The vector scattering amplitude $\mathbf{X}$ for an isotropic dipole with polarizability α is (5.17):

$$\mathbf{X} = \frac{ik^3}{4\pi}\alpha\hat{\mathbf{e}}_r \times (\hat{\mathbf{e}}_r \times \hat{\mathbf{e}}_x).$$

Since $\hat{\mathbf{e}}_r \times \hat{\mathbf{e}}_x = \hat{\mathbf{e}}_y$ and $\hat{\mathbf{e}}_r \times \hat{\mathbf{e}}_y = -\hat{\mathbf{e}}_x$ when $\theta = 0°$, we have

$$(\hat{\mathbf{e}}_x \cdot \mathbf{X})_{\theta=0} = \frac{-ik^3}{4\pi}\alpha,$$

which can be substituted into (9.17) to obtain

$$\mathbf{E}_t = \mathbf{E}_i\left(1 + i\frac{\mathfrak{N}kh\alpha}{2}\right) = \mathbf{E}_i + \mathbf{E}_s. \tag{9.18}$$

It follows from (9.5) and (9.6) that the polarizability can be written $\alpha = |\alpha|e^{i\Theta}$, where Θ is the phase difference between the displacement of an oscillator and the field that excites it. Thus, the total scattered field is

$$\mathbf{E}_s = \tfrac{1}{2}\mathbf{E}_i\mathfrak{N}hk|\alpha|e^{i(\Theta+\pi/2)}, \tag{9.19}$$

where we have used $i = e^{i\pi/2}$. The significance of (9.19) is that the oscillators scatter waves that interfere at P to give a total scattered wave $\mathbf{E}_s$ which is 90° out of phase with the oscillators and $\Theta + 90°$ out of phase with the incident wave. Figure 9.3a shows the incident wave i, the 90° phase-shifted scattered wave s, and the transmitted wave t for low frequencies where $\Theta \simeq 0$; phasor diagrams are included for those who are enlightened by such. Note that the transmitted wave is retarded (i.e., it *lags* the incident wave) because of interactions with all the oscillators in the slab. Interaction with successive slabs of oscillators as the wave moves through the medium results in a phase velocity less than $c : n > 1$. Figure 9.4 shows the phase relations that lead to $n > 1$ at low frequencies; on the left side of the figure the vertical scale shows the oscillator phase relative to the incident wave, and on the right side the vertical scale shows the phase of the total scattered wave relative to the incident wave; phasor diagrams are also included. At frequencies well above the resonant frequency the oscillator phase is almost 180° and the scattered wave has a phase of almost 270°. Figure 9.3*b* shows that the scattered wave *leads* the incident wave, giving a transmitted wave that is *advanced* relative to the incident wave because of interaction with the oscillators in the slab. Interaction with successive slabs of oscillators as the wave moves through the medium results in a phase velocity greater than $c : n < 1$. Such a seemingly strange effect as the speeding up of light is thus no stranger than the slowing down of light; both effects depend on the relative phase of the waves radiated by oscillators that have been set in motion by the electromagnetic field. It should be carefully noted, however, that this entire discussion has been based on the *steady-state* response of the oscillators. Consideration of the transient response would give quite different results, including "precursors" which travel through the medium at exactly *c* regardless of the oscillator response (see, e.g., Stratton, 1941, pp. 333–340, for a discussion of the transients leading up to the steady-state response). Thus, a pulse of electromagnetic energy would not follow the behavior we have just sketched because the steady state would not

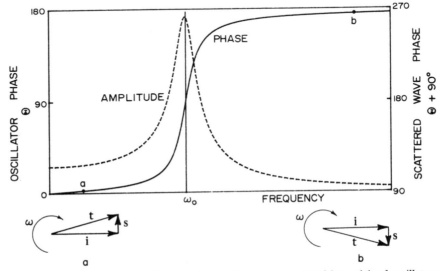

Figure 9.4 Phase of a single oscillator, and that of the wave scattered by a slab of oscillators, relative to that of the wave exciting them.

have time to build up. In particular, the signal carried by the pulse could not be propagated faster than the free-space speed of light.

9.1.4 An Example of Electronic Excitations: MgO

For all solids and liquids there is a frequency region in the ultraviolet where light is strongly absorbed because the photon energy matches energy differences between filled and empty electron energy levels. Since there is a continuum of both empty and filled levels (bands), the result is an extensive region of high absorption. The gross properties of this complicated quantum system can be represented by a single classical oscillator with sufficient strength and damping to give it strong absorption over a broad frequency range. Of course, such a model cannot be expected to give the details of the electronic absorption structure, but it can give the general features. Thus, it is useful in guiding intuition and sometimes in coarse modeling of optical constants.

Our first illustrative example of a real material is the electronic excitations of MgO, an insulating crystalline solid, which will be one of the representative solids of Chapter 10. The reflectance measured at near-normal incidence over the photon energy range from about 1 to 30 eV (2 eV corresponds to a wavelength of about 0.62 μm) is shown in Fig. 9.5; these data and the corresponding optical constants are from Roessler and Walker (1967), who give a table of n and k. Peaks and valleys in the reflectance curve are caused mostly by variations in the density of occupied and vacant electron states. The optical constants derived from a Kramers–Kronig analysis of the reflectance data (see Section 2.7) are shown in the bottom part of the figure; the $1/\omega^4$ limiting behavior of the reflectance (9.15) was used to extrapolate the data to infinite energy. The dashed curves are the result of calculations based on a one-oscillator model using the parameters given in the figure caption. With these oscillator parameters the dielectric functions were calculated from (9.8) and (9.9); the corresponding optical constants n and k follow from (9.2), and the normal-incidence reflectance from (2.58).

If we disregard the fine structure, we see that the gross behavior of R, n, and k is given correctly by this simple one-oscillator model. The similarity to Fig. 9.2 is evident. At energies well below the resonance peak, which includes the visible region from about 2 to 3 eV, the material is nonabsorbing (small k), while n is almost constant, increasing only slightly toward higher energy. This behavior of n is the normal dispersion shown by all common transparent materials at visible wavelengths; it is the sort of frequency dependence one finds in elementary texts for the refractive index of glass, water, ice, and so on. In fact, this nearly constant refractive index that we have grown up with undoubtedly contributes to our collective prejudice that optical constants are really essentially constant, except for a small increase of n toward the blue, which causes separation of white light into a rainbow of colors upon passing through a prism or a water drop. Even the one-oscillator model clearly shows, however, that this prejudice is an all too simplistic view of optical properties;

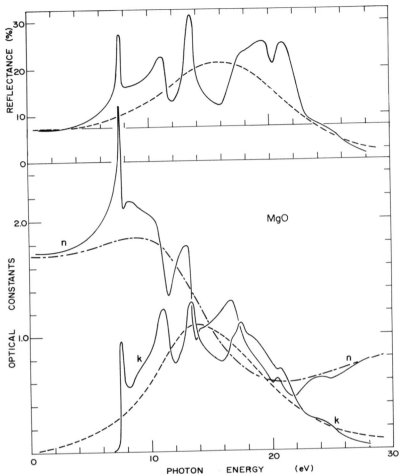

Figure 9.5 Measured reflectance and derived optical constants for MgO are shown by solid lines (Roessler and Walker, 1967). Calculations for a one-oscillator model with $\omega_p^2 = 321$ (eV)^2, $\gamma = 8$ eV, and $\omega_0 = 13$ eV are shown by dashed lines.

normal dispersion of n in the nonabsorbing region is inseparably connected with, and in a sense caused by, the strong absorption band centered at $\mathcal{E}_0 = \omega_0/\hbar$. All visibly transparent solids and liquids must therefore have strong absorption in the ultraviolet where they are opaque.

In the energy region broadly centered about 14 eV ($\lambda \simeq 880$ Å), light is strongly absorbed in MgO after penetrating the surface of the crystal. We can estimate the penetration depth from (2.52): $I/I_0 = \exp(-4\pi kz/\lambda)$. For $k \simeq 1$ and $\lambda = 0.88 \times 10^{-5}$ cm, the $1/e$ penetration depth is about 10^{-6} cm, or 100 Å; thus, it would be hopeless to attempt cutting and polishing or cleaving a crystal thin enough so that transmission measurements could be made in this energy region. On the other hand, a 100-Å penetration depth corresponds to many atomic layers in the crystal; hence, reflection is not just a phenomenon confined to surface layers of atomic dimensions, as is, for example, low-energy electron diffraction by solids.

High absorption in the far ultraviolet is accompanied by high reflectance, as was pointed out for the general oscillator model. Figure 9.5 shows that MgO reflects over 20% of normally incident light in the vicinity of 20 eV ($\lambda \simeq 600$ Å); this compares well with platinum—a common far-ultraviolet reflectance coating—which has a reflectance of about 12% at 18 eV (Hass and Tousey, 1959). Many nonconducting solids have higher reflectances in the far ultra-violet than metals. At still higher energies, of course, the reflectance goes to zero, as does k, and n approaches 1 from below; all these trends are shown in Fig. 9.5 by the measured values, which are in accord with the limiting behavior (9.14) and (9.15) predicted by the model calculations. Low reflectance in the extreme ultraviolet (~ 500 Å or less) is common for all solids: there are no efficient normal-incidence mirrors in the far ultraviolet and soft x-ray regions.

9.1.5 An Example of Lattice Vibrations: α-SiC

It was shown in the preceding section that the optical properties corresponding to electronic transitions in condensed matter are qualitatively described correctly by a one-oscillator model, although one would hesitate to use such a simple model to, for example, analyze reflectance data quantitatively. In contrast, the lattice vibrational modes in some crystals are so accurately described by simple oscillator theory that it is commonly used to determine the optical constants from measurements. Since the lattice vibrations and electronic excitations are well separated in energy for many solids, the effect of the electrons can be well approximated by the low-frequency limit (9.16). That is, at frequencies low compared with characteristic electronic excitation frequencies, the dielectric function is taken to be a real constant; with this assumption, the modified one-oscillator model appropriate to lattice vibrations is

$$\epsilon = \epsilon_{0e} + \frac{\omega_p^2}{\omega_t^2 - \omega^2 - i\gamma\omega}. \tag{9.20}$$

The notation ϵ_{0e} indicates that this is the dielectric function at frequencies low compared with *electronic* excitation frequencies. We have also replaced ω_0 with ω_t, the frequency of the *transverse* optical mode in an ionic crystal; micro-scopic theory shows that only this type of traveling wave will be readily excited by a photon. Note that ω_p^2 in (9.20) corresponds to $\mathfrak{N}e^2/m\epsilon_0$ for the lattice vibrations (ionic oscillators) rather than for the electrons. The mass of an electron is some thousands of times less than that of an ion; thus, the plasma frequency for lattice vibrations is correspondingly reduced compared with that for electrons.

A solid for which this one-oscillator model fits extremely well is α-SiC; this is a "textbook" example, although it is a solid of considerable technological importance. Furthermore, it seems to be one of those few solids that have been identified in the far reaches of interstellar space by means of just such vibrational absorption features presented here. Since α-SiC is not completely isotropic we should actually treat it according to the anisotropic model of Section 9.3. However, its infrared properties are almost the same for the two principal directions. For normal incidence reflectance from platelets of SiC, which tend to grow with flat faces perpendicular to the hexagonal axis, the optical properties are isotropic.

Measured normal incidence reflectances of α-SiC for incident electric field perpendicular to the hexagonal axis are shown in Fig. 9.6; these are unpublished measurements made in the authors' laboratory, but they are similar to those published by Spitzer et al. (1959). Also included in this figure are both sets of optical constants—n, k and ϵ', ϵ''—calculated from the best fit of a one-oscillator model to the experimental data. Note that the model curve is almost a perfect representation of the data over the entire range shown; for this solid, the technique of fitting data with a one-oscillator model is both a simple and accurate method for extracting optical constants.

Many of the features exhibited by the simple model (Fig. 9.2) are also seen in Fig. 9.6: normal dispersion of n on either side of the resonance frequency; a narrow region of anomalous dispersion in the neighborhood of the absorption band; and high reflectance around the absorption band with lower reflectance away from the resonance frequency where k is small. A new feature appears because ϵ_{0e} has replaced 1 in the simple oscillator equation (9.7): the region of negative ϵ' ($n < k$) is confined to a narrower region near the resonance; in contrast, ϵ' in Fig. 9.2 is negative in a greater range above the frequency near ω_0 where ϵ' crosses the ω axis. The high-frequency edge of the negative ϵ' region, where $\epsilon' = 0$, is denoted by ω_l, the *longitudinal* optical mode frequency; this is the frequency at which longitudinal oscillations of the ions are resonant. Without going into detail we can obtain some insight into this by considering the Maxwell equation

$$\nabla \cdot \mathbf{D} = \epsilon \nabla \cdot \mathbf{E} = 0.$$

If $\epsilon \neq 0$, then $\nabla \cdot \mathbf{E} = 0$ and the electric field is *transverse*. However, if $\epsilon = 0$ at some frequency, then $\nabla \times \mathbf{H} = 0$. This in turn, together with $\nabla \cdot \mathbf{B} = 0$, implies that $\mathbf{B} = 0$ and, consequently, $\nabla \times \mathbf{E} = 0$; that is, the electric field is *longitudinal*. Note then that ω_l is the frequency at which ϵ vanishes or nearly does so.

In the region between ω_t and ω_l, which for SiC is between about 800 and 1000 cm^{-1}, the reflectance is high not because of large k but because of small n. If $n \simeq 0$, the normal incidence reflectance is nearly 100%; only for the undamped oscillator ($\gamma = 0$) is the reflectance actually 100%, but solids like SiC approach this rather closely. If the damping constant γ in (9.20) is set equal to zero, the real part of the dielectric function becomes

$$\epsilon' = \epsilon_{0e} + \frac{\omega_p^2}{\omega_t^2 - \omega^2}, \tag{9.21}$$

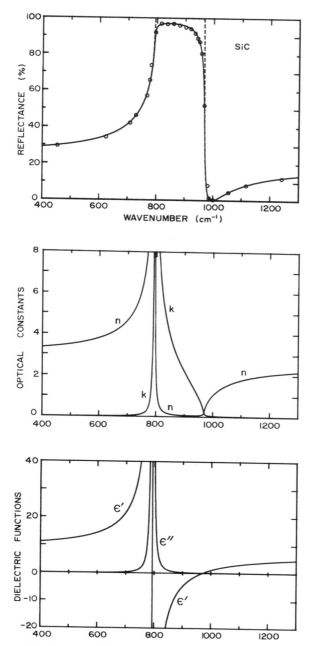

Figure 9.6 Measured reflectance (circles) of α-SiC. The solid curves are from (9.20) with $\omega_t = 793$ cm^{-1}, $\gamma = 4.76$ cm^{-1}, $\omega_p^2 = 2.08 \times 10^6$ cm^{-2}, and $\epsilon_{0e} = 6.7$; the dashed curve is for the same model with $\gamma = 0$. The wave number is $1/\lambda$.

and the imaginary part ϵ'' is an infinitely sharp spike centered at ω_t. R calculated for this undamped model is shown by dashed lines in Fig. 9.6. Note that SiC is well approximated by neglecting γ, which is small compared with ω_t; in particular, the 100% theoretical reflectance between ω_t and ω_l is almost reached by the actual reflectance, which exceeds 97% in this region. Such high reflectances of ionic crystals in narrow infrared spectral regions provide means for making band-pass reflection filters, which are used in some commercial infrared spectrophotometers. The contrast between reflectances in and away from the high-reflectance region can be increased by multiple reflection: a beam of light with a continuum of infrared frequencies, upon being multiply reflected by a series of ionic crystals, will emerge with mostly frequencies in the high-reflectance region remaining. Discovery of this effect gave rise to the terminology *Reststrahlen* (residual ray) mode for this type of crystal oscillation.

An interesting and useful relation can be derived between the frequencies and the real parts of the dielectric function on either side of the *Reststrahlen* band. The frequency ω_l at which ϵ' vanishes is, from (9.21), given approximately by

$$\omega_l^2 = \omega_t^2 + \frac{\omega_p^2}{\epsilon_{0e}}. \tag{9.22}$$

For frequencies low compared with the transverse optical frequency ω_t, the dielectric function (9.21) approaches the limiting value ϵ_{0v}:

$$\epsilon_{0v} = \epsilon_{0e} + \frac{\omega_p^2}{\omega_t^2}. \tag{9.23}$$

The physical significance of ϵ_{0e} and ϵ_{0v} should be clear from Fig. 9.16. If (9.22) and (9.23) are combined we obtain the *Lyddane–Sachs–Teller relation*:

$$\frac{\omega_l^2}{\omega_t^2} = \frac{\epsilon_{0v}}{\epsilon_{0e}}. \tag{9.24}$$

9.2 THE MULTIPLE-OSCILLATOR MODEL

The one-oscillator model is quite useful in describing many optical excitations, especially when modified to include—in the low-frequency limit—the effect of all oscillators removed to higher frequencies [e.g., (9.20)]. The model becomes even more useful over a broader range of frequencies if it is extended to a multiplicity of oscillators. A naive pictorialization of the multiple-oscillator model is that it is a collection of weights and springs, as in Fig. 9.1, but of more than one type; that is, there is more than one resonant frequency. A rigorous approach to treating lattice vibrations, for example, would be to separate the motions of the electrons and the lattice ions, write down the equations of motion for the ions moving in some kind of effective potential, expand this potential to terms of second order in the displacement of the ions from equilibrium (the *harmonic approximation*), and transform the equations of motion to a system of normal coordinates; the result is a set of independent equations of motion for effective harmonic oscillators—a collection of weights and springs. Because of the superposition principle for amplitudes in the harmonic approximation, the complex polarizability is an additive quantity, that is, summable over the effective oscillators. In general, therefore, the dielectric function for a collection of oscillators is just the sum over the various oscillators

$$\epsilon = \epsilon_0 + \sum_j \frac{\omega_{pj}^2}{\omega_j^2 - \omega^2 - i\gamma_j\omega},\tag{9.25}$$

where the parameters of the jth oscillator are the resonant frequency ω_j, the damping constant γ_j, and the plasma frequency ω_{pj}. As in (9.20), ϵ_0 represents the effect of all oscillators well removed to higher frequencies. This is a division for convenience; if *all* oscillators are included in the summation, then $\epsilon_0 = \epsilon(\infty) = 1$. Note that n and k are not summable: they must be obtained from (9.25).

A simple application of the multiple-oscillator theory is to fit measured reflectance data for MgO in the *Reststrahlen* region. In Section 9.1 we considered the electronic excitations of MgO, whereas we now turn our attention to its lattice vibrations. A glance at the far-infrared reflectance spectrum of MgO in Fig. 9.7 shows that it does not completely exhibit one-oscillator behavior: there is an additional shoulder on the high-frequency side of the main reflectance peak, which signals a weaker, but still appreciable, second oscillator. The solid curves in Fig. 9.7 show the results of a two-oscillator calculation using (9.25); the reflectance data were taken from Jasperse et al. (1966), who give the following parameters for MgO at 295°K:

$$\epsilon_0 = \epsilon_{0e} = 3.01$$

$$\omega_1 = 401 \text{ cm}^{-1} \qquad \gamma_1 = 7.62 \text{ cm}^{-1} \qquad \frac{\omega_{p1}^2}{\omega_1^2} = 6.6$$

$$\omega_2 = 640 \text{ cm}^{-1} \qquad \gamma_2 = 102.4 \text{ cm}^{-1} \qquad \frac{\omega_{p2}^2}{\omega_2^2} = 0.045$$

Notice the second, weaker oscillator at 640 cm^{-1}; from a quantum viewpoint, this oscillator is interpreted as the excitation of two phonons by a photon. In order for momentum to be conserved, two phonons with almost equal and opposite momenta must be created: a photon has negligible momentum compared with phonon momenta. Such two-phonon transitions usually occur to a noticeable extent in ionic crystals, which necessitates a multiple-oscillator correction to the main one-oscillator *Reststrahlen* band.

We note that in MgO there is a rather large difference between the transverse optical mode frequency at about 400 cm^{-1}, where ϵ'' is a maximum,

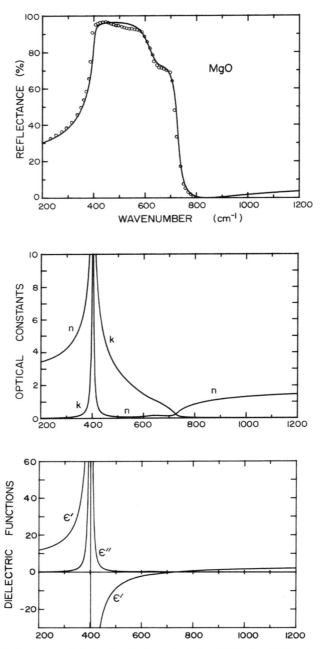

Figure 9.7 Reflectance and derived optical constants for MgO; the measurements (circles) are from Jasperse et al. (1966).

and the longitudinal optical mode frequency, where $\epsilon' = 0$. This in turn implies [see (9.24)] a large difference between the real part of the dielectric functions in the high- and low-frequency limits: $\epsilon_{0e} = 3.01$ and $\epsilon_{0v} = 9.64$. More important for absorption by small particles, the negative ϵ' region extends between ω_t and ω_l. We shall show in Chapter 12 that small particles of various shapes can absorb strongly throughout the negative ϵ' region. Because of the extent of this frequency region in MgO, its small-particle absorption spectrum can look quite different from its absorption spectrum in thin films, which is similar to the k curve.

9.3 THE ANISOTROPIC OSCILLATOR MODEL

Common liquids are optically isotropic, and the solids that physicists seem to like most are cubic and therefore isotropic. As a consequence, treatments of optical properties, particularly from a microscopic point of view, usually favor isotropic matter. Among the host of naturally occurring solids, however, most are *not* isotropic. This somewhat complicates both theory and experiment; for example, measurements of optical constants must be made with oriented crystals and polarized light. But because of the prevalence of optically anisotropic solids, we are compelled to extend the classical models to embrace this added complexity.

In the preceding sections the optical response of matter has been described by a *scalar* dielectric function ϵ, which relates the electric field $\mathbf{E}$ to the displacement $\mathbf{D}$. More generally, $\mathbf{D}$ and $\mathbf{E}$ are connected by the *tensor* constitutive relation (5.46), which we write compactly as $\mathbf{D} = \varepsilon_0 \bar{\epsilon} \cdot \mathbf{E}$. The dielectric tensor $\bar{\epsilon}$ is often symmetric, so that a coordinate system can be found in which it is diagonal:

$$\bar{\epsilon} = \begin{pmatrix} \epsilon_1 & 0 & 0 \\ 0 & \epsilon_2 & 0 \\ 0 & 0 & \epsilon_3 \end{pmatrix},$$

where $\epsilon_1, \epsilon_2, \epsilon_3$, are the *principal* dielectric functions. Thus, if the electric field is aligned along one of the principal axes, then $\mathbf{D}$ and $\mathbf{E}$ are parallel.

To give physical meaning to the principal dielectric functions, we consider propagation of plane waves $\mathbf{E}_0 \exp(i\mathbf{k} \cdot \mathbf{x} - i\omega t)$ in an anisotropic medium; that is, we ask: What kind of plane waves can propagate in such a medium without change of polarization? If we follow the same reasoning as in Section 2.6, we obtain from the Maxwell equations

$$\mathbf{k}(\mathbf{k} \cdot \mathbf{E}_0) - \mathbf{E}_0(\mathbf{k} \cdot \mathbf{k}) = -\frac{\omega^2}{c^2}\bar{\epsilon} \cdot \mathbf{E}_0, \qquad \mathbf{k} \cdot (\bar{\epsilon} \cdot \mathbf{E}_0) = 0,$$

where we have assumed that $\mu = \mu_0$. Let us consider the special case where $\mathbf{k}$ is along one of the principal axes, which we may take to be the z axis without loss of generality. Then if $\mathbf{E}_0$ is referred to principal axes, the equations above reduce to

$$\left(k^2 - \frac{\omega^2 \epsilon_1}{c^2} \right) E_{0x} = 0, \qquad \left(k^2 - \frac{\omega^2 \epsilon_2}{c^2} \right) E_{0y} = 0, \qquad \epsilon_3 E_{0z} = 0.$$

If $\epsilon_3 \neq 0$, then $E_{0z} = 0$: the wave is transverse. There are two solutions to these equations:

$$k^2 = \frac{\omega^2 \epsilon_1}{c^2}; \qquad E_{0x} \neq 0, \quad E_{0y} = 0,$$

$$k^2 = \frac{\omega^2 \epsilon_2}{c^2}; \qquad E_{0x} = 0, \quad E_{0y} \neq 0.$$

Thus, plane waves can propagate along the z axis without change of polarization provided that they are either x-polarized or y-polarized. The complex refractive indices for these two types of waves, however, are different:

$$n_1 + ik_1 = \sqrt{\epsilon_1}, \qquad n_2 + ik_2 = \sqrt{\epsilon_2}.$$

If $\epsilon_1 = \epsilon_2$, then the z axis—the direction of propagation—is called the *optic axis* or the *c axis*. In this analysis we have tacitly assumed that the coordinate transformation to principal axes diagonalizes both the real and imaginary parts of the dielectric tensor.

In frequency regions where absorption is small the two indices of refraction n_1 and n_2 give rise to the phenomenon of double refraction. One of the most common uses for this property is in making wave retarders such as quarter-wave plates: incident light linearly polarized with equal x and y field components is phase shifted upon transmission because of the two different phase velocities c/n_1 and c/n_2. An entire field, usually referred to as crystal optics, arises out of this and further applications of crystal anisotropy.

Different refractive indices n for different linear polarization states of light propagating along a principal axis is called (linear) *birefringence*. Crystals with this property are said to be birefringent or doubly refracting. If there is appreciable absorption, the attenuation of a wave will also depend on polarization; this is referred to as *dichroism* or, more specifically, linear dichroism, to distinguish it from absorption differences in circularly polarized light.

The classical picture that describes these anisotropic effects based on the Lorentz model is illustrated in Fig. 9.8, which is a generalization of the spring model of Fig. 9.1; note that the spring stiffness depends on direction.

To obtain expressions for the principal dielectric functions we need only write three equations similar to (9.20) with three different sets of oscillator parameters appropriate to $\epsilon_1, \epsilon_2, \epsilon_3$. Each set of parameters corresponds to one

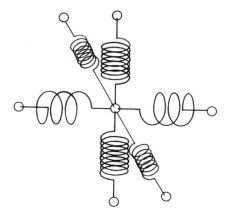

Figure 9.8 Anisotropic oscillator.

of the three different springs in Fig. 9.8. A more accurate description of the principal dielectric functions than can be given with one oscillator for each principal axis is a multiplicity of such oscillators (see Section 9.2).

The spring model suggests that the symmetry of the crystal lattice determines the different forms of the dielectric tensor; that is, they are related to the seven types of crystalline solid (amorphous solids and most liquids are isotropic). This is summarized as follows:

Isotropic. $\epsilon_1 = \epsilon_2 = \epsilon_3$

Amorphous solids
Most liquids
Cubic crystals
$$\begin{pmatrix} \epsilon' + i\epsilon'' & 0 & 0 \\ 0 & \epsilon' + i\epsilon'' & 0 \\ 0 & 0 & \epsilon' + i\epsilon'' \end{pmatrix}$$

Uniaxial Crystals. $\epsilon_1 = \epsilon_2 \neq \epsilon_3$

Tetragonal
Hexagonal
Trigonal
$$\begin{pmatrix} \epsilon_1' + i\epsilon_1'' & 0 & 0 \\ 0 & \epsilon_1' + i\epsilon_1'' & 0 \\ 0 & 0 & \epsilon_3' + i\epsilon_3'' \end{pmatrix}$$

Biaxial Crystals. $\epsilon_1 \neq \epsilon_2 \neq \epsilon_3$

Orthorhombic
$$\begin{pmatrix} \epsilon_1' + i\epsilon_1'' & 0 & 0 \\ 0 & \epsilon_2' + i\epsilon_2'' & 0 \\ 0 & 0 & \epsilon_3' + i\epsilon_3'' \end{pmatrix}$$

Biaxial Crystals

Triclinic
Monoclinic
$$\begin{pmatrix} \epsilon_1' & 0 & 0 \\ 0 & \epsilon_2' & 0 \\ 0 & 0 & \epsilon_3' \end{pmatrix} \quad \text{or} \quad \begin{pmatrix} \epsilon_1'' & 0 & 0 \\ 0 & \epsilon_2'' & 0 \\ 0 & 0 & \epsilon_3'' \end{pmatrix}$$

For the low-symmetry triclinic and monoclinic crystals the principal axes for the real and imaginary parts of the dielectric tensor are different. This makes life very complicated, and we—along with most other authors—will avoid such complications.

An example of a solid with anisotropic optical properties in the lattice vibration region is crystalline quartz, SiO_2, often found naturally as large, clear, single hexagonal crystals. Quartz is sometimes used for wave-retarding plates and has other commercial and scientific uses. Two different infrared reflectance spectra, taken from Spitzer and Kleinman (1961), are shown in Fig. 9.9. Polarized light was used in this experiment, and the crystal was oriented and cut properly for determination of the optical properties along the two principal axes. The positions of the *Reststrahlen* bands are quite different for the two polarization directions (parallel and perpendicular to

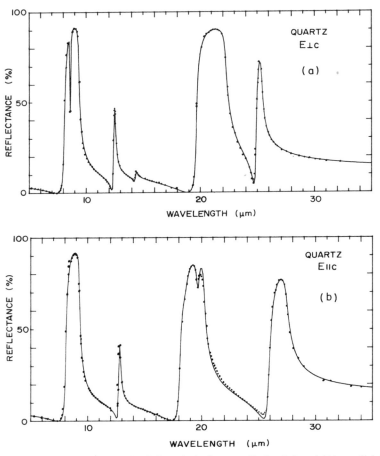

Figure 9.9   Reflectance of quartz for light polarized perpendicular (a) and (b) parallel to the c axis. Measurements (dots) and a theoretical fit (solid lines) are from Spitzer and Kleinman (1961).

Table 9.1 Oscillator Parameters Used to Fit the Reflectance Data Shown in Fig. 9.9 (From Spitzer and Kleinman, 1961)

$\omega_j \ (\text{cm}^{-1})$	ω_{pj}^2/ω_j^2	γ_j/ω_j	
1227	0.009	0.11	
1163	0.01	0.006	
1072	0.67	0.0071	Electric field
797	0.11	0.009	perpendicular
697	0.018	0.012	to c axis
450	0.82	0.0090	
394	0.33	0.007	

$$\epsilon_{0e} = 2.356$$

1220	0.011	0.15	
1080	0.67	0.0069	
778	0.10	0.010	Electric field
539	0.006	0.04	parallel
509	0.05	0.014	to c axis
495	0.66	0.0090	
364	0.68	0.014	

$$\epsilon_{0e} = 2.383$$

the c axis), especially at wavelengths greater than about 15 μm; this shows the difference in the effective spring constants for different crystallographic directions. Spitzer and Kleinman fit the reflectance data for each polarization direction with seven oscillators, the parameters of which are given in Table 9.1. With these parameters it is an easy matter to calculate optical constants from (9.25). Comparison of calculated reflectances (solid lines) with measured reflectances (dots) shows the success of the fitting procedure; this provides an excellent quantitative illustration of the anisotropic multiple-oscillator model.

9.4 THE DRUDE MODEL

There is a marked difference between the optical properties of conductors and nonconductors of electricity; this is shown schematically in Fig. 9.10 by simple energy band diagrams. Because of the large number of electrons in a solid, there is nearly a continuum of energy states, or levels, that these electrons can occupy. However, a consequence of the periodicity of the crystal lattice is that the energy levels are grouped into *bands*. If there is a forbidden energy gap, the *band gap*, between completely filled and completely empty energy bands, the material is a nonconductor (i.e., an insulator or a semiconductor). If, on the other hand, a band of electron states is incompletely filled, or if an otherwise

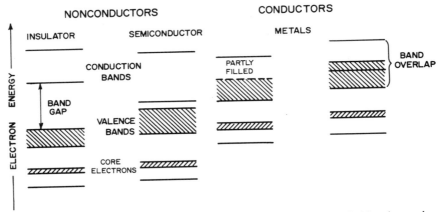

Figure 9.10 Electron energy bands in nonconductors and conductors. Filled bands are shown hatched.

filled band overlaps in energy with an empty band, the material is a conductor: electrons at the top of the energy distribution can be excited into adjacent unoccupied states by an applied electric field, which results in an electric current. This availability of vacant electron states in the same energy band provides a mechanism, *intraband* absorption, for absorption of low-energy photons. Absorption in nonconductors, *interband* absorption, is only likely for photon energies greater than the band gap. This difference between conductors and nonconductors gives rise to substantial optical differences: insulators tend to be transparent and weakly reflecting for photons with energies less than the band gap, whereas metals tend to be highly absorbing and reflecting at visible and infrared wavelengths.

Electrons in metals at the top of the energy distribution (near the Fermi level) can be excited into other energy and momentum states by photons with very small energies; thus, they are essentially "free" electrons. The optical response of a collection of free electrons can be obtained from the Lorentz harmonic oscillator model by simply "clipping the springs," that is, by setting the spring constant K in (9.3) equal to zero. Therefore, it follows from (9.7) with $\omega_0 = 0$ that the dielectric function for free electrons is

$$\epsilon = 1 - \frac{\omega_p^2}{\omega^2 + i\gamma\omega}, \tag{9.26}$$

with real and imaginary parts

$$\epsilon' = 1 - \frac{\omega_p^2}{\omega^2 + \gamma^2},$$

$$\epsilon'' = \frac{\omega_p^2\gamma}{\omega(\omega^2 + \gamma^2)}. \tag{9.27}$$

This is the *Drude model* for the optical properties of a free-electron metal. The plasma frequency is given by $\omega_p^2 = \mathfrak{N}e^2/m\varepsilon_0$, where $\mathfrak{N}$ is the density of free electrons and m is the effective mass of an electron. We have used the symbol ω_p before, but in the present context the plasma frequency has a simple physical interpretation. Let us take as our classical model of a metal a collisionless gas of free electrons moving against a fixed background of immobile positive ions. The number density $\mathfrak{N}$ of positive charges is therefore constant in space and time. In equilibrium, the density of electrons is also $\mathfrak{N}$. But if the electrons are disturbed slightly from equilibrium by some unspecified means, the nonuniform charge distribution will set up an electric field that will tend to restore charge neutrality. The electrons, having acquired momentum from the field, will overshoot the equilibrium configuration: there will be an oscillation. If we denote the electron number density by $\mathfrak{N} - \delta\mathfrak{N}$, the electric field is given by

$$\nabla \cdot \mathbf{E} = \frac{e\,\delta\mathfrak{N}}{\varepsilon_0}. \tag{9.28}$$

If we consider only small departures from equilibrium ($|\delta\mathfrak{N}/\mathfrak{N}| \ll 1$), the equation of continuity is approximately

$$\nabla \cdot \mathbf{u} = \frac{\partial}{\partial t}\frac{\delta\mathfrak{N}}{\mathfrak{N}}, \tag{9.29}$$

where $\mathbf{u}$ is the velocity field of the electron gas, which is taken to be a continuous charged fluid. The equation of motion of this charged fluid is

$$\frac{\partial\mathbf{u}}{\partial t} + (\mathbf{u}\cdot\nabla)\mathbf{u} = -\frac{e}{m}\mathbf{E}, \tag{9.30}$$

where we neglect magnetic and pressure gradient forces. The first term on the left side of (9.30) is of order U/τ, where U is a characteristic velocity associated with the electron motion and τ is a characteristic time; the second term is of order U^2/L, where L is a characteristic length. If we assume that $1/\tau \gg U/L$, then (9.30) is approximately

$$\frac{\partial\mathbf{u}}{\partial t} = -\frac{e}{m}\mathbf{E}.$$

From these equations we obtain

$$\frac{\partial^2}{\partial t^2}\frac{\delta\mathfrak{N}}{\mathfrak{N}} + \omega_p^2\frac{\delta\mathfrak{N}}{\mathfrak{N}} = 0. \tag{9.31}$$

The existence of plane-wave solutions $\mathbf{E} = \mathbf{E}_0\exp(i\mathbf{k}\cdot\mathbf{x} - i\omega t)$, $\delta\mathfrak{N}/\mathfrak{N} = C\exp(i\mathbf{k}\cdot\mathbf{x} - i\omega t)$ to (9.28) and (9.31) requires that

$$i\mathbf{k}\cdot\mathbf{E} = \frac{e\,\delta\mathfrak{N}}{\varepsilon_0}, \tag{9.32}$$

$$C\left(\omega_p^2 - \omega^2\right) = 0. \tag{9.33}$$

There are two solutions to (9.33): (1) $\omega^2 \neq \omega_p^2$, $C = 0$, which implies that the plane wave is transverse ($\mathbf{k} \cdot \mathbf{E} = 0$); and (2) $C \neq 0$, $\omega^2 = \omega_p^2$. The Drude dielectric function (9.26) vanishes at $\omega = \omega_p$ (neglecting damping); but we showed in Section 9.1 that the field is *longitudinal* at frequencies where $\epsilon(\omega) = 0$. Thus, the second solution corresponds to a longitudinal oscillation. This collective oscillation of the electron gas is called a *plasma oscillation*; it originates from long-range correlation of the electrons caused by Coulomb forces. Such plasma oscillations in gaseous discharges were investigated theoretically and experimentally by Tonks and Langmuir (1929). Further refinements were added to the classical theory by Bohm and Gross (1949ab); this work, in turn, led to a series of seminal papers on the quantum theory of plasma oscillations (Bohm and Pines, 1951, 1953; Pines and Bohm, 1952).

In the more general—and realistic—case of nonzero damping, the Drude dielectric function vanishes at the *complex* frequency $\omega_p - i\gamma/2$, provided that $\omega_p^2 \gg \gamma^2/4$. In the quantum-mechanical language of elementary excitations we refer to the excitation of a plasma oscillation as the creation (or production, or excitation) of a *plasmon*, the quantum of plasma oscillation, with energy $\hbar\omega_p$ and lifetime $\tau = 2/\gamma$. For a plasmon to be a well-defined entity, its lifetime must be sufficiently long ($\omega_p\tau \gg 1$). Although a plasmon is made up of electrons, it is *not* an electron: it is a gang, or collection, of electrons that get together under the urging of the long-range Coulomb force and decide to act in concert. Hence, for the purpose of discussing their behavior, they may, like an orchestra or a choir, be considered as a single entity following the same (Coulombic) conductor. In a choir, members are sometimes called upon to sing solos; so it is also in a collection of electrons. Individual particle behavior is exhibited, for example, in the electronic excitation spectrum of MgO (Fig. 9.5).

An elementary treatment of the free-electron motion (see, e.g., Kittel, 1962, pp. 107–109) shows that the damping constant is related to the average time τ between collisions by $\gamma = 1/\tau$. Collision times may be determined by impurities and imperfections at low temperatures but at ordinary temperatures are usually dominated by interaction of the electrons with lattice vibrations: *electron–phonon scattering*. For most metals at room temperature γ is much less than ω_p. Plasma frequencies of metals are in the visible and ultraviolet: $\hbar\omega_p$ ranges from about 3 to 20 eV. Therefore, a good approximation to the Drude dielectric functions at visible and ultraviolet frequencies is

$$\epsilon' \simeq 1 - \frac{\omega_p^2}{\omega^2},$$
$$\qquad (\omega \gg \gamma). \qquad (9.34)$$
$$\epsilon'' \simeq \frac{\omega_p^2 \gamma}{\omega^3},$$

These equations are identical with the high-frequency limit (9.13) of the Lorentz model; this indicates that at high frequencies all nonconductors behave like metals. The interband transitions that give rise to structure in optical properties at lower frequencies become mere perturbations on the free-electron type of behavior of the electrons under the action of an electromagnetic field of sufficiently high frequency.

The reflectance, dielectric functions, and refractive indices, together with calculations based on the Drude theory, for the common metal aluminum are shown in Fig. 9.11. Aluminum is described well by the Drude theory except for the weak structure near 1.5 eV, which is caused by bound electrons. The parameters we have chosen to fit the reflectance data, $\hbar\omega_p = 15$ eV and $\hbar\gamma = 0.6$ eV, are appreciably different from those used by Ehrenreich et al. (1963), $\hbar\omega_p = 12.7$ eV and $\hbar\gamma = 0.13$ eV, to fit the low-energy ($\hbar\omega < 0.2$ eV) reflectance of aluminum. This is probably caused by the effects of band transitions and the difference in electron scattering mechanisms at higher energies. The parameters we use reflect our interest in applying the Drude theory in the neighborhood of the plasma frequency.

For aluminum at low frequencies ϵ' is large and negative and ϵ'' is also large; the magnitude of both decreases monotonically with increasing frequency. The corresponding n and k are both large at low frequencies, but decrease toward higher frequencies; k is greater than n, which is less than 1 in much of the region below ω_p (2–15 eV). Because of the small value of n, the normal

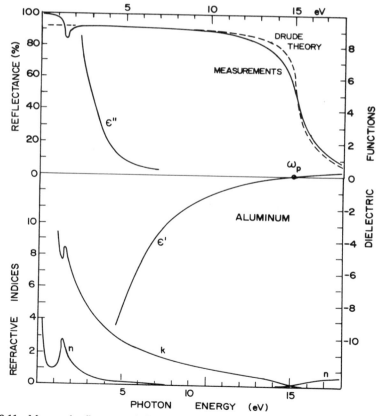

Figure 9.11 Measured reflectance of aluminum compared with the Drude theory. The dielectric function and refractive index are from Hagemann et al. (1974).

incidence reflectance is nearly 100% throughout this region, and even thin films of aluminum would transmit little light. At frequencies well above ω_p, $\epsilon' \simeq 1$ and $\epsilon'' \simeq 0$, which means that $n \simeq 1$ and $k \simeq 0$: at these frequencies, aluminum is transparent. The change from opacity to transparency near the plasma frequency is an example of *ultraviolet transparency*; it occurs in all free-electron type of metals: the alkali metals Li, Na, K, and Rb, and multivalent metals such as Mg, Al, and Pb. Ultraviolet transparency is a cut-on optical filter effect, which can be used to block light of wavelengths greater than λ_p.

We have mentioned that the region of negative ϵ' is of special importance for the optics of small particles, which can absorb and scatter strongly at frequencies that depend on their shape. In particular, strong absorption by spheres occurs at the frequency where $\epsilon' = -2$. Figure 9.11 shows that there can be an extensive region of shape-dependent absorption and scattering by small metallic particles; this will be discussed more fully in Chapter 12.

The Drude free-electron model is not limited to metals; as mentioned previously, nonconductors show a free-electron type of behavior at sufficiently high frequencies. For example, if one looks at the reflectance and dielectric functions of silicon (Philipp and Ehrenreich, 1963), one sees strong absorption bands between about 3 and 6 eV. Above about 7 eV, however, the reflectance falls smoothly from high to low values; ϵ'' goes smoothly toward zero; and ϵ' increases from a large negative value, going through zero at the plasma frequency of about 17 eV. All this is characteristic of free electrons.

Impurities in semiconductors, which release either free electrons or free holes (the absence of an electron in an otherwise filled "sea" of electrons), also give rise to optical properties at low energies below the minimum band gap (e.g., 1.1 eV for Si) that are characteristic of the Drude theory. Plasma frequencies for such doped semiconductors may be about 0.1 eV.

An interesting qualitative application of the Drude theory can be made to the ionosphere, the region of the atmosphere lying between about 50 and 500 km, in which the number density of free electrons is sufficient to affect radio-wave propagation. An electron density of 10^6 cm^{-3} gives rise to a plasma frequency of order 10^7 sec^{-1}, which corresponds to a wavelength of about 30 m. At lower frequencies (longer wavelengths) than ω_p the ionosphere is highly reflecting, just as a metal is, which has an important effect on low-frequency radio-wave propagation. The ionosphere becomes transparent to radio waves greater than ω_p for the same reason that free-electron metals become transparent in the ultraviolet. Therefore, high-frequency radio waves are not so greatly affected by "skipping" or reflection from the ionosphere. Obviously, radio communication with satellites or spaceships above the ionosphere must rely on frequencies greater than its maximum plasma frequency. The electron density is not uniform with height and varies considerably because of such changing effects as sunspot activity; also, the earth's magnetic field complicates the propagation of radio waves in the ionosphere. Nevertheless, the Drude model gives a qualitative description that helps in understanding the observed phenomena.

Table 9.2   Plasma Frequencies and Corresponding Wavelengths for a Range of Free-Electron Densities

$\mathfrak{N}$ (cm^{-3})	ω_p (sec^{-1})	λ_p	Type of Plasma
10^{24}	5.7×10^{16}	330 Å	
10^{22}	5.7×10^{15}	3300 Å } metals	$\text{Al}(\lambda_p = 830 \text{ Å})$
10^{20}	5.7×10^{14}	3.3 μm	$\text{Cs}(\lambda_p = 3620 \text{ Å})$
10^{18}	5.7×10^{13}	33 μm	
10^{17}	1.8×10^{13}	105 μm } electron–hole droplets	doped semiconductors
10^{16}	5.7×10^{12}	330 μm	
10^{6}	5.7×10^{7}	33 m	
10^{5}	1.8×10^{7}	105 m } ionosphere (F layer)	

Plasma frequencies and the corresponding wavelengths for a range of free-electron densities are given in Table 9.2; the plasma effects mentioned in this section are noted beside the wavelengths (electron–hole droplets will be discussed in Chapter 12).

9.4.1   Free and Bound Electrons in Metals

Despite its applicability to metals such as aluminum, Drude theory alone does not accurately describe the optical characteristics of many other metals. Silver is a good example of a metal that exhibits some free-electron type of behavior, which can be treated with the Drude theory; but it also has a substantial bound-electron component, which appreciably alters the free-electron optical properties. The normal-incidence reflectance of silver (Irani et al., 1971) is shown in Fig. 9.12a; a logarithmic scale was used to emphasize the unusually large variation of reflectance over a narrow range around ω_p. The low-energy part of the spectrum suggests a free-electron type of reflectance: it is almost 100% and falls precipitously to quite low values near the plasma frequency at about 3.9 eV. Beyond the plasma frequency, however, results differing markedly from Drude theory occur as the reflectance rises abruptly and vacillates a little before again falling to low values; this behavior above 4 eV is interpreted as a bound charge effect. In Fig. 9.12b the experimentally determined ϵ' for silver is decomposed into components contributed by free and bound charges. That is, polarizabilities and, hence, susceptibilities are additive; thus, we may write a composite dielectric function $\epsilon = \epsilon_f + \delta\epsilon_b$ that includes a contribution $\delta\epsilon_b$

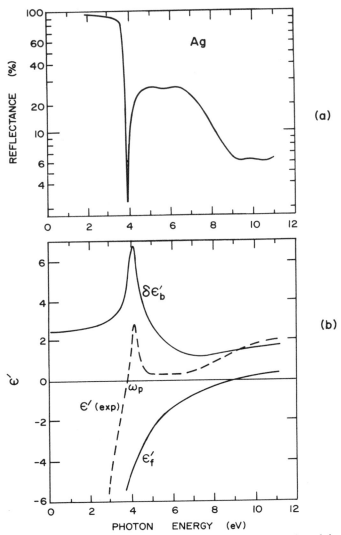

Figure 9.12 (*a*) Reflectance of silver (from Irani et al., 1971). (*b*) Separation of the measured ϵ' into free and bound contributions (from Ehrenreich and Philipp, 1962).

from Lorentz oscillators and a free-electron contribution ϵ_f:

$$\epsilon = 1 - \frac{\omega_{pe}^2}{\omega^2 + i\gamma_e\omega} + \sum_j \frac{\omega_{pj}^2}{\omega_j^2 - \omega^2 - i\gamma_j\omega},$$

where the subscript e is appended to free-electron parameters.

The free-electron contribution to the dielectric function in Fig. 9.12*b* is obtained from the Drude theory with parameters determined from the low-frequency limit of the data ($\hbar\omega < 0.1$ eV); the bound-charge contribution then follows simply by subtraction: $\delta\epsilon'_b = \epsilon' - \epsilon'_f$. These results clearly show the strong effect of bound charges on the plasma frequency. From the equation $\omega_p^2 = \mathfrak{N}e^2/m\epsilon_0$ with the electron density $\mathfrak{N}$ appropriate to silver, the plasma frequency is calculated to be 9.2 eV, the frequency at which ϵ' would be approximately zero if only free electrons determined the optical properties of silver. However, the appreciable positive contribution of the bound charges at energies below about 6 eV pulls up the ϵ' curve so that it crosses the axis at 3.9 eV rather than 9 eV. The plasma frequency, defined as the frequency at which $\epsilon' = 0$, is shifted from the far to the near ultraviolet because of the bound charges, resulting in the onset of ultraviolet transparency near 3200 Å instead of 1300 Å. Almost as soon as the reflectance drops at the plasma frequency, however, the peak in $\delta\epsilon'_b$ near 4 eV causes a rise in reflectance that would not occur were it not for the bound charges. We thus see how competing bound and free charges in metals can greatly alter their optical properties.

An interesting metal to compare with silver is copper: its reddish-brown color obviously signals considerable differences in optical constants in the visible; yet the electron energy band structures of the two metals are qualitatively quite similar. The difference in their appearance lies in the position of the ϵ' peak caused by bound charges, which is near 4 eV in silver and near 2 eV in copper. Because the free-electron contribution to ϵ' is so strongly negative, the positive contribution of the bound charges is not sufficient to raise ϵ' to positive values and thereby decrease the plasma frequency, as it does in silver. Although the reflectance does not drop because the plasma frequency is shifted, the bound charges do have an important effect. The onset of their effects near 2 eV causes a sharp decrease in reflectance from nearly 100% at lower energies; a weaker example of the reflectance decrease caused by bound charges can be seen near 1.5 eV in the aluminum reflectance spectrum of Fig. 9.11. At 2 eV, which corresponds to wavelengths in the red part of the visible spectrum, the reflectance of copper is still high; at the other end of the visible spectrum (about 3 eV), the bound charges greatly reduce the reflectance. It is this higher reflectance in the red than in the blue that gives the metal its characteristic "copper" color.

9.5   THE DEBYE RELAXATION MODEL

There is another important mechanism for polarizing matter containing *permanent*, as opposed to induced, electric dipoles: partial alignment of the dipoles along the electric field against the counteracting tendency toward disorientation caused by thermal buffeting. The restoring "force" that tries to return a polarized region to an unpolarized state is thus the statistical tendency toward random orientation of the dipoles; this is quite different from the conceptual springs of the Lorentz theory. The dipole restoring tendency does not cause overshoot leading to oscillation of the electric polarization; it is as if permanent electric dipoles are *overdamped*, whereas Lorentz oscillators are *underdamped*. Because the permanent dipoles merely relax to equilibrium, this mode of polarization is referred to as *relaxation* and is usually associated with Debye, who did the early definitive work (one of the best references continues to be *Polar Molecules*, 1929, by Debye). Since many liquids are composed of molecules that have permanent dipole moments, Debye relaxation by molecular rotation plays an important role in determining the optical constants of these liquids at certain frequencies, usually in the microwave region. Many solids also show Debye relaxation because of the presence of charged defects or impurities, which may have nonequivalent positions in the crystal lattice, leading to the possibility of reorientation of the resulting dipole moments.

The return to equilibrium of a polarized region is quite different in the Debye and Lorentz models. Suppose that a material composed of Lorentz oscillators is electrically polarized and the static electric field is suddenly removed. The charges equilibrate by executing damped harmonic motion about their equilibrium positions. This can be seen by setting the right side of (9.3) equal to zero and solving the homogeneous differential equation with the initial conditions $x = x_0$ and $\dot{x} = 0$ at $t = 0$; the result is the damped harmonic oscillator equation:

$$x = \text{Re}\left\{ x_0 e^{-i\omega_0 t} e^{-\gamma t} \right\}.$$

The decay of the initial polarization is therefore

$$P_l(t) = P_l(0)e^{-\gamma t}\cos(\omega_0 t);$$

we append l to the polarization to indicate a collection of Lorentz oscillators. In contrast with these damped harmonic oscillators, a collection of Debye oscillators (permanent dipoles), if polarized initially, returns to equilibrium with the time dependence

$$P_d(t) = P_d(0)e^{-t/\tau}. \tag{9.35}$$

There is no oscillation: the polarization merely relaxes toward zero with a time constant τ. In the following paragraphs, we shall use (9.35), the basic assumption of the Debye theory, to derive an expression for the dielectric function of a collection of permanent dipoles.

If at time t_0 a constant electric field E_0 (i.e., a step function) is suddenly applied to a sample of polarizable matter, the polarization will follow the time evolution illustrated schematically in Fig. 9.13. There are contributions to the polarization from electrons, lattice ions, and permanent dipoles. We shall assume that the response of the permanent dipoles is much slower than that of the electrons and ions; that is, the response of the latter may be considered to occur instantaneously compared with the time required for the matter to reach

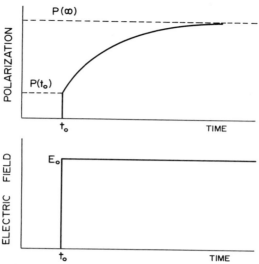

Figure 9.13   Schematic diagram of the time response of polarization following application of a field E_0 at time t_0.

its final equilibrium polarization. When the field is applied at time t_0, therefore, a polarization $P(t_0) = \varepsilon_0 \chi_{0v} E_0$ is instantaneously induced; as time increases, the polarization approaches the limiting value $P(\infty) = \varepsilon_0 \chi_{0d} E_0$. We remind the reader that in our notation χ_{0v} is the susceptibility at frequencies low compared with characteristic lattice vibrational frequencies, which are in turn low compared with electronic frequencies; and χ_{0d} is the susceptibility at frequencies low compared with dipole frequencies (i.e., the static, or dc, susceptibility). By assuming an exponential approach to equilibrium as in (9.35) we can write a physically plausible expression for the time-dependent polarization:

$$P(t) = P(\infty) - Ce^{-(t-t_0)/\tau} \qquad (t > t_0). \qquad (9.36)$$

The constant C is determined from the requirement that $\lim\limits_{t \downarrow t_0} P(t) = P(t_0)$:

$$C = P(\infty) - P(t_0) = P(t_0)(\chi_{0d} - \chi_{0v})/\chi_{0v}.$$

With a bit of rearranging (9.36) can be put in a form which is easier to interpret and which will enable us to generalize it to a series of step functions:

$$P(t) = \varepsilon_0 \chi_{0v} E_0 + \varepsilon_0 (\chi_{0d} - \chi_{0v})[1 - e^{-(t-t_0)/\tau}]E_0. \qquad (9.37)$$

The first term on the right side of (9.37) is the contribution to the total polarization from the lattice ions; the second term $P_d(t)$, is the contribution from the permanent dipoles.

Suppose now that two different two-step fields are applied: (1) at t_0 the electric field $2E_1$ is turned on and at t_1 the field suddenly changes to E_1; and (2) the field at t_1 is the same as in (1) but between t_0 and t_1 it is $E_1/2$. It is clear from Fig. 9.14 that the polarization in these two cases is different, even for $t > t_1$ when the applied field is the same; that is, the polarization at time t depends on the history of the applied field and not merely its instantaneous value. This is a specific example of a general conclusion that we made about the response of a linear medium to a time-dependent electric field (see Section 2.3). The polarization at all times greater than t_1 can be obtained by following reasoning similar to that which led to (9.37): we write $P(t)$ in the form (9.36) and require that $\lim_{t \downarrow t_1} P(t) = P_d(t_1) + \varepsilon_0 \chi_{0v} E_1$; the result is

$$P(t) = \varepsilon_0 \chi_{0v} E_1 + \varepsilon_0 (\chi_{0d} - \chi_{0v}) E_1 [1 - e^{-(t - t_1)/\tau}]$$

$$+ \varepsilon_0 (\chi_{0d} - \chi_{0v}) E_0 [e^{-(t - t_1)/\tau} - e^{-(t - t_0)/\tau}] \qquad (t > t_1). \quad (9.38)$$

It is not difficult to show that (9.38) can be written

$$P(t) = \varepsilon_0 \chi_{0v} E_1 + \varepsilon_0 (\chi_{0d} - \chi_{0v}) \int_{t_0}^{t} E(t') \frac{d}{dt'} [e^{-(t - t')/\tau}] \, dt'. \quad (9.39)$$

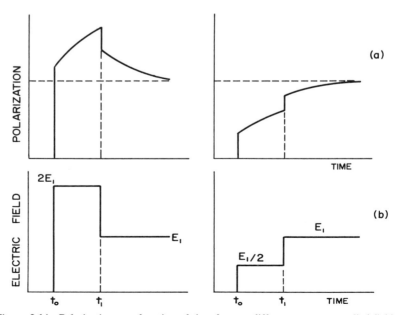

Figure 9.14 Polarization as a function of time for two different two-step applied fields.

Although this was obtained for a series of two step functions, it should be clear that it is valid for any number of such functions. Indeed, any field can be approximated by a series of step functions, where the error can be made arbitrarily small by choosing sufficiently small steps; thus, (9.39) is valid for any field. The technique of representing an arbitrary forcing function by a series of steps (or, more correctly, impulses) was first developed by Green, and a good account of this is given by Marion (1970, p. 139); the exponential function multiplying $E(t')$ in (9.39) is therefore the *Green's function* for the system of permanent dipoles in the Debye theory. The first term in (9.39) is the lattice polarization, which we assume follows the instantaneous field; the second term is the sum of all polarization changes caused by the slow dipole relaxation mechanism. Note that (9.39) is a particular example of a linear, causal constitutive relation (Section 2.3).

To obtain the frequency-dependent susceptibility $\chi(\omega)$, we need the polarization in response to a time-harmonic field $E_0 e^{-i\omega t}$:

$$P(t) = \varepsilon_0 \chi E_0 e^{-i\omega t} = \varepsilon_0 \chi_{0v} E_0 e^{-i\omega t}$$

$$+ \varepsilon_0 (\chi_{0d} - \chi_{0v}) \int_{-\infty}^{t} E_0 e^{-i\omega t'} \frac{d}{dt'} [e^{-(t-t')/\tau}] \, dt'. \quad (9.40)$$

Implicit in (9.40) is the assumption that ω is small compared with lattice vibrational frequencies. The susceptibility in the frequency region where Debye relaxation is the dominant mode of polarization is therefore

$$\chi = \chi_{0v} + \frac{\Delta}{1 - i\omega\tau}, \qquad \Delta = \chi_{0d} - \chi_{0v}$$

from which follows the dielectric function

$$\epsilon = \epsilon_{0v} + \frac{\Delta}{1 - i\omega\tau} \quad (9.41)$$

with real and imaginary parts

$$\epsilon' = \epsilon_{0v} + \frac{\Delta}{1 + \omega^2\tau^2}, \qquad \epsilon'' = \frac{\omega\tau\Delta}{1 + \omega^2\tau^2}. \quad (9.42)$$

Our derivation of (9.41) follows closely that of Gevers (1946) and is similar to that of Brown (1967, pp. 248–255). Because of the nature of this derivation it should hardly be necessary to do so, but it can be shown directly by integration—more easily than for the Lorentz oscillator—that the real and imaginary parts of the Debye susceptibility satisfy the Kramers–Kronig relations (2.36) and (2.37).

The imaginary part of the dielectric function (9.41) is a maximum at $\omega = 1/\tau$ and behaves similarly to ϵ'' for the Lorentz oscillator. The real part, however, behaves quite differently: it has no maxima and no minima but decreases monotonically from ϵ_{0d} at low frequencies to ϵ_{0v} at high frequencies; the transition occurs in the neighborhood of $\omega = 1/\tau$. At low frequencies, the permanent dipoles easily follow the electric field changes, and the dc dielectric function can be quite large; at high frequencies, however, they are unable to keep up with the oscillating field, and ϵ' falls to a value that does not include a contribution from the permanent dipoles.

The Debye equations (9.42) are particularly important in interpreting the large dielectric functions of polar liquids; one example is water, the most common liquid on our planet. In Fig. 9.15 measured values of the dielectric functions of water at microwave frequencies are compared with the Debye theory. The parameters ϵ_{0d}, ϵ_{0v}, and τ were chosen to give the best fit to the experimental data; $\tau \simeq 0.8 \times 10^{-11}$ sec follows immediately from the frequency at which ϵ'' is a maximum; $\epsilon_{0d} - \epsilon_{0v}$ is $2\epsilon''_{max}$.

On physical grounds, relaxation of permanent dipoles is expected to be highly dependent on temperature; this is in contrast with Lorentz oscillators, the dielectric behavior of which is relatively insensitive to changes in temperature. Debye (1929, Chap. 5) derived a simple classical expression for the relaxation time of a sphere of radius a in a fluid of viscosity η:

$$\tau = \frac{4\pi\eta a^3}{k_B T}, \qquad (9.43)$$

where T is the absolute temperature and k_B is Boltzmann's constant. Two

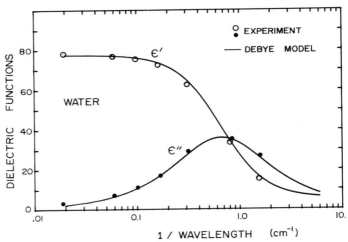

Figure 9.15 Dielectric function of water at room temperature calculated from the Debye relaxation model with $\tau = 0.8 \times 10^{-11}$ sec, $\epsilon_{0d} = 77.5$, and $\epsilon_{0v} = 5.27$. Data were obtained from three sources: Grant et al. (1957), Cook (1952), and Lane and Saxton (1952).

factors conspire to affect the relaxation time: the numerator of (9.43) arises from the viscous restoring torque on a small sphere; the denominator arises from the thermal buffeting. The greater is T, the lower the viscosity (in liquids) and the shorter the relaxation time; the greater is T, the more quickly will thermal motion randomize an oriented collection of permanent dipoles after the orienting field is removed. The room temperature viscosity of water is about 0.01 g/cm-sec, and if we take $a \simeq 10^{-8}$ cm, (9.43) yields a relaxation time of about 0.3×10^{-11} sec; thus, the simple theory is in good agreement with the relaxation time deduced from the experimental data of Fig. 9.15.

A naive interpretation of the phase transition from liquid to solid water as the temperature is decreased is that the water undergoes a large discontinuous increase in viscosity. Thus, the permanent electric dipoles that were free to rotate in the liquid are suddenly immobilized. Consistent with this interpretation is the expectation that the relaxation time for ice should be much greater than for water, with a corresponding decrease in the frequency at which ϵ'' is a maximum. This is indeed so: if water were to freeze suddenly from room temperature, then ϵ' at frequencies in the region shown in Fig. 9.15 would plunge from a high of about 80 to a low of about 3.2. Such large changes can be observed, for example, in radar backscattering from melting ice particles or freezing water droplets; this large difference between the dielectric functions of ice and liquid water is the major contributor to the "bright band" in vertical radar reflectivity profiles observed by meteorologists (see, e.g., Battan, 1973, p. 190).

Dipole relaxation is often described by a superposition of exponentially decaying functions with different relaxation times in place of the single relaxation time in (9.35). This is analogous to the need for several oscillator parameters in the Lorentz model. Both ϵ' and ϵ'' are functions of frequency; therefore, frequency can be eliminated to give a functional relationship between them. Experimental measurements of the dielectric functions in the dipole frequency region are often displayed as a plot of ϵ'' versus ϵ', called the *Cole–Cole plot*. From a glance at such a plot it is immediately obvious whether or not a single relaxation time is sufficient to determine the frequency dependence of the dielectric function; it is not difficult to show from (9.42) that for a single relaxation time, the Cole–Cole plot is a semicircle centered on the ϵ' axis at $(\epsilon_{0d} + \epsilon_{0v})/2$ and with radius $(\epsilon_{0d} - \epsilon_{0v})/2$:

$$\left[\epsilon' - \tfrac{1}{2}(\epsilon_{0d} + \epsilon_{0v})\right]^2 + \epsilon''^2 = \left[\tfrac{1}{2}(\epsilon_{0d} - \epsilon_{0v})\right]^2.$$

9.6   GENERAL RELATIONSHIP BETWEEN ϵ' AND ϵ''

We must reemphasize that the real and imaginary parts of the complex dielectric function (and the complex refractive index) are not independent. Arbitrary choices of ϵ' and ϵ'' (or n and k) do not necessarily correspond to any real solid or liquid at some wavelength. The interdependence of ϵ' and ϵ'' has been illustrated in each of the classical models of this chapter: the Lorentz model, the Drude model, and the Debye model. These connections between real and imaginary parts of the same complex function are specific examples of the Kramers–Kronig relations discussed in Section 2.3. For any solid or liquid the real part of the dielectric function at any frequency ω is related to the imaginary part at all frequencies by

$$\epsilon'(\omega) = 1 + \frac{2}{\pi} P \int_0^\infty \frac{\Omega \epsilon''(\Omega)}{\Omega^2 - \omega^2} d\Omega. \qquad (9.44)$$

We are now in a position to better understand and, we hope, appreciate, the sometimes mysterious Kramers–Kronig relations.

First, we note that the consequence of no absorption ($\epsilon'' = 0$) at all frequencies is that the integral in (9.44) vanishes and $\epsilon' = 1$. Optically, such a material does not exist: there is no way that it can be distinguished from a vacuum by optical means. The Kramers–Kronig relations also tell us that it is a contradiction to assert that either the real or imaginary parts of the dielectric function can be independent of frequency; the frequency dependence of the one implies the frequency dependence of the other. These consequences of the Kramers–Kronig relations are almost trivial, but it is disturbing how often they are blithely ignored.

The high- and low-frequency behavior of ϵ' can be inferred from (9.44) regardless of the nature and number of absorption features in ϵ''. As ω becomes arbitrarily large, all contributions to the integral go to zero and

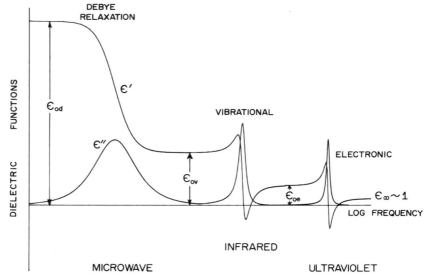

Figure 9.16 Schematic diagram of the frequency variation of the dielectric function of an ideal nonconductor.

$\lim_{\omega \to \infty} \epsilon'(\omega) = 1$. Thus, it is a general result that at frequencies far above all absorption bands ϵ' approaches the free-space value 1; the frequency is so high that none of the polarization mechanisms can respond. At low frequencies, on the other hand, ϵ' is the sum of contributions from each polarization mechanism—electronic, vibrational, and dipolar—with the lowest-frequency mechanism contributing most:

$$\epsilon'(0) = 1 + \frac{2}{\pi} \int_0^\infty \frac{\epsilon''(\Omega)}{\Omega} d\Omega. \tag{9.45}$$

The consequence of (9.45) is that for any material other than free space, somewhere at higher frequencies there must be an absorption mechanism lurking.

A schematic diagram of the frequency dependence of ϵ' and ϵ'' for an ideal nonconductor (no Drude term) is given in Fig. 9.16; this summarizes several of the points made in this chapter. Possible fine structure in the optical constants is ignored for simplicity. The notation we have used for the high- and low-frequency dielectric functions in the various models will become clearer after consulting this figure. At $\omega = 0$ the dielectric function is composed of contributions from permanent dipoles, vibrational oscillators, and electronic oscillators. As the frequency is increased, the permanent dipoles cannot respond and ϵ' drops from the static value ϵ_{0d} to ϵ_{0v}, the value at frequencies low compared with characteristic vibrational frequencies. As ω increases through the vibrational region, ϵ' oscillates and then settles down to ϵ_{0e}, the low-frequency limit for electronic modes. Finally, as the frequency increases beyond the point where all the electronic modes are exhausted, ϵ' approaches 1. In each frequency interval where ϵ' is changing there is an associated peak in ϵ'', the absorptive part of the dielectric function. There are other polarization modes associated with core electrons in the atoms, which give rise to much weaker variations in ϵ' and ϵ'' at x-ray frequencies; we have arbitrarily chosen to exclude these from our discussion of "optical properties."

NOTES AND COMMENTS

Hodgson (1970) and Christy (1972) have treated optical properties of matter in a spirit similar to that of this chapter. And so has Garbuny (1965), although within a book of broader scope.

We direct the reader to the book by Davies (1969) for further reading about wave propagation in the ionosphere.

Crystal optics—the optics of anisotropic media—is treated at an elementary level by Wood (1977). More advanced treatises are those by Nye (1957) and by Ramachandran and Ramaseshan (1961).

Fröhlich and Pelzer (1955) determined the frequencies of longitudinal waves in matter described by the three simple dielectric functions—Lorentz, Drude, and Debye—discussed in this chapter.

Chapter 10

Measured Optical Properties

In the preceding chapter we discussed classical theories of optical constants based on simple hypothetical models. Lest we mislead the reader, we must emphasize that the illustrative materials and frequency ranges in that chapter were carefully selected; everything does not, alas, conform in detail to such simple models. Ultimately, therefore, recourse must be had to measured frequency-dependent optical constants, and to fully understand scattering and absorption by small particles one must be aware of how the optical constants of real materials vary. In this chapter we give synopses of measured optical constants of an insulating solid, a metal, and a liquid: magnesium oxide, aluminum, and water. These materials illustrated simple theoretical dielectric functions in Chapter 9; the emphasis in this chapter is on measurements and their interpretation for a wide range of frequencies, from microwave to far ultraviolet. We also briefly comment on the magnitude of k, the imaginary part of the refractive index, and the validity of using bulk dielectric functions in small-particle calculations.

10.1 OPTICAL PROPERTIES OF AN INSULATING SOLID: MgO

Although not a common solid, magnesium oxide has been available for many years as pure single crystals, which have been studied over a broad wavelength range. Single crystals are cleaved easily for reflection and transmission studies, and small particles are produced by simply burning magnesium ribbon in air. Ultraviolet and infrared optical properties of MgO were used in Sections 9.1 and 9.2 to illustrate the Lorentz model applied to electronic and vibrational absorption. In Fig. 10.1 the real and imaginary parts of the dielectric function are shown for photon energies from far infrared to far ultraviolet. Because investigators of optical properties are usually interested in limited spectral regions, it was necessary to piece together results from several sources which are cited in the figure caption (unfortunately, there is not yet a compendium of optical constants of solids). Note that the imaginary part of the dielectric function is plotted on a logarithmic scale, whereas the real part is shown on a

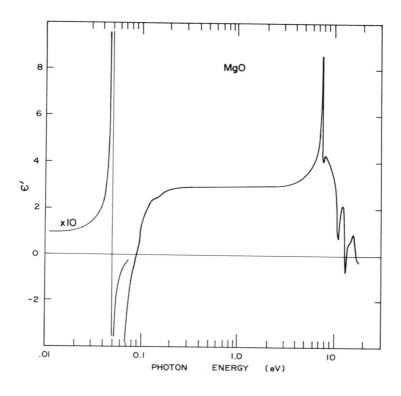

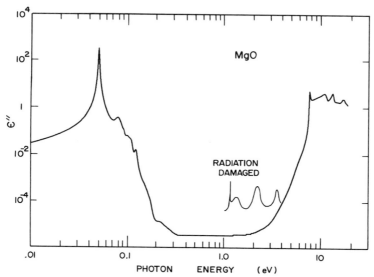

Figure 10.1 Dielectric functions of magnesium oxide. Infrared data are from Jasperse et al. (1966); visible and ultraviolet data are from Roessler and Walker (1967).

linear scale; therefore, structure in the real part appears more prominent than associated structure in the imaginary part.

The imaginary part of the dielectric function of MgO ranges over eight decades, an enormous variation [and recall that ϵ'' enters into the exponent of the transmission equation (2.52)]. There are two distinct absorption regions: absorption by electrons occurs above about 1 eV and that by lattice vibrations below about 0.3 eV. The sharp peak below the minimum band gap energy ($\sim$ 7.6 eV), especially prominent in the ϵ' curve (and in Fig. 9.5), is evidently the result of creation of an *exciton*. When a photon with energy slightly below the band gap is absorbed by an electron near the top of a filled band, the electron leaves behind a positively charged hole. Interaction of the hole and the electron can bind the pair together, much like an electron and a proton are bound in a hydrogen atom. The bound electron–hole pair—an excitation which cannot transport charge—is called an exciton. In ionic solids such as the alkali halides and MgO the exciton is tightly bound, and the electron is confined mostly to the vicinity of the negative ion (O^{2-} in MgO). A sequence of progressively closer spaced absorption bands leading up to the continuous transitions is sometimes observed; the electron–hole pair may be thought of as having a sequence of bound-state energy levels like a one-electron atom (e.g., hydrogen). In weakly bound excitons, the electron orbit ranges over many lattice parameters; in this instance the permittivity of free space, which is appropriate for the hydrogen atom, must be replaced by the permittivity of the solid in which the electron moves. The larger permittivity of solids compared with free space has the twofold effect of compressing the hydrogen atom energy levels and increasing the orbital radius of the electron. A notable example of a material showing hydrogen-like energy-level structure near the band gap is CuO_2, which has a number of exciton bands at low temperatures.

Beyond the exciton peak in the far ultraviolet there is a series of other absorption features in MgO, which are more distinct in Fig. 9.5. These features, a theoretical treatment of which requires electron energy band calculations, are caused by maxima in the combined density of available states for electrons in the ground and excited states.

On the low-energy side of the transparent region there is strong absorption by lattice vibrations. The huge peak near 0.05 eV is the *Reststrahlen* band, which was discussed in Section 9.2. For both energy and momentum to be conserved, only the long-wavelength transverse optical mode phonon can couple strongly to the photon; this interaction produces the dominant infrared peak. A second peak in ϵ'', some three orders of magnitude less intense, is on the high-energy side of the *Reststrahlen* peak at about 0.08 eV. This is interpreted as the result of excitation of two phonons by a photon, a much less probable event. In the region 0.1–0.2 eV there is weak residual absorption, which is probably related to multiple phonon excitation.

Between the regions of strong absorption by electronic and vibrational transitions there is a region of high transparency where absorption is dominated by impurities and imperfections. Artificial crystals of MgO are thus quite transparent to visible light. This transparent region can be made more interesting by a bit of stimulation. We have inset a separate spectrum of ϵ'' for radiation-damaged MgO in Fig. 10.1 (the variations of ϵ' are too small to be seen in the figure); the dielectric function was obtained by a four-oscillator fit to transmission data of Chen and Sibley (1969). These additional bands in MgO are examples from the rich field of *color centers*, or *F-centers* (from the German *Farbenzentrum*), in transparent crystals. Because the bands absorb differently in different parts of the visible spectrum the originally clear crystals are colored. In the simplest color center, for example, an electron is trapped in the vicinity of a missing negative ion site, which has an effective positive charge. The trapped electron has energy levels to which it can be excited by photons of approximately visible wavelengths. Although the energy levels of the trapped electron can be quite narrowly spaced, absorption of a photon is usually accompanied by production of several phonons along with the electronic excitations; this gives the possibility of photon absorption over a rather broad band of energies, which accounts for the 0.05-eV width of the three MgO absorption bands (Fig. 10.1). The sharp spike on the low-energy side of the first band seems to be a no-phonon band where (as the name implies) the electronic transition is not accompanied by phonon emission and attendant band broadening. Only in rare instances, of which this is one, is the no-phonon band strength comparable with that of broad phonon-assisted bands. There are many other kinds of color centers in crystals. The most extensive work has been done on alkali halide crystals, which can be colored either by irradiation or by heating the solid in its alkali metal vapor. Natural radioactivity is thought to be the agent producing color centers in some crystals highly prized for their color—amethyst, for example. Old bottles left in the sun for many years owe their purple hue to color centers induced by low-level ultraviolet photon bombardment. Impurities must sometimes be present to make radiation-induced energy levels possible. Much work remains to be done before the exact formation mechanisms of all color centers are understood.

Impurity atoms in transparent crystals also frequently have electronic energy levels that contribute to absorption bands in the transparent region. In most instances the electronic transitions are coupled to multiphonon processes, as in F-centers, causing broadening of the absorption bands; sometimes there are no-phonon bands. Not all impurity atoms produce such effects. Particularly important coloring agents are the transition metal atoms, which can yield a variegated collection of absorption bands; the colors of gemstones such as ruby, saphire, and the garnets are the result of transition metal impurities in otherwise clear crystals, and many colored minerals owe their hues to this same cause. Even the reds and oranges of the Grand Canyon can be traced to iron impurities in crystals of its rocks.

10.2   OPTICAL PROPERTIES OF A METAL: ALUMINUM

In modern times aluminum has become one of the most common metals in the industrialized world; every day we encounter aluminum as cooking ware, foil, and building materials. Aluminum is highly reflecting at all visible wavelengths

MEASURED OPTICAL PROPERTIES

and into the far ultraviolet, making it the commonly used coating for high-quality mirrors, including reflecting telescope mirrors. Although aluminum oxidizes readily, the oxide coating inhibits further deterioration by oxidation; this preserves the quality of the surface for most optical applications because the oxide is transparent to visible light.

The dielectric function and reflectance of aluminum, which were displayed in Fig. 9.11, are shown over a broader energy range in Fig. 10.2; note that two

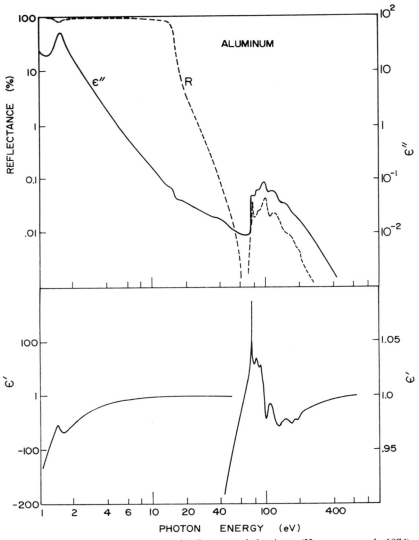

Figure 10.2 Dielectric functions and reflectance of aluminum (Hagemann et al., 1974).

logarithmic scales are needed to properly show structure at both high and low energies. These optical properties were taken from Hagemann et al. (1974), who did experiments in the range 13–150 eV and also surveyed the literature for determinations of optical constants at lower energies; these authors give very useful tables of optical constants for aluminum as well as magnesium, copper, silver, gold, bismuth, carbon, and aluminum oxide.

Below the plasma frequency at about 15 eV the only appreciable deviation from Drude theory occurs near 1.5 eV, where interband electronic transitions produce a peak in ϵ'' and associated structure in ϵ'; with this exception, ϵ' for aluminum goes monotonically toward negative infinity and ϵ'' toward positive infinity as the energy approaches zero.

Features abruptly appear at about 60 eV, where ϵ' is nearly 1 and ϵ'' has fallen to about 0.01. This signals the onset of absorptive effects associated with deeper lying core electrons in the aluminum atoms; note carefully that the change in scale for ϵ' greatly exaggerates the relative importance of the core transitions. The electron configuration of the aluminum atom is $[Ne]3s^2 3p$; that is, two $3s$ electrons and one $3p$ electron outside a closed neon-like shell. The three outer electrons only partially fill the electron energy band that is formed when the atoms come together to make up the solid; it is this partially filled band that gives continuous free-electron absorption. Electrons in the neon-like cores are so tightly bound that photons of energy greater than about 60 eV are required to excite them. It is these core electrons that give rise to structure in the dielectric function of aluminum at high energies. Although such core transitions were not mentioned in the chapter on classical theories, they could be modeled reasonably well by a superposition of Lorentzian oscillators.

At energies high compared with the core electron excitation energies, ϵ' again approaches 1 and ϵ'' tends toward zero. In the x-ray region, of course, further excitation of deep-lying electrons will occur, giving absorption structure and attendant variation in ϵ'. These transitions are more properly regarded as atomic transitions: they are little affected by the fact that the atoms are grouped in a solid.

10.3 OPTICAL PROPERTIES OF A LIQUID: WATER

Optical properties of liquids are similar in many ways to those of solids. Electrically, there are metallic liquids such as mercury and molten iron, but the majority of common liquids are nonmetallic. As an illustration of a liquid we have chosen H_2O, a ubiquitous substance on our planet; water dominates not only atmospheric processes but the chemistry of life.

The optical properties of water have been studied for centuries; the modern results are scattered throughout the literature of many scientific fields. Fortunately, several authors have critically surveyed the literature on H_2O and have assembled collections of optical constants for broad wavelength ranges. These reviews include those by Irvine and Pollack (1968), Hale and Querry

(1973), and Ray (1972); extensive tabulations of n and k, which are very useful for small-particle calculations, are given in the first two papers. Because of more recent data included in the paper by Hale and Querry, and because they ensured that n and k are connected by Kramers–Kronig relations, we have based our figures for liquid H_2O on their paper.

Dielectric functions calculated from refractive indices tabulated by Hale and Querry are plotted in Fig. 10.3; optical constants at wavelengths shorter than 0.2 μm were taken directly from the work of Kerr et al. (1972). In deference to the many workers who do not commonly use electron volts, we include a wavelength scale. As was done for MgO, ϵ' is plotted linearly and ϵ'' is plotted logarithmically. We also include on the ϵ'' plot some results for solid H_2O from Irvine and Pollack; comparison of ice and water gives insight into the similarities and differences between a solid and its liquid. The insulating liquid H_2O is not unlike the insulating solid MgO: the electronic and vibrational excitation regions for both materials are well separated by a highly transparent region. Electronic absorption by H_2O begins to be appreciable at about 0.2 μm; strong vibrational absorption bands rise above the continuum between about 1 and 100 μm. The major difference between MgO and liquid H_2O occurs longward of the vibrational absorption region: dipole relaxation in H_2O (discussed in Section 9.5) maintains a high level of absorption from the microwave well into the infrared; this also is the cause of the major differences between the optical properties of liquid and solid water.

10.3.1 The Transparent Region

The lowest point in the entire absorption spectrum lies in the visible, where ϵ'' plunges to less than 10^{-8}. A simple calculation shows that water is quite transparent to visible light for path lengths of tens of centimeters. If we ignore reflection, which is only a few percent, then the transmission T by a slab of thickness d is $\exp(-\alpha d)$, where to good approximation α is $2\pi\epsilon''/n\lambda$. In the visible, therefore, α is about 10^{-3} cm^{-1}. Even in a large glass full of water ($d \simeq 10$ cm) there is no discernible absorption. However, $T \simeq 0.6$ for $d = 500$ cm; thus, there is appreciable absorption in a home swimming pool of moderate size. Moreover, there is appreciable differential absorption over the visible spectrum. If we take as blue $\lambda = 4250$ Å, as red $\lambda = 6000$ Å, and use k values from Hale and Querry of 1.3×10^{-9} and 10.9×10^{-9}, respectively, the transmission percentages for a 5-m path length of pure water are the following: blue, 82.5%; red, 32%. Thus, white light transmitted over such a distance should take on a distinctly bluish hue. This apparently can be observed in the clear water of swimming pools, although competing effects such as reflection of the blue sky, color of the pool bottom and sides, impurities and particles in the water, and scattering by the water molecules themselves can render the interpretation complicated. Because absorption in ice and water are similar, light transmitted by pure ice over distances of several meters should also

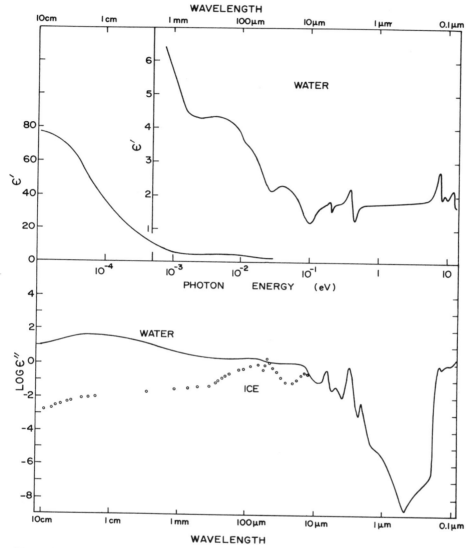

Figure 10.3 Dielectric functions of water (Hale and Querry, 1973). ϵ'' for ice is taken partly from Irvine and Pollack (1968) and partly from an unpublished compilation of the optical constants of ice, from far ultraviolet to radio wavelengths, by Stephen Warren (to be submitted to *Applied Optics*).

appear bluish; this seems to be an important factor in the blue color of glacial crevasses and other large-scale natural ice formations.

10.3.2 The Debye Relaxation Region

At microwave frequencies Debye relaxation is the dominant absorption mechanism in water. The broad maximum in ϵ'' near $\lambda = 1$ cm is shown in Fig. 9.15, which also shows how well the simple Debye theory fits experimental data. This suggests that H_2O is not a linear molecule with, say, two hydrogen atoms at equal distances from and on opposite sides of the oxygen atom. Instead, the hydrogen atoms are grouped toward the same side of the oxygen atom; the angle between them is about $105°$. The resulting electron distribution favors a higher electron density around a hydrogen atom; this imparts to the molecule a slightly ionic binding character and gives it a permanent dipole moment which can respond to microwave electric fields. Damping of the dipoles, which determines the relaxation time, is highly temperature dependent. Moreover, the molecules abruptly lose their orientational mobility as liquid water freezes to ice, and the Debye relaxation vanishes. The result is a huge change in the dielectric function at microwave frequencies, with effects extending well into the infrared. Because of the frequent coexistence of ice and water and the prevalent use of radar to probe the earth's environment, the difference between the dielectric functions of water and ice can have important consequences. For example, radar reflections from thunderstorms are quite different for rain and hail. Radio waves have been used to map the topography of polar ice sheets by means of reflected signals (Robin et al., 1969); however, it may not be possible to use such techniques for glaciers and other ice masses containing appreciable liquid water.

10.3.3 The Molecular Vibration Region

The infrared absorption bands of water can be grouped into two regions: the molecular bands between 1 and 10 μm, and the intermolecular bands between 10 and 100 μm. In the first region the absorption bands are similar to those of the free molecule; they occur in the liquid at 1.45, 1.94, 2.95, 4.7, and 6.05 μm. The fundamental vibration frequencies of the free (vapor phase) H_2O molecule help elucidate the vibrational modes of the H_2O molecule in the liquid phase. The nonlinear triatomic molecule H_2O has nine degrees of freedom: three describing the translational motion of the center of mass; three rotational degrees of freedom; and three vibrational degrees of freedom. The normal modes of vibration of the free H_2O molecule are illustrated schematically in Fig. 10.4 together with the notation commonly used to denote the three normal frequencies. We may now surmise that the absorption band near 3 μm in liquid and solid H_2O is related to the $\bar{\nu}_1$ and $\bar{\nu}_3$ vibrational modes of the free molecule, which are the O—H stretching modes. The 6.05-μm band in the liquid is the $\bar{\nu}_2$ bending mode of the molecule, which occurs at 6.27 μm in the

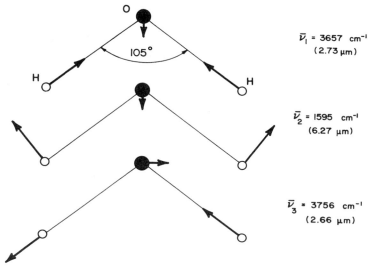

Figure 10.4 Normal modes of vibration of the free water molecule.

free molecule. The other absorption bands in water are more difficult to interpret precisely, but they are evidently overtones and combination bands of the three fundamental vibrations.

There are two broad absorption bands in water at 17 and 62 μm superimposed on the tail of the Debye absorption. Since there are no broad absorption bands in H_2O vapor in this region these bands must be the result of interaction between molecules—intermolecular vibrations. The fact that these bands show larger differences between the liquid and solid phases than the shorter-wavelength molecular vibrations is consistent with this interpretation. There are two kinds of intermolecular vibrations: rotational and translational. The shorter wavelength band, at 17 μm in water and 12 μm in ice, is the rotational band. Note the appreciable shift of this band toward higher frequencies in passing from water to ice, a consequence of the expected "tightening" of the rotational restoring forces as the liquid freezes. The 62-μm band in water is the intermolecular translational band. In ice it is made up of two bands, at 44.8 μm and 62 μm. Largely as a result of much less Debye absorption at these wavelengths, the intermolecular bands in ice are much more dominant than in water. In Table 10.1 we summarize the absorption bands in H_2O that are the result of molecular motions.

There is an important difference between an ionic crystal like MgO and a molecular substance such as liquid or solid H_2O. In the molecular substance there is a clear distinction between molecular and intermolecular modes because the atoms are bound into molecules much more tightly than the molecules are bound to one another. In the ionic crystal there is no such

Table 10.1   Molecular Absorption Bands in Liquid and Solid H_2O

Band Positions (μm)		Type of Molecular Motion
Water	Ice	
1. 45	1. 52	
1. 94	2. 0	
2. 95	3. 08	O—H molecular stretching mode ($\bar{\nu}_1$ and $\bar{\nu}_3$)
4. 7	4. 5	
6. 05	6. 05	H—O—H molecular bending mode ($\bar{\nu}_2$)
17	12	Intermolecular rotational
62	45	Intermolecular translational
	62	
1 cm	—	Debye relaxation

distinction; only one dominant absorption mode exists associated with lattice waves in which positive and negative ions move in opposition (the optic mode). One cannot even distinguish a unique cluster of atoms or ions to call a molecule; in a sense, the entire crystal is a molecule. An important consequence of this distinction between ionic and molecular substances is that vibrational absorption in the former is packed into one small frequency range, whereas in the latter there are a number of isolated absorption bands. It follows from the Kramers–Kronig relations, therefore, that ionic crystals usually have much stronger vibrational absorption bands than do molecular substances.

10.3.4   The Electronic Absorption Region

Features in the far-ultraviolet absorption spectrum of H_2O are the result of electronic excitations. The most prominent features in ϵ'' for water occur at 8.2 eV (8.5 eV for ice), 9.8 eV, and 13.5 eV. It is clear from inspection of the dielectric function that the electronic excitations are not exhausted beyond 14.6 eV, the region for which there is no published data; for if this were so, then ϵ'' would be approaching zero and ϵ' would be approaching one from lower values (see Section 9.2 and the high-energy data in Figs. 9.5 and 10.2). Interpretation of the high-energy peaks for H_2O relies heavily on intercomparisons among the liquid, vapor, and solid phases (Kerr et al., 1972). For example, the water band at 8.2 eV is associated with an electronic excitation in the free molecule at 7.5 eV; the energy is shifted in going from the vapor to the liquid or solid because of bonding between molecules. The other two bands are similarly related to corresponding structure in the electronic absorption spectrum of H_2O vapor. This close correspondence between electronic absorption spectra of free molecules and condensed phases is similar to that in the vibrational region. Interpretation of electronic absorption in the ionic crystal MgO, however, requires a complete energy band calculation for guidance, and molecular spectroscopy is of little help.

10.4   A COMMENT ON THE MAGNITUDE OF k

There are two fields in which great effort has been expended to infer the imaginary part of the complex refractive index of remote particles: the study of interstellar dust and the study of atmospheric aerosols. Determination of k for atmospheric and interstellar particles is a stringent test of methods for inverting data obtained by remote sensing. In both fields values of k in the range 10^{-3}–10^{-2} at visible wavelengths have frequently been inferred. However, it is rare to find a solid or liquid with these values of k in the visible, and for good physical reasons. Most solids and liquids either have very low values of k (e.g., 10^{-6}) or k of order 1. Intermediate values, such as 10^{-3}, are usually found only at wavelengths where k is rapidly changing from low to high values. For example, k in the visible for MgO is less than 10^{-5} and for water is less than 10^{-8} (see Figs. 10.1 and 10.3); for both materials k is 10^{-3} at about 2000 Å near a rapidly rising absorption edge. The reason for the propensity for either high or low values of k in solids at visible wavelengths lies in the electron energy band structure (see Fig. 9.10). In nonmetallic solids there is very little absorption for photon energies less than the band gap and, consequently, k is small; for photon energies greater than the band gap, however, k is large. Absorption usually rises rapidly as the photon energy increases through the band gap region. Metals, on the other hand, have large values of k at all frequencies below the plasma frequency, which is characteristically in the ultraviolet.

Our discussion so far has been confined to pure materials. It might be supposed that any value of k could be achieved by introducing controlled amounts of impurities or defects; however, this is generally not so. A solid cannot be doped to arbitrary levels; in addition, the impurity must produce absorption in the band gap region. Although some atoms, most notably transition metal atoms, do give absorption bands when incorporated into some solids, it is usually difficult to have enough impurities to give $k = 0.01$; we illustrate this with a story.

A Dirty Silicate Story A friend of one of the authors is an astronomer—as well as a professional mineral dealer—who became interested in "dirty silicates" as candidates for interstellar dust. He therefore selected for determination of k the blackest natural silicate mineral in his possession, the coal-black mineral hornblende, which contains a high concentration of impurities such as iron. A slice about 100 μm thick was polished, and transmission was measured in a recording spectrophotometer. The fact that appreciable light was transmitted for all near-infrared and visible wavelengths indicated that k was rather small. Calculations indeed confirmed that k was less than 10^{-4} between about 6 and 0.3 μm. And yet this was the blackest silicate in the possession of a professional collector. It is not easy to find $k = 0.01$ in the band gap region of a solid; on the other hand, where interband or intraband transitions occur, it is not easy to find such a low value.

The example in the preceding paragraph also illustrates that a macroscopic sample that looks black does not have a k of order 1; if it did have such a high k value, it would probably look shiny. Let us emphasize this with a quantitative example. Suppose that a 1-mm-thick wafer of a black solid transmits 1% of the light incident on it. If we neglect reflection losses (typically small for nonconductors), then k is approximately (see Section 2.8) $(\lambda/4\pi h)\ln(I_i/I_t)$. For $\lambda = 0.5$ μm, $h = 0.1$ cm, and $I_i/I_t = 100$, k is about 2×10^{-4}. *Thus, black materials need not have k values in the visible much greater than 10^{-4}.* A k of 10^{-4}, which might seem small at first glance, is in fact huge: it gives rise to appreciable absorption of visible light in all but very thin samples.

10.5 VALIDITY OF BULK OPTICAL CONSTANTS IN SMALL-PARTICLE CALCULATIONS

Before continuing into Chapters 11–13, where we use bulk optical constants in small-particle calculations, it is well to pause and remind ourselves that bulk optical constants, without modification, may not always apply to small particles. Up to this point we have tacitly assumed that the interaction of electromagnetic waves with particles can be treated in the continuum approximation—the electromagnetic theory of Part 1—with constitutive relations independent of size; dielectric functions, such as those presented in this chapter and the preceding one, are for bulk matter. It requires no great insight, however, to realize that matter cannot be subdivided indefinitely without change in optical properties. For example, optical absorption by atomic aluminum is clearly not the same as that by bulk metallic aluminum, and the absorption spectrum of molecular MgO bears little resemblance to that of crystalline MgO. The question to ask, therefore, is not if, but at what size, optical constants appropriate to small particles deviate appreciably from those of bulk matter. This question cannot be answered simply and precisely because there may be several reasons for the breakdown of bulk optical constants, the relative importance of each of which depends on the type of material, particle size, and frequency. For example, in sufficiently small metallic particles the electron mean free path is limited by the particle boundary, with consequences for their optical properties; this is discussed in Section 12.1. Other effects are increased spacing between electronic and vibrational energy levels and the increased importance of surface states in particles with large surface-to-volume ratios; these have been surveyed by Huffman (1977).

Although we admit that these effects must surely appear in small particles, we cannot countenance the uncritical invocation of vague and esoteric "quantum size effects," as is sometimes done, to dispose of optical phenomena in small particles that are not understood. Often the correct interpretation may be much simpler: particle shape, for example, which we shall discuss in Sections 12.3 and 12.4. Our best advice, based on calculations (Martin, 1973; Chen et al., 1978, for example) and those experiments which have been done with sufficient care to allow unambiguous interpretation, is that quantum size effects are unimportant except for particle sizes well below 1000 Å, in most instances below 100 Å. We shall bring evidence to bear in support of this assertion in Chapter 12.

10.6 SUMMARY OF ABSORPTION MECHANISMS AND TEMPERATURE EFFECTS

There are many ways in which electromagnetic waves can interact with matter in its condensed phases, liquid and solid. Some of these have been treated with simple models in Chapter 9, and examples are given in this chapter. Lest we leave the reader with an oversimplified view of optical constants we list in Table 10.2 several absorption mechanisms in solids together with the spectral regions in which they are important. References, primarily review articles and monographs, are also included to guide the reader in further study.

Most optical studies of condensed matter have been made near room temperature, but there are applications in which optical properties are required at high temperatures (e.g., particles as solar radiation absorbers) or at low temperatures (e.g., interstellar dust grains). Although we cannot discuss all possibilities, the following brief comments may be helpful.

In general, temperature tends to affect more greatly low-frequency (infrared and far infrared) than high-frequency (visible and ultraviolet) absorption mechanisms. At room temperature ($\sim 300°K$) $k_B T$ corresponds to about 0.025 eV ($\lambda \simeq 10 \ \mu m$); thermal energies are therefore quite small compared with ultraviolet photon energies, so temperature changes only slightly perturb ultraviolet absorption mechanisms. Of the two major types of absorption mechanisms, vibrational and electronic, the former are far more temperature dependent. Infrared absorption bands of ionic solids, associated with collective lattice oscillations, commonly increase in width and decrease in height with increasing temperature; the frequency of peak absorption tends to shift to lower values, although the shift is small except for large temperature changes ($100–1000°K$). Excellent examples are given by Jasperse et al. (1966), who studied the *Reststrahlen* band in LiF from 7.5 to 1060°K and that in MgO from 8 to 1950°K. Far-infrared ($> 100 \ \mu m$) absorption in crystalline solids may change by several orders of magnitude as temperature increases, although absorption in amorphous solids tends to be independent of temperature (Mitra and Nudelman, 1970).

Temperature changes do not appreciably affect ultraviolet optical properties of both metals and insulators, although at low temperatures absorption bands associated with excitons and electron band transitions are usually sharper, and frequencies of peak absorption may shift slightly. In the soft x-ray region, transitions of core electrons buried in the interior of atoms hardly notice temperature changes.

Table 10.2 Summary of Absorption Mechanisms

Spectral Region	Type of Absorption	References
Microwave > 1 mm	Debye relaxation in polar matter	Debye (1929)

Far infrared
  ~ 1 mm (1000 μm)
  to ~ 100 μm
	Electron energy splitting
	  in rare earth ions
	Phonon-difference processes
	Superconducting band gap
	Defect-induced vibrations
	Impurity-induced vibrations
		Hadni (1970a)
		
		Hadni (1970b)
		Tinkham (1970b)
		Genzel (1969)
		McCombie (1970)

Infrared
  ~ 100
  to ~ 2 μm
	Intramolecular vibrations
	Intermolecular vibrations
	Reststrahlen absorption
	  in ionic crystals
	Two-phonon and multiphonon
	  processes
	Magnetic excitation
	  in magnetic insulators
	Intraband electron
	  transitions in metals
	Free carriers in
	  semiconductors
		Sherwood (1972)
		Sherwood (1972)
		Mitra (1969);
		  Plendl (1970)
		Bendow (1978)
		
		Tinkham (1970a)
		
		Nilsson (1974);
		  Glicksman (1971)
		Dixon (1969);
		  Moss (1959)

Near visible
  ~ 2 μm to
  ~ 3000 Å
	Electron states of impurities
	
	No-phonon lines and
	  phonon sidebands
	No-magnon lines and
	  magnon sidebands
	Electron defect states
	  (color centers)
	Transition-metal-ion
	  energy levels
	
	Electron band gaps
	  in semiconductors
	Electron band tailing
	  in amorphous semiconductors
		McClure (1959b);
		  Burns (1970)
		Silsbee (1969);
		  Rebane (1974)
		McClure (1974)
		
		Markham (1966)
		
		DiBartolo (1974);
		  Crosswhite and
		  Moos (1967)
		Harbeke (1972)
		
		Tauc (1972)

Far ultraviolet
  ~ 3000 to
  ~ 500 Å
	Exciton bands (single
	  and series)
	Band-to-band transitions
	  of electrons
	Molecular electron
	  transitions
	Electron plasma
	  oscillations
	  in metals
		Nikitine (1969);
		  Reynolds (1969)
		Phillips (1966)
		
		McClure (1959a)
		
		Steinmann (1968);
		  Glicksman (1971)

Extreme ultraviolet,
  soft x-ray
  ~ 500 to
  ~ 20 Å
	Core electron transitions	Brown (1974)

Recall, however, that temperature changes often give rise to phase changes such as solid-to-liquid, magnetic, and superconducting phase transitions, with associated appreciable changes in optical properties at all wavelengths. We mentioned in Section 9.5 that the Debye relaxation mechanism, which dominates the radio-frequency behavior of liquid water, practically vanishes in the transition to ice. Similarly, superconducting band gap absorption disappears above the transition temperature.

NOTES AND COMMENTS

Optical properties of solids encompasses a field of study too extensive to be covered adequately in a single volume. Monographs that emphasize different aspects of the subject, together with the topics they treat especially well, are listed below.

Wooten (1972): Metals.

Greenaway and Harbeke (1968): Semiconductors.

Sherwood (1972): Infrared lattice vibrations.

Burns (1970): Ions in crystals.

Some of the most comprehensive coverages of optical properties are found in the following proceedings of summer schools and conferences: Tauc (1966), Nudelman and Mitra (1969), Mitra and Nudelman (1970), Abelès (1972), Mitra and Bendow (1975); the first two are especially recommended.

The following articles in the series *Solid State Physics* offer comprehensive reviews of various topics.

McClure (1959ab): Molecules and ions in crystals.

Stern (1963): Optical properties in general.

Markham (1966): Color centers.

Phillips (1966): Fundamental electronic spectra.

Glicksman (1971): Metals.

Brown (1974): Extreme ultraviolet spectra of solids.

Nilsson (1974): Metals and alloys.

Bendow (1978): Transparent insulators.

The first two volumes in a compilation of optical data on all the metallic elements, in the photon energy range 0.1–500 eV, have been recently published by Weaver et al. (1981ab).

Two highly recommended papers on the color of water are those by Bancroft (1919) and by Raman (1922). The first of these contains lengthy quotes from earlier work.

Molecular vibrations are treated in great detail in the books by Herzberg (1945) and by Wilson et al. (1955).

Part 3

Optical Properties
of Particles

Chapter 11

Extinction

Extinction is the attenuation of an electromagnetic wave by scattering and absorption as it traverses a particulate medium. In homogeneous media the dominant attenuation mechanism is usually absorption. Comparison of extinction spectra for small particles of various sizes with absorption spectra for the bulk parent material reveals both similarities and differences; we shall make such comparisons in the early part of this chapter. After reviewing the meaning of extinction and the various ways it is commonly presented (Section 11.1), we survey extinction effects in particles of the three illustrative materials of Chapter 10 (Section 11.2). Following this survey, the major features of spectral extinction (except those reserved for Chapter 12) are discussed in more detail: extinction dominated by scattering, including size and size distribution effects (Section 11.3); ripple structure (Section 11.4); and the effects of absorption (Section 11.5). Calculations of extinction by spheroids and infinite cylinders in Section 11.6 are followed in Section 11.7 by measurement techniques and experimental results for both spherical and nonspherical particles. We conclude with a brief summary in Section 11.8.

11.1 EXTINCTION = ABSORPTION + SCATTERING

If multiple scattering is negligible the irradiance of a beam of light is exponentially attenuated from I_i to I_t in traversing a distance h through a particulate medium (see Section 3.4):

$$\frac{I_t}{I_i} = \exp(-\alpha_{\text{ext}} h).$$ (11.1)

Extinction is the result of both absorption and scattering:

$$\alpha_{\text{ext}} = \mathfrak{N}(C_{\text{abs}} + C_{\text{sca}});$$ (11.2)

$\mathfrak{N}$ is the number of particles per unit volume; C_{abs} and C_{sca} are the absorption and scattering cross sections. Although both processes occur simultaneously, there are instances where one or the other dominates. For example, visible light passing through a fog is attenuated almost entirely by scattering, whereas light passing along the shaft of a coal mine might be attenuated primarily by absorption. A simple but effective demonstration of these two extremes, particularly well suited to large audiences, is shown schematically in Fig. 11.1. Two transparent containers (Petri dishes serve quite well) are filled with water, placed on an overhead projector, and their images focused on a screen. To one container, a few drops of milk are added; to the other, a few drops of India ink. The images can be changed from clear to a reddish hue to black by increasing the amount of milk or ink. Indeed, both images can be adjusted so that they appear equally dark; in this instance it is not possible, judging solely by the light transmitted to the screen, to distinguish one from the other: the amount of extinction is about the same. But the difference between the two suspensions immediately becomes obvious if one looks directly at the containers: the milk is white whereas the ink is black. Milk is a suspension of very weakly absorbing particles which therefore attenuate light primarily by scattering; India ink is a suspension of very small carbon particles which attenuate light primarily by absorption. Although this demonstration is not meant to be quantitative, and its complete interpretation is complicated somewhat by multiple scattering, it clearly shows the difference between extinction by scattering and extinction by absorption. Moreover, it shows that merely by observing transmitted light it is not possible to determine the relative contributions of absorption and scattering to extinction; to do so requires an additional independent observation.

Extinction of light by particles is a fairly commonly observed phenomenon: sunlight through a dust storm or a polluted layer of air; automobile headlights in fog; a skin diver's light in murky water. All these examples of extinction, however, are not necessarily described by the theoretical expressions (11.1) and (11.2). Multiple scattering aside, underlying (11.2) is the assumption that *all* light scattered by the particles, however small the scattering angle, is excluded from the detector. But the eye, or any other detector, collects light scattered in

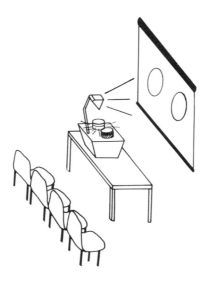

Figure 11.1 A demonstration of how extinction may be dominated either by scattering or by absorption: one Petri dish contains a milk suspension; the other contains an India ink suspension.

some possibly small but finite set of angles around the forward direction; the larger the particles the more the scattering diagram is peaked in the forward direction and the greater the possible discrepancy between measured and calculated extinction. Thus, a detection system may have to be carefully designed to measure extinction that can be legitimately compared with theoretical extinction, particularly by large particles. Extinction by interstellar dust is an example where the requirements of theory are most likely to be strictly satisfied: distances to particles are thousands of light years and a detector coupled to a telescope receives almost entirely light that has not been scattered.

There is not a unique way of representing extinction. Physicists, because they are accustomed to cross sections for various atomic and nuclear processes, are probably most comfortable with the extinction cross section $C_{ext} = C_{abs} + C_{sca}$. On the other hand, the extinction efficiency $Q_{ext} = C_{ext}/$(cross-sectional area) is more likely to bring a smile to the face of an astronomer or an atmospheric scientist. In the books by van de Hulst (1957) and Kerker (1969), extinction efficiencies are presented almost exclusively; consequently, this has become widespread. But we argued in Section 3.4 that the extinction cross section per unit particle volume (or mass) may be a more appropriate measure of how efficiently a particle attenuates light. These three ways of representing extinction, which are listed below, may be looked upon as merely different ways of normalizing the extinction cross section.

Cross section	C_{ext}	$Q_{ext}\pi a^2$ (sphere)
Cross section per unit area	C_{ext}/A	Q_{ext}
Cross section per unit volume	C_{ext}/V	$3Q_{ext}/4a$ (sphere)

Plots of each of these quantities as a function of particle size would look quite different and, therefore, would tell different stories. Except for a scale factor, each of them plotted as a function of wavelength for the same particle size would be identical. In our first example of extinction (Fig. 4.6) we displayed the efficiency Q_{ext}, as we shall often do in this chapter. In Chapter 12, however, our preference switches to the extinction cross section per unit particle volume. Unnormalized extinction cross sections (strictly speaking, the differential scattering cross section integrated over the acceptance angle of the detector) are more appropriate in Section 13.5 on particle sizing.

11.2 EXTINCTION SURVEY

Curves of extinction as a function of size parameter show a wealth of features, even when calculated with uninteresting constant optical constants, as has often been done. When realistic optical constants are used in calculations the different types of extinction effects become even more numerous. In this section we incorporate optical constants of the three illustrative materials of

Chapter 10 with Mie theory in a survey of extinction effects which we shall discuss in more detail in subsequent sections.

11.2.1 Magnesium Oxide

Volume-normalized extinction is plotted in Fig. 11.2 as a function of photon energy for several polydispersions of MgO spheres; both scales are logarithmic. For comparison of bulk and small-particle properties the bulk absorption coefficient $\alpha = 4\pi k/\lambda$ is included. Some single-particle features, such as ripple structure, are effaced by the distribution of radii. The information contained in these curves is not assimilated at a glance: they require careful study.

At and near visible wavelengths, where bulk MgO is highly transparent, extinction is dominated by scattering. As a consequence, attenuation by MgO particles is quite different from that by the bulk solid. If the particles are much smaller than the wavelength, scattering varies as the fourth power of the photon energy $(1/\lambda^4)$; at such wavelengths extinction is a linear function of photon energy on a logarithmic plot. Greater extinction of higher than lower energy light gives rise to reddening, which may be observed in light transmitted through an MgO-smoked microscope slide. If the particles are large compared with the wavelength, extinction dominated by scattering is nearly independent of photon energy. Note in particular the broad range of neutral extinction for the particles with mean radius 1.0 μm. White light would undergo little discernible change in color upon transmission by such particles.

At energies where MgO electronic absorption sets in, structure in the bulk optical constants appears in extinction (both scattering and absorption), but only for the smallest particles (0.01 μm). No structure is apparent in the extinction spectrum of particles with mean radius 0.05 μm at energies greater than about 7.6 eV. At these energies the particles are "black": no appreciable light that penetrates the particles reemerges carrying the spectral information. This effect could be called "saturation", although this term is used in many different ways.

Because the particles must be very small, not more than a few hundred angstroms for most substances, there have been few laboratory observations of structure in ultraviolet extinction spectra of small, nonmetallic particles. Particles of the required size are difficult to make. Grinding them from the bulk, with the attendant problem of separation, is almost hopeless. Various smokes can be made, MgO particularly easily, but not of all substances that might be of interest.

Extinction and absorption spectra for 0.01 μm particles with optical constants appropriate to radiation-damaged MgO (see Fig. 10.1) are also shown in Fig. 11.2. The lowest-energy absorption band among the three broad bands shows clearly in extinction but not the highest-energy band. In this instance obscuration of structure is caused by the dominance of scattering over absorption, not "saturation" of absorption. This is obvious from the absorption spectrum, in which all three bands are evident. To observe the band at 3.5 eV

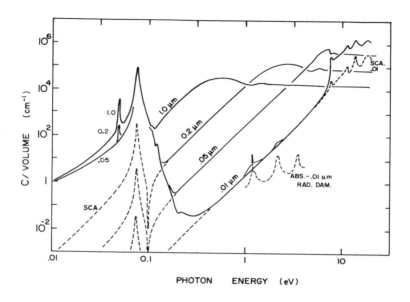

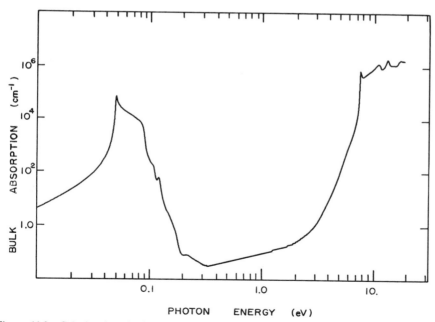

Figure 11.2 Calculated extinction spectrum of MgO spheres below which is the absorption spectrum of bulk MgO.

would require measuring absorption in the presence of a large amount of scattering. Diffuse reflectance techniques (Kortüm, 1969) or the increasingly popular photoacoustic spectroscopy (Rosencwaig, 1980) could be used. These techniques, however, would be applied less successfully to absorption bands between 7 and 20 eV for 0.05-μm particles or larger.

At infrared wavelengths extinction by the MgO particles of Fig. 11.2, including those with radius 1 μm, which *can* be made by grinding, is dominated by absorption. This is why the KBr pellet technique is commonly used for infrared absorption spectroscopy of powders. A small amount of the sample dispersed in KBr powder is pressed into a pellet, the transmission spectrum of which is readily obtained. Because extinction is dominated by absorption, this transmission spectrum should follow the undulations of the intrinsic absorption spectrum—but not always. Comparison of Figs. 10.1 and 11.2 reveals an interesting discrepancy: calculated peak extinction occurs at 0.075 eV, whereas absorption in bulk MgO peaks at the transverse optic mode frequency, which is about 0.05 eV. This is a large discrepancy in light of the precision of modern infrared spectroscopy and could cause serious error if the extinction peak were assumed to lie at the position of a bulk absorption band. This is the first instance we have encountered where the properties of small particles deviate appreciably from those of the bulk solid. It is the result of surface mode excitation, which is such a dominant effect in small particles of some solids that we have devoted Chapter 12 to its fuller discussion.

The scattering cross section in Fig. 11.2 also peaks at about 0.075 eV, although it is much less than the corresponding absorption cross section. Note that the scattering cross section drops sharply near 0.1 eV, where the real part of the refractive index is about 1 and the imaginary part is low: the particles are almost optically identical with free space. This is the *Christiansen effect*, which can be exploited to make a band-pass filter (see, e.g., Smith et al., 1968, pp. 395–399). Transmission through a collection of particles is greatest near the Christiansen frequency (where $n \simeq 1$ and $k \simeq 0$) because scattering nearly vanishes; at neighboring frequencies scattering greatly diminishes the transmitted light. Larger particles than we have discussed in connection with Fig. 11.2 are required, however, so that absorption does not dominate over scattering.

11.2.2   Water

Mie calculations with the optical constants of water given in Fig. 10.3 are shown in Fig. 11.3; extinction and absorption are plotted logarithmically, photon energy linearly. The bulk absorption coefficient of water is shown in Fig. 11.3c. Because many of the extinction features of water and MgO, both of which are insulators, are similar, we present calculations for a single water droplet (in air) with radius 1.0 μm. Size-dependent spectral features are therefore not obscured as they are for a distribution of radii.

Perhaps the most conspicuous features in extinction appear where water is transparent, between about 0.5 and 7 eV. Interference of incident and

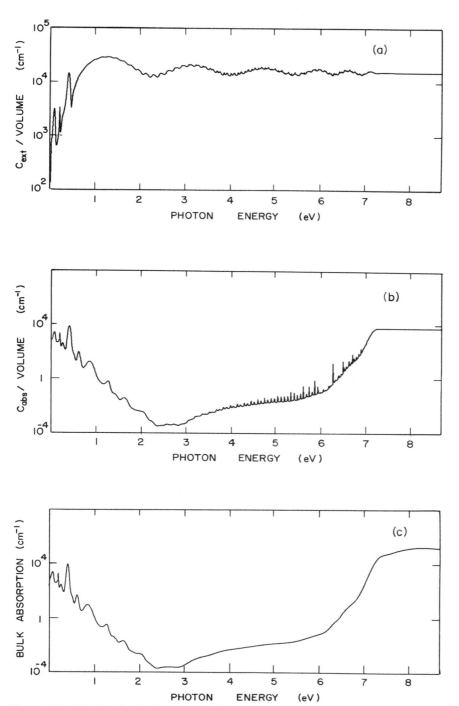

Figure 11.3 Calculated extinction (*a*) and absorption (*b*) by a water droplet of radius 1.0 μm. The absorption spectrum of water is shown in (*c*).

forward-scattered light (see Section 4.4) gives rise to *interference structure*: a series of broad, regularly spaced extinction maxima and minima. Superimposed on this is sharp and highly irregular *ripple structure*, which originates in resonant electromagnetic modes of a sphere. We shall discuss interference and ripple structure in more detail in Sections 11.3 and 11.4. Since the positions of these features are size dependent (especially ripple structure) they tend not to appear in calculations for polydispersions. This is why the extinction curves for MgO (Fig. 11.2) are so different from the extinction curve in Fig. 11.3. Also included in this figure is the absorption cross section per unit particle volume. Interference structure does not appear in absorption but ripple structure does, although at energies where absorption is very small. Comparison of the curves for a water droplet (*a* and *b*) with that for water (*c*) shows that interference and ripple structure give to a droplet spectral features which contrast markedly with those of its parent material.

The extinction features at energies where water is transparent are rapidly squelched in the ultraviolet as the onset of electronic transitions greatly increases bulk absorption. In the infrared, however, vibrational absorption bands in water are carried over into similar bands in extinction (dominated by absorption if $a \ll \lambda$) by a water droplet. Unlike MgO there are no appreciable spectral shifts in going from the bulk to particulate states. The reason for this lies in the strength of bulk absorption and will be discussed more thoroughly in Chapter 12.

11.2.3 Aluminum

The calculated extinction spectrum of a polydispersion of small aluminum spheres (mean radius 0.01 μm, fractional standard deviation 0.15) is shown in Fig. 11.4; both scales are logarithmic. In some ways spectral extinction by metallic particles is less interesting than that by insulating particles, such as those discussed in the preceding two sections. The free-electron contribution to absorption in metals, which dominates other absorption bands, extends from radio to far-ultraviolet frequencies. Hence, extinction features in the transparent region of insulating particles, such as ripple and interference structure, are suppressed in metallic particles because of their inherent opacity. But extinction by metallic particles is not without its interesting aspects.

Note that there is *no bulk absorption band* in aluminum corresponding to the prominent extinction feature at about 8 eV. Indeed, the extinction maximum occurs where bulk absorption is monotonically decreasing. This feature arises from a resonance in the collective motion of free electrons constrained to oscillate within a small sphere. It is similar to the dominant infrared extinction feature in small MgO spheres (Fig. 11.2), which arises from a collective oscillation of the lattice ions. As will be shown in Chapter 12, these resonances can be quite strongly dependent on particle shape and are excited at energies where the real part of the dielectric function is negative. For a metal such as aluminum, this region extends from radio to far-ultraviolet frequencies. So the

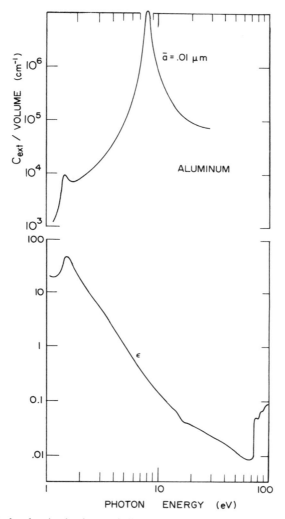

Figure 11.4   Calculated extinction by a polydispersion of aluminum spheres (top) compared with the bulk absorption spectrum of aluminum (bottom).

possible range of shape-dependent features in extinction spectra of small metallic particles is correspondingly broad.

11.3   SOME EXTINCTION EFFECTS IN INSULATING SPHERES

In the preceding section and elsewhere we have touched on a few points that deserve further elaboration. For example, we have emphasized that extinction calculations should, in general, be done with wavelength-dependent optical constants; in this section we examine the consequences of not doing so. Then we go on to consider the progressive obliteration of features in extinction by a polydispersion of spheres as the spread of their radii is increased.

11.3.1 Two Views of Extinction

Two curves showing extinction by a sphere are given in Fig. 11.5. The refractive index for the top curve is $1.33 + i10^{-8}$, which corresponds to water at about 5600 Å. This is the type of extinction curve most commonly encountered, although it is easy to misinterpret. It strictly represents the dependence of extinction on radius at a *fixed* wavelength, not its dependence on wavelength, despite the fact that the abscissa is proportional to $1/\lambda$. The reason for this is that the optical constants of water—indeed, any substance—depend on wavelength. Only if a suitable refractive index $m(\lambda)$ is incorporated with Mie theory can a calculated curve faithfully show how extinction varies with wavelength, and then only for a single size. This type of curve, for a particle of radius 1.0 μm, is shown in Fig. 11.5b.

Comparison of the two curves in Fig. 11.5 reveals both similarities and differences. In the transparent region, between about $x = 3$ and $x = 37$, the extinction features are qualitatively similar but the positions of the major (interference) peaks may be appreciably displaced. The positions of the first peaks (near $x = 7$) are almost identical. But the second peaks are slightly shifted and the third- and higher-order peaks are considerably shifted. This progressively increasing displacement of peak positions is a result of the increase of the real part of the refractive index of water toward the ultraviolet. In the regions where water is absorbing—infrared and ultraviolet—the two curves show little similarity. At $x = 37$ ripple structure abruptly vanishes from the extinction curve of Fig. 11.5b; this results from the strong electronic absorption that sets in shortward of about 0.2 μm, the absorption edge. A similar effect occurs for all nonmetallic solids and liquids, although the position of the absorption edge—in or near the ultraviolet—is different for each substance. At values of x less than about two extinction features in Fig. 11.5b do not have counterparts in Fig. 11.5a; these features arise from the infrared absorption bands of water.

11.3.2 Size Distribution Effects

Extinction features that strongly depend on particle size will be obscured, if not totally obliterated, in a polydispersion. Many analytical expressions for the radius probability distribution have been used in Mie calculations. For purposes of illustration we have chosen the Gaussian distribution, according to which the probability that a sphere has radius between a and $a + da$ is

$$\frac{1}{\sigma\sqrt{2\pi}} \exp\left[-\frac{1}{2}\left(\frac{a - \bar{a}}{\sigma} \right)^2 \right] da,$$

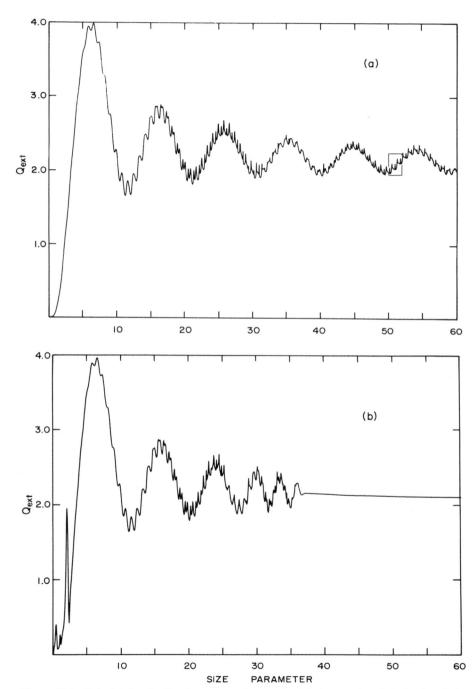

Figure 11.5 Calculated extinction by a water droplet. In (*a*) the wavelength is fixed and the radius is varied; in (*b*) the radius (1.0 μm) is fixed and the wavelength is varied.

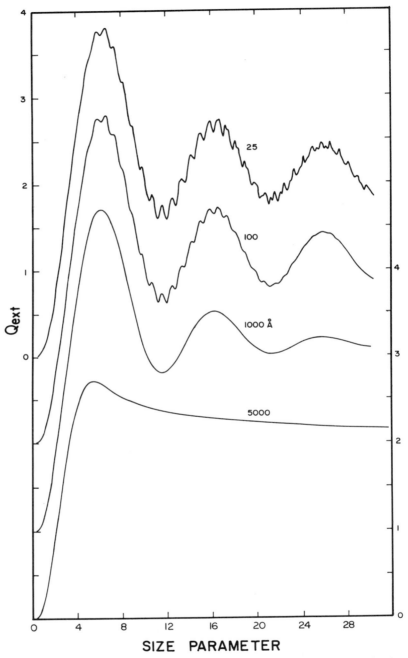

Figure 11.6 The effect of size dispersion on extinction of visible light by water droplets. Each curve is labeled with σ, the standard deviation in the Gaussian size distribution.

298

where $\bar{a}$ is the mean radius and σ^2 is the variance. Although the Gaussian distribution probably does not represent natural aerosols and other particulate systems very well its advantages are simplicity and familiarity.

To show the effect of increasing size dispersion on extinction a series of calculations for water droplets is given in Fig. 11.6. The topmost curve reproduces the calculations of Fig. 11.5a for a single sphere; the standard deviation σ is increased in successively lower curves.

Ripple structure, beginning with the sharpest at large size parameters, is the first to disappear as σ is increased. As the distribution is further widened, the interference structure fades away. For the widest distribution the only remaining features are reddening at small size parameters, and, at the other extreme, an asymptotic approach to the limiting value 2.

Without an appreciation for the possible spread of sizes in real particulate systems the values of σ in Fig. 11.6 are merely those of an adjustable parameter. We therefore give distribution widths for some natural and artificial aerosols and hydrosols in Table 11.1; we excluded from this list broad distributions, such as raindrops, to which the notion of a width is not really applicable.

Table 11.1 Approximate Size Distribution Parameters for Some Natural and Artificial Aerosols and Hydrosols

Particles	Mean Radius (μm)	Fractional Standard Deviation	Reference
Polystyrene spheres in water	0.05–2	0.005–0.01	
Bergland–Liu generator		0.01	Bergland and Liu (1973)
Sinclair–LaMer generator		0.1–0.2	Kerker (1969, p. 320)
Ultrasonic nebulizer		0.3	Perry et al. (1978)
Nozzles and sprays	20–300	1.1	Corn and Esmen (1976)
Corn rust spores	3.4	0.038	Corn and Esmen (1976)
Lycopodium spores	15	0.01	Corn and Esmen (1976)
Blood cells (dried)	8	0.08	Corn and Esmen (1976)
Clover pollen	26	0.04	Corn and Esmen (1976)
Giant ragweed pollen	9.8	0.05	Corn and Esmen (1976)
Orchard grass pollen	15.5	0.07	Corn and Esmen (1976)
Upper clouds of Venus	1.0	0.3	Hansen and Hovenier (1974)

11.4 RIPPLE STRUCTURE

Ripple structure—sharp, irregularly spaced features in extinction by weakly absorbing spheres—is evident in Figs. 4.6, 11.3, 11.5, and 11.6. In some ways ripple structure is merely a nuisance. It is, after all, unlikely to be observed in extinction by many polydispersions, natural and artificial. Yet calculations for a single sphere are of necessity done at a finite number of size parameters, one of which may just happen to correspond to a sharp peak unrepresentative of the generally smooth variation of extinction. Ways have been devised to get around this problem (see, e.g., Penndorf, 1958). Often the best approach in these days of fast computers is to do calculations for a distribution of sizes rather than for a single size (see Fig. 11.6). Before the digital computer was commonplace it was prohibitively expensive to do calculations for a very fine size parameter grid. Consequently, many of the older extinction curves do not exhibit as much structure as that shown in Fig. 11.5a. But even this curve is not indicative of the extreme fineness of extinction structure. A closer look at a small part of Fig. 11.5a (enclosed by a rectangle) reveals even more structure. Figure 11.7 shows results of extinction calculations for a water droplet similar to those of Chýlek et al. (1978a). Structure that was barely perceptible in Fig. 11.5a is now well resolved. Even on this expanded scale, however, there are unresolved extinction peaks; these are indicated by the vertical lines at about $x = 50.33$, 50.68, 51.12, and 51.90. To resolve these peaks requires an even finer size parameter grid. For example by decreasing Δx to 10^{-6} the peak near $x = 50.33$ is resolved in Fig. 11.8. Thus, the closer we look the more structure we find. One might well wonder if there is an end to this: is it a matter of dogs having fleas, which themselves have fleas, and so on *ad infinitum*? Chýlek et al. (1978a) have made a careful study of this question and claim that no more

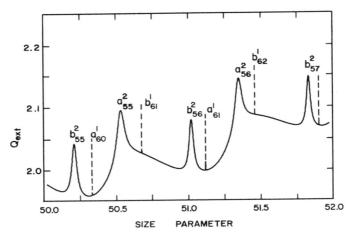

Figure 11.7 High-resolution ($\Delta x = 10^{-4}$) calculation of the ripple structure in extinction by a water droplet ($m = 1.33 + i10^{-8}$). After Chýlek et al. (1978a).

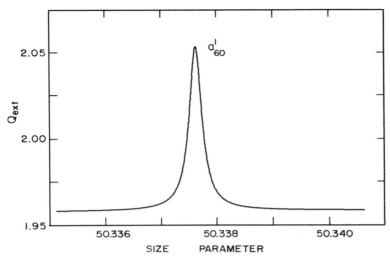

Figure 11.8   The unresolved extinction peak near $x = 50.33$ in Fig. 11.7 is resolved by decreasing Δx to 10^{-6}.

structure is revealed by decreasing Δx below 10^{-7}; therefore, the peak shown in Fig. 11.8 is an example of the narrowest to be found. Ripple structure is not limited to the scattering component of extinction but is present in absorption as well (Bennett and Rosasco, 1978).

We discussed in Section 4.3 the electromagnetic normal modes, or virtual modes, of a sphere, which are resonant when the denominators of the scattering coefficients a_n and b_n are minima (strictly speaking, when they vanish, but they only do so for complex frequencies or, equivalently, complex size parameters). But Q_{ext} is an infinite series in a_n and b_n, so ripple structure in extinction must be associated with these modes. The coefficient c_n (d_n) of the internal field has the same denominator as $b_n(a_n)$. Therefore, the energy density, and hence energy absorption, inside the sphere peaks at each resonance: there is ripple structure in absorption as well as scattering.

For each index n there is a sequence of values of x for which the mode associated with a_n or b_n is excited. We may therefore label each (nonoverlapping) extinction peak with the type of mode [electric (a_n) or magnetic (b_n)], the index n, and the sequential order of x (Chýlek, 1976): for example, a_{60}^1, a_{60}^2, and so on, where the superscript indicates the order of x; the extinction peaks between $x = 50$ and $x = 52$ for a water droplet are so labeled in Fig. 11.7.

Three peaks, labeled a_{60}^1, a_{60}^2, a_{60}^3, in the real part of $a_{60}(x)$ for a water droplet are shown in Fig. 11.9; the real part of the refractive index is fixed but the imaginary part is varied. These optical constants could be obtained by, for example, adding a little dye (food coloring) to water; this would increase k without appreciably changing n. Note that the horizontal scale is different for

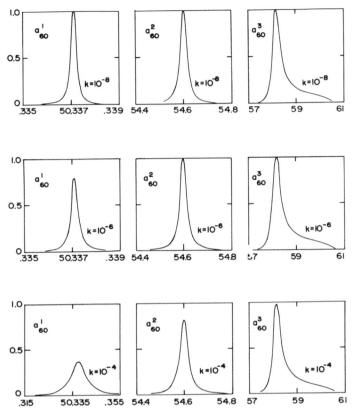

Figure 11.9 Three peaks in the real part of a_{60}. The imaginary part of the refractive index is progressively increased from top to bottom but the real part is fixed (1.33). After Chýlek et al. (1978b).

each set of curves. Similar curves, for different optical constants, have been given by Chýlek et al. (1978b).

The peaks appear to be Lorentzian, and if we were to plot $\mathrm{Im}\{a_{60}(x)\}$ this would be even more obvious. That this is indeed so, to good approximation, was shown by Fuchs and Kliewer (1968), who discussed in detail the normal modes of an ionic sphere.

For given k, the peaks broaden with increasing order. As absorption is increased for a fixed order each peak is broadened at the expense of its height, particularly the a_{60}^{1} peak; the broader, higher-order peaks are not nearly as severely damped by absorption. This accounts for the rapid loss of the sharpest extinction peaks with increased absorption (Fig. 11.12).

Ripple structure was observed in scattering at 90° by water droplets as they nucleated and grew in a cloud chamber (Dobbins and Eklund, 1977). We shall show in Section 11.7 that ripple structure is easily observed in extinction by

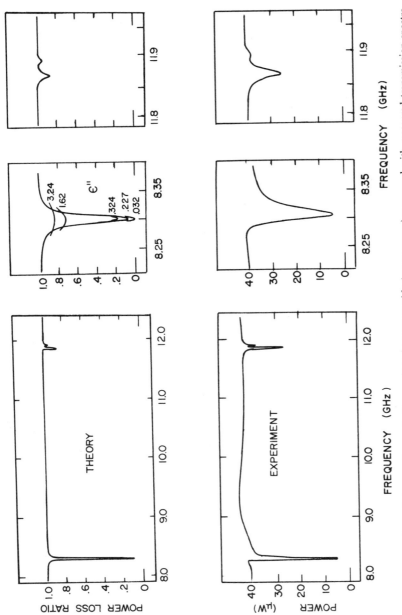

Figure 11.10 Theoretical power loss ratio for a sphere in a waveguide (upper curves) compared with measured transmission spectra (lower curves). The curves on the right are enlargements of the two bands; the effect of absorption on the lowest frequency band is shown in the upper right. From P. Affolter and B. Eliasson, *IEEE Trans. Microwave Theory Tech.*, **MTT-21** (1973), 573–578; © 1973 IEEE.

polystyrene spheres suspended in water. In both these examples the ripples are fairly broad and therefore can be observed despite the distribution of particle sizes. Single spheres are required, however, if one is to observe the very sharp structure associated with larger, more refractive spheres. Such observations have been reported for the microwave and visible regions; these are discussed below.

11.4.1 Sphere in a Microwave Waveguide

The possibility of making use of ripple structure in microwave devices has been discussed for years (e.g., Richtmyer, 1939). For example, a "notch" filter can be made by combining a sphere with a waveguide. This was done by Affolter and Eliasson (1973): they suspended a 0.2-cm $SrTiO_3$ ($\epsilon' = 324.4$) sphere symmetrically in a waveguide and measured the transmitted power as a function of frequency. Their results are shown in the lower part of Fig. 11.10; calculations of the power loss ratio (equivalent to I/I_0 in our notation) are shown in the upper part of this figure. Attenuation of microwave power is almost complete in a narrow frequency band. Calculations show that the shape of this band strongly depends on ϵ'', which is also evident in Fig. 11.12; Affolter and Eliasson suggested that this could be used to measure ϵ'' of weakly absorbing materials.

11.4.2 Ripples in Radiation Pressure

It might be thought that ripple structure, particularly the sharp features at large x (Fig. 11.5), would be primarily of theoretical interest because it could not be observed in a distribution of sizes. Even with a single particle a small dispersion of wavelengths would obliterate such structure. In recent years, however, many features of the complex ripple structure pattern have been observed in a set of elegant and beautiful experiments. Ashkin and Dziedzic (1971, 1976, 1977) have developed a method of levitating transparent particles in a strongly focused laser beam. Not only is a particle balanced against gravity by radiation pressure, it is also laterally stabilized in the beam. An electronic feedback system adjusts the laser power so that a particle is kept at a fixed height. If the wavelength is varied, then the radiation pressure, and hence the laser power necessary for stable levitation, also varies; this technique has been called radiation-pressure-force spectroscopy.

The power required to levitate an oil drop as its size parameter is varied by tuning the dye laser wavelength is shown in the lower curves of Fig. 11.11. The calculated radiation pressure efficiency (plotted as $1/Q_{pr}$) is shown in the middle curve and Q_{ext} in the upper curve; the refractive index $m = 1.47 + i10^{-6}$ is approximately constant over the small wavelength interval. This figure is taken from Chýlek et al. (1978b), who identified the peaks in the upper curve. Curve a of the experimental results is for values of x calculated from the drop size determined microscopically with an accuracy of $\pm 5\%$. The ripple structure

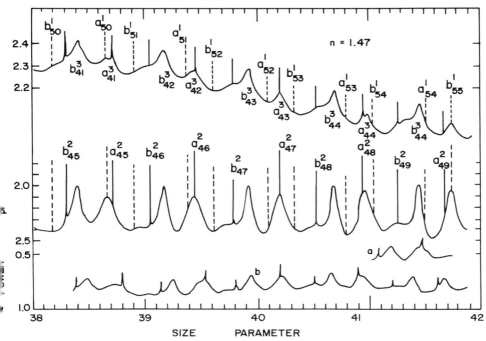

Figure 11.11 Calculated extinction and radiation pressure for a sphere compared with the power necessary to levitate the sphere. From Chýlek et al. (1978b), who used the experimental results of Ashkin and Dziedzic (1977).

features are so distinctive, however, that a much more accurate determination of x was made by matching experimental and theoretical curves; curve b is based on the more accurate drop size. Experiment and theory now agree very well except for the small displacements attributed to drop evaporation during the scan. All features of the calculated spectrum are observed in the measurements except those with width Δx of order 10^{-6}, shown as dashed lines in the calculations. These correspond to first-order resonances of the normal modes (e.g., b_{50}^1, a_{52}^1, etc.) and are too narrow to be resolved despite the laser's high spectral purity.

The experiments of Ashkin and Dziedzic impressively demonstrate the complexity of the electromagnetic modes of a sphere and the high degree of accuracy with which Mie theory describes them; they also provide a means for measuring the sizes of single spheres to within 1 part in 10^5 or 10^6 and for sensitively monitoring small size changes.

11.5 ABSORPTION EFFECTS IN EXTINCTION

At this point the reader should be well aware that all solids and liquids are strongly absorbing in some spectral regions and that this has consequences for extinction, some of which were discussed in the survey section (11.1). We examine the effects of absorption on extinction further in this section.

11.5.1 Ripple and Interference Structure

Absorption is progressively increased in the sequence of extinction curves in Fig. 11.12 for a narrow Gaussian distribution of radii. Because the refractive index (wavelength) is fixed for each curve the abscissa is proportional to the mean radius. Although the optical constants for the upper curves are similar to those of water at some infrared wavelengths, $m = 1.33 + i0.1$ is not; in this particular instance our point is better made by varying only k despite our general aversion to this sort of arbitrary and unrealistic fiddling with optical constants.

It is obvious at a glance that ripple structure is the first to fade as absorption is increased, beginning at the larger values of x. Next, the interference structure is damped, and again, it is more pronounced for large x. In Section 4.4 we interpreted this structure as arising from interference between the undeviated beam and the central ray through a sphere. Consistent with this interpretation is the expectation that these extrema should vanish whenever k is sufficiently large ($kx > 1$) that no appreciable light penetrates through the particle; that this is so is evident from Fig. 11.12.

11.5.2 Effect of an Absorption Edge

What is at first sight a paradox emerges from careful study of Fig. 11.12: extinction at fixed x does not always increase with increasing k. For example, at $x = 2$ extinction increases with increasing absorption; this is most evident in the bottom two curves. But at $x = 6$ it decreases with increasing absorption. At $x = 11$, however, there is again a reversal. Thus, the effect of a rapid onset of strong absorption, such as that occurring in the ultraviolet for all insulating solids and liquids, will depend on the particle size. For example, near the absorption edge of MgO (~ 7 eV) extinction by a 0.01-μm sphere sharply increases (Fig. 11.2). But when the radius is increased to 0.05 μm, extinction sharply drops just beyond the absorption edge. These effects are more clearly apparent in the experimental extinction curves for polystyrene (Fig. 11.19). The physical reason why increasing absorption can decrease extinction is that interference is thwarted if the light does not penetrate through the particle. At size parameters corresponding to interference peaks—destructive interference—an absorption edge decreases extinction; conversely, at size parameters corresponding to interference valleys, thwarting of constructive interference increases extinction.

11.5.3 Asymmetry Associated with Narrow Absorption Bands

Suppose that a material has a narrow, symmetric absorption band in the bulk state. It seems no more than common sense to expect that the corresponding

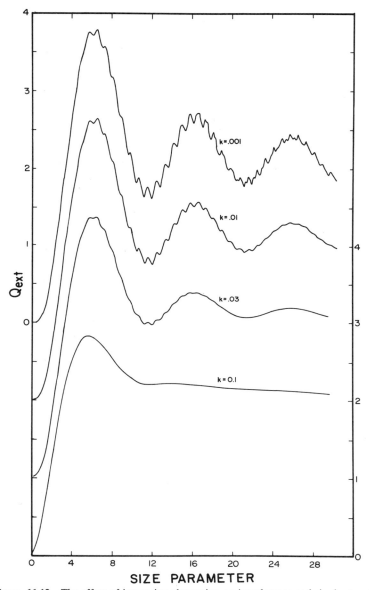

Figure 11.12 The effect of increasing absorption on interference and ripple structure.

extinction band for a particle of this material should also be symmetric. Yet when the particle is sufficiently large that the extinction band has an appreciable scattering component, it may be not at all symmetric. This type of asymmetry was first pointed out by van de Hulst (1957) in connection with small particles of the interstellar medium. Since then various calculations have been made for both coated and uncoated particles (Wickramasinghe and

Nandy, 1970; Greenberg and Stoeckly, 1971; Greenberg and Hong, 1974).

We have chosen the sharp absorption band at 1.18 eV in radiation-damaged MgO (see the inset curves in Figs. 10.1 and 11.2) to illustrate the asymmetry effect. The band strength derived from measurements is not sufficient to clearly illustrate the effect, so we increased it, along with the strengths of the three broader peaks that accompany it, by a factor of 100. Optical constants were calculated from the multioscillator model (9.25) with parameters chosen to best fit the data; then the strengths ω_{pj}^2 were increased a hundredfold. The resulting band strengths might be thought of as representing much higher levels of radiation damage, although it is doubtful that they could be realized in this way. The maximum ϵ'' of the enhanced 1.18-eV band is 0.082, corresponding to a k of about 0.024. Note that this is still some hundreds of times lower than intrinsic absorption in MgO between 7 and 20 eV.

Calculations for a range of particle sizes are shown in Fig. 11.13. Note that the scales have not been shifted for the different sizes: extinction increases with size because of scattering. The extinction band for the 0.1-μm particle faithfully reflects the characteristics of the intrinsic absorption band. But asymmetries develop for particles larger than about 0.2 μm; indeed, at a radius of 0.3 μm the absorption band looks like an "emission" band relative to the continuum. The explanation for this strange extinction behavior near an absorption band lies in the preceding section: extinction is not a steadily increasing function of bulk absorption. A narrow absorption band is similar to a small absorption edge that falls just as rapidly as it rises, which can thus cause extinction peaks, dips, or both.

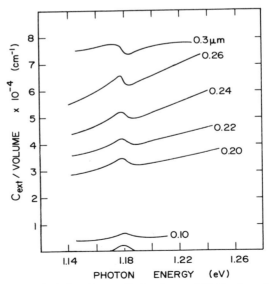

Figure 11.13 Extinction calculations for a radiation-damaged MgO sphere showing how increasing size gives asymmetric bands. The lowest curve (unlabeled) is for a sphere of radius 0.01 μm. From Huffman (1977).

11.5.4   Dominance of Absorption in the Rayleigh Limit

Extinction is proportional to the square of the volume of a particle small compared with the wavelength—but only if absorption is negligible. If it is not it will always dominate over scattering for sufficiently small sizes [see (5.12)], in which instance extinction is proportional to particle volume. The volume attenuation coefficient $\alpha_v = C_{ext}/v$ of water droplets is shown as a function of radius in Fig. 11.14a for a set of wavelengths. Water is very weakly absorbing at visible wavelengths ($k < 10^{-8}$) and extinction is therefore dominated by scattering; interference structure is evident but ripple structure has been suppressed for clarity. As the wavelength is increased the first interference peak shifts to larger radii; and α_v is constant for the smaller sizes, which indicates that extinction is predominantly absorption. At wavelengths where absorption is sufficiently strong (e.g., $\lambda = 12.5$ μm) α_v is constant until the radius is so large that almost no light penetrates into the particle's interior; this has practical consequences: at such wavelengths extinction by a polydispersion is independent of the details of the size distribution provided that the volume fraction of larger particles is small. Thus, Carlon et al. (1977) suggested that liquid water content along a path in the atmosphere could be determined by measuring attenuation of an infrared beam along that path. Most solids and liquids are somewhat absorbing in the infrared; moreover, larger particles have short atmospheric residence times. So extinction of infrared radiation by atmospheric aerosols is expected to be nearly independent of size (but not necessarily shape, as we shall see in Chapter 12), and attenuation measurements could therefore be used to remotely monitor their total volume.

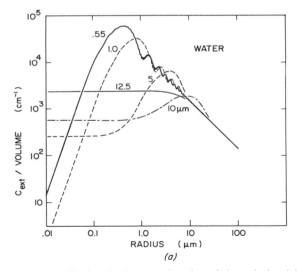

Figure 11.14 Volume-normalized extinction as a function of size calculated for spheres of (a) water (after Carlon et al., 1977) and (b) aluminum (from Rathmann, 1981). The optical constants used are those appropriate to the wavelengths noted on the curves.

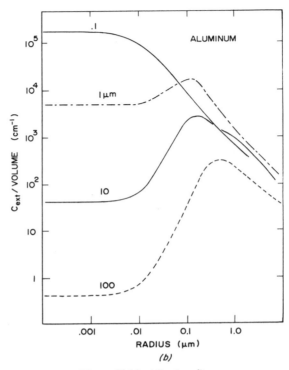

Figure 11.14 (*Continued*)

There are some notable differences apparent in Fig. 11.14 between the extinction curves for aluminum spheres and those for water droplets. For example, α_v is still constant for sufficiently small aluminum particles but the range of sizes is more restricted. The large peak is *not* an interference maximum: aluminum is too absorbing for that. Rather it is the dominance of the magnetic dipole term b_1 in the series (4.62). Physically, this absorption arises from eddy current losses, which are strong when the particle size is near, but less than, the skin depth. At $\lambda = 0.1 \ \mu m$ the skin depth is less than the radius, so the interior of the particle is shielded from the field; eddy current losses are confined to the vicinity of the surface and therefore the volume of absorbing material is reduced.

11.6 EXTINCTION CALCULATIONS FOR NONSPHERICAL PARTICLES

To this point we have dealt only with spheres, which have the advantage that their extinction properties are easily calculated while still giving substantial guidance into extinction by small particles in general. And there are many particles that are indeed spherical: cloud droplets; sulfuric acid droplets (apparently present in the upper clouds of Venus); sulfur colloids; and the much-used polystyrene spheres. For such particles Mie theory is admirably suited.

A few particles, such as spores, seem to be rather well approximated by spheroids, and there are many examples of elongated particles which may fairly well be described as infinite cylinders. Our next step toward understanding extinction by nonspherical particles is to consider calculations for these two shapes. To a limited extent this has already been done: spheroids small compared with the wavelength in Chapter 5 and normally illuminated cylinders in Chapter 8. We remove these restrictions in this section; measurements are presented in the following section. Because calculations for these shapes are more difficult than for spheres, we shall rely heavily on those of others.

11.6.1 Spheroids

Asano and Yamamoto (1975) solved the problem of absorption and scattering by an isotropic, homogeneous spheroid of arbitrary size by expanding the fields in vector spheroidal harmonics and determining the coefficients of the scattered field in a manner similar to that for a sphere (Chapter 4). A few results of calculations were given in this first paper; Asano (1979) subsequently published a more extensive set of calculations of scattering and absorption by spheroids of varying size, shape, and refractive index under various conditions of illumination. We now borrow a few examples from this work to extend our understanding of extinction by nonspherical particles. We note immediately that three factors add complications beyond those for spheres. First, a spheroid may be either prolate or oblate; second, the direction and state of polarization of the incident beam must be specified; and third, a spheroid has two characteristic lengths, its major and minor axes. As a consequence, the number of possible sets of calculations in a complete treatment of extinction by spheroids is vastly greater than that for spheres. Out of this multitude we have chosen only a few for illustration.

Figure 11.15 shows Asano's calculations of extinction by nonabsorbing spheroids for an incident beam parallel to the symmetry axis, which is the major axis for prolate and the minor axis for oblate spheroids. Because of axial symmetry extinction in this instance is independent of polarization. Calculations of the scattering efficiency Q_{sca}, defined as the scattering cross section divided by the particle's cross-sectional area projected onto a plane normal to the incident beam, are shown for various degrees of elongation specified by the ratio of the major to minor axes (a/b); the size parameter $x = 2\pi a/\lambda$ is determined by the semimajor axis a.

A glance at the curves in Fig. 11.15 reveals extinction characteristics similar to those for spheres: at small size parameters there is a Rayleigh-like increase of Q_{sca} with x followed by an approximately linear region; broad-scale interference structure is evident as is finer ripple structure, particularly in the curves for the oblate spheroids. The interference structure can be explained

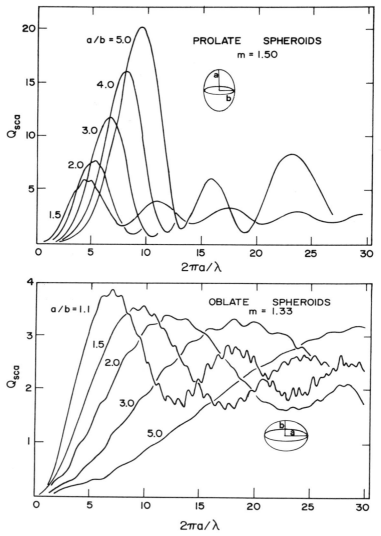

Figure 11.15 Calculated extinction by spheroids; the incident light is parallel to the symmetry axis. From Asano (1979).

qualitatively, as was done for spheres in Section 4.4, as alternate destructive and constructive interference between the ray passing through the center of the particle, thereby undergoing a phase shift of $4\pi a(m - 1)/\lambda$, and the undeviated beam. For the prolate spheroids, with $m = 1.5$, this phase shift is simply x, and we might expect the positions of the extinction maxima to be independent of elongation. This does not, however, occur, although the separation Δx

between peaks is approximately 2π. Note in particular that the position of the first extinction maximum (the first destructive interference) shifts to larger x with increasing elongation a/b. Asano has attributed this to edge phenomena, which become more important as the curvature at the point where the incident beam grazes the particle decreases, that is, with increasing a/b (for fixed a). In the limit $a/b \rightarrow \infty$ the prolate spheroid illuminated end on becomes an infinite cylinder with the wave propagating along its axis (i.e., a cylindrical waveguide); there is indeed some similarity between the extinction curve in Fig. 11.15 for the largest a/b and that for an infinite cylinder illuminated at nearly grazing incidence (Fig. 11.17). Additional evidence in support of this interpretation is provided by the curves in Fig. 11.15 for oblate spheroids; the curvature of such spheroids at the point of grazing *increases* with increasing a/b (for fixed a). If we were to plot Q_{sca} for the oblate spheroids as a function of the phase shift parameter $4\pi b(m - 1)/\lambda$, the positions of the first maxima would be more nearly congruent than they were for the prolate spheroids.

Perhaps the greatest difference between the extinction calculations for prolate and oblate spheroids is in the ripple structure, which is much more obvious for the latter and even persists to the largest a/b ratios shown, although with reduced amplitude.

Asano's calculations of absorption and scattering efficiencies for an absorbing prolate spheroid (which are not reproduced here) are qualitatively similar to those for spheres. For example, absorption dampens both large- and small-scale oscillations. As x increases Q_{abs} and Q_{sca} approach the respective asymptotic limits $1 - Q_{refl}$ and $1 + Q_{refl}$; Q_{refl}, the average reflectance over the particle's illuminated face (see Section 7.1), is small for the given refractive index ($1.5 + i0.1$), so in this instance the asymptotic efficiencies are approximately 1. The extinction efficiency therefore oscillates about two with decreasing amplitude as x increases.

Extinction calculations for obliquely incident light, also taken from Asano (1979), are shown in Fig. 11.16. The symmetry axis is parallel to the z axis and the direction of the incident beam, which makes an angle ζ with the symmetry axis, lies in the xz plane, the plane of incidence. The incident light is polarized either with its electric field or its magnetic field perpendicular to the plane of incidence; these two polarization states are denoted by TE (transverse electric) and TM (transverse magnetic).

Straight lines connect the extinction efficiencies computed at size parameter intervals $\Delta x = 1$; this coarse spacing attests to the difficulties of doing computations for obliquely incident light. In general, TE and TM waves are scattered similarly, although there are slight differences which lead to polarization (in the forward direction) of unpolarized obliquely incident light. Extinction of light incident along the symmetry axis of the oblate spheroid is broad and rippled, but the maximum becomes more peaked and shifts to smaller values of x as the obliquity is increased (i.e., as ζ is increased from $0°$ to $90°$). For the prolate spheroid a similar, but opposite, effect is evident: the first extinction maximum decreases and shifts to larger values of x with increasing

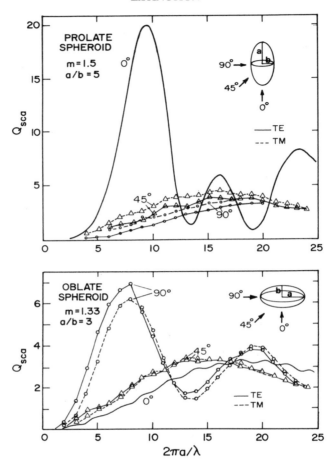

Figure 11.16   Calculated extinction of obliquely incident light by spheroids. From Asano (1979).

obliquity. It should be kept in mind, however, in interpreting these results that the distance through the center of the prolate spheroid *along the direction of incidence* decreases with increasing obliquity, whereas the opposite is true for the oblate spheroid. Extinction by the prolate spheroid, which is rather elongated ($a/b = 5$), is quite similar to that by an infinite cylinder, which we discuss next.

11.6.2   Infinite Cylinder

An infinite right circular cylinder is another particle shape for which the scattering problem is exactly soluble (Section 8.4), although it might be thought that such cylinders are so unphysical as to be totally irrelevant to real problems. But microwave extinction measurements (discussed in Section 11.7) indicate that finite cylinders with length-to-diameter ratios as low as about 5 can be closely approximated as infinite cylinders. Thus, many problems involving particles such as textile fibers, asbestos fibers, and even smoke particles aggregated into chain-like structures, can be treated adequately within the framework of infinite cylinder theory.

We have already given examples of extinction by infinite cylinders illuminated normally to their axes (Section 8.4), and Appendix C contains a

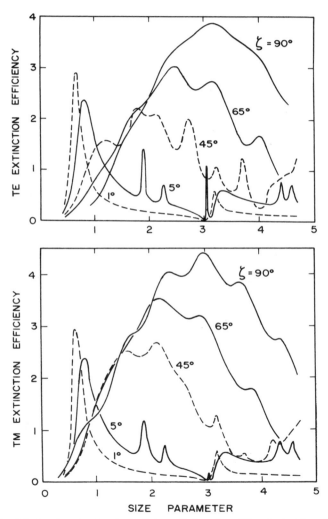

Figure 11.17 Calculated extinction by infinite cylinders for obliquely incident light; $\zeta = 90°$ corresponds to normally incident light. TE and TM denote light with the electric and magnetic vectors, respectively, perpendicular to the xz plane. From Lind and Greenberg (1966).

program for such computations. The same type of computations are required to determine extinction of obliquely incident light; although we omit the details here, they may be found in the paper by Lind and Greenberg (1966), from which we have taken Fig. 11.17. The refractive index $m = 1.6$ is appropriate to Lucite at microwave frequencies. As the angle of incidence ζ is varied from 90° (normal) to 1° (nearly grazing), the extinction efficiencies for TE and TM waves approach each other, as required by symmetry. The curve of extinction as a function of size is broad and rippled when the light is normally incident but narrows as ζ is increased until at nearly grazing incidence it is a series of sharp peaks dominated by the one at smallest x; this was also evident in the extinction curves for the elongated prolate spheroid (Fig. 11.16). At grazing incidence the efficiencies vanish, but in this instance the cylinder is more properly considered as a waveguide.

It seems fairly obvious from inspection of Fig. 11.17 that the extinction curve for a collection of randomly oriented cylinders would be similar to that for a polydispersion of spheres.

11.7   EXTINCTION MEASUREMENTS

Extinction is determined by measuring the ratio of transmitted to incident irradiance (11.1). Many laboratories are equipped with recording spectrophotometers which can measure this quantity very quickly for liquid or solid samples. In principle this same type of instrument may be used for measuring extinction by particulate samples. The results, however, may be unreliable unless the detector is designed to reject forward-scattered light, which may be the major contributor to extinction by particles larger than the wavelength.

The essentials of a proper experimental arrangement for measuring extinction are shown in Fig. 11.18. Light from a point source is collimated by, for example, a lens, transmitted through the particulate sample, and focused through a small pinhole onto a detector. In principle, light scattered in other than the forward direction is not detected. Of course, the pinhole cannot be infinitesimally small; in practice, therefore, the acceptance angle is determined by the size of the pinhole and the focal length of the second lens. Figure 11.18 is only schematic, however; the effective size of the source may be reduced by additional apertures and a monochromator may be inserted in the system.

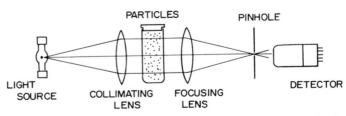

Figure 11.18   Schematic diagram of an experiment to measure extinction.

Although conventional spectrophotometers are not usually designed to reject light scattered at small angles, such instruments may be adequate for extinction measurements if the particles are not too large. But it is well to be aware of the potential pitfalls of measuring extinction this way. Several of the extinction curves in Chapter 12 for very small smoke particles (< 0.1 μm) were obtained with a commercial spectrophotometer. Infrared scattering by such particles is small and nearly uniform in all directions, even for rather large refractive indices. In this instance the simple measurement technique is quite accurate.

11.7.1 Polystyrene Spheres

Measured extinction spectra for aqueous suspensions of polystyrene spheres—the light scatterer's old friend—are shown in Fig. 11.19. Water is transparent only between about 0.2 and 1.3 μm, which limits measurements to this interval. These curves were obtained with a Cary 14R spectrophotometer, a commonly available double-beam instrument which automatically adjusts for changing light intensity during a wavelength scan and plots a continuous, high-resolution curve of optical density. To reproduce the fine structure faithfully, the curves were traced exactly as they were plotted by the instru-

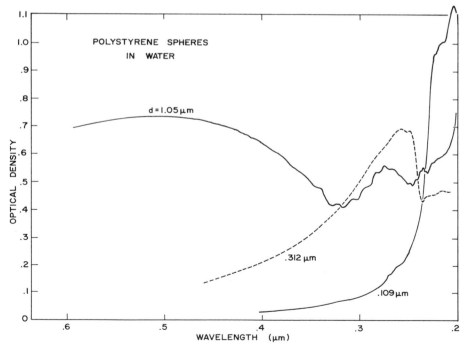

Figure 11.19 Measured extinction by aqueous suspensions of polystyrene spheres with three different mean diameters.

ment. Note that in contrast with previous extinction curves wavelength is the abscissa in Fig. 11.19. Although the measurements were not made using the "proper" experimental arrangement shown in Fig. 11.18, the particles are not too large and are suspended in water, so measured extinction is not greatly different from what would be obtained if forward-scattered light were excluded from the detector.

Almost all the calculated extinction features discussed previously may be found in the measured extinction curves: reddening when a/λ is small; interference peaks, the first of which shifts to longer wavelengths with increasing size; ripple structure; and the effects of an ultraviolet absorption edge. Reddening, as a result of greater scattering of shorter wavelengths, is evident in the spectrum for the 0.312-μm spheres. But the 1.05-μm spheres exhibit a tendency toward bluing, which is rare in natural aerosols. Even small-scale ripple structure, which sets in on the short-wavelength side of the interference peak for the 1.05-μm spheres, is obvious. Although such ripple structure is not usually observed in extinction spectra of polydispersions, it is in this instance because of the small size dispersion ($\sim 1\%$) and the high resolution of the spectrophotometer. Absorption dominates extinction by the smallest spheres and rises sharply near $\lambda = 0.23$ μm, the absorption edge of polystyrene; the barely perceptible features near 0.22 μm and 0.27 μm are also observed in bulk polystyrene (Inagaki et al., 1977). With the absorption edge so clearly marked it is apparent that the extinction band near 0.26 μm for the 0.312-μm spheres is a result of abrupt suppression of the first interference peak by strong absorption; this is an excellent experimental example of an effect that was discussed in Section 11.2 in connection with calculations for MgO and in Section 11.5.

Extinction spectra could be used to size particles by matching measured features with those calculated from Mie theory provided that the size distribution is narrow and the particles are nearly spherical.

11.7.2   Irregular Quartz Particles

Disappearance of prominent extinction features for spheres with an increase in the spread of sizes was illustrated in Fig. 11.6; similar effects occur for irregular particles. For example, Fig. 11.20 shows measurements for various size-graded fractions of irregularly shaped quartz particles suspended in water (Hodkinson, 1963). Measurements were made at three different wavelengths and the results plotted against size parameter (the refractive index changes only slightly). Data were normalized by assuming that the constant extinction efficiency at the largest size parameters was equal to the asymptotic limit 2. None of the prominent features calculated for spheres appear in these results; extinction merely rises steadily at small size parameters (reddening) to a constant value at large size parameters. Although some size information is present near values of x where extinction levels off it seems that accurate sizing of such a distribution of irregular particles by extinction holds little promise of success.

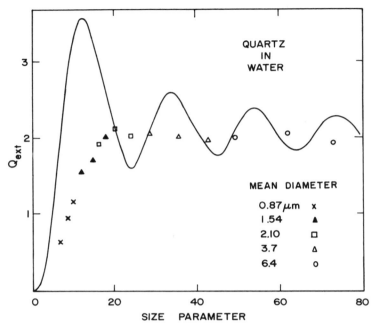

Figure 11.20 Measured extinction by five aqueous suspensions of irregular quartz particles (Hodkinson, 1963) at the wavelengths 0.365, 0.436, and 0.546 μm.

11.7.3 Additional Measurements

The measurements presented in the preceding two sections represent extremes: a highly monodisperse suspension of spheres with much extinction structure, and a broad size distribution of irregularly shaped particles with no prominent features. Measured extinction intermediate between these two extremes has been reported. For example, DeVore and Pfund (1947) segregated different size fractions of several fine powders on glass substrates. The first extinction maximum was clearly apparent for each sample, in contrast with the results in Fig. 11.20. These authors were even able to infer refractive indices of the powder solids by noting the shift of the first interference peak as the particles were surrounded by different liquids. Procter and Barker (1974) and Procter and Harris (1974) determined extinction by segregated size fractions of quartz and diamond dust. For these irregular particles there was evidence of the first interference maximum, after which extinction fell steadily toward the asymptotic limit.

11.7.4 Direct Measurement of Absorption

Extinction is not difficult to measure—in principle. But if it is to be separated into its components an independent measurement of either scattering or

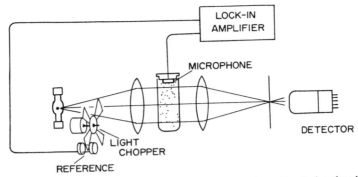

Figure 11.21 Schematic diagram of an experiment to measure absorption (using the photoacoustic method) and extinction.

absorption is required, which may be difficult. To determine total scattering either differential scattering measurements must be integrated over all directions or some special optical arrangement, such as an integrating sphere, must be used to collect all the scattered light. Errors in Q_{sca} are carried over into Q_{abs} inferred from $Q_{ext} - Q_{sca}$; and if $Q_{sca} \simeq Q_{ext}$ the relative error in Q_{abs} can be quite large. It is possible, however, to measure absorption directly—rather than extinction less scattering—by a method variously referred to as optoacoustic, photoacoustic, or spectrophone. Particles illuminated by light chopped at audio frequencies are periodically heated; this causes pressure oscillations (sound) which are detected by a sensitive microphone. Only absorbed light contributes to the heating, so absorption can be measured directly even in the presence of appreciable scattering. A schematic arrangement for simultaneously measuring absorption by the photoacoustic method and extinction is shown in Fig. 11.21. A closed cell contains particles, which in this instance are merely suspended in air. The signal from the microphone is amplified by a lock-in amplifier referenced to the frequency and phase of the light chopper. By rotating the detector arm in the scattering plane angular scattering could also be measured. If a focused laser beam were used as the light source, absorption by single particles could conceivably be measured, and scattering combined with this absorption to give extinction. An advantage of the photoacoustic method is that the sensitivity of the detector, a microphone, in no way depends on the wavelength of the light.

Absorption by particles, for example acetylene smoke (Roessler and Faxvog, 1979a) and diesel emissions (Faxvog and Roessler, 1979), has been measured by the photoacoustic method.

11.7.5 Microwave Extinction Measurements

There undoubtedly have been many studies—many of which are probably classified—of absorption and scattering of microwaves by all kinds of objects, particularly large objects such as aircraft and missiles. We have not thoroughly searched for all papers relevant to microwave scattering because our main interest has been in infrared to ultraviolet radiation. There is, however, a series of microwave experiments undertaken primarily to understand scattering of much shorter wavelength light by particles of comparable size, but by exploiting the inherent advantages of microwaves.

Extinction and scattering of light by a particle of given shape depend only on the *ratio* of a characteristic dimension (for a sphere, its radius) to the wavelength and the refractive index at that wavelength. Because of this similarity principle, scattering of visible light by small particles ($\sim$ 1 μm) may be studied by using microwaves and much larger particles ($\sim$ 10 cm) with the same shape and refractive index as the particles of interest. This *microwave analog technique* is discussed further in Section 13.3.

The essential ingredients of a microwave extinction experiment are shown in Fig. 11.22. The particle is suspended by threads so that its orientation can be changed and it can be moved in and out of the beam. Extinction is most conveniently determined in this instance by measuring the magnitude and phase of the forward-scattering amplitude and obtaining the cross section from the optical theorem (3.24). The experiment consists of nulling the detector (both amplitude and phase) with no scatterer present and then measuring the additional amplitude and phase of the forward-scattered wave with the particles inserted in the beam; calibration is accomplished with spheres of known size and refractive index.

The obvious advantage of the microwave experiment is that oriented single particles of arbitrary shape and, within limits, arbitrary refractive index, can be studied easily. Multilayered and other inhomogeneous particles pose no particular problems.

Microwave (λ = 3 cm) extinction measurements for beams incident parallel ($\zeta = 0°$) and perpendicular ($\zeta = 90°$) to the symmetry axis of prolate spheroids

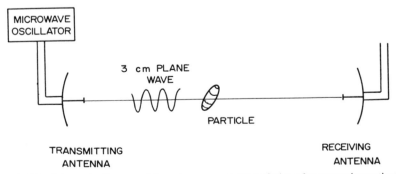

Figure 11.22 Schematic diagram of the microwave analog technique for measuring extinction by single oriented particles.

of varying semimajor axis a, but constant elongation $a/b = 2$, are shown in Fig. 11.23. These curves were taken from the paper by Greenberg et al. (1961), although a few notational changes were made for the sake of consistency. The general characteristics of this measured extinction are similar to those calculated for spheroids (Figs. 11.15 and 11.16). When the beam is incident perpendicular to the symmetry axis the extinction peaks are broad, and the ripple structure for TM polarization appears somewhat more pronounced than that for TE polarization. When the beam is incident parallel to the symmetry axis the extinction peak narrows and shifts to a smaller size parameter. The same kind of behavior is evident in the extinction curve for infinite cylinders (Fig. 11.17).

Extinction by a cylinder of the same material and with the same elongation as the spheroid, also measured by Greenberg et al., is shown in Fig. 11.24. Not surprisingly, extinction by the cylinder and spheroid are qualitatively similar. Infinite cylinder calculations agree rather well with measurements even for such a short cylinder. Additional experiments by Greenberg et al. suggest that such calculations tend to agree better with measurements if the length-to-diameter ratio is greater than about 5. Although the two extinction efficiencies (TE and TM) do not separately approach the infinite cylinder limits uniformly with increasing elongation (they oscillate about these values), the *difference* between the efficiencies more nearly does. Thus, there is some experimental evidence to support the assertions near the end of Section 8.4, made on the basis of

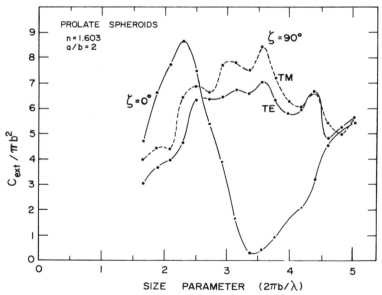

Figure 11.23 Measured extinction of microwave radiation by prolate spheroids. From Greenberg et al. (1961).

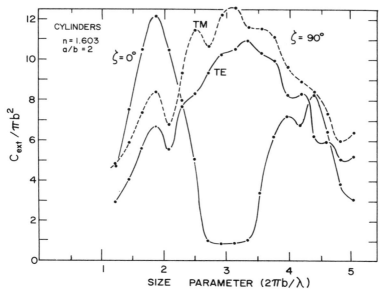

Figure 11.24   Measured extinction of microwave radiation by cylinders. From Greenberg et al. (1961).

diffraction theory, about when a finite cylinder may be considered effectively infinite. Of course, differences will always remain between finite and infinite cylinders with the same diameter and composition. The extent to which the former may be approximated by the latter will depend on the physical quantity of interest (extinction, angular scattering) and the desired accuracy.

11.8   EXTINCTION: A SYNOPSIS

Mie theory does an admirable job of predicting extinction by spherical particles with known optical constants: even the finest details it predicts—ripple structure—have been observed in extinction by single spheres. Several different causes—a distribution of sizes or shapes, and absorption—have the same effect of effacing the ripple structure or even the broader interference structure.

Absorption dominates over scattering for sufficiently small absorbing particles. Volumetric extinction by such particles is independent of their size but not of their shape; we shall discuss shape effects further in the following chapter.

Extinction is frequently dominated by scattering if the particle is about the same size as or larger than the wavelength. But absorption, which is usually manifested by absorption bands or absorption edges, can strongly affect extinction in unexpected ways: extinction may either increase or decrease with increasing absorption, and symmetric absorption bands in bulk matter may be transformed into highly asymmetric or even inverted extinction bands in small particles.

Extinction is easy to measure in principle but may be difficult in practice, especially for large particles where it becomes difficult to discriminate between incident and forward-scattered light. Spheres and ensembles of randomly oriented particles do not linearly polarize unpolarized light upon transmission. But single elongated particles or oriented ensembles of such particles can polarize unpolarized light by differential extinction.

NOTES AND COMMENTS

Asymptotic expressions for extinction and absorption efficiencies of spheres averaged over a size parameter interval $\Delta x \sim \pi$ (i.e., with no ripple structure) have been derived by Nussenzveig and Wiscombe (1980).

Further discussion of extinction by oriented particles is given in van de Hulst (1957, Chaps. 15 and 16).

Chapter 12

Surface Modes
in Small Particles

There is a class of electromagnetic modes in small particles, called *surface modes*, which give rise to interesting—perhaps puzzling at first sight—absorption spectra: small-particle absorption spectra can have features where none exist in the bulk and several features where only a single absorption band exists in the bulk. Particle shape and the variation of the dielectric function determine the number, position, and width of such features, which are manifestations of surface modes. Despite their dominant influence on absorption and scattering, and the ease with which they emerge from simple theory, surface modes have received little attention, particularly from applied scientists.

The extinction curves for magnesium oxide particles (Fig. 11.2) and aluminum particles (Fig. 11.4) show the dominance of surface modes. The strong extinction by MgO particles near 0.07 eV($\sim 17 \ \mu m$) is a surface mode associated with lattice vibrations. Even more striking is the extinction feature in aluminum that dominates the ultraviolet region near 8 eV: no corresponding feature exists in the bulk solid. Magnesium oxide and aluminum particles will be treated in more detail, both theoretically and experimentally, in this chapter.

In Sections 12.1 and 12.2 we discuss the theory of surface modes in spherical and nonspherical particles, respectively; in Sections 12.3 and 12.4 comparisons between theory and experiment are given, first for insulators and then for metals and metal-like materials.

There are several persistent misconceptions about the interaction of light with particles small enough to be described by Rayleigh theory. One is that they are very "inefficient" absorbers and scatterers of light. By any measure of efficiency, including the conventional efficiency Q (a somewhat unusual efficiency as it is not restricted to be less than 1), the small particles discussed in this chapter are very efficient. Q_{abs} is often much greater than 1, and small particles are, per unit mass, among the most efficient of absorbing materials. Another common misconception is that if a particle is small compared with the wavelength, an electromagnetic wave does not probe the details of its structure and, consequently, shape does not appreciably affect its absorption spectrum:

small particles obediently follow in the footsteps of the parent material. In fact, one of the most interesting aspects of small particles is that they can exhibit absorption features that are totally dominated by shape and that bear little resemblance to those of the bulk material. When first confronted with intense shape-dependent absorption in very small particles there is an unfortunate tendency to assume that bulk dielectric functions have become inapplicable, perhaps because of the importance of quantum-level spacings. We emphasize, however, that these effects can usually be explained satisfactorily with classical electromagnetic theory (indeed, classical electrostatics!) and bulk optical constants.

12.1   SURFACE MODES IN SMALL SPHERES

The conditions for the vanishing of the denominators of the scattering coefficients a_n and b_n for a homogeneous sphere are (4.54) and (4.55). We now consider these conditions in the limit of vanishingly small x. From the series expansions (5.1) and (5.2) of the spherical Bessel functions of order n, together with a bit of algebra, we can show that the denominator of a_n vanishes in the limit $x \to 0$ (finite $|m|$) provided that

$$m^2 = -\frac{n+1}{n}, \qquad n = 1, 2, \ldots \qquad (12.1)$$

where we have taken the sphere to be nonmagnetic. In the limit $x \to 0$ there is no solution to (4.55), the condition that the denominator of b_n vanish, for any n. At frequencies where (12.1) is satisfied, the corresponding scattering coefficient a_n is infinite. These frequencies are complex and, consequently, the associated normal modes are said to be *virtual*. Nevertheless, at real frequencies close to these complex frequencies, the scattering coefficients will be large. If (12.1) is approximately satisfied at some frequency, there will be a maximum, or resonance, in the cross sections. We shall refer to the normal modes (i.e., the vector spherical harmonics) associated with these frequencies as *surface modes*; they are characterized by internal electric fields with no radial nodes. The radial variation of the radial component of the electric field inside the sphere for a given normal mode is, from (4.40), (4.50), and (5.1),

$$E_{1r} \propto \left(\mathbf{N}_{e1n}^{(1)}\right)_r \propto r^{n-1} \qquad (x, |m|x \ll 1). \qquad (12.2)$$

The greater the order of the normal mode, the more the field is localized near the surface of the sphere, hence the designation surface modes. The lowest-order mode ($n = 1$) is uniform throughout the sphere, and this mode is sometimes called the *mode of uniform polarization*.

For sufficiently small spheres, a_1 will be the dominant coefficient; for $n = 1$ the condition (12.1) is

$$m^2 = -2. \qquad (12.3)$$

If, for the moment, we take the sphere to be in free space, then (12.3) in component form becomes

$$n^2 - k^2 = -2; \qquad 2nk = 0, \qquad (12.4)$$

where $n + ik$ is the complex refractive index of the particle. The solution to (12.4) is

$$n = 0; \qquad k = \sqrt{2}. \qquad (12.5)$$

The conditions (12.5) have been stumbled upon from time to time and then dismissed as "unphysical": n cannot be 0! But the reader who has faithfully waded through Chapters 9 and 10 should by now be somewhat hardened to refractive indices less than 1—or even 0. Indeed, one of our objectives in Chapter 9 was to clear the way for the introduction of (12.5), knowing full well that it is often unpalatable. Prejudices about what the dielectric function can or cannot be are not nearly so deeply rooted as those surrounding the refractive index; thus, (12.3) can be cast in a more palatable form in terms of the complex dielectric function of the particle $\epsilon = \epsilon' + i\epsilon''$:

$$\epsilon = -2\epsilon_m, \qquad (12.6)$$

where ϵ_m is the dielectric function of the surrounding medium (assumed to be nonabsorbing). The solution to (12.6) is

$$\epsilon' = -2\epsilon_m; \qquad \epsilon'' = 0. \qquad (12.7)$$

We shall call the frequency at which $\epsilon' = -2\epsilon_m$ and $\epsilon'' \simeq 0$ the *Fröhlich frequency* ω_F; the corresponding normal mode—the mode of uniform polarization—is sometimes called the *Fröhlich mode*. In his excellent book on dielectrics, Fröhlich (1949) obtained an expression for the frequency of polarization oscillation due to lattice vibrations in small dielectric crystals. His expression, based on a one-oscillator Lorentz model, is similar to (12.20). The frequency that Fröhlich derived occurs where $\epsilon' = -2\epsilon_m$. Although he did not explicitly point out this condition, the frequency at which (12.6) is satisfied has generally become known as the Fröhlich frequency. The oscillation mode associated with it, which is in fact the lowest-order surface mode, has likewise become known as the Fröhlich mode. Whether or not Fröhlich's name should be attached to these quantities could be debated; we shall not do so, however. It is sufficient for us to have convenient labels without worrying about completely justifying them.

The condition (12.6) has on occasion been attributed to Mie, presumably because it can be obtained from the Mie theory. But it is sobering to realize that it follows from simple electrostatics. For we showed in Section 5.2 that the absorption efficiency in the electrostatics approximation is

$$Q_{\text{abs}} = 4x \, \text{Im}\left\{ \frac{\epsilon - \epsilon_m}{\epsilon + 2\epsilon_m} \right\}, \qquad (12.8)$$

from which (12.6) follows almost trivially. (We have omitted the subscript 1 from ϵ for convenience; it will be reintroduced when we discuss coated particles.) Prediction of a resonance in the cross sections at the frequency where $\epsilon = -2\epsilon_m$ underscores our assertion at the beginning of Chapter 9 that it is often more enlightening to express various quantities in terms of the dielectric function rather than the refractive index. If we write (12.8) in terms of the relative refractive index m,

$$Q_{abs} = 4x \, \mathrm{Im}\left\{ \frac{m^2 - 1}{m^2 + 2} \right\},$$

then the resonance is almost certain not to be noticed: there is a deep psychological resistance to looking upon the square of the refractive index as anything other than an inherently positive number. So we encourage a kind of bilingualism when considering the optics of small particles.

The origin of the misconception that the absorption spectrum of particles in the Rayleigh limit is not appreciably different from that of the bulk parent material is easy to trace. Again, for convenience, let us take the particles to be in free space. In Chapter 3 we defined the volume attenuation coefficient α_v as the extinction cross section per unit particle volume; if absorption dominates extinction, then α_v for a sphere is $3Q_{abs}/4a$, where a is the radius. If we assume that $n \gg k$, which is true for most insulating solids at *visible* wavelengths, then

$$\alpha_v \simeq \frac{9n}{(n^2 + 2)^2} \alpha \qquad (n \gg k), \qquad (12.9)$$

where $\alpha = 4\pi k / \lambda$ is the bulk absorption coefficient. The factor multiplying α in (12.9) does not vary greatly over spectral regions in which n does not vary greatly and in many instances is not much different from 1. But this is not always true; in particular, n can vary greatly in the negative ϵ' region where, moreover, the assumption that k is much less than n, which is necessary for the validity of (12.9), completely breaks down. Thus, (12.9) is not always a reliable guide to spectral effects in small spheres (or, indeed, small particles of any shape); there will be many illustrations of this in succeeding paragraphs.

Let us return now to the general expression (12.8) for absorption by small spheres and write it as a function of ϵ' and ϵ'':

$$Q_{abs} = 12x \frac{\epsilon_m \epsilon''}{(\epsilon' + 2\epsilon_m)^2 + \epsilon''^2}. \qquad (12.10)$$

The absorption efficiency at the Fröhlich frequency is therefore

$$Q_{abs}(\omega_F) = \frac{12x\epsilon_m}{\epsilon''(\omega_F)}, \qquad (12.11)$$

which runs completely counter to intuition: the maximum absorption is *inversely* proportional to the absorptive part of the complex dielectric function. Note that although x is restricted to small values, say $x < 0.1$, Q_{abs} is not necessarily small at the Fröhlich frequency. For example, if x is 0.1 and $\epsilon''(\omega_F)/\epsilon_m$ is 0.1 (and there is no physical reason why it cannot take on this or even smaller values), then $Q_{abs} = 12$!

We tacitly assumed in the preceding paragraph that the absorption maximum in the region where ϵ' is negative occurs at the Fröhlich frequency. Although this is not strictly correct, the Fröhlich frequency is usually approximately equal to the frequency of maximum absorption; the precise position of the maximum depends on the behavior of the dielectric function and can be determined only by detailed calculations. This is analogous to the position of absorption peaks in the bulk material: we usually assume that such peaks occur at the maxima of ϵ'', whereas this is only approximately correct.

12.1.1   The Effect of Finite Size on the Fröhlich Frequency

Equation (12.6) is strictly valid only in the limit of vanishingly small x. We can obtain a better approximate condition for small but finite-sized particles by retaining more terms in the series expansions (5.1) and (5.2). The condition that the denominator of a_1 vanish is, from (4.54),

$$m\psi_1(mx)\xi_1'(x) - \xi_1(x)\psi_1'(mx) = 0. \tag{12.12}$$

If ψ_1 is expanded to terms of order x^4 and ξ_1 to terms of order x, then (12.12) correct to terms of order x^2 is

$$\epsilon = -\left(2 + \tfrac{12}{5}x^2\right)\epsilon_m. \tag{12.13}$$

For small x, this is not appreciably different from (12.6). However, (12.13) gives us an indication of how the Fröhlich frequency shifts as the size of the sphere increases. If in the neighborhood of the frequency where $\epsilon = -2\epsilon_m$ the real part of the dielectric function is an increasing function of frequency (this will almost always be so), an increase in particle size shifts the Fröhlich frequency to lower values (i.e., to longer wavelengths).

12.1.2   The Effect of a Coating

In the preceding paragraphs we considered a homogeneous sphere. Let us now examine what happens when a homogeneous core sphere is uniformly coated with a mantle of different composition. Again, the condition for excitation of the first-order surface mode can be obtained from electrostatics. In Section 5.4 we derived an expression for the polarizability of a small coated sphere; the condition for excitation of the Fröhlich mode follows by setting the denominator of (5.36) equal to zero:

$$(\epsilon_2 + 2\epsilon_m)(\epsilon_1 + 2\epsilon_2) + f(2\epsilon_2 - 2\epsilon_m)(\epsilon_1 - \epsilon_2) = 0, \tag{12.14}$$

where ϵ_1 and ϵ_2 are the dielectric functions of core and mantle, respectively, and f is the fraction of the total particle volume occupied by the core. If we take the dielectric function ϵ_m of the surrounding medium to be 1, then (12.14) becomes

$$\epsilon_1 = -2\epsilon_2 \left[\frac{\epsilon_2(1-f) + (2+f)}{\epsilon_2(2f+1) + 2(1-f)} \right]. \tag{12.15}$$

As a check on (12.15) we note that $\epsilon_1 = -2$ when $f = 1$, as required. If the core volume is small compared with that of the mantle ($f \ll 1$), then (12.15) becomes $\epsilon_1 \simeq -2\epsilon_2$, which is the Fröhlich mode condition for a homogeneous sphere with dielectric function ϵ_1 in a medium with dielectric function ϵ_2 (this is not a particularly startling result). Thus, the effect of coating a small, homogeneous sphere is to shift its Fröhlich frequency; the magnitude of this shift depends on the behavior of ϵ_1 as well as the kind and amount of coating.

12.1.3 Fröhlich Modes of Voids and Bubbles

In previous chapters we have always taken particles to be in a nonabsorbing medium. We now briefly remove this restriction. The notion of extinction by particles in an absorbing medium is not devoid of controversy: more than one interpretation is possible. But Bohren and Gilra (1979) showed that if the extinction cross section is interpreted as the reduction in area of a detector because of the presence of a particle [see Section 3.4, particularly the development leading up to (3.34)], then the optical theorem for a spherical particle in an absorbing medium is formally similar to that for a nonabsorbing medium:

$$C_{\text{ext}} = 4\pi \, \text{Re} \left[\frac{S(0°)}{k^2} \right], \tag{12.16}$$

where the wave number k is now *complex*. For a nonabsorbing medium k is real and may be taken outside the brackets in (12.16). As before, we may *define* the extinction efficiency Q_{ext} as $C_{\text{ext}}/\pi a^2$, which in the small-particle limit is

$$Q_{\text{ext}} = 4 \, \text{Im} \left\{ \frac{x(\epsilon - \epsilon_m)}{\epsilon + 2\epsilon_m} \right\}. \tag{12.17}$$

Note that the size parameter x for a sphere in an absorbing medium is complex.

Consider now a spherical *void* ($\epsilon = 1$) in an otherwise homogeneous medium. Light is not absorbed by such a void, but it can influence the absorption of light in the surrounding medium. The condition for a resonance in the extinction efficiency of a small spherical void follows readily from (12.17):

$$\epsilon_m = -\tfrac{1}{2}. \tag{12.18}$$

The resonance condition for a hollow sphere, or *bubble*, in air is, from (12.14),

$$(\epsilon + 2)(1 + 2\epsilon) + f(2\epsilon - 2)(1 - \epsilon) = 0, \tag{12.19}$$

where ϵ is the dielectric function of the solid part of the bubble. There are two roots to (12.19):

$$\epsilon_+ = \frac{-(5 + 4f) + 3\sqrt{1 + 8f}}{4 - 4f}, \qquad \epsilon_- = \frac{-(5 + 4f) - 3\sqrt{1 + 8f}}{4 - 4f}.$$

If f is small (a nearly solid bubble), the two roots are approximately $\epsilon_+ = -\frac{1}{2}$ and $\epsilon_- = -2$; these are the resonance conditions for a spherical void in a medium with dielectric function ϵ and a solid sphere with dielectric function ϵ in free space, respectively. As f is increased ϵ_+ increases monotonically and ϵ_- decreases monotonically, where $\lim_{f \to 1} \epsilon_+ = 0$ and $\lim_{f \to 1} \epsilon_- = -\infty$. (Of course, when $f = 1$, the bubble bursts!)

Up to this point we have considered only the conditions for resonances in the cross sections of small spherical particles of various kinds; we have said nothing quantitative about their strengths and the frequencies at which they might occur other than brief introductory remarks about ionic crystals in the infrared and metals in the ultraviolet. To determine if a resonance is realizable, where it occurs, and its strength, we need to know how the dielectric function varies with frequency. Therefore, in the following sections we shall examine some of the preceding resonance conditions in the light of simple, but realistic, dielectric functions.

12.1.4 Crystals with Simple Vibrational Modes

Detailed calculations of surface modes in small spherical particles are best carried out using the exact theory and measured optical constants. Even though simple models may fit measured data reasonably well, there can be considerable differences between calculations based on model and measured optical constants. This has been stressed by Hunt et al. (1973), who illustrated their point with NiO. Nevertheless, the usefulness of back-of-the-envelope calculations is not to be gainsaid provided their limitations are kept firmly in mind. Cross-section resonances for spheres are sharp and, consequently, are often missed in calculations done in a state of ignorance about their existence; it is wise to chart the approximate position of resonances before beginning a series of calculations. In the following paragraphs, therefore, we examine the consequences of (12.6) for crystals that are well described by the simple one-oscillator model of Section 9.1. We shall sometimes loosely refer to such crystals as "ionic," which is not strictly correct. The reason for this terminology is that strongly ionic crystals have been given prominence in work on lattice vibrational modes. But ionic particles are not the only ones that can support surface modes. Indeed, an ionic crystal is a limiting case, an idealization, in which bonding is entirely Coulombic and covalent bonding is negligible.

If we ignore damping, the Fröhlich frequency follows readily from (12.6), (9.21), and (9.23):

$$\omega_F^2 = \omega_t^2 \left(\frac{\epsilon_{0v} + 2\epsilon_m}{\epsilon_{0e} + 2\epsilon_m} \right). \tag{12.20}$$

A consequence of (9.23) and (9.24) is that $\omega_t < \omega_F < \omega_l$. This is no more than a statement that (12.6) is satisfied only in the region where ϵ' is negative, which, for simple one-oscillator materials, lies between ω_t and ω_l; we shall call this the surface mode region.

Laboratory measurements of extinction spectra are often done with particles suspended in some nonvacuous medium. It is usually taken for granted that features in such spectra are insensitive to this medium provided that it is weakly absorbing, but in the surface mode region this assumption can be greatly in error. For if we differentiate (12.20) with respect to ϵ_m, which we take to be greater than or equal to 1, then $d\omega_F/d\epsilon_m < 0$; in going from air to some nonvacuous medium, the Fröhlich frequency shifts to a lower value. Note that the magnitude of this shift depends on the width of the surface mode region: it can at most be $\omega_l - \omega_t$. This is just one example of an important general rule about surface modes: their characteristics strongly depend on how the dielectric function varies with frequency.

We noted in Section 9.1 that for a one-oscillator model, $\epsilon''(\omega)$ falls to one-half its maximum value $\epsilon''(\omega_t)$ at $\omega = \omega_t \pm \gamma/2$ provided that $\gamma/\omega_t \ll 1$. Under the same condition on γ/ω_t [see (12.33)] the half-width of the sphere absorption spectrum (12.10) is also γ. That is, the *width* of the absorption band is preserved in going from the bulk to particulate states, although the *position* of the band can be appreciably shifted.

We now pause briefly in the theoretical discussion to consider a specific example, silicon carbide, the infrared optical constants of which are given to good approximation by a one-oscillator model (Fig. 9.6). Extinction efficiencies for small silicon carbide spheres in air calculated from Mie theory and the parameters of Fig. 9.6 are shown in Fig. 12.1. For the 0.1-μm-radius sphere, which is sufficiently small that Rayleigh theory is adequate, the single extinction feature at 930 cm^{-1} is the Fröhlich mode, or lowest-order surface mode; its position is very near that predicted by (12.20). The frequency at which absorption is a maximum is appreciably shifted from what it would be in thin films, the transverse optic mode frequency $\omega_t = 793$ cm^{-1}. The small-particle extinction efficiency at ω_F is orders of magnitude greater than that at ω_t, which does not even show on the linear plot. As the size is increased, the surface mode peak shifts slightly to lower frequencies, in accordance with (12.13), and broadens as higher-order surface modes, which are not resolved on this plot, are excited. Simultaneously, an increasingly complicated series of modes

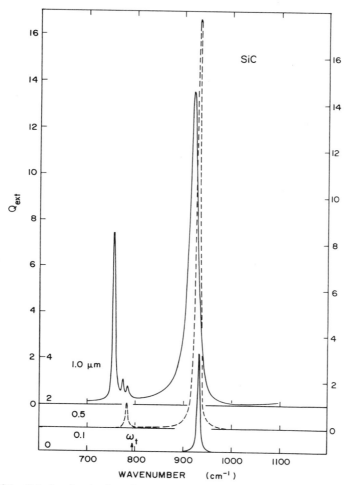

Figure 12.1   Calculated extinction efficiencies of silicon carbide spheres in air. The wave number denotes the inverse of the wavelength.

appears just below ω_t. These modes are of the same nature as the ripple structure modes discussed in Chapter 11. They appear at rather small size parameters in this instance because the refractive index rises to very large values just below ω_t (Fig. 9.6). Fuchs and Kliewer (1968) have given a detailed discussion of the positions and widths of these modes, which they call "low-frequency modes." It is worth noting again that the extinction efficiency Q_{ext} can be much greater than 1 in the surface mode region even for spheres small compared with the wavelength.

Extinction spectra for a 0.1-μm-radius SiC sphere in potassium bromide ($\epsilon_m = 2.33$) and in air ($\epsilon_m = 1$) are shown in Fig. 12.2. The Fröhlich mode

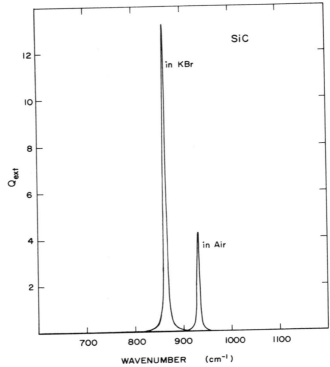

Figure 12.2 Calculated extinction efficiencies of a silicon carbide sphere (0.1 μm) in air and in potassium bromide.

shifts to a lower frequency in going from air to KBr, in agreement with (12.20). Note also that the extinction efficiency at the Fröhlich frequency is greater for the particle in a KBr matrix than in air even though KBr is nonabsorbing over the SiC surface mode region.

A number of experiments over the past 10 years have qualitatively confirmed that the general aspects of absorption by small ionic particles are in accord with the preceding discussion for spheres. In particular, the dominant absorption peak is shifted toward higher frequencies in going from bulk to particulate states and there is a predictable shift depending on the surrounding medium. Some notable examples are the alkali halides KCl, NaCl, and KBr (Martin, 1969, 1970, 1971; Bryksin et al., 1971), MgO (Genzel and Martin, 1972, 1973), and UO_2 and ThO_2 (Axe and Pettit, 1966). Notably lacking in the comparison between theory and experiment has been agreement about the width and strength of absorption bands. To understand these discrepancies requires that we consider the effect of departures from sphericity; this will be undertaken in Section 12.2 as a prelude to further discussion of vibrational modes in insulators.

12.1.5   Simple Metals

We showed in the preceding section that for solids with strong vibrational bands the position of features in absorption spectra can be shifted appreciably in going from the bulk to particulate states. Metallic particles can deviate even more markedly from the behavior of the bulk parent material: they can have absorption features over broad frequency regions where *none* appear in the bulk. For a simple metal—one that is well described by the Drude formula (9.26)—the imaginary part of the dielectric function has no maximum: it merely decreases monotonically with increasing frequency. But there will be a peak in the absorption cross section of a small spherical particle of such a metal near the frequency ω_F where $\epsilon = -2\epsilon_m$. If $\gamma^2 \ll \omega_p^2$, it follows from (9.27) that

$$\omega_F = \frac{\omega_p}{\sqrt{1 + 2\epsilon_m}}. \tag{12.21}$$

In air $\epsilon_m = 1$ and (12.21) reduces to

$$\omega_F = \frac{\omega_p}{\sqrt{3}}. \tag{12.22}$$

Unlike ionic materials, the negative ϵ' region for a simple metal is not confined to a relatively narrow band of frequencies: the surface mode region extends from ω_p down to zero frequency. As a consequence, metallic particles can be richer in surface modes than ionic particles. This will become apparent in Section 12.2 when we discuss the effect of shape on surface modes; for the moment, we content ourselves with spheres.

Following (9.27) we discussed the physical interpretation of the plasma frequency for a simple metal and introduced the concept of a plasmon, a quantized plasma oscillation. It may help our understanding of the physics of surface modes in small particles and the terminology sometimes encountered in their description if we expand that discussion.

It was tacitly assumed in Chapter 9 that the plasma was unbounded; that is, we had in mind *bulk* plasmons. But because of the long-range nature of the organizing forces in a plasma oscillation, it is reasonable to expect that for a sufficiently small system, the electrons will sense the presence of the boundaries and modify their collective behavior accordingly. Indeed, following hard on the heels of the acceptance of bulk plasmons in metals came the realization that *surface plasmons* were possible in thin films (Ritchie, 1957; Stern and Ferrell, 1960). Whereas the energy of a bulk plasmon is $\hbar\omega_p$, that of a surface plasmon in a thin film (in air) is $\hbar\omega_p/\sqrt{2}$. The next member of this family of plasmons is the surface plasmon in a sphere (in air) with energy $\hbar\omega_p/\sqrt{3}$ [see (12.22)]. Thus, surface modes in small metallic particles are often called *surface plasmons*. All of this illustrates a general rule, which we can state but not prove: if there is an interesting effect in a thin film, there will be a corresponding effect, albeit with possibly a few new twists, in small particles. Both are examples of systems with at least one small dimension.

Ions in the lattice of a solid can also partake in a collective oscillation which, when quantized, is called a *phonon*. Again, as with plasmons, the presence of a boundary can modify the characteristics of such lattice vibrations. Thus, the infrared surface modes that we discussed previously are sometimes called *surface phonons*. Such surface phonons in ionic crystals have been clearly discussed in a landmark paper by Ruppin and Englman (1970), who distinguish between *polariton* and pure phonon modes. In the classical language of Chapter 4 a polariton mode is merely a normal mode where no restriction is made on the size of the sphere; pure phonon modes come about when the sphere is sufficiently small that retardation effects can be neglected. In the language of elementary excitations a polariton is a kind of hybrid excitation that exhibits mixed photon and phonon behavior.

The choice of quantum-mechanical or classical language to describe surface modes in small particles is dictated more by taste than by necessity. However, there is an unfortunate tendency among physicists to consider that "quantum mechanics is intrinsically better than classical mechanics, and that classical mechanics is something real physicists ought to grow out of"; we agree with Pippard (1978, p. 3) that this is a "disputable proposition." Indeed, much mischief has been done—and is still being done—by incorrectly applying quantum theory to "explain" the strange optical behavior of small particles. Surface modes in small particles are adequately and economically described in their essentials by simple classical theories. Even, however, in the classical description, quantum mechanics is lurking unobtrusively in the background; but it has all been rolled up into a handy, ready-to-use form: the dielectric function, which contains all the required information about the collective as well as the individual particle excitations. The effect of a boundary, which is, after all, a macroscopic concept, is taken care of by classical electromagnetic theory.

We must again emphasize, even more strongly than we did at the beginning of this chapter, that surface plasmons and surface phonons are not examples of the failure of the bulk dielectric function to be applicable to small particles. Down to surprisingly small sizes—exactly how small is best stated in specific examples, as in Sections 12.3 and 12.4—the dielectric function of a particle is the same as that of the bulk parent material. But this dielectric function, which is the repository of information about elementary excitations, manifests itself in different ways depending on the size and shape of the system.

12.1.6 Limitation of the Mean Free Path

There is one clear exception to the rule that bulk dielectric functions tend to be applicable to very small particles: in metal particles smaller than the mean free path of conduction electrons in the bulk metal, the mean free path can be dominated by collisions with the particle boundary. This effect has been invoked by many authors, including Doyle (1958), Doremus (1964), Kreibig and von Fragstein (1969), Kreibig (1974), and Granqvist and Hunderi (1977).

The dielectric function of a metal can be decomposed into a free-electron term and an interband, or bound-electron term, as was done for silver in Fig. 9.12. This separation of terms is important in the mean free path limitation because only the free-electron term is modified. For metals such as gold and copper there is a large interband contribution near the Fröhlich mode frequency, but for metals such as silver and aluminum the free-electron term dominates. A good discussion of the mean free path limitation has been given by Kreibig (1974), who applied his results to interpreting absorption by small silver particles. The basic idea is simple: the damping constant in the Drude theory, which is the inverse of the collision time for conduction electrons, is increased because of additional collisions with the boundary of the particle. Under the assumption that the electrons are diffusely reflected at the boundary, γ can be written

$$\gamma = \gamma_{\text{bulk}} + \frac{v_F}{L},$$

where γ_{bulk} is the bulk metal damping constant, v_F is the electron velocity at the Fermi surface, and L is the effective mean free path for collisions with the boundary. Kreibig used $L = 4a/3$ for a sphere of radius a, although there is slight disagreement among various authors about the constant of proportionality between L and a.

Near the plasma frequency in metals $\omega^2 \gg \gamma^2$; therefore, to good approximation, the imaginary part of the Drude dielectric function (9.26) is

$$\epsilon''(\omega, a) = \frac{\omega_p^2}{\omega^3} \gamma = \frac{\omega_p^2}{\omega^3} \left(\gamma_{\text{bulk}} + \frac{3v_F}{4a} \right)$$

$$= \epsilon''_{\text{bulk}} + \frac{3}{4} \frac{\omega_p^2}{\omega^3} \frac{v_F}{a}. \tag{12.23}$$

Although the effect of the mean free path limitation on the real part of the dielectric function is slight, the effect on the imaginary part is often substantial. For small silver particles Kreibig (1974) found that ϵ'' near the Fröhlich frequency is given by

$$\epsilon'' = 0.23 + \frac{26.4}{a},$$

where a is in angstroms. Thus, ϵ'' is enhanced by more than 10% for particles of radius 1000 Å, and for a radius of about 115 Å, ϵ'' is twice the bulk value. The effect of the decreased mean free path is to increase the width and lower the peak height of the surface plasmon absorption. A combination of Mie theory and the free-electron contribution to the dielectric function suitably modified to include the mean free path limitation gives good agreement with experiments in which the metal particles are spherical and well isolated from one another.

12.1.7   Surface Modes in Small Aluminum Spheres

As an example of extinction by spherical particles in the surface plasmon region, Fig. 12.3 shows calculated results for aluminum spheres using optical constants from the Drude model taking into account the variation of the mean free path with radius by means of (12.23). Figure 9.11 and the attendant discussion have shown that the free-electron model accurately represents the bulk dielectric function of aluminum in the ultraviolet. In contrast with the Q_{ext} plot for SiC (Fig. 12.1), we now plot volume-normalized extinction. Because this measure of extinction is independent of radius in the small size

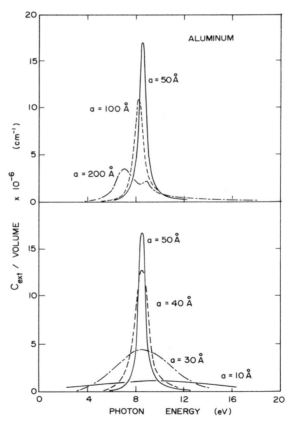

Figure 12.3   Calculated extinction per unit volume of aluminum spheres.

limit deviations from Rayleigh theory are clearly evident. The upper curve of Fig. 12.3 shows the shift of the Fröhlich mode toward lower energies as the size increases (12.13). A higher-order surface mode is also obvious in the curve for the 200-Å particle. The net result of increasing size is thus a shift of the maximum toward lower energies, broadening of the band, and a decrease in volume-normalized extinction. The lower part of the figure shows calculated results for radii smaller than 50 Å, where the mean free path limitation becomes an appreciable effect. Increased damping in the Drude formula (9.26) gives rise to an increased width of the dominant, lowest-order surface mode; at the same time the peak height is reduced. Both the excitation of higher-order surface modes with increasing size and greater damping for very small particles have been verified experimentally; this will be discussed in Section 12.4.

12.1.8 Field Lines of the Poynting Vector

Small spheres can absorb more than the light incident on them. The truth of this assertion follows from simple calculations using (12.11). But analytical proofs have less force, to some minds at least, than geometrical proofs. For this reason, therefore, we consider the interaction of light with a small sphere in a way which, as far as we know, has not been done before. The result is not new knowledge but rather new evidence supporting and firmly implanting in our minds what we already know.

We showed in Section 3.3 that the total Poynting vector $\mathbf{S}$ in the region surrounding an arbitrary particle can be written as the sum of three terms:

$$\mathbf{S} = \mathbf{S}_i + \mathbf{S}_s + \mathbf{S}_{ext}.$$

$\mathbf{S}_i$ is the Poynting vector of the incident field and $\mathbf{S}_s$ that of the scattered field; we may interpret $\mathbf{S}_{ext}$ as the term that arises because of interaction between the incident and scattered fields. Of greater interest here, however, is the flow of electromagnetic energy exclusive of that scattered. Thus, the Poynting vector under consideration (normalized by I_i, the magnitude of $\mathbf{S}_i$) is

$$\mathbf{A} = \frac{\mathbf{S}_i + \mathbf{S}_{ext}}{I_i}.$$

Were it not for the particle, of course, $\mathbf{A}$ would just be a unit vector parallel to the direction of propagation of the incident plane wave, and the field lines would be parallel lines. At sufficiently large distances from the particle the field lines are nearly parallel, but close to it they are distorted. It is the nature of this distortion in the neighborhood of a small sphere and its relation to the optical properties of the sphere that we now wish to investigate.

If the incident wave is x-polarized, the ϕ-component of $\mathbf{A}$ is zero in the xz plane ($\phi = 0$). In this plane, therefore, the field lines are solutions to the differential equation

$$\frac{dr}{d\theta} = \frac{rA_r}{A_\theta}. \tag{12.24}$$

For a sufficiently small sphere, a_1 is the dominant scattering coefficient in the series (4.45) and is given by (5.4). All the ingredients are at hand, therefore, for writing (12.24) in explicit form, a laborious task the details of which are best omitted; the result is

$$\frac{d\rho}{d\theta} = -\rho \frac{\cos\theta}{\sin\theta}$$

$$\times \frac{\left[\rho^3 + \{(x^2\rho^2\cos\theta + x^2\rho^2 - 1)(K_r\cos\xi + K_i\sin\xi) + (x\rho\cos\theta + x\rho)(K_r\sin\xi - K_i\cos\xi)\}\right]}{\left[\rho^3 + \{(x^2\rho^2\cos\theta + 2)(K_r\cos\xi + K_i\sin\xi) + (x\rho\cos\theta - 2x\rho)(K_r\sin\xi - K_i\cos\xi)\}\right]}, \quad (12.25)$$

where $K = K_r + iK_i = (\epsilon - 1)/(\epsilon + 2)$, $\xi = x\rho(\cos\theta - 1)$, and $\rho = r/a$. Subject to restrictions on the size of the sphere, (12.25) is completely general: it gives the field lines of the Poynting vector right up to the boundary of the sphere.

This equation was solved numerically with a fourth-order Runge–Kutta scheme. It was usually more convenient to recast (12.25) as a differential equation in the rectangular Cartesian coordinates; sometimes, however, the advantage was tipped in favor of the polar coordinates. The results shown in Fig. 12.4 were obtained with a mixture of the two approaches.

At a photon energy of about 8.8 eV—the surface plasmon energy—the real part of the dielectric function of aluminum is -2; the corresponding imaginary part is about 0.2. It follows from (12.11), therefore, that the absorption efficiency of a small aluminum sphere (in air) with size parameter 0.3 is about 18: such a sphere presents to incident photons a target area 18 times greater than its geometrical cross-sectional area. More palpable evidence of the sphere's great size in this instance is provided by Fig. 12.4a, which shows the field lines of **A** in the region surrounding the sphere. Note the strong convergence of field lines near the sphere; light that, according to geometrical optics, would have passed the sphere without impediment, is deflected toward it.

An absorption cross section 18 times greater than the geometrical cross section implies that the absorption radius—to coin a term—is about 4.2 times greater than the geometrical radius. This follows from the analytical expression (12.11), but it should also emerge from purely geometrical reasoning. And indeed it does: note in Fig. 12.4a that those field lines extending to about 3.9 times the particle radius converge onto the particle.

At energies on either side of 8.8 eV a small aluminum sphere presents a much smaller target to incident photons. At 5 eV, for example, the absorption efficiency of a sphere with $x = 0.3$ is about 0.1; as far as absorption is concerned, the sphere is much smaller than its geometrical cross-sectional area. The field lines of the Poynting vector, shown in Fig. 12.4b, are what are to be

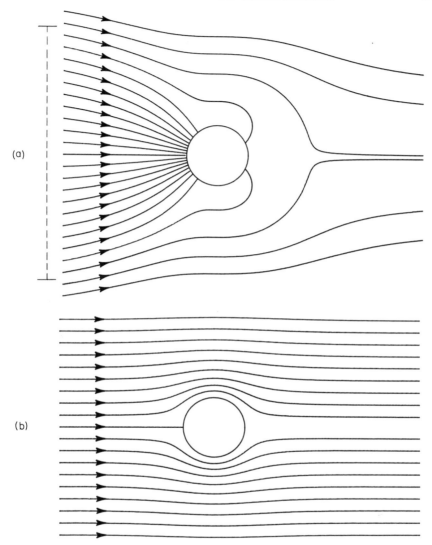

Figure 12.4 Field lines of the total Poynting vector (excluding that scattered) around a small aluminum sphere illuminated by light of energy 8.8 eV (*a*) and 5 eV (*b*). The dashed vertical line in (*a*) indicates the effective radius of the sphere for absorption of light.

expected for such a small target: a few lines intersect the sphere but most are deflected around it.

The imaginary part of the dielectric function of SiC at its Fröhlich frequency in the infrared (about 932 cm^{-1}) is close to that of aluminum at 8.8 eV. So Fig. 12.4*a* also shows the field lines of the Poynting vector around a small SiC sphere illuminated by light of frequency 932 cm^{-1}. At nearby frequencies, 900

cm^{-1} for example, the field lines around a SiC sphere ($x = 0.3$) are similar to those shown in Fig. 12.4b for aluminum at 5 eV.

No textbook on electromagnetic theory would be complete without a figure showing the field lines around a sphere in an electrostatic field. The reason, of course, is that this is a very effective way of presenting an idea—the sphere distorts the otherwise uniform field—in such a way that it can be grasped at a glance. But a small sphere illuminated by a plane wave also disturbs the flow of electromagnetic energy in its neighborhood. So the field lines of the Poynting vector (excluding that of the scattered field) around the sphere help to elucidate how a particle can absorb more than the light incident on it.

12.2 SURFACE MODES IN NONSPHERICAL PARTICLES

In many ways, surface modes in nonspherical particles are more interesting than those in spherical particles. Moreover, they are more likely to be observed in laboratory investigations: it is not easy to prepare samples of particles that are small compared with the wavelength, spherical to a high degree, and unagglomerated. Yet this is what is required if predictions based on the theory for small particles are to be legitimately compared with measurements. Unless special care is taken (we shall have more to say about this in the following section), solid particles prepared in the laboratory are likely to be nonspherical; even if spherical at their moment of birth, they can quickly coagulate into irregular clumps. When we leave the laboratory and seek out solid particles in natural environments—the earth's atmosphere, interplanetary space, the interstellar medium—it is almost certain that such particles are not spherical. It is often assumed that shape is irrelevant to extinction spectra of small particles: a collection of randomly oriented irregular particles is "equivalent" somehow to a collection of spheres. As we shall see, this assumption is demonstrably false in the surface mode region. Therefore, if data are analyzed, or computations are based, on the assumption that irregularly shaped particles can be adequately approximated by spheres, the results can be greatly in error.

Notable progress in analyzing nonspherical particles has been made by Fuchs (1975), who calculated absorption by cubes in the electrostatics approximation and applied the results to experimental data for MgO and NaCl. We shall discuss Fuchs's results at the end of Section 12.3. Langbein (1976) also did calculations for rectangular parallelepipeds, including cubes, which give valuable insights into nonspherical shape effects. Because the cube is a common shape of microcrystals, such as MgO and the alkali halides, these theoretical predictions have been used several times to interpret experimental data. We shall do the same for MgO. Our theoretical treatment of nonsphericity, however, is based on ellipsoids. Despite its simplicity, this method predicts correctly many of the nonspherical effects.

12.2.1 Ellipsoids

There is no "exact" theory for irregularly shaped particles; nor is there an approximate theory suitable for our purposes. Indeed, the very notion of what precisely is an irregular particle is clouded by ambiguity; it is difficult to parameterize nonsphericity in general. Except for infinite cylinders, the exact theory for other regular shapes (e.g., spheroids and ellipsoids) is quite complicated. Fortunately, many of the interesting surface mode effects occur in particles small compared with the wavelength; therefore, we can appeal to electrostatics (i.e., the Rayleigh theory of Chapter 5) for guidance, if not for exact quantitative results. Ellipsoidal particles, which include spheres and long cylinders (needles) as special cases, represent perhaps the simplest departure from sphericity. If the incident electric field is parallel to a principal axis of a small, homogeneous ellipsoid of volume v, then its polarizability (5.32) may be written

$$\alpha = v \frac{\epsilon - \epsilon_m}{\epsilon_m + L(\epsilon - \epsilon_m)}, \tag{12.26}$$

where the geometrical factor L may take any value from 0 to 1, and we have omitted the subscript from the dielectric function ϵ of the particle for convenience. The absorption and scattering cross sections corresponding to (12.26) are

$$C_{abs} = k \, \text{Im}\{\alpha\}; \qquad C_{sca} = \frac{k^4}{6\pi} |\alpha|^2,$$

where k is the wave number. Thus, there will be a resonance in *both* cross sections (i.e., a surface mode will be excited) at the frequency where the denominator of α vanishes:

$$\epsilon = \epsilon_m \left(1 - \frac{1}{L} \right). \tag{12.27}$$

For any real material, the frequency at which (12.27) is satisfied is complex—the surface modes are virtual. However, its real part is approximately the frequency where the cross sections have maxima, provided that the imaginary part is small compared with the real part. We shall denote this frequency by ω_s. For a sphere, ω_s is the Fröhlich frequency ω_F. If used intelligently, always keeping in mind its limitations, (12.27) is a guide to the whereabouts of peaks in extinction spectra of small ellipsoidal particles; but it will not necessarily lead to the exact frequency.

There is only one distinct geometrical factor L for a sphere; there are two for a spheroid, and three for the general ellipsoid. Thus, there is the possibility of one, two, or three distinct extinction peaks depending on the shape of the particle. The width, height, and separation of these peaks depends, of course, on the behavior of the dielectric function. Because of the wide range of $1/L$, (12.27) may be satisfied over a correspondingly wide range of frequencies; but again, this depends on the shape and magnitude of the dielectric function.

The geometrical factors for spheroids are given by (5.33) and (5.34), together with the relations $L_2 = L_3$ (prolate) or $L_1 = L_2$ (oblate) and $L_1 + L_2$

$+ L_3 = 1$. In order to show how ϵ' for spheroid resonances depends on the ratio of axis lengths, (12.27) is plotted in Fig. 12.5. For any value of L_j, a surface mode resonance will occur at the frequency where ϵ'/ϵ_m has the value given by a point on the curve. The positions of ϵ'/ϵ_m for a sphere and for several spheroids, including the limiting cases of disks and circular cylinders, are shown for different polarizations of the incident electric field. For a sphere, of course, all L_j are equal, and the resonance condition is $\epsilon'/\epsilon_m = -2$ regardless of the polarization of the incident light. The resonance condition for a prolate spheroid is split into two branches: ϵ'/ϵ_m moves down the curve toward $-\infty$ with increasing elongation for the electric field parallel to the long axis and up the curve toward -1 for the electric field perpendicular to this axis. Similarly, the resonance condition for an oblate spheroid has two branches, the end points of which are 0 and $-\infty$. There should thus be two peaks in the absorption spectrum of randomly oriented spheroids and dichroism (absorption depending on polarization) for aligned spheroids; experimental verification of this for metallic particles will be discussed in Section 12.4.

In the preceding paragraphs we discussed only the conditions for surface mode resonances in the cross sections of small ellipsoidal particles. We now turn to specific examples to further our understanding of these resonances.

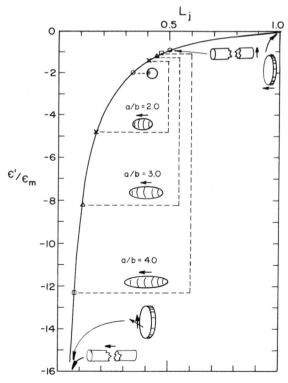

Figure 12.5 Effect of shape on the position of the lowest-order surface mode of small spheroids. Arrows next to the various shapes show the direction of the electric field.

12.2.2 Metallic Ellipsoids

In this section we try to extract as much physics as possible from a combination of electrostatics and the Drude theory of metals. It must be kept in mind, however, that our conclusions are rigorously correct only to the extent that both of these theories are valid. But by sacrificing rigor, we gain in understanding and insight.

The frequency-dependent absorption cross section of a metallic ellipsoid with dielectric function (9.26) is

$$C_{abs}(\omega) = \frac{v\gamma\omega_p^2}{c} f(\epsilon_m, L) \frac{\omega^2}{\left(\omega^2 - \omega_s^2\right)^2 + \gamma^2\omega^2}, \qquad (12.28)$$

where c is the speed of light *in vacuo* and

$$\omega_s^2 = \frac{L\omega_p^2}{\epsilon_m - L(\epsilon_m - 1)}, \qquad f(\epsilon_m, L) = \frac{\epsilon_m^{3/2}}{\left[\epsilon_m - L(\epsilon_m - 1)\right]^2}.$$

By means of this combination of the cross section for an ellipsoid with the Drude dielectric function we arrive at resonance absorption where there is no comparable structure in the bulk metal absorption. The absorption cross section is a maximum at $\omega = \omega_s$ and falls to approximately one-half its maximum value at the frequencies $\omega = \omega_s \pm \gamma/2$ (provided that $\gamma^2 \ll \omega_s^2$). That is, the surface mode frequency is ω_s or, in quantum-mechanical language, the surface plasmon energy is $\hbar\omega_s$. We have assumed that the dielectric function of the surrounding medium is constant or weakly dependent on frequency.

When $L = 1$, the surface mode frequency is the plasma frequency, which for most metals lies in the ultraviolet; when $L = 0$, ω_s vanishes. So there is an enormous range of possible collective excitations in small, ellipsoidal, metallic particles: their frequencies can be anywhere from the ultraviolet to the radio. For a given shape, the surface mode frequency is a monotonically decreasing function of ϵ_m; so in going from free space to a denser medium, the surface mode frequencies shift to lower values.

The maximum absorption cross section is

$$C_{abs}(\omega_s) = \frac{v\omega_p^2}{c\gamma} f(\epsilon_m, L). \qquad (12.29)$$

Note that for particles in air ($\epsilon_m = 1$), the maximum absorption is independent of shape. But if the particles are embedded in some nonvacuous medium ($\epsilon_m > 1$), the high-frequency peaks are greater than the low-frequency peaks. The maximum ratio of the height of peaks is $f(\epsilon_m, 1)/f(\epsilon_m, 0) = \epsilon_m^2$.

Up to this point we have considered only a single ellipsoidal particle oriented so that the electric field of the incident wave is parallel to one of its principal axes. A more realistic configuration is a collection of identical particles that are randomly oriented. The average absorption cross section $\langle C_{abs} \rangle$ of such a collection is merely the arithmetic average of the three principal cross sections:

$$\langle C_{abs} \rangle = k \, \mathrm{Im} \left\{ \frac{\alpha_1 + \alpha_2 + \alpha_3}{3} \right\}, \tag{12.30}$$

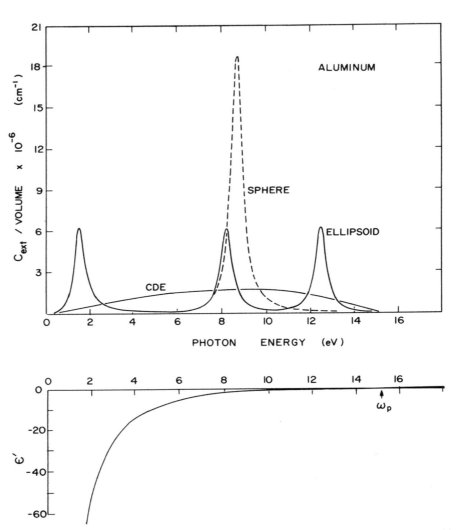

Figure 12.6 Calculated absorption spectra of aluminum spheres, randomly oriented ellipsoids (geometrical factors 0.01, 0.3, and 0.69), and a continuous distribution of ellipsoidal shapes (CDE). Below this is the real part of the Drude dielectric function.

where α_j is given by (12.26) with $L = L_j$. Thus, the absorption spectrum corresponding to (12.30) is characterized, in general, by three distinct peaks of approximately equal width (provided, of course, that the separation of surface mode frequencies is large compared with the width of the peaks). Moreover, if the particles are in air, the height of all three peaks is approximately the same. This is shown in Fig. 12.6, where the absorption spectrum of ellipsoidal aluminum particles is given. Also shown in this figure is the absorption spectrum for spheres, which has a single peak; the absorption spectrum for spheroids would exhibit two peaks. We shall explain the significance of the curve labeled CDE in a later section.

Let us return now to a single oriented ellipsoid. Integrated absorption is sometimes of interest:

$$\int_0^\infty C_{abs}(\omega)\,d\omega = \frac{v\gamma\omega_p^2}{c}f(\epsilon_m, L)\int_0^\infty \frac{\omega^2}{\left(\omega^2 - \omega_s^2\right)^2 + \gamma^2\omega^2}\,d\omega, \quad (12.31)$$

where the upper limit of integration should not be interpreted too rigidly. Obviously, the Rayleigh theory fails to be valid for indefinitely large frequencies. So the symbol infinity in (12.31) indicates a frequency sufficiently large that the absorption cross section is negligible, but not so large that Rayleigh theory is inapplicable. It may be shown by applying the residue theorem that

$$\int_0^\infty \frac{\omega^2}{\left(\omega^2 - \omega_s^2\right)^2 + \gamma^2\omega^2}\,d\omega = \frac{\pi}{2\gamma},$$

for all values of γ/ω_s; therefore, the integrated absorption is

$$\int_0^\infty C_{abs}(\omega)\,d\omega = \frac{\pi}{2}\frac{v\omega_p^2}{c}f(\epsilon_m, L). \quad (12.32)$$

There are several interesting observations that can be made about (12.32). Integrated absorption is independent of the damping constant γ; the only bulk parameter that affects it is the plasma frequency. If the particles are in air, then integrated absorption is independent of the shape; this is true not only for a single oriented ellipsoid but also for a collection of randomly oriented ellipsoids. It is instructive to rewrite (12.32) using (12.29):

$$\int_0^\infty C_{abs}(\omega)\,d\omega = \frac{\pi}{2}\gamma C_{abs}(\omega_s),$$

which shows that there is a simple proportionality between peak and integrated absorption.

12.2.3 Vibrational Surface Modes in Ellipsoids

The frequency-dependent absorption cross section of an ellipsoid with dielectric function (9.20) is

$$C_{\text{abs}}(\omega) = \frac{v\gamma\omega_p^2}{c\sqrt{\epsilon_m}} f(\xi, L) \frac{\omega^2}{\left(\omega^2 - \omega_s^2\right)^2 + \gamma^2\omega^2},$$

$$\omega_s^2 = \omega_t^2 + \frac{L\omega_p^2}{\epsilon_m + L(\epsilon_{0e} - \epsilon_m)}, \qquad (12.33)$$

$$f(\xi, L) = \frac{1}{\left[1 + L(\xi - 1)\right]^2}, \qquad \xi = \frac{\epsilon_{0e}}{\epsilon_m}.$$

Equation (12.33) is similar in form to (12.28), the absorption cross section of a metallic ellipsoid: the maximum absorption is at ω_s and the half-width of the absorption peak is approximately γ. There are some important differences between absorption spectra of ionic and metallic ellipsoids, however. If we use the approximate relation (9.23), then ω_s may be written

$$\omega_s^2 = \omega_t^2 \left[\frac{\epsilon_{0v} + \epsilon_m(1/L - 1)}{\epsilon_{0e} + \epsilon_m(1/L - 1)}\right],$$

from which it follows that ω_s lies between ω_t and ω_l. This is in marked contrast with ω_s for metallic ellipsoids, which ranges from 0 to the plasma frequency. Note also that the maximum absorption cross section

$$C_{\text{abs}}(\omega_s) = \frac{v\omega_p^2}{\gamma c\sqrt{\epsilon_m}} f(\xi, L)$$

is not independent of particle shape except in the special case where $\epsilon_{0e} = \epsilon_m$. In general, the low-frequency peaks are higher than the high-frequency peaks; the maximum ratio is $f(\xi, 0)/f(\xi, 1) = \xi^2$.

The average cross section of identical, but randomly oriented ellipsoids will, in general, exhibit three peaks in the frequency range between ω_t and ω_l. An example of this is given in Fig. 12.7, where C_{abs} for a silicon carbide ellipsoid is shown as a function of frequency.

12.2.4 Randomly Oriented Disks, Needles, and Spheres

In the two preceding sections we considered features in the absorption spectra of idealized ellipsoids. Because of the simple form of the dielectric functions we

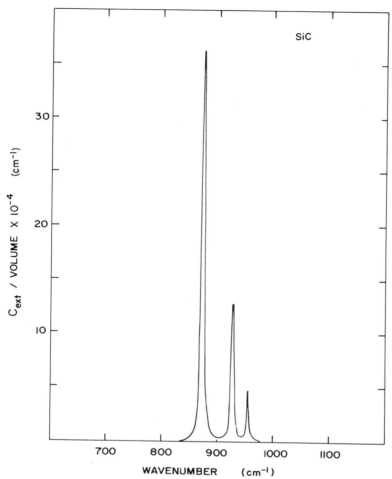

Figure 12.7 Calculated extinction cross section per unit volume of a silicon carbide ellipsoid with geometrical factors 0.1, 0.3, and 0.6. $C_{ext} \simeq C_{abs}$ for sufficiently small absorbing particles.

were able to obtain explicit expressions for the frequency-dependent cross section from which the position, width, and height of spectral features were obvious almost at a glance. But nature is not so cooperative as to provide us with only such simple materials; in general, recourse must be had to experimentally determined dielectric functions. In this section, therefore, we emphasize how the cross sections depend on the real and imaginary parts of the dielectric function, keeping in mind, of course, that these quantities are frequency dependent, but without actually specifying this dependence.

Corresponding to each point in the triangular region of the $L_1 L_2$ plane shown in Fig. 12.8 there is a unique ellipsoid, and conversely. It would be an unmanageable task, therefore, to consider all possible ellipsoids; but it also is hardly necessary to do so. Spheres, disks, and needles represent extreme forms of ellipsoids; they more or less bracket the range of possibilities. Moreover, they are readily visualized. So let us restrict ourselves for the moment to these three shapes. The average cross sections for identical, but randomly oriented particles are

$$\langle C_{abs} \rangle_{sphere} = \frac{kv}{3} \left[\frac{27}{(\epsilon' + 2)^2 + \epsilon''^2} \right] \epsilon'',$$

$$\langle C_{abs} \rangle_{needle} = \frac{kv}{3} \left[\frac{8}{(\epsilon' + 1)^2 + \epsilon''^2} + 1 \right] \epsilon'',$$

$$\langle C_{abs} \rangle_{disk} = \frac{kv}{3} \left[\frac{1}{\epsilon'^2 + \epsilon''^2} + 2 \right] \epsilon'',$$

$$\langle C_{sca} \rangle_{sphere} = \frac{k^4 v^2}{18\pi} |\epsilon - 1|^2 \left[\frac{27}{(\epsilon' + 2)^2 + \epsilon''^2} \right],$$

$$\langle C_{sca} \rangle_{needle} = \frac{k^4 v^2}{18\pi} |\epsilon - 1|^2 \left[\frac{8}{(\epsilon' + 1)^2 + \epsilon''^2} + 1 \right],$$

$$\langle C_{sca} \rangle_{disk} = \frac{k^4 v^2}{18\pi} |\epsilon - 1|^2 \left[\frac{1}{\epsilon'^2 + \epsilon''^2} + 2 \right],$$

$$(12.34)$$

where ϵ is the dielectric function of the particle relative to that of the surrounding medium and k is the wave number in this medium. It is clear that, at a given frequency, the only difference between the cross sections for the three shapes is the term in brackets. We are now in a position to examine further the origin of the misconception that there are no shape effects in extinction spectra of small particles. For most nicely behaved materials (insulators at visible wavelengths, for example), ϵ'' is small and ϵ' lies between 2 and 3. If $\epsilon'' \ll 1$ and $\epsilon' = 2$, for example, the cross sections are in the ratio (sphere : needle : disk) 1 : 1.12 : 1.33. So for such a material there are no strong shape effects. If, on the other hand, $|\epsilon'|$ or ϵ'' is large, then the cross section for a disk or a needle can be appreciably greater than that for a sphere. Of course, in the region where ϵ' is negative there can be strong shape effects.

More insight into shape effects in absorption spectra of small particles can be acquired from contour plots in the complex ϵ plane; lines of constant dimensionless cross section $3\langle C_{abs} \rangle / kv$ are shown in Fig. 12.9a, b, c. Note that the curves are symmetric about the lines $\epsilon' = -2$, $\epsilon' = -1$, and $\epsilon' = 0$ for the sphere, needle, and disk, respectively. Three points representing certain solids

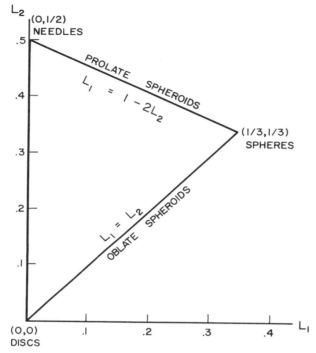

Figure 12.8 Each point of the triangular region corresponds to a unique ellipsoid, and conversely.

at specific wavelengths are labeled on the contour plots: i at $(2.3, 0.0)$ represents a typical insulator at visible wavelengths; c at $(3.0, 4.0)$ corresponds approximately to carbon in the visible; m at $(-2.0, 0.3)$ corresponds to magnesium oxide at its Fröhlich frequency (~ 620 cm^{-1}) in the infrared. Although the resolution of these contour maps is not sufficient for estimating shape effects in the vicinity of i, the calculations in the preceding paragraph indicate only slight differences ($\sim 30\%$) among the three shapes. For carbon, which is highly absorbing in the visible, estimation from the contour plots gives approximately 3, 5, and 8 for spheres, needles, and disks, respectively. But the values for MgO at its Fröhlich frequency are estimated to be greater than 50, about 3, and about 1, which shows the extreme shape dependence of small particle absorption in this instance.

Figure 12.9d shows the dielectric function of several metals that either have been discussed in Chapter 9 or will be discussed in connection with small particle extinction in Section 12.4. The energy dependence of the dielectric function is given in the form of trajectories in the complex ϵ plane, similar to the Cole–Cole plots (1941) that are commonly used for polar dielectrics; the numbers indicated on the trajectories are photon energies in electron volts.

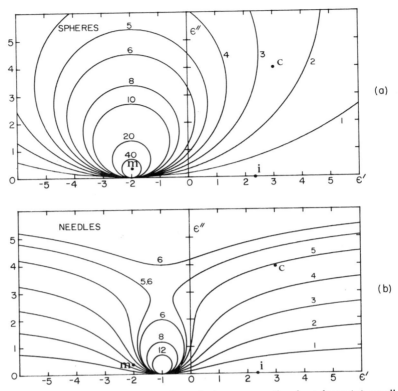

Figure 12.9 Contour plots of constant dimensionless cross section for spheres (*a*), needles (*b*), and disks (*c*). Cole–Cole plots are shown in (*d*) for various metals.

When used in conjunction with the contour plots these trajectories enable one to quickly estimate the magnitude of particle absorption and its dependence on shape. The degree to which a metal behaves like a free-electron (Drude) metal can also be determined at a glance. For example, ϵ'' of a free-electron metal goes monotonically to zero with increasing frequency as ϵ' approaches 1 from below (see Section 9.4). Aluminum is a good illustration; its trajectory is typical of a free-electron metal. From 3.0 to 3.6 eV the trajectory of silver is similar to that of aluminum, but at higher energies veers sharply from the goal of (1.0, 0.0) because of the onset of interband electronic transitions, which are discussed in connection with Fig. 9.12. The trajectory of copper nowhere looks very much like that of a Drude metal, although it terminates at the point (1.0, 0.0) as it must. Similarly, gold does not behave like a free-electron metal above 2.2 eV.

Differences in surface plasmon absorption among various metals are clearly revealed by imagining the trajectories to be superposed onto the contour plots. Spherical silver and aluminum particles have intense surface plasmon absorption peaks because ϵ'' is small at the frequency where ϵ' is -2, whereas gold

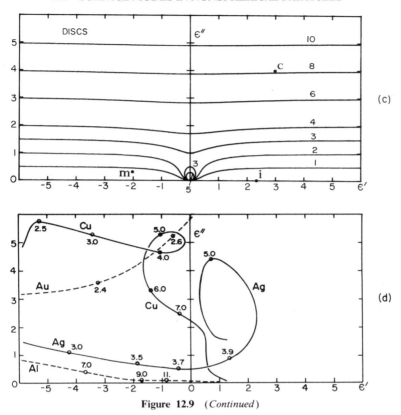

Figure 12.9 (*Continued*)

and copper particles are less absorbing because of much greater values of ϵ''. Because their trajectories closely approach the ϵ' axis, where the poles of the absorption cross section lie, absorption by silver and aluminum particles is much more dependent on shape than that by copper and gold particles.

12.2.5   Distribution of Ellipsoidal Shapes

At this point the reader who has studied the preceding sections may well wonder why we are so interested in, if not obsessed with, ellipsoidal particles: most real particles are no more ellipsoidal than they are spherical. One reason for devoting so much space to ellipsoids is that they are a means for dispelling widespread misconceptions about the nonexistence of shape effects in small-particle absorption spectra. For if there are strong shape effects in spectra of ellipsoidal particles, then there are certainly such effects in the spectra of other, less well-defined nonspherical particles. But there is at least one other reason, which may prove to be of greater practical utility: the hope that spectra of irregular particles can be approximated somehow by suitably averaging over all ellipsoidal shape parameters to obtain simple expressions for the average absorption cross section. This idea derives from conversations with D. P. Gilra, who did calculations for distributions of spheroids. Gilra's unpublished calculations were used by Treffers and Cohen (1974) in an attempt to identify particles in the space surrounding cool stars by means of features in infrared emission spectra. In this section we derive some expressions for average cross sections under various assumptions; in later sections we shall offer experimental evidence to support the validity of these expressions in describing absorption spectra of irregular particles.

The average absorption cross section of a randomly oriented collection of *identical* homogeneous ellipsoids (12.30) may be written

$$\langle C_{abs} \rangle = \frac{kv}{3} \, \mathrm{Im} \left\{ \sum_{j=1}^{3} \frac{1}{\beta + L_j} \right\},$$

where $\beta = 1/(\epsilon - 1)$ and ϵ is the dielectric function of the ellipsoid relative to that of the surrounding medium. Suppose that, in addition to being randomly oriented, the collection consists of ellipsoidal particles of all possible shapes; that is, the geometrical factors L_1, L_2 are not restricted to a single set of values but are distributed according to some *shape probability function* $\mathscr{P}(L_1, L_2)$. Because of the requirement that $L_1 \leq L_2$, $\mathscr{P}(L_1, L_2)$ is strictly defined only on the hatched triangular region in the $L_1 L_2$ plane shown in Fig. 12.10. However, it is convenient to extend the domain of definition of $\mathscr{P}(L_1, L_2)$ onto the larger

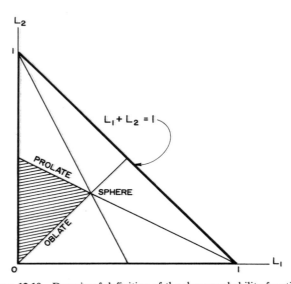

Figure 12.10 Domain of definition of the shape probability function.

triangular region Δ shown in the figure. This region may be subdivided into six equivalent regions of equal area, each of which corresponds to one of the six possible ways of choosing the relative lengths of the axes a, b, c of the ellipsoid (in Chapter 5 we required that $a > b > c$). The shape probability function is normalized to unity on Δ:

$$\iint_\Delta \mathcal{P}(L_1, L_2) \, dL_1 \, dL_2 = 1.$$

The absorption cross section averaged over the shape distribution and over all orientations is, therefore,

$$\langle\langle C_{\text{abs}} \rangle\rangle = \iint_\Delta \langle C_{\text{abs}} \rangle \mathcal{P}(L_1, L_2) \, dL_1 \, dL_2$$

$$= \frac{k v}{3} \, \text{Im}\{\mathcal{I}_1 + \mathcal{I}_2 + \mathcal{I}_3\},$$

$$\mathcal{I}_1 = \iint_\Delta \frac{\mathcal{P}(L_1, L_2)}{\beta + L_1} dL_1 \, dL_2, \tag{12.35}$$

$$\mathcal{I}_2 = \iint_\Delta \frac{\mathcal{P}(L_1, L_2)}{\beta + L_2} dL_1 \, dL_2,$$

$$\mathcal{I}_3 = \iint_\Delta \frac{\mathcal{P}(L_1, L_2)}{\beta + 1 - L_1 - L_2} dL_1 \, dL_2.$$

The integral of a function $f(L_1, L_2)$ over Δ may be written as an iterated integral:

$$\iint_\Delta f(L_1, L_2) \, dL_1 \, dL_2 = \int_0^1 dL_1 \int_0^{1 - L_1} f(L_1, L_2) \, dL_2.$$

We have assumed that all particles have the same volume v; however, if there is no correlation between shape and volume, the *total* absorption cross section of the collection is

$$\mathfrak{N} \frac{k \langle v \rangle}{3} \, \text{Im}\{\mathcal{I}_1 + \mathcal{I}_2 + \mathcal{I}_3\},$$

where $\mathfrak{N}$ is the total number of particles per unit volume and $\langle v \rangle$ is the average particle volume. It has also been implicitly assumed that $\mathcal{P}(L_1, L_2)$ is continuous; this is not a necessary restriction, however, and we can take into account discrete distributions by replacing the integrals above with summations over a discrete set of points (L_1, L_2) in Δ.

12.2.6   Uniform Distribution of Ellipsoidal Shapes

Perhaps the simplest conceivable distribution is one for which all shapes are equally probable, in which instance $\mathcal{P}(L_1, L_2) = 2$ and the integrals in (12.35) are readily evaluated:

$$\mathcal{G}_1 = \mathcal{G}_2 = \mathcal{G}_3 = \frac{2\epsilon}{\epsilon - 1} \, \text{Log} \, \epsilon - 2.$$

Therefore, the average cross section is

$$\langle\langle C_{\text{abs}} \rangle\rangle = kv \, \text{Im} \left\{ \frac{2\epsilon}{\epsilon - 1} \, \text{Log} \, \epsilon \right\}. \tag{12.36}$$

In (12.36) Log z denotes the *principal value* of the logarithm of a complex number $z = re^{i\Theta}$ (Churchill, 1960, p. 56):

$$\text{Log} \, z = \text{Log} \, r + i\Theta \qquad (r > 0, \, -\pi < \Theta < \pi),$$

where Log $r = \ln r$ if r is real.

12.2.7   Summary of Shape Effects

Before presenting experimental data on surface mode absorption by small particles, we briefly summarize shape effects calculated in the Rayleigh approximation. Figure 12.11 is a schematic of surface mode absorption for the two idealized classes of solids with ϵ' negative at some frequencies: an insulator described by a one-oscillator (Lorentz) dielectric function (left) and a free-electron (Drude) metal (right); the particles are in free space. Spheres of both solids absorb strongly in the single narrow band around the frequency where ϵ' is -2; spheroids have two bands, and ellipsoids three, at frequencies determined by the relative lengths of their principal axes. Continuous distributions of ellipsoids have the broad absorption spectra sketched for both materials. Absorption bands for other nonspherical particles are expected at frequencies that depend on their shape factors (which may be laborious to calculate); for example, the six most important surface mode frequencies of an insulating cube, calculated by Fuchs (1975), are shown in the figure.

There are several important generalizations to be gleaned from this summary: (1) small particles of any shape can absorb strongly at frequencies where ϵ' is negative; (2) a distribution of shapes broadens the absorption bands at the expense of maximum absorption; and (3) although shape effects in insulating particles are confined to the region between the transverse and longitudinal optical mode frequencies, strong absorption by metallic particles may occur at any frequency from the bulk plasma frequency (commonly in the far ultraviolet) down through the visible and infrared to radio frequencies; thus, shape effects can be much more pronounced in metals than in insulators.

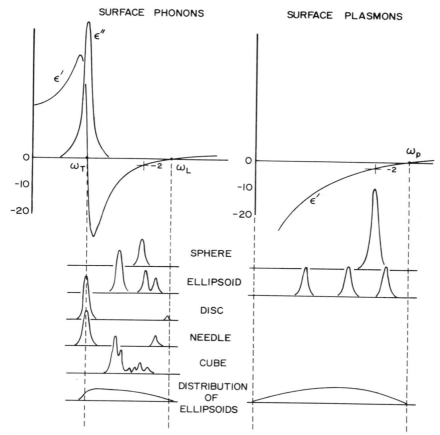

SURFACE PHONONS   SURFACE PLASMONS

Figure 12.11 Surface mode frequencies for insulating and metallic particles of various shapes.

12.3 VIBRATIONAL MODES IN INSULATORS

In this section we compare the theory of the preceding two sections with experimental measurements of infrared extinction by small particles. Comparisons between experiment and theory for spheres of various solids, most notably alkali halides and magnesium oxide, have been published in the scientific literature; many of these papers are cited in this chapter. In most of this work, however, there is an arbitrary normalization of theory and experiment, which tends to hide discrepancies. For this reason, most theoretical calculations in this section are compared with mass-normalized extinction measurements. The new measurements presented here were made in the Department of Physics at the University of Arizona. A group of solids was selected to illustrate different aspects of surface modes. Results on amorphous quartz (SiO_2) particles, for example, illustrate the agreement between experiment and theory for very small spheres described by the bulk dielectric function of the parent material. This demonstrated agreement gives confidence in the procedure of using bulk dielectric functions for small-particle calculations. Results are next shown for crystalline quartz particles, which are definitely nonspherical as well as optically anisotropic, necessitating the use of shape distribution theory and orientational averaging. Next, extinction data for SiC and MgO particles are discussed; we have used these solids for illustrative purposes in Chapters 9 and 10 and in preceding sections of this chapter. Also, there have been numerous papers relating to surface mode absorption in MgO smoke particles, which are nearly cubical in contrast with the wide distribution of shapes found in particulate samples of some of the other solids.

12.3.1 Important Experimental Considerations

For theory to be legitimately compared with experiment it is necessary that samples be prepared in which the particles are quite small (usually submicrometer), well isolated from one another, and that the total mass of particles be accurately known; also, reliable optical constants obtained from measurements on bulk samples must be at hand. These requirements are, of course, easy to state but often difficult to meet; however, if they are not met, then comparison between theory and experiment—agreement or disagreement—is likely to be specious.

In order to comply with the assumptions underlying the theory in this chapter—single, isolated, homogeneous particles—it is desirable to disperse the particles in a solid matrix. Although it might seem at first thought that dispersal in air or another gas would be more convenient and equally satisfactory, such a procedure yields a dynamic system in which the particulate characteristics continually change because of coagulation and settling. Dispersal in the zero-gravity, high-vacuum environment of an orbiting spacecraft might prevent these undesirable effects. But a considerably less expensive experimental technique is to disperse the particles in a solid matrix; this ensures that their isolation and independence are maintained. A common matrix technique, which has been used in infrared spectroscopy for many years by chemists, is the KBr pellet technique: small quantities of the particulate sample are mixed thoroughly with powdered KBr; because of the softness of KBr and its bulk transparency between about 40 and 0.2 μm, the KBr and particle mixture can be pressed into a clear pellet. Transmission measurements in conventional infrared spectrophotometers then yield extinction spectra for the particles in the KBr matrix. Although KBr has been by far the most popular matrix, other materials, such as TlBr and KI, can be used. Polyethylene powder with a dispersed particulate sample also can be pressed into infrared-transparent samples for use at wavelengths longer than about 20 μm; quite satisfactory samples can be made by pressing the powder mixture between two glass plates on a hot plate. An alternative procedure is to allow particles to collect as smoke, or by settling in air, onto thin sheets of polyethylene film, then cut the sheets into small squares (~ 1 cm) which can be stacked and fused together between glass plates on a hot plate. A more sophisticated method, used for many years to isolate molecules and molecular clusters, is to inject the sample (vapor or particles) into a flowing gas stream of, for example, argon, which is then solidified. Martin and Schaber (1977) have used this technique in recent years to isolate small solid particles, as has Welker (1978), who also studied aggregates of silver atoms. All these matrix techniques merely serve to isolate the particles from one another and to maintain them in such a state for spectroscopic study.

The KBr technique used in most of the work reported in this section has the additional advantage that samples can be stored for many years in a desiccator without change for future study. A disadvantage is the possibility of altering some of the particles by the grinding and pressing process. To avoid this possibility some workers have opted for a loose collection of particles on an infrared-transparent substrate (see, e.g., Genzel and Martin; 1972). Although this is a perfectly acceptable approach, it usually violates our assumption of single scattering by independent particles, which thereby necessitates a modified theoretical treatment, such as the Maxwell Garnett theory, to account for the interaction between particles.

Measurements of extinction by small particles are easier to interpret and to compare with theory if the particles are segregated somehow into a population with sufficiently small sizes. The reason for this will become clear, we hope, from inspection of Fig. 12.12, where normalized cross sections using Mie theory and bulk optical constants of MgO, SiO_2, and SiC are shown as functions of radius; the normalization factor is the cross section in the Rayleigh limit. It is the *maximum* infrared cross section, the position of which can shift appreciably with radius, that is shown. The most important conclusion to be drawn from these curves is that the mass attenuation coefficient (cross section per unit particle mass) is independent of size below a radius that depends on the material (between about 0.5 and 1.0 μm for the materials considered here). This provides a strong incentive for dealing only with small particles: provided that the total particle mass is accurately measured, comparison between theory and experiment can be made without worrying about size distributions or arbitrary normalization.

There are two different ways to obtain submicrometer particles: (1) grind bulk material as finely as possible and disperse the resultant particles in air or water for segregation by settling; and (2) use a technique that generates only submicrometer particles, such as vaporization in an electric arc and subsequent condensation in a gas. Some of the particles discussed in succeeding paragraphs—amorphous SiO_2 and SiC—were prepared by arc vaporization; MgO particles were obtained by burning magnesium ribbon in air. These processes yield mostly particles that are very small (less than about 0.1 μm). Particles made from the bulk solid (quartz, for example) were ground vigorously for several hours in steel and agate mortars, dispersed both in air and in water, and settled for times long enough to leave only particles less than about 1 μm

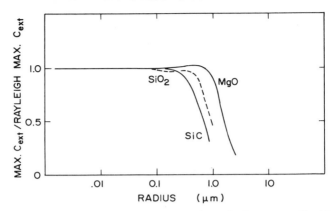

Figure 12.12 Maximum infrared extinction cross sections of spheres normalized by the value in the Rayleigh limit.

in suspension; the suspensions were then filtered to collect the particles. In all instances, about 100 μg of particles was dispersed in about 0.5 g of KBr powder. The powder–KBr mixtures were agitated in a glass vial with steel balls for periods ranging from a few hours to a few days before being pressed into pellets (~ 1 cm diameter) under a force of about 10 tons. This method of preparation seems to produce samples of particles reasonably well isolated from one another and sufficiently small that their volumetric extinction is independent of size.

12.3.2 Amorphous Quartz Spheres

A continually recurring question is the range of applicability of bulk optical constants to small particles: below what size are bulk properties no longer valid in small-particle calculations? Because *some* small-particle optical effects may be interpreted as the failure of bulk properties to be valid (see the paragraphs in Sections 12.1 and 12.4 on the limitation of the mean free path), it has become common, unfortunately, to doubt the correctness of using bulk dielectric functions even for micrometer-size particles and larger. Too often inexplicable effects exhibited by small particles—inexplicable, that is, within the framework of isolated sphere theory—are interpreted as resulting from the inapplicability of bulk properties without exploring other alternatives, such as shape effects and interactions between particles. But measurements on *nonspherical* particles cannot be compared with *sphere* calculations to decide for or against the validity of using bulk optical constants in such calculations; the only proper experimental test requires measurements on spheres. Small solid spheres are not, as a rule, easily generated; an exception is SiO_2 smoke, which can be produced readily enough by striking an arc (ac or dc) in air between silicon electrodes or carbon electrodes embedded with pieces of silicon or quartz. These smokes consist of nearly perfect spheres of amorphous SiO_2 with diameters in the range between about 100 and 1000 Å; they satisfy the requirements of being spherical, isotropic, composed of a material with accurately measured infrared optical constants, and well within the range where volumetric extinction is independent of size. The biggest problem to be overcome is that upon formation in air the spheres link together in clusters and chains which, in electron micrographs, resemble strings of pearls. Steyer et al. (1974) published infrared extinction data on this system of particles because of its approach to ideality; however, they found a factor of 2.2 between calculated and measured peak extinction. Subsequently, experiments have been undertaken in which a more concerted effort has been made to disrupt clusters and produce isolated spheres in a KBr matrix. Results of the newer experiments are shown in Fig. 12.13, where measured extinction is compared with that calculated from sphere theory with no adjustable parameters. Bulk properties used

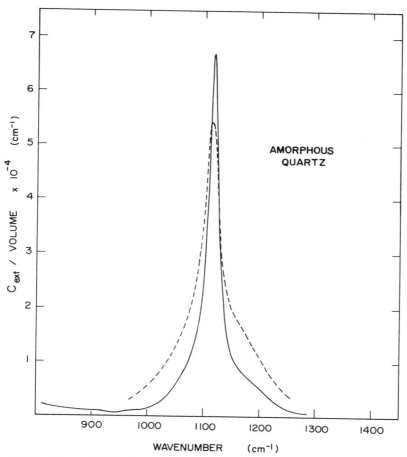

Figure 12.13   Measured (dashed curve) and calculated (solid curve) infrared extinction by amorphous quartz spheres.

in the calculations, taken from the paper by Steyer et al. (1974), agree well with independent measurements reported by Neuroth (1956) and Zolotarev (1970). Evidence that the silica spheres have been dispersed more than in earlier efforts is that the discrepancy between measured and calculated peak extinction has been reduced from a factor of 2.2 to about 20%.

Several predicted features of infrared surface mode absorption by small spheres are verified by the experimental results shown in Fig. 12.13. The frequency of peak absorption by spheres is shifted an appreciable amount from what it is in the bulk solid: the ϵ'' curve peaks at 1070 cm^{-1}, whereas the peak of the small-sphere absorption is at 1111 cm^{-1}, very close to the frequency where ϵ' is $-2\epsilon_m$ (-4.6 for a KBr matrix). The absorption maximum (absorption is nearly equal to extinction for these small particles) is very strong: Q_{abs} for a 0.1-μm particle is about 7 at the Fröhlich frequency.

We feel that the remaining discrepancy between sphere calculations and measurements can be attributed to residual clumping of particles, which causes slight broadening of the small-particle absorption bands at the expense of decreased peak height. Even without invoking such residual clumping, however, the comparison based on bulk optical constants seems favorable: the predicted absorption band—position and shape—is very close to that measured. In view of possible experimental uncertainties in the bulk dielectric function, the agreement is sufficiently close to convince us that, in this instance at least, bulk optical constants are appropriate to particles averaging considerably less than 0.1 μm in diameter.

12.3.3 Crystalline Quartz

Crystalline quartz is one of the earth's most common solid substances, bestrewing the surface as common sand. Its optical properties are anisotropic and it has strong infrared absorption bands near 9 μm. In Section 9.3, quartz exemplified an anisotropic solid the optical constants of which have been successfully extracted from infrared reflectance measurements by use of a multiple-oscillator model. Extinction measurements are presented in Fig. 12.14 for a collection of submicrometer quartz particles segregated from a finely ground powder by suspending it in water for a sufficient time to allow the larger particles to settle; these results have been published by Huffman and Bohren (1980). Comparisons of the measured volume-normalized extinction with calculations based on optical constants measured by Spitzer and Kleinman (1961) are shown in the figure. In the upper part, sphere calculations incorporating the treatment of anisotropy given in Section 5.6 are compared with measurements; without arbitrary normalization, the agreement is poor: the measured band width is much greater and the peak height much smaller than that calculated. Note that adjusting the measured and calculated curves to bring the peaks into congruence, as is commonly done, would make the agreement more favorable, but this is merely a cosmetic device that masks the discrepancies. Greatly improved agreement with experiment is exhibited by

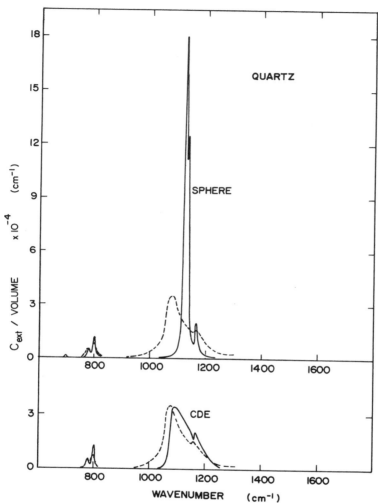

Figure 12.14 Measured infrared extinction by crystalline quartz particles (dashed curves) compared with calculations for spheres (top) and a continuous distribution of ellipsoids (bottom).

the theoretical spectrum calculated from (12.36) for a continuous distribution of ellipsoids, which is shown in the bottom part of the figure. Although all small-particle infrared spectra that we have measured have not shown such good agreement, the results do show a marked improvement in the treatment of shape effects, particularly for a sample containing particles widely distributed in shape. Moreover, the large differences between sphere calculations and those for a distribution of ellipsoids, together with good experimental agreement with the latter, bear witness to the importance of shape effects in the vicinity of strong infrared absorption bands.

12.3.4   Silicon Carbide

Although silicon carbide occurs in a variety of crystalline forms, the infrared optical properties of the major allotropes are very similar (Spitzer et al., 1959). In Section 9.1 the lattice absorption band in SiC near 11 μm was displayed as our canonical example of a real solid that conforms to a simple one-oscillator model; SiC was also invoked to illustrate surface modes in small spheres (see Figs. 12.1 and 12.2) and the effect of particle shape in the surface mode region (Fig. 12.7). Because of this prominence given to SiC we present in Fig. 12.15 experimental data for SiC particles size-segregated from powder by settling in water and then dispersed in KBr. Although we previously used the one-oscillator approximation for cubic SiC, the calculations shown in Fig. 12.15 are based on the anisotropic optical constants of hexagonal α-SiC measured by Spitzer et al. (1959). As with crystalline quartz, measurements and sphere calculations

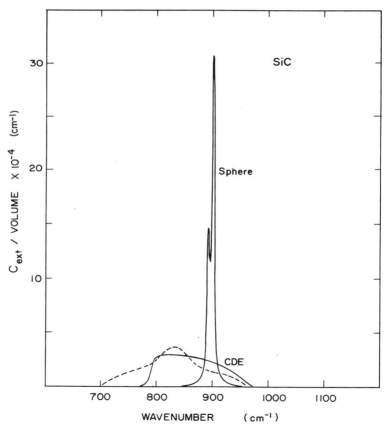

Figure 12.15 Measured infrared extinction by silicon carbide particles (dashed curve) compared with calculations for spheres and a continuous distribution of ellipsoids (CDE).

are in serious conflict. The more favorable agreement of calculations for a continuous distribution of ellipsoids (CDE) with experiment underscores the consequences of particle shape to absorption spectra. Measured integrated absorption as well as the approximate magnitude of absorption in the region between 800 and 950 cm^{-1} compare rather well with the CDE calculations. There are some discrepancies, however. Unlike the calculated band, which dips abruptly at ω_t (793 cm^{-1}), the measured absorption band extends to frequencies lower than ω_t, with just a hint of a shoulder near 780 cm^{-1}. This is the frequency region where *spheres* of SiC begin to exhibit bulk absorption modes as they depart from the behavior dictated by Rayleigh theory. These bulk modes in spheres of SiC were discussed in connection with Fig. 12.1. They would not, of course, appear in the calculations of Fig. 12.15, but they should begin to make their presence felt, although distorted in a complicated way by nonsphericity, in the larger particles (0.5–1.0 μm) of the experimental sample. Measured absorption also tends to be greater than that calculated from the CDE theory, but the peak is at a lower frequency than the sphere mode. This may be a consequence of a preferred shape among the particles, such as platelets, which would presumably not exist if the SiC particles, like the quartz particles, were ground from the bulk; the particles used in our measurements were segregated from the manufacturer's sample without further grinding because of the extreme hardness of SiC.

Further experimental evidence of shape effects in absorption spectra of SiC particles is found in the data of Pultz and Hertl (1966), who investigated infrared absorption by SiC fibers with and without SiO$_2$ coatings. Although these measurements were not mass-normalized, they show a strong absorption band at 795 cm^{-1} and a weaker band at 941 cm^{-1}. If the fibers are approximated as ellipsoids with $L_2 = L_3 = \frac{1}{2}$ and $L_1 = 0$ (i.e., a cylinder), then the ellipsoid equation (12.27) predicts absorption peaks for particles in air at frequencies where $\epsilon' = -1$ and $\epsilon' = -\infty$. This corresponds to absorption bands at 797 and 945 cm^{-1} for the dielectric function of isotropic SiC, in excellent agreement with the experimental peak positions for the fibers.

12.3.5 Magnesium Oxide

Possibly no other solid has been studied so much for the purpose of understanding small-particle infrared surface modes than MgO. The reason for this may in part be the ease with which small, highly crystalline cubes of MgO are generated by simply burning magnesium ribbon in air. Genzel (1974) has surveyed much of the experimental and theoretical work.

Genzel and Martin (1972, 1973) measured extinction by MgO smokes loosely packed on transparent substrates both in air and covered with the transparent oil Nujol. Their results showed absorption bands appreciably shifted from the bulk absorption band; the peak frequencies agreed with calculations, but the widths were consistently greater than predicted by sphere theory. In addition, a narrower absorption feature always appeared at the approximate bulk absorption maximum (the maximum of ϵ''). Enhanced damping in small particles was invoked to explain the broadening. Fuchs (1975) derived an expression for absorption by cubes in the electrostatics approximation which he applied to MgO with optical constants given by a one-oscillator model, but with a damping factor larger than for bulk MgO. Fuchs (1978) has also compared results from his continuum model with the lattice-dynamical calculations of Chen et al. (1978) for an MgO microcrystal of 900 atoms (10 × 10 × 9); there was good agreement between the two theories, which suggests that relatively few atoms are necessary before an MgO particle may be properly regarded as macroscopic. Luxon et al. (1969) examined infrared absorption by several different kinds of MgO samples and, in the interpretation of their results, discussed the role of particle shape. Matumura and Cho (1981) measured emissivities in the infrared surface mode region, as did Kälin and Kneubühl (1976) for alkali halides. Despite all this effort, agreement between experiment and theory has not been completely satisfactory, particularly since arbitrary normalization is present in much of this work.

Results of our recent measurements of extinction by MgO particles are shown in Fig. 12.16. Solid curves represent experimental data, where the particles are successively more dispersed as one progresses downward; dashed curves are theoretical calculations based on the bulk optical constants measured by Jasperse et al. (1966). In the upper part of the figure volume-normalized extinction is shown for a sample prepared by burning magnesium ribbon in air and collecting the particles on a clean KBr pellet. The middle curve is for the same smoke but dispersed by grinding with KBr and shaking the mixture for 3 hours in a glass vial with steel balls before pressing a pellet; the bottom curve is for a similarly prepared sample, but shaken for 3 days before pressing a pellet.

Particles dispersed on the KBr substrate are not strictly isolated, and perhaps this should be taken into account by basing calculations on the Maxwell Garnet theory (or a similar theory) as Genzel and Martin did. The particle volume fraction is small, however, so we are not misrepresenting the experiment too badly by comparing it with calculations for isolated spheres and a continuous distribution of isolated ellipsoids. Our sphere calculations are in poor agreement with measurements, even the position of peak absorption. But Genzel and Martin obtained good theoretical agreement with measured peak absorption using the optical constants of Häfele (1963). This illustrates the sensitivity of surface mode calculations to optical constants and the consequent difficulties one often faces in deciding which among a possibly contradictory set are "best" for such calculations.

Comparison of measurements for particles dispersed *on* and *in* KBr is quite revealing. The extinction curve for particles on a KBr substrate shows a peak at approximately 400 cm^{-1}, the transverse optical mode frequency for bulk MgO. This feature has been observed a number of times and it is discussed in some of the references already cited. Its explanation now appears to be the tendency of MgO cubes to link together into chains, which more closely

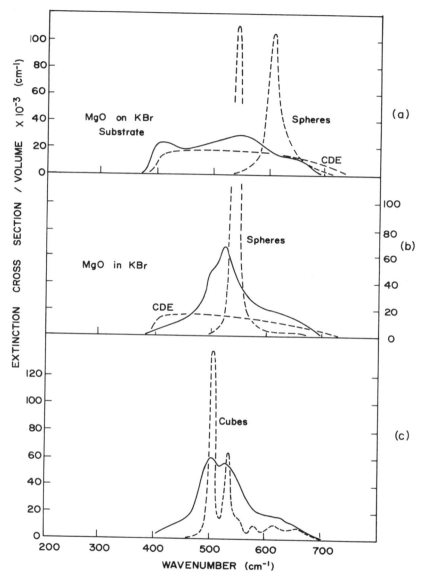

Figure 12.16 Measured infrared extinction by magnesium oxide cubes (solid curves); the particles are progressively more dispersed going from (*a*) to (*c*). Calculations for various particle shapes are shown by dashed curves.

resemble cylinders or elongated spheroids than spheres. A similar effect was observed in NiO smoke particles by Hunt et al. (1973), who analyzed their results on the basis of calculations for cylinders as well as spheres. Further evidence that the chain formation is responsible for the 400-cm^{-1} peak is that it disappears upon thorough dispersal of the particles in KBr (Fig. 12.16b, c); this was pointed out by Dayawansa and Bohren (1978).

In Fig. 12.16b sphere and CDE calculations are compared with measurements on MgO cubes well dispersed in KBr; neither is very satisfactory. The calculated position of peak absorption by spheres is fairly close to that measured but not coincident with it; the CDE calculations show appreciable absorption over approximately the same frequency range as the measurements but no structure. If the optical constants we have used accurately place the Fröhlich frequency, Fig. 12.16b suggests that neither spheres nor a broad distribution of shapes are good approximations for MgO particles. This is hardly surprising because electron micrographs reveal that MgO smoke is composed of cubes. These cubes are so nearly perfect that they have been used to quickly determine the resolution of electron microscopes: degraded resolution results in apparently rounded corners.

Fuchs's (1975) result for volume-normalized absorption by randomly oriented cubes in the electrostatics approximation is

$$k \, \text{Im} \left\{ \sum_{j=1}^{6} \frac{C(j)}{\epsilon_m / (\epsilon - \epsilon_m) + n_j} \right\}. \qquad (12.37)$$

The sum over all strength factors $C(j)$ is 1, and the n_j are analogous to the geometrical factors L_j in the expression (12.30) for the absorption cross section per unit volume of randomly oriented ellipsoids:

$$k \, \text{Im} \left\{ \sum_{j=1}^{3} \frac{\frac{1}{3}}{\epsilon_m / (\epsilon - \epsilon_m) + L_j} \right\}.$$

We used the dielectric function ϵ of bulk MgO calculated from oscillator parameters determined by Jasperse et al. (1966), together with the dielectric function ϵ_m of the KBr matrix given by Stephens et al. (1953) (corrected by June, 1972), to calculate the absorption spectrum (12.37) of a dilute suspension of randomly oriented MgO cubes. These theoretical calculations are compared with measurements on well-dispersed MgO smoke in Fig. 12.16c. Superimposed on a more or less uniform background between about 400 and 700 cm^{-1}, similar to the CDE spectrum, are two peaks near 500 and 530 cm^{-1}, the frequencies of the two strongest cube modes. It appears that for the first time these two modes have been resolved experimentally. If this is indeed so we conclude that the widths of individual cube modes are not much greater than the width of the dominant bulk absorption band. Genzel and Martin (1972) used an eightfold increase in the damping factor and Fuchs (1975) invoked a factor of 2.5 to bring theory into congruence with measurements. Such enhanced damping suggested by the broad and featureless absorption spectra previously published for MgO cubes does not seem to be necessary now in view of the apparent resolution of the cube modes. The MgO results and the agreement between theory and experiment for amorphous quartz spheres (Sections 12.3) imply that bulk optical constants do not have to be modified appreciably—by increasing the damping factor, for example—for infrared surface modes in particles in the size range 100–1000 Å.

12.4   ELECTRONIC MODES IN METALS

The first discussions of surface plasmon modes, although not designated as such, concerned colloidal silver and gold. From the classical descriptions of Faraday and the theoretical treatment of Mie to the present, the vivid colors of metal colloids have attracted serious scientific attention. In more recent years many observations have been made on surface modes in various metal and metal-like particles. These include colloidal dispersions in water, photographic emulsions and photosensitive glasses, island films formed by evaporating small quantities of metal onto smooth surfaces, cermets and granular films, aggregated color centers in alkali halide crystals, metallic smokes, electron–hole droplets in semiconductors at low temperatures, and even particles in interstellar space. Rather than survey this vast field spanning various disciplines within both pure and applied science, we have chosen a few examples that illustrate the more salient points. Gold is first, partly because of its historical interest, and because it continues to be a good example of surface mode absorption by small spherical particles at visible wavelengths. Next, silver is discussed because shape effects, which are much more pronounced in this nearly ideal free-electron metal than in gold, have been nicely demonstrated by experiment. New measurements on aluminum particles widely distributed in shape, which causes surface plasmon absorption to extend into the far infrared, is then discussed. Finally droplets of electron–hole plasma condensed in semiconductors at low temperatures are briefly treated as a modern example of surface modes in a plasma of greatly different density than common metals.

12.4.1   Gold

In his classical paper of 1908 Mie interpreted quantitatively the vivid colors of colloidal gold that had been discussed qualitatively much earlier by Faraday (1857). Despite his lack of a computer, Mie's calculations were in rather good agreement with measurements by Kirchner and Zsigmondy (1904). In recent years much theoretical and experimental work on gold particles has been published: Granqvist and Hunderi (1977) list over 50 papers published during the past 20 years. Some of this effort has been directed toward remeasuring optical constants of gold, which are (presumably) better known now than in

Mie's time. Various types of particulate systems have been studied; however, gold particles in aqueous solutions, in gelatin, and in glass are most likely to be well isolated from one another, and for this reason we emphasize these systems. To illustrate the size dependence of extinction by colloidal gold, data from Turkevich et al. (1954) and from Doremus (1964) are combined in Fig. 12.17. They investigated different size ranges, but in the region of overlap their data agree well. Because they presented their measurements differently, the curves for radii less than 26 Å are for a constant number of particles, whereas those for radii greater than 26 Å are for a constant mass. We adjusted the data of Doremus so that peak absorption is the same for the 26- and 100-Å particles. The median sizes of the larger particles studied by Turkevich et al. are the same as those for which calculations were reported by Mie (1908); they also reproduced his calculations.

The curves of Fig. 12.17 nicely illustrate the varied optical effects exhibited by small metallic particles in the surface mode region, both those explained by Mie theory with bulk optical constants and those requiring modification of the electron mean free path (see Section 12.1). Absorption by particles with radii between about 26 and 100 Å peaks near the Fröhlich frequency ($\lambda_F \simeq 5200$ Å), which is independent of size. Absorption decreases markedly at longer

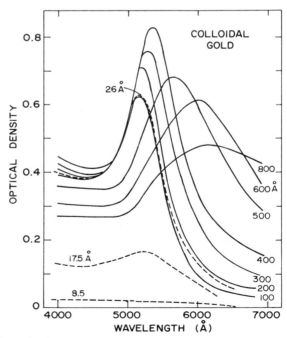

Figure 12.17 Absorption by gold particles of different radii. The solid curves are from Turkevich et al. (1954); the dashed curves are from Doremus (1964).

wavelengths (note that this is completely contrary to the absorptive behavior of thin gold films); this gives rise to the ruby red color observed in transmission of white light by small gold sols and "ruby" glass. As the radius increases beyond 100 Å the absorption peak broadens and, in accordance with (12.13), shifts to longer wavelengths; thus, the observed color changes from ruby red through purple and violet to pale blue for the largest particles (800 Å). Scattering of white light by the smaller particles is weak but becomes noticeable for particles larger than about 250 Å. The color of this scattered light is reddish brown because the scattering cross section maximum lies in the long-wavelength part of the visible spectrum; blue light by transmission is a result of both preferential scattering and absorption of longer wavelengths.

The shapes of the absorption band cease to be independent of size for particles smaller than about 26 Å, which suggests that the bulk dielectric function is inapplicable. Indeed, the broadening and lowering of the absorption peak can be explained by invoking a reduced mean free path for conduction electrons (Section 12.1). Thus, the major features of surface modes in small metallic particles are exhibited by this experimental system of nearly spherical particles well isolated from one another. But when calculations and measurements with no arbitrary normalization are compared, some disagreement remains. Measurements of Doremus on the 100-Å aqueous gold sol, which agree with those of Turkevich et al., are compared with his calculations in Fig. 12.18; the two sets of calculations are for optical constants obtained

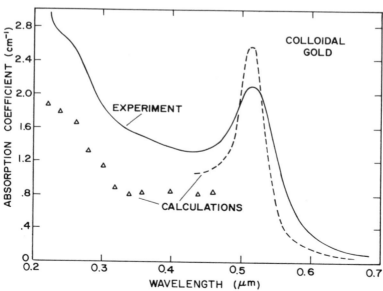

Figure 12.18 Measured and calculated absorption spectra for gold particles of 100 Å radius in aqueous suspension. From Doremus (1964).

from different sources. There is a residual discrepancy, seemingly ubiquitous: the calculated peak is sharper and higher than that measured; this was also noted by Doremus and by Granqvist and Hunderi. Doremus felt that its explanation lies in differences between bulk and small-particle absorption because of imperfections and impurities trapped by the gold particles during growth. After exploring several hypotheses, including nonsphericity and oxide coatings, Granqvist and Hunderi concluded that unavoidable clumping of particles into clusters and chains would result in effective shape factors similar to single nonspherical particles, which would broaden and decrease the absorption peak (see the summary of shape effects at the end of Section 12.2).

Despite some disagreement, it is surprising how well the combination of bulk optical constants, especially when suitably modified, and Mie theory succeeds in explaining the observed optical properties of very small particles. Doremus concluded that even without invoking mean free path limitations, the classical theory is adequate down to diameters of 85 Å; Granqvist and Hunderi found good agreement with classical calculations for particles in the size range 30–40 Å when the mean free path effect and nonsphericity are accounted for; they specifically remark that they found no evidence for a quantum size effect.

12.4.2 Silver

Surface plasmon absorption has been observed for silver particles in various media, including aqueous solutions (Kreibig, 1974), gelatin (Skillman and Berry, 1968), and glass (Stookey et al., 1978). Size effects exhibited by nearly spherical silver particles are similar to those for gold: limitation of the mean free path by the particle boundary broadens and decreases peak absorption at very small sizes, while at larger sizes the peak shifts to longer wavelengths and broadens as higher-order modes are excited. As pointed out in the discussion of Fig. 12.9, the low value of ϵ'' for silver near the Fröhlich frequency gives rise to an intense absorption band which is more sensitive to shape than the highly damped bands of gold. The Fröhlich mode for silver spheres in air is at about 3600 Å and shifts to about 4100 Å in a medium such as gelatin or glass. Because peak absorption by very small spheres occurs at the blue edge of the spectrum, increasing the size causes the transmission colors to be swept through the visible. This size effect, together with shape broadening and shifting, provides a mechanism for coloring glass (Weyl, 1951, Chaps. 24 and 25). Wiegel (1954), for example, has described colors in silver colloids ranging from yellow in the smallest sizes through the sequence red, purple-red, violet, dark blue, light blue, and gray-green as the size increases from about 100 to 1300 Å. Silver in various photosensitive materials has been widely studied because of its importance in the photographic process. This process enables particles to be generated in definite amounts by exposure and subsequently grown by photographic development. Because the number of particles remains constant as the size is increased, samples with both controlled number densities and sizes can be prepared; some control over shape has also been achieved.

Efforts to observe the effect of controlled departures from sphericity (see Section 12.2) on absorption spectra of small particles have been particularly successful for silver. Absorption spectra of three samples of silver particles produced in gelatin by photographic exposure and development are shown in Fig. 12.19 (Skillman and Berry, 1968); the average axial ratios (a/b) of the approximately spheroidal particles were determined from electron micrographs. For nearly spherical particles ($a/b = 1.18$) only one band is apparent, close to the Fröhlich mode wavelength 4200 Å; mode splitting (see the discussion of Fig. 12.5) occurs for more elongated particles ($a/b = 2.5$ and 3.35): one band moves toward longer and the other toward shorter wavelengths. The bottom half of Fig. 12.19 shows, in addition to ϵ' for silver, the two sets of wavelengths at which (12.27) is satisfied for the shape factors L_1

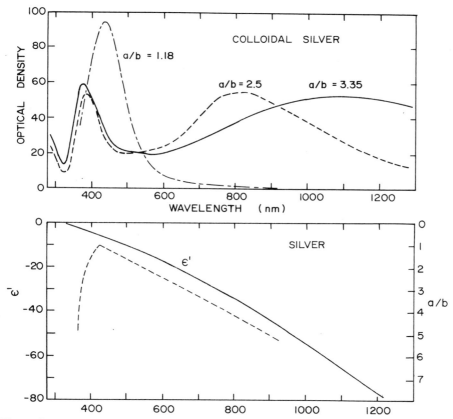

Figure 12.19   Absorption spectra of silver colloids (top); particle shape is specified by the average axial ratio a/b (from Skillman and Berry, 1968). Below this are ϵ' for silver (solid curve) and the surface mode wavelengths (dashed curves) for the two shape factors of a spheroid.

and L_2 ($L_3 = L_2$). Although there is some quantitative disagreement, measurements and calculations are qualitatively similar: the greater the particle elongation, the greater the separation of the two modes. However, the position of the short-wavelength band is predicted better than that of the long-wavelength band; this discrepancy may be the result of a distribution of spheroidal shapes, which was approximated simply by an average axial ratio, and by the fact that the particles were not perfect spheroids.

Additional measurements of absorption spectra for nonspherical silver particles were made by Stookey and Araujo (1968). Using a combination of heat treatment to grow particles in photosensitive glass and stretching to align them, they produced samples with polarization-dependent absorption; the band lies on the long-wavelength side of the Fröhlich mode if the electric vector is parallel to the long axis of the particles and on the short-wavelength side if it is perpendicular to the axis. Further experiments on photochromic glasses by Stookey et al. (1978) showed that silver is deposited selectively in the narrow tips of elongated pyramids. Although these particles are not ellipsoids, Rayleigh theory was used successfully to interpret the experimental results. However, the measured spectral shifts were somewhat *less* than those calculated, in contrast with the results of Skillman and Berry. Stookey and co-workers point out, as we did at the beginning of this section, that as a result of the low values of ϵ'' over the surface plasmon region, vitreous colloidal silver has unusually high coloring power; because of this and the possibility of spanning the visible spectrum by changing particle size and shape, various applications of this full-color photographic medium are anticipated.

Absorption resonances resulting from excitation of surface modes are accompanied by scattering resonances at approximately the same frequencies; this was pointed out following (12.26). In most experiments transmission is measured to determine extinction, which is nearly equal to absorption for sufficiently small particles. However, surface mode resonances have been observed in spectra of light *scattered* at 90° by very small particles of silver, copper, and gold produced by nucleation of vapor in an inert gas stream (Eversole and Broida, 1977). The scattering resonance peak was at 3670 Å, near the expected position of the Fröhlich mode, for the smallest silver particles. Although peak positions were predictable, differences in widths and shapes of the bands were concluded to be the result of nonsphericity.

12.4.3   Aluminum

Extinction calculations for aluminum spheres and a continuous distribution of ellipsoids (CDE) are compared in Fig. 12.6; the dielectric function was approximated by the Drude formula. The sum rule (12.32) implies that integrated absorption by an aluminum particle in air is nearly independent of its shape: a change of shape merely shifts the resonance to another frequency between 0 and 15 eV, the region over which ϵ' for aluminum is negative. Thus, a distribution of shapes causes the surface plasmon band to be broadened, the extent of which is most apparent in extinction spectra plotted logarithmically. Figure 12.20 shows calculations, using optical constants of aluminum measured by Hagemann et al. (1974), for particles in air (solid curves) and in a medium with a dielectric function approximately equal to that of KBr over its transparent region (dashed curves). Note that there is little difference between the CDE calculations for particles in air and in a solid medium. Because the low-frequency wing of the sphere extinction spectrum is so steep, the difference between the CDE and sphere calculations becomes ever larger toward longer wavelengths: they differ by a factor of about 10^2 at 1 μm, 10^3 at 20 μm, and 10^4 at 300 μm; and at millimeter wavelengths the difference reaches 10^5. Experimental data from two sources are shown in the figure: the circles are

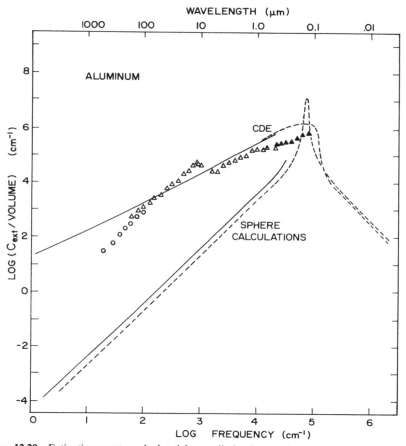

Figure 12.20 Extinction spectra calculated for small aluminum spheres and a continuous distribution of ellipsoids (CDE) in air (——) and in a medium with $\epsilon = 2.3$ (---). The circles show data from Granqvist et al. (1976), the triangles from Rathmann (1981); solid triangles are for particles on a substrate and open triangles are for particles in a matrix of KBr or polyethylene.

far-infrared measurements of Granqvist et al. (1976); the triangles are measurements of Rathmann (1981), which will be discussed shortly. Although the shape distribution of aluminum particles produced experimentally is certainly not a continuous distribution of ellipsoids, the CDE calculations are considerably closer to the data points than those for spheres; this was also true for SiC and crystalline quartz particles (Section 12.3).

The large discrepancy between sphere calculations and far-infrared extinction measurements (Fig. 12.20) has attracted much attention recently. The original impetus for such measurements was a search for superconducting band gaps of small particles. Although the telltale superconducting feature was not found, the unexpectedly high extinction inspired a series of explanations: Tanner et al. (1975) discussed the possibility of quantum size effects as the responsible mechanism; Glick and Yorke (1978) suggested a mechanism by which incomplete screening of conduction electrons near the particle surface would result in increased coupling of photons to lattice vibrations; Šimánek (1977) proposed that an amorphous oxide coating would greatly enhance absorption by particles aggregated into clusters; Maksimenko et al. (1977) and Lushnikov et al. (1978) reinvestigated the quantum-mechanical approach of Gor'kov and Eliashberg (1965) to explain the extinction; Ruppin (1979) considered both oxide coatings and spheroidal shapes incorporated into effective medium theories. Various of these ideas were reviewed and discussed at the Optical Society of America Topical Meeting on "Optical Properties Peculiar to Matter of Small Dimensions" held in Tucson, Arizona, March 1980. None of the proposed explanations seem to be completely satisfactory, particularly in view of the factor-of-10^5 discrepancy between measurements and calculations of far-infrared extinction by small platinum particles reported by Sievers.

We propose that a distribution of particle shapes broadens the surface plasmon band sufficiently that far infrared extinction is about 1000 times greater than would be obtained for spheres; this is strongly suggested by the CDE calculations of Fig. 12.20. To test this hypothesis further, extinction measurements were made over the wavelength interval between about 50 and 0.12 μm for aluminum particles prepared in a manner similar to that of Granqvist et al. (1976). Aluminum was evaporated from a tantalum boat in a helium atmosphere (~ 5 torr) and collected as very thin coatings on LiF and quartz substrates for ultraviolet, visible, and near-infrared transmission measurements. For measurements at longer wavelengths, where more mass was needed, aluminum smoke particles were dispersed in KBr and polyethylene matrices in the manner described at the beginning of Section 12.3. Volume-normalized extinction determined from transmission and mass measurements is represented by the triangles in Fig. 12.20. Agreement between measurements and calculations is considerably better for the distribution of ellipsoidal shapes than for spheres.

The peak near 18 μm is evidence for an oxide coating, which is unavoidable with the experimental technique used, on the particles. Although oxide absorption, especially in an amorphous coating on elongated particles, could give the enhanced far-infrared absorption (Šimánek, 1977), such an oxide should be weakly absorbing between about 10 and 0.3 μm. Indeed, calculated absorption by coated aluminum particles using measured optical constants of both crystalline and amorphous aluminum oxide (Rathmann, 1981) is much too low in this spectral region to account for the measurements shown in Fig. 12.20.

The considerably better agreement of CDE theory with recent measurements on aluminum smoke in the near infrared, together with similar improved agreement for nonspherical insulating particles (Section 12.3), is evidence that shape effects are indeed responsible for large far-infrared absorption by aluminum and other metallic particles. The individual particles may be highly nonspherical. But if they are nearly spherical, as they sometimes appear in electron micrographs, surface plasmon absorption would not be shifted into the far infrared; in this instance, it is likely that small clusters have not been disrupted by dispersal, which gives rise to effective shapes sufficiently different from spherical to cause the observed shifts. A similar cluster effect in gold particles on a substrate was discussed in detail by Granqvist and Hunderi (1977), who suggested this as a possibly important factor in far-infrared absorption by metallic particles.

12.4.4   Electron–Hole Droplets in Germanium

At low temperatures a pure semiconductor is a perfect insulator with no free carriers. Upon laser irradiation at a frequency greater than the semiconducting band gap, a high density of electron–hole pairs can be excited which, at liquid-helium temperatures, condense into small droplets of electron–hole plasma. These electron–hole (e–h) droplets have been discussed thoroughly in a dedicated volume of *Solid State Physics* that contains reviews of theoretical aspects (Rice, 1977) and experiments (Hensel et al., 1977).

It has been possible to measure the angular dependence of scattering by e-h droplets in germanium because the host is quite transparent between 1.66 μm (the band gap) and 25 μm (the lattice absorption band). From these measurements it was ascertained that the condensate indeed existed in the form of droplets with radii between about 1 and 10 μm.

Far-infrared absorption measurements gave an independent determination of the electron density; from the position of the Fröhlich mode near 9 meV ($\sim$ 140 μm) a density of 2.3×10^{17} cm^{-3} was inferred. Other experiments on Sb-doped Ge and pure germanium irradiated at different powers showed appreciable changes in absorption band positions and shapes. Rose et al. (1978) interpreted the shift of peak absorption from 9 meV (intrinsic Ge) to about 5.5 meV (doped Ge) as the result of an increase in droplet size under the assumption that the other e–h plasma parameters remained constant. Their (arbitrarily normalized) Mie calculations for a set of droplet sizes are shown in Fig. 12.21. There is good agreement with measurements of far-infrared absorption by e–h droplets in Sb-doped Ge (Timusk and Silin, 1975), which allows

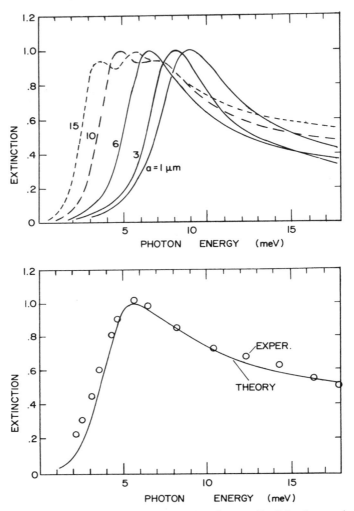

Figure 12.21 Calculated far infrared extinction (arbitrarily normalized) by electron–hole droplets in Sb-doped germanium (from Rose et al., 1978) below which are experimental data (circles) of Timusk and Silin (1975).

the droplet size to be determined where light scattering measurements are difficult because of impurity-induced opacity of the germanium host. Thus, the sizing of e–h droplets is a simple application of Mie theory coupled with Drude theory to a somewhat exotic particulate medium.

12.4.5 Survey of Other Plasmons

Surface plasmon absorption has been observed for small particles of several other metals, and many calculations have been published; these are too numerous to discuss in detail. But it is worthwhile to summarize the properties of surface plasmons for various metallic particles. In Table 12.1 the most important parameters characterizing surface plasmons are listed: the position of the Fröhlich mode for spheres in air; and the value of ϵ'' at the frequency where $\epsilon' = -2$, which determines the width and height of the absorption band. Also listed are energies where $\epsilon' = 0$ (bulk plasmon energies) to show the possible high-energy extent of shape effects. The sources of this experimental data are indicated in the table.

The alkali metals, with only one free electron per atom, have lower plasmon energies than those of divalent free-electron metals such as Mg and Al because the plasma frequency decreases with decreasing electron density. Thus, surface plasmon energies for alkali metals are in or near the visible, whereas they are in the far ultraviolet for Mg, Al, and Pb. Surface plasmon energies of the divalent metals Ag, Au, and Cu are shifted toward and into the visible because of interband transitions (see Fig. 12.9d); this is also the cause of the large values of ϵ'' for Au and Cu.

Another example listed in the table is graphite, the Fröhlich mode of which is near 5.5 eV (2200 Å); the boundaries of the negative ϵ' region are about 4 and 6.5 eV. The graphite surface plasmon has been tentatively identified as responsible for a feature in the interstellar extinction spectrum (see Section 14.5).

For photon energies large compared with the band gap in semiconductors, electronic transitions are only slightly perturbed by the presence of the gap; the valence electrons in this instance act like free electrons. Thus, the high-energy optical properties of semiconductors are similar to those of free-electron metals. But there are marked differences in their low-energy optical properties, which bear on the importance of particle shape to extinction spectra. Whereas ϵ' for metals approaches large negative values with decreasing photon energy,

Table 12.1 Characteristics of Bulk and Surface Plasmons

Solid	Bulk Plasmon Energy (eV)	Surface Plasmon Energy (eV)	ϵ'' where $\epsilon' = -2$	Reference
Lithium	6.6	3.4	1.0	Rasigni and Rasigni (1977)
Sodium	5.4	3.3	0.12	Palmer and Schnatterly (1971)
Potassium	3.8	2.4	0.13	Palmer and Schnatterly (1971)
Magnesium	10.7	6.3	0.5	Hagemann et al. (1974)
Aluminum	15.1	8.8	0.2	Hagemann et al. (1974)
Iron	10.3	5.0	5.1	Moravec et al. (1976)
Copper	—	3.5	4.9	Hagemann et al. (1974)
Silver	3.8	3.5	0.28	Huebner et al. (1964)
Gold	—	2.5	5.0	Hagemann et al. (1974)
Graphite	—	5.5	2.7	Taft and Phillip (1965)

ϵ' for semiconductors becomes positive again; at photon energies below the band gap energy the optical properties of semiconductors are similar to those of insulators. Thus, the surface plasmon region—where ϵ' is negative—for semiconducting particles is bounded from above *and* from below; this contrasts with aluminum particles, for example, in which, depending on their shape, surface plasmons are possible at all energies below the plasma frequency. Consequently, shape effects in metallic particles are, in general, more pronounced than in semiconducting particles.

NOTES AND COMMENTS

Gilra (1972ab) has very thoroughly discussed absorption by small ellipsoidal particles, including those with coatings, in the surface mode region.

The question of which shape maximizes and which minimizes the absorption cross section of a particle with fixed volume and composition has been answered by Bohren and Huffman (1981) within the framework of the Rayleigh ellipsoid theory.

An independent derivation of (12.36), preceding ours, was published by Aronson and Emslie (1975).

Chapter 13

Angular Dependence of Scattering

Within the framework of this book all the information about elastic scattering by small particles is contained in the 4 × 4 scattering matrix (3.16); each of its 16 elements is an angle-dependent function of wavelength, particle size, shape, and composition. There is so much information in the complete scattering matrix, however, that only recently have the properties of all its elements begun to be investigated. Most theoretical and experimental work has been limited to scattering of unpolarized or linearly polarized light. Because of this, the present chapter is arranged as follows: in Sections 13.1 through 13.5 the incident light is unpolarized or linearly polarized; this restriction is removed in Sections 13.6 through 13.8. In Section 13.1 we discuss calculated scattering by spheres to give the reader some feeling for the various effects that can occur. Measurement techniques are then treated briefly in Sections 13.2 and 13.3. This is followed in Section 13.4 by calculations interspersed with measurements on spherical and nonspherical systems. Section 13.5 treats an important application: the sizing of particles by light scattering. Section 13.6 is somewhat of a digression: symmetry properties of the scattering matrix. Measurement techniques for the complete scattering matrix are given in Section 13.7, and results of measurements in Section 13.8. Section 13.9, the concluding section, summarizes the differences and similarities between spherical and nonspherical particles, partly in an attempt to answer the important practical question: To what extent is Mie theory applicable to nonspherical particles?

13.1 SCATTERING OF UNPOLARIZED AND LINEARLY POLARIZED LIGHT

The maximum amount of information about scattering by any particle or collection of particles is contained in all the elements of the 4 × 4 scattering matrix (3.16), which will be treated in more generality later in this chapter. Most measurements and calculations, however, are restricted to unpolarized or linearly polarized light incident on a collection of randomly oriented particles with an internal plane of symmetry (no optical activity, for example). In such instances, the relevant matrix elements are those in the upper left-hand 2 × 2 block of the scattering matrix, which has the symmetry shown below (see, e.g.,

Perrin, 1942, and Section 13.6):

$$
\begin{pmatrix} I_s \\ Q_s \\ U_s \\ V_s \end{pmatrix} = \frac{1}{k^2 r^2} \begin{pmatrix} S_{11} & S_{12} & \vdots & \\ S_{12} & S_{22} & \vdots & 0 \\ \cdots & \cdots & \cdots & \cdots \\ & 0 & \vdots & \\ & & \vdots & \end{pmatrix} \begin{pmatrix} I_i \\ Q_i \\ U_i \\ V_i \end{pmatrix}. \tag{13.1}
$$

Although we have chosen to emphasize the more common Stokes parameters (I, Q, U, V), the system $(I_\parallel, I_\perp, Q, V)$ is more suited to measurements in which linear polarizers are interposed in the incident and scattered beams. In the latter system, the Stokes parameters are defined by

$$
I_\parallel = \langle E_\parallel E_\parallel^* \rangle, \qquad I_\perp = \langle E_\perp E_\perp^* \rangle,
$$

in place of (2.84); U and V are the same in both systems. In the $(I_\parallel, I_\perp, U, V)$ system the scattering matrix corresponding to (13.1) is

$$
\begin{pmatrix} I_{\parallel s} \\ I_{\perp s} \\ U_s \\ V_s \end{pmatrix} = \frac{1}{k^2 r^2} \begin{pmatrix} Q_{11} & Q_{12} & \vdots & \\ Q_{12} & Q_{22} & \vdots & 0 \\ \cdots & \cdots & \cdots & \cdots \\ & 0 & \vdots & \\ & & \vdots & \end{pmatrix} \begin{pmatrix} I_{\parallel i} \\ I_{\perp i} \\ U_i \\ V_i \end{pmatrix}.
$$

The matrix elements in the two systems are related by

$$ S_{11} = \tfrac{1}{2}(Q_{11} + 2Q_{12} + Q_{22}), \qquad Q_{11} = \tfrac{1}{2}(S_{11} + 2S_{12} + S_{22}), $$

$$ S_{12} = \tfrac{1}{2}(Q_{11} - Q_{22}), \qquad Q_{12} = \tfrac{1}{2}(S_{11} - S_{22}), $$

$$ S_{22} = \tfrac{1}{2}(Q_{11} - 2Q_{12} + Q_{22}), \qquad Q_{22} = \tfrac{1}{2}(S_{11} - 2S_{12} + S_{22}). $$

13.1.1   A Few Definitions

The scattered irradiances per unit incident irradiance (dimensionless irradiances) for incident light parallel and perpendicular to the scattering plane are (omitting $k^2 r^2$)

$$ i_\parallel = S_{11} + S_{12} = Q_{11} + Q_{12}, $$

$$ i_\perp = S_{11} - S_{12} = Q_{22} + Q_{12}, $$

and the dimensionless scattered irradiance for incident unpolarized light is

$$ i = \frac{i_\parallel + i_\perp}{2} = S_{11}. $$

Other quantities commonly measured are the *degree of linear polarization P* of the scattered light for incident unpolarized light

$$ P = -\frac{S_{12}}{S_{11}} = \frac{Q_{22} - Q_{11}}{Q_{11} + 2Q_{12} + Q_{22}}, $$

and the *cross polarization*

$$Q_{12} = \tfrac{1}{2}(S_{11} - S_{22}),$$

which is measured by inserting linear polarizers fore and aft of the scattering medium, one with its axis parallel and the other with its axis perpendicular to the scattering plane; the order of the polarizers is unimportant for the special scattering matrix (13.1).

With the programs in the appendixes one can compute elements of the scattering matrix for spheres and, consequently, all the quantities defined in the preceding paragraph. The cross polarization vanishes for spherical particles, and the following relations hold:

$$P = \frac{Q_{22} - Q_{11}}{Q_{22} + Q_{11}}, \qquad i_{\parallel} = Q_{11}, \quad i_{\perp} = Q_{22}.$$

There are many other angle-dependent scattering functions in the scientific literature, which is a source of endless confusion. In the hope that it will help the confused—among whom we count ourselves—to reconcile the notation and terminology of various authors, some of the more commonly encountered functions are expressed in our notation.

The *differential scattering cross section* $dC_{sca}/d\Omega$, a familiar quantity in atomic physics, is defined as the energy scattered per unit time into a unit solid angle about a direction $\hat{\Omega}$—which may be specified by two angles, the scattering angle θ and the azimuthal angle ϕ (see Fig. 3.3)—for unit incident irradiance. It is expressed in terms of the scattered irradiance $I_s(\theta, \phi)$, the incident irradiance I_i, and the distance r to the detector as

$$\frac{dC_{sca}}{d\Omega} = \frac{r^2 I_s}{I_i}. \tag{13.2}$$

($dC_{sca}/d\Omega$ should not be interpreted as the derivative of a function of Ω; it is *formally* written as a derivative merely as an aid to memory). Although the quantity on the right side of (13.2) is often referred to as the "Rayleigh ratio," we shall avoid this term in favor of the more descriptive term (for physicists at least) differential scattering cross section. If the incident light is unpolarized

$$\frac{dC_{sca}}{d\Omega} = \frac{S_{11}}{k^2} = \frac{i_{\parallel} + i_{\perp}}{2k^2}.$$

For an isotropic medium such as a collection of many randomly oriented particles, which may themselves be anisotropic, the scattered irradiance and hence the differential scattering cross section is independent of ϕ.

The differential scattering cross section divided by the total scattering cross section

$$p = \frac{1}{C_{\text{sca}}} \frac{dC_{\text{sca}}}{d\Omega}$$ (13.3)

is called the *phase function*, a term originating in the phases of astronomical bodies. The phase function (13.3) agrees with that of van de Hulst (1957), but differs by a factor 4π from phase functions of other authors (e.g., Rozenberg, 1960; Hansen and Travis, 1974).

13.1.2   Scattering by Spheres

One example of calculated angular scattering by a sphere has been given already (Fig. 4.9). To further develop understanding of scattering by spheres, we display in various ways $i_\parallel$, $i_\perp$, and P for a sequence of spheres of increasing size in Figs. 13.1–13.3. Readers are encouraged to use the program in Appendix A to devise their own examples.

Calculations for spheres with optical constants appropriate to water at visible wavelengths are shown in Fig. 13.1; $i_\parallel$ and $i_\perp$ are plotted on the left, P on the right. Similar results for spheres with refractive index $m = 1.55 + i0.0$, which corresponds approximately to fused quartz in the visible region, are shown in Fig. 13.2.

For small size parameters, the familiar Rayleigh (see Section 5.1) scattering patterns are obtained: perpendicularly polarized light is scattered isotropically, while light polarized parallel to the scattering plane vanishes at a scattering angle of 90°; as a consequence, incident unpolarized light is completely polarized at 90°. For all sizes, $i_\parallel = i_\perp$ at 0° and 180°: the two polarizations are indistinguishable in these directions because of symmetry; in other directions, the scattering plane enforces a distinction. Polar plots of scattering functions are shown in Fig. 13.3 for various water droplets; no new information is contained in these plots, but they can evoke sharper physical images.

The first deviations from Rayleigh theory appear as forward–backward asymmetry, with more light being scattered in forward directions; also, peak polarization decreases and shifts to larger angles. As the size is increased further, the asymmetry becomes more pronounced and the dominant forward-scattering lobe narrows. A concomitant of an increase in size is more undulations, as if new peaks appear in the backscattering direction and move forward. For large particles, the increased complexity of scattering is an indicator of its extreme sensitivity to size. Thus, comparison of measured scattering with sets of calculations is a possible means of accurately sizing spheres. For a given refractive index, the number of peaks in the scattering pattern gives a fairly good measure of the sphere radius. This sensitivity to radius, however, causes structure to be obliterated as the size dispersion in a collection of particles increases, giving rise to much smoother patterns.

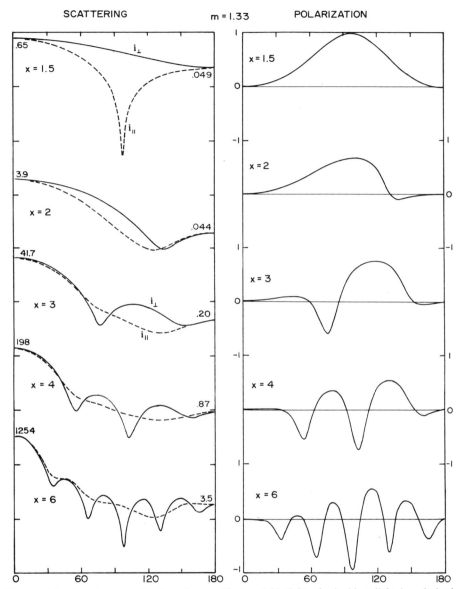

SCATTERING $m = 1.33$ POLARIZATION

Figure 13.1 Angular scattering by spheres with $m = 1.33$ (left); the incident light is polarized parallel (---) or perpendicular (——) to the scattering plane. On the right is the degree of polarization of scattered light for incident unpolarized light.

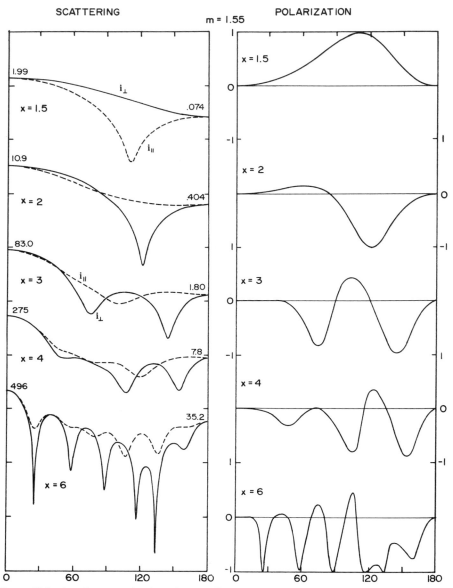

Figure 13.2 Angular scattering by spheres with $m = 1.55$ (left); the incident light is polarized parallel (---) or perpendicular (——) to the scattering plane. On the right is the degree of polarization of scattered light for incident unpolarized light.

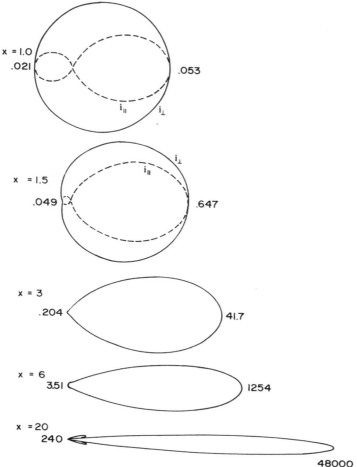

Figure 13.3   Polar plots of angular scattering by spheres with $m = 1.33$. Note the great change in scale: as x increases by 20 forward-to-backward scattering increases by about 1000.

Calculations of Hansen and Travis (1974) for a distribution of sizes are shown in Fig. 13.4; the size parameters are distributed according to $n(x) = x^6 \exp(-9x/x_{\text{eff}})$. At visible wavelengths the three effective size parameters x_{eff} correspond to radii of about 20 μm, 80 μm, and 0.3 mm; this covers a range of sizes found in fogs and clouds. All curves show the relatively smooth behavior characteristic of a distribution of sizes, in contrast with the highly structured curves for single spheres. Scattering is strongly peaked in the forward direction; the forward (diffraction) lobe becomes increasingly confined to smaller angles as x_{eff} increases from 37.5 to 600. For $m = 1.33$ (water), two peaks develop in the region between 130 and 140°; these are the primary ($\theta \simeq 137°$)

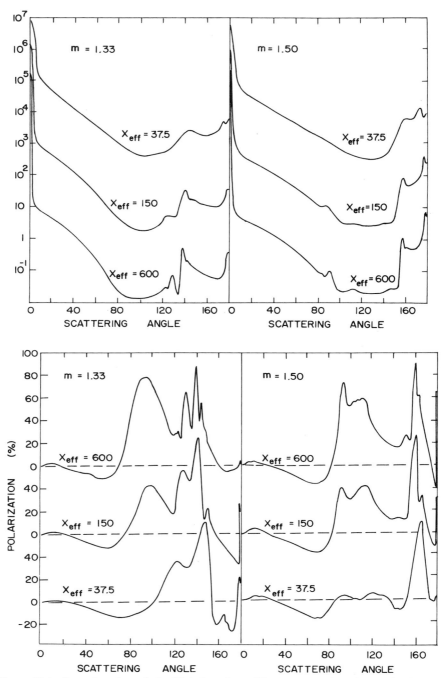

Figure 13.4 Scattering of unpolarized light by spheres. The top curves show the angular distribution of the scattered light and the bottom curves its degree of polarization. From Hansen and Travis (1974); copyright © 1974 by D. Reidel Publishing Company, Dordrecht, Holland.

and secondary ($\theta \simeq 129°$) rainbows, which may be attributed to one and two internal reflections, respectively, in a water droplet (see Section 7.2). Geometrical optics accounts well for the positions of the rainbows, which are rather insensitive to size for the larger droplets. The rainbow features are also prominent in the polarization curves.

Note the sharp increase in scattering near the backward direction, particularly for the largest spheres; this is the origin of the *glory*, one of the more spectacular natural phenomena. Observant passengers in airplanes flying above clouds may see the glory, a series of colored rings, around the shadow of the airplane. Unlike the rainbow the glory is not easy to explain, other than to say that it is a consequence of all the thousands of terms in the scattering series, a correct but unsatisfying statement. After all, the same statement could be made about the rainbow, which nevertheless has a simple physical explanation. A similar explanation of the glory, universally understood and accepted, has not yet been achieved, although there have been several admirable attempts toward this end, particularly recently. Rather than enter into a protracted digression on this interesting topic, we refer the reader to several relevant papers: those by van de Hulst (1947), Bryant and Cox (1966), Khare and Nussenzveig (1977), and Nussenzveig (1979).

13.2   TECHNIQUES OF MEASUREMENT AND PARTICLE PRODUCTION

Until about 10 years ago angular scattering measurements at visible wavelengths were limited to collections of *many* particles, either natural dispersions as in the atmosphere or those generated in the laboratory. With the advent of high-power lasers it has become feasible to measure light scattering by *single* particles. There are good reasons for measuring scattering by both single and many particles.

Most angular scattering measurements have been made at visible and near-ultraviolet wavelengths where detectors (primarily the photomultiplier tube) are sensitive, sources are intense, and good polarizing filters and other optical elements are readily available. These advantages of visible light diminish when we turn to other wavelength regions, although there is no lack of interest in them. The other principal wavelength region in which laboratory studies of angular scattering have been made is the microwave. As with extinction (see Chapter 11), scattering by single nonspherical and inhomogeneous particles in various orientations can be studied.

13.2.1   POLAR NEPHELOMETERS

An instrument for angular light scattering measurements is often called a *nephelometer*, from the Greek work *nephele* for cloud. To be more precise, the instrument shown schematically in Fig. 13.5 is a *polar* nephelometer, so named because of its angular detection capability. Its essential elements are a collimated light source and an arm that can be rotated about the sample (scattering cell); mounted on the arm is a detector system, which includes optical elements to collect light scattered within a small solid angle. The nephelometer shown in Fig. 13.5 is somewhat idealized, and each part of the instrument may be more complicated; descriptions of actual nephelometers are given, for example, by Stacey (1956), Pritchard and Elliot (1960), Holland and Gagne (1970), and Hunt and Huffman (1973).

The light source may be a lamp (tungsten–halogen, high-pressure mercury or xenon) with suitable collimators or, alternatively, a laser. Although lasers are easy to use and have seen wide use since their commercialization, it should not be assumed that a laser is always the best source; it certainly is not the most economical. A 300- to 500-watt tungsten–halogen lamp for home slide projectors is inexpensive and available at most photography shops; although only a small fraction of the rated power is available as visible light, the cost of a lamp is only a small fraction of that of a 1-watt continuous wave (CW) laser! Particularly for the study of collections of particles, where neither small beam size nor a high degree of monochromaticity is required, the tungsten–halogen lamp should be considered. In studies of scattering by a single particle, a laser is usually the best source.

The telescope on the detector arm, consisting of a lens followed by an aperture, limits the angular acceptance of the detector; this is accomplished at the expense of less detection sensitivity. The various factors determining resolving power and sensitivity are discussed by Pritchard and Elliot (1960). Intersection of the incident beam with the detector field of view determines the scattering volume (illuminated volume), which consequently changes with angle; therefore, the measured signal must be corrected by the multiplicative factor $\sin \theta$. It is not possible to make measurements at scattering angles near

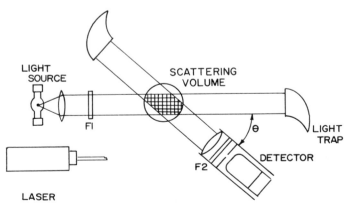

Figure 13.5 Schematic diagram of a polar nephelometer for measuring angular scattering. F1 and F2 are possible polarizing filters.

180° with conventional nephelometers because the arm interferes with the incident beam. Scattering at large angles is important, however, and this frequently requires that modifications be made so that these angles are accessible to measurement.

13.2.2   Absolute and Relative Measurements

Angular light scattering measurements are sometimes classified as either *absolute* or *relative*. In an absolute measurement I_s/I_i, which is directly related to the differential scattering cross section (13.2), is determined; in a relative measurement the irradiance is referred to some arbitrary scattering angle, say 10°, so that (assuming azimuthal symmetry)

$$\frac{I_s(\theta)}{I_s(10°)} = \frac{I_s(\theta)/I_i}{I_s(10°)/I_i} = \frac{dC_{sca}(\theta)/d\Omega}{dC_{sca}(10°)/d\Omega}.$$

Relative measurements are considerably easier to make and are the type most commonly reported. However, absolute measurements are of importance, for example, in comparing measured scattering cross sections of nonspherical particles with calculations for equivalent spheres. Note that "absolute" as we are using the term here means that scattering is not normalized to some arbitrary reference angle; it does not mean that absolute *irradiances* are measured, as with calibrated detectors. In both "relative" and "absolute" measurements, it is relative (i.e., dimensionless) *irradiances* that are determined.

Absolute measurements can be made by swinging the detector arm from 0° to θ, thereby obtaining I_i and $I_s(\theta)$; easy to say but less easily done: I_i may be thousands of times greater than I_s, and appreciable errors are likely because of the lack of detector linearity over such a range. One method for overcoming this is to attenuate I_i with neutral density filters to a level comparable with I_s; if the optical density is known, the unattenuated incident irradiance can be determined. Another technique is to use a diffusing surface, such as opal or MgO-smoked glass, in the incident beam to scatter light uniformly in all directions. Nonuniform illumination of the photocathode, arising from nonuniform illumination of the scattering volume, may also be a problem. Pritchard and Elliot (1960) and Holland and Gagne (1970) calibrated their instruments to account for these and other effects by moving a calibrated diffusing plate through the scattering volume and integrating the measured signal over the traverse for each angle and for all filter combinations. Perhaps the simplest technique is to use spheres (e.g., polystyrene) with known properties—concentration, size distribution, and refractive index—together with Mie calculations. For a given light source and optical elements such as filters, the detector signal D_r is measured with the reference sample in the scattering cell

$$D_r = KI_s = K\frac{I_i}{r^2}\left(\frac{dC_{sca}}{d\Omega}\right)_r,$$

where K is a dustbin (angle dependent) into which we deposit all our ignorance about instrument calibration factors. Under the same conditions another angular scan is made of the detector signal D with the "unknown" sample in the scattering cell

$$D = K \frac{I_i}{r^2} \frac{dC_{sca}}{d\Omega}.$$

$dC_{sca}/d\Omega$ is determined from the measured ratio of signals and $(dC_{sca}/d\Omega)_r$ calculated from Mie theory for the reference sample:

$$\frac{dC_{sca}}{d\Omega} = \frac{D}{D_r} \left(\frac{dC_{sca}}{d\Omega} \right)_r.$$

This technique is quite adaptable to instruments under the control of a microcomputer: the reference signal as well as calculated differential scattering cross sections are stored in the computer memory for a set of scattering angles; as the data scan proceeds the digital signal at each angle is divided by the reference signal and multiplied by the reference differential scattering cross section to give the absolute differential scattering cross section of the sample.

Because of the small amount of light scattered by dilute suspensions of particles it is necessary to carefully exclude extraneous light from the detector. Such light may originate from the surroundings (ambient light) or within the instrument itself. To exclude ambient light, measurements can be made in a darkroom; or the entire nephelometer can be enclosed in a dark box; or the scattering cell can be enclosed. The last option complicates the instrument by requiring a rotating light-tight aperture for the detector arm. Another possibility is to selectively reject the ambient light either with a light chopper and lock-in amplifier tuned to the chopper frequency or with a filter in the detector arm at the frequency of the source if it is nearly monochromatic. To exclude extraneous light originating from within the detector, light traps are used to stop the incident beam and as a backdrop for the detector (Fig. 13.5). Similarly, careful attention must be given to the design of the scattering cell, which may reflect forward-scattered light back into the detector. This is a particularly serious problem in the backscattering direction for large particles: even a very small amount of specular reflection from the scattering cell can easily dominate over scattering by the sample. Design of scattering cells becomes even more difficult for measurements of small amounts of circular polarization: strain-induced birefringence can give rise to appreciable errors.

13.2.3   Particle Production

Not only a nephelometer is needed to systematically study light scattering in the laboratory but also means for producing particles of known composition, size distribution, and shape. A very useful device, therefore, would be a "magic particle maker" having a set of dials with which to specify the type of distribution and its parameters (mean size and width), chemical composition (to set the optical constants), and shape. When set for particle shape "sphere," for example, such a device used in conjunction with a calibrated nephelometer could give data similar to the curves of Figs. 13.1–13.3 calculated from Mie theory. Unfortunately, even for spheres, the simplest shape, such a magic box is not available. The experimenter who wishes to change a single parameter, such as mean size or refractive index, may have to work for several years to develop a different method of particle production; his theoretical counterpart need only change one number at a computer terminal! Thus, making well-defined collections of particles is a difficult task which has been undertaken in many laboratories for a long time. Only a few selected examples are given here.

For production of highly monodisperse aerosols, the Sinclair–LaMer generator (described by Kerker, 1969, p. 319) has often been used with success. Liquid droplets of controlled size in the approximate range 0.01–1 μm can be produced by condensing vapor onto nuclei created for the purpose. Careful control of temperature and temperature gradient permits the production of liquid aerosols with fractional standard deviations in particle radius of about 0.1–0.2. Various nebulizers have been used to generate liquid droplets which may be dried to yield particles of soluble solids. Size distributions from conventional nebulizers are often quite broad. An ultrasonic nebulizer, however, can give narrow distributions: the ultrasonic transducer excites surface waves in the liquid which can be driven to crest and break above the surface, releasing droplets of similar size at each crest (Lobdell, 1968); fractional standard deviation in size is about 0.3. Changing the frequency by using different transducers or higher harmonics enables droplet size to be varied. A vibrating orifice has been used to generate even narrower droplet size distributions (Bergland and Liu, 1973; Pinnick et al., 1973). A liquid is forced at high pressure through a small orifice. The orifice is driven into oscillation by an ultrasonic transducer, causing the emerging liquid to break off into droplets with size determined by the properties of the liquid, the applied pressure, the orifice size, and the vibration frequency. Relative size deviation is typically a few percent.

Liquid droplets with a dissolved component can be dried to leave a residue of solid aerosol particles with a size distribution similar to that of the parent liquid aerosol. Irregular solid particles have been produced beginning with droplets from an ultrasonic nebulizer (Perry et al., 1978) and droplets from a vibrating orifice (Pinnick et al., 1973; Pinnick et al., 1976; Pinnick and Auvermann, 1979). It is also possible to make aerosol particles that are not soluble in common liquids by means of chemical reactions in a flame. For example, following Nielsen et al. (1963) we have produced α-Fe_2O_3 (hematite) particles by nebulizing an aqueous solution of $FeCl_3$ and passing the resulting droplets into an air–hydrogen flame; α-Fe_2O_3 particles, with a mean size and standard deviation calculable from the original droplet size distribution, are formed by the reaction

$$2FeCl_3 + 3H_2O \underset{flame}{\rightarrow} Fe_2O_3 + 6HCl.$$

The vibrating orifice method for producing monodisperse liquid aerosols, combined with drying or chemical reaction in a flame, comes about as close to being a magic particle maker as any technique we are aware of.

13.3 MEASUREMENTS ON SINGLE PARTICLES

In 1961, before lasers were a common laboratory instrument, Gucker and Egan published angular light-scattering measurements for single isolated particles. Since the advent of high-power collimated lasers many measurements on single particles have been published, particularly in the last decade, and instrument designs have proliferated. To successfully obtain angular light scattering information from single particles it is usually necessary to take one of two approaches: stably suspend the particles and use a conventional "take-your-time" nephelometer such as the one in Fig. 13.5, or make rapid measurements on single particles in flow. We shall discuss each of these in turn.

13.3.1 Particle Suspension Methods

There are several methods for levitating single particles. One of them uses a modified Millikan oil-drop apparatus in which the particle is charged and balanced against gravity by an electric field between parallel plates. To stabilize the particle laterally the electric field is deformed by a charged needle; an electronic servomechanism maintains the applied voltage at a level necessary to ensure vertical stability. This technique is described by Wyatt and Phillips (1972); it has been used for measurements on polystyrene spheres (Phillips et al., 1970), bacteria (Wyatt and Phillips, 1972), NaCl and NaCl–H_2O particles (Tang and Munkelwitz, 1978), and to determine diffusion coefficients (Davis and Ray, 1977). Particles suspended by electrostatic levitation tend to tumble randomly; this is an advantage in studies of scattering averaged over all particle orientations, but a disadvantage if one is interested in oriented particles.

A vertical laser beam has been used by Ashkin (1970) and Ashkin and Dziedzic (1971) to levitate weakly absorbing spherical particles by radiation pressure. Lateral stability results from the dominance of refracted over reflected components of the scattered light (see Table 7.1). Unequal reflection on opposite sides of the particle, which is caused by beam nonuniformity, produces a net force that drives the particle toward lower light levels; this instability is countered by refraction, which produces a reaction that drives the particle toward higher light levels. The particle is thus laterally stabilized in the most intense part of the beam. Laser levitation has the disadvantage that it cannot be used with strongly absorbing particles, which are likely to vaporize in beams sufficiently powerful for levitation.

Perhaps the simplest, but least used method for supporting single particles is to attach them to very fine fibers such as drawn glass fibers or spider threads (Saunders, 1970, 1980). The fiber diameter can be made small compared with particles larger than about 1 μm. Furthermore, the characteristics of scattering by long cylinders (see Section 8.4, particularly Fig. 8.5) can be used to advantage: scattering by fibers oblique to the scattering plane is confined to directions only one or two of which lie in this plane. This simple method provides a means for fixing the orientation of a particle; its disadvantages are the need to consider scattering by the support fiber and the possibility of electromagnetic interaction between the fiber and the particle.

13.3.2 Rapid Light Scattering Measurements

Rather than suspend a particle in the scattering volume for a time sufficient to make a scan with a conventional nephelometer, scattering data can be recorded rapidly during the short time it takes a particle to transit the scattering volume. This can be done in two ways: (1) with a *rotating* detector or aperture (or both), or (2) with an array of *fixed* detectors.

Marshall et al. (1976) used an annular segment of an ellipsoidal mirror to direct light scattered by a particle at one focus to a detector fixed at the other focus; an aperture (5°) rotating at 3000 rpm then gives a 360° scan in 20 msec. Morris et al. (1979) mounted a detector on a turntable that rotates at 1 Hz; in 1 sec this gives a scan from 0 to 180° and from 180 to 360°; these scans are not necessarily identical. There are at least two reasons for making 360° scans instead of the usual 180° scans: single particles (other than spheres) are likely to be azimuthally asymmetric; and such scans show instrument misalignment.

Fixed detector arrays were devised by Diehl et al. (1979) and by Bartholdi et al. (1980) to measure scattering by single particles. The former authors mounted detectors at ±45°, ±90°, ±135°, combinations of either two or three of which sampled scattered light at 16.7-msec intervals (60-Hz sampling rate). The instrument of Bartholdi et al. has an annular segment of an ellipsoidal reflector to focus scattered light onto a circular array of 60 photodiode detectors. This instrument was designed for applications in which biological cells flow in single file (at flow rates of up to 1000 particles per second) through the scattering volume; by analyzing the scattered light, cells from heterogeneous populations can be identified and possibly separated downstream.

13.3.3 Microwave Analog Scattering

Microwave scattering is an important analog technique for investigating scattering of visible light by single nonspherical particles. Because of the ratio of about 10^5 between microwave and visible wavelengths, arbitrarily oriented particles of many shapes and with microwave size parameters of less than 1 can be readily fashioned (e.g., by machining); materials can be chosen with microwave optical constants closely matching those at visible wavelengths for particles of interest. Microwave extinction ($\theta = 0°$) has already been discussed in Section 11.7; to adapt the instrument shown in Fig. 11.22 to angular scattering measurements, the receiving antenna need only be mounted so that it can move about the particle. Extraneous radiation reflected from walls, floor, and ceiling is usually eliminated by the liberal use of absorbing material.

Materials with microwave optical constants similar to those of common solids in the visible have been made from plastics of varying density; an absorber, such as fine carbon dust, can be added to increase the imaginary part of the refractive index. For example, acrylic plastic (Dupont trade name, Lucite) has a microwave refractive index of about 1.6, which is similar to that of some silicate materials in the visible. Polystyrene beads impregnated with a volatile material can be expanded by heating to form a foam with an effective refractive index which depends on its density. In this way a material with optical constants similar to those of ice or water at visible wavelengths can be made. The microwave optical constants of such composite media are determined by techniques similar to those that are used for homogeneous media (Chapter 2); the fact that these optical constants can be used in Mie calculations for homogeneous spheres which agree with experiment is a partial justification for treating some composite media as homogeneous with effective refractive indices, although this does not necessarily follow for all combinations of materials (see Section 8.5 for a discussion of theories of optical constants for inhomogeneous media). There are practical limits to the range of microwave optical constants obtainable with common materials: for example, optical constants close to those that give strong shape-dependent scattering and absorption (see Chapter 12), such as $n = 0$ and $k = \sqrt{2}$, would probably be difficult to obtain.

Microwave angular scattering measurements for incident light polarized parallel and perpendicular to the scattering plane have been made by Zerull and Giese (1974) and by Zerull et al. (1977, 1980) for particles of various shapes, including spheres, spheres with roughened surfaces, cubes, octahedrons, irregular particles with both convex and concave surfaces, and "fluffy" particles consisting of loose aggregates of smaller compact particles; n was mostly in the range from about 1.5 to 1.7 and k between about 0.005 and 0.015. Greenberg et al. (1961) investigated scattering by spheroids and finite cylinders; they chose materials with microwave optical constants similar to those of water and silicate materials at visible wavelengths.

The microwave analog technique has the great advantage that scattering by single particles of any shape and orientation can be studied with relative ease; within limits, the optical constants can also be varied arbitrarily by using composite media. Therefore, microwave experiments should be the best test of theories of scattering by irregular particles, such as those discussed in Section 8.6. For collections of randomly oriented particles, however, microwave measurements can be quite laborious: measurements for many orientations must be made at each scattering angle and for many particles. Examples of angular scattering of microwave radiation are given in the following section.

13.4 SOME THEORETICAL AND EXPERIMENTAL RESULTS

Measurements of angular scattering by spheres agree well with Mie calculations, both for single spheres (Marshall et al., 1976; Bartholdi et al., 1980; Phillips et al., 1970) and for monodisperse collections of spheres (e.g., Pinnick et al., 1976). Sphere measurements are now used primarily to check instrument operation and calibration or to infer some physical property from accurate determinations of size and refractive index (Davis and Ray, 1977, for example). The present thrust in light scattering, both theoretical and experimental, is toward a better understanding of nonspherical particles, although the standard of comparison is likely to be Mie calculations for equivalent spheres.

Numerical approximation techniques can now handle almost any nonspherical and inhomogeneous particle, but the computational demands are high, especially when a distribution of shapes, sizes, and orientations must be considered. An additional complication is that scattering by nonspherical particles may depend on the azimuthal angle ϕ as well as the scattering angle θ, which makes scattering diagrams less easy to display.

13.4.1 Scattering by Spheroids

A spheroid is perhaps the simplest (finite) nonspherical particle. Many approximate treatments of spheroids, valid for certain limiting cases, have been given (see e.g., Section 5.3). Two approaches to the problem of scattering by spheroids of arbitrary shape and composition, which have been taken in recent years, are (1) constructing separable solutions to the scalar wave equation in spheroidal coordinates and expanding the fields in vector spherical harmonics in a manner similar to that for spheres (Section 4.1), and (2) using the T-matrix method of Waterman (1965, 1971) (Section 8.6). The former approach was initiated by Asano and Yamamoto (1975); the latter adopted by Barber and Yeh (1975). Both of these methods are exact. Scattering by spheroids has also been investigated within the framework of various approximate methods (Latimer et al., 1978).

Calculated scattering of unpolarized light by spheroids is shown in Figs. 13.6 and 13.7. In the first figure polar scattering diagrams of Latimer et al. (1978) are displayed for equal-volume spheroids (1 μm^3)—sphere, prolate, oblate—with symmetry axes parallel to the scattering plane and in various orientations relative to the incident beam; the refractive index $m = 1.05$, which approximately corresponds to that of biological cells in water at visible wavelengths, reflects the authors' interest in particles of biological origin. Scattering diagrams from Asano (1979) are shown in Fig. 13.7 for two different

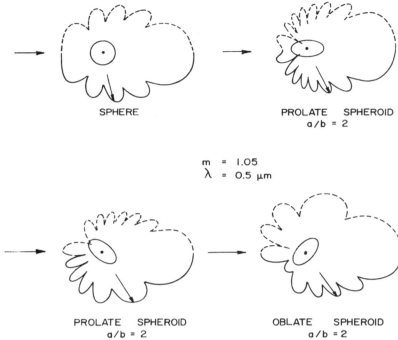

Figure 13.6 Polar scattering diagrams for equal-volume spheroids. The incident light is un-polarized. From Latimer et al. (1978).

orientations of a spheroid with axial ratio 5.0 and refractive index 1.5; the symmetry axis is again parallel to the scattering plane.

Note first that spheroids oriented with their symmetry axes oblique to the incident beam scatter asymmetrically about the forward direction; this is more evident in the polar plots of Fig. 13.6. Latimer et al. pointed out that there tend to be more lobes in the scattering diagram near directions toward which the particle presents a large width than near directions for which the projected width is smaller; this is somewhat analogous to scattering by spheres: the larger the sphere, the more lobes there are. They also concluded that for sufficiently large particles (radius of equal volume sphere greater than about 1 μm), small-angle (1.5°) scattering is strongly dependent on shape and orientation. In general, the larger the spheroid, the greater the number of lobes in the scattering diagram and the more it is peaked in the forward direction. Note in Fig. 13.7 that forward scattering by a prolate spheroid is greater when its axis is parallel (0°) than when it is oblique (45°) to the incident beam even though it presents a smaller area to the beam in the parallel orientation. The effect of increasing absorption (not shown in the figure but discussed by Asano) is to dampen oscillations in the scattering diagram.

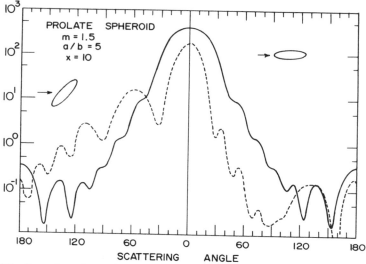

Figure 13.7 Scattering of unpolarized light incident parallel (——) and oblique (---) to the symmetry axis of a spheroid. From Asano (1979).

Asano also gives scattering diagrams for a large, slender prolate spheroid illuminated obliquely (45°), but in scattering planes that do not contain the symmetry axis; scattering resembles that by infinite circular cylinders (Section 8.4). Scattering by single coated spheroids is discussed by Wang and Barber (1979).

Scattering depends on azimuthal angle for arbitrarily oriented spheroids but not for collections of randomly oriented spheroids. Therefore, if theory is to be compared with measurements either on collections of randomly oriented particles or on single particles randomized by tumbling, calculations must be done for many orientations. This has not been done in most calculations. Exceptions are the work of Wang et al. (1979) on coated as well as homogeneous spheroids and that of Asano and Sato (1980). Because of the range of sizes and refractive indices considered, the latter paper is particularly valuable for comparing theory with experimental results; the physical mechanisms responsible for departures from scattering by spheres are also discussed at length by Asano and Sato.

13.4.2 Scattering by Collections of Particles

Several examples of scattering by spherical and by nonspherical particles are collected in Fig. 13.8: calculations for randomly oriented prolate and oblate spheroids; measured scattering of microwave radiation by a polydispersion of nonspherical particles; and measured scattering of visible light by irregular

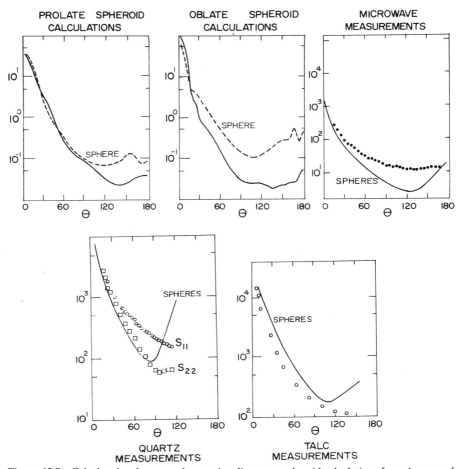

Figure 13.8 Calculated and measured scattering diagrams: spheroid calculations from Asano and Sato (1980); microwave measurements from Zerull et al. (1980); quartz measurements from Holland and Gagne (1970); and talc measurements from Holland and Draper (1967).

quartz and by talc particles. Accompanying each of these are calculations for equivalent spheres. The meaning of equivalent, as well as size parameters and refractive indices, is given in the papers (cited in the figure caption) from which these curves were taken (redrafted for uniformity).

There are both similarities and differences between scattering by spherical and by nonspherical particles. Near the forward direction, where scattering may be associated primarily with diffraction, external reflection, and twice-refracted transmission (Hodkinson and Greenleaves, 1963), nonspherical particles scatter similarly to area-equivalent spheres, in general. Forward scattering by large particles is dominated by diffraction, which depends on particle area and is independent of refractive index; for this reason, forward scattering may be used to size nonspherical particles with unknown optical properties. Features in scattering diagrams for spheres, such as rainbows, are not exhibited by nonspherical particles. In directions greater than about 90°, scattering diagrams for nonspherical particles tend to be flatter than those for spheres; in particular, scattering does not increase sharply near the back direction (i.e., glories are not associated with nonspherical particles). It is difficult to generalize about differences between scattering by nonspherical and equivalent spherical particles at intermediate scattering angles (in the range 45–135°, say), which may reflect the absence of a general trend: equivalent spheres scatter less at these angles than irregular quartz particles and cubes but more than talc particles and spheroids. These differences also appear in calculated asymmetry parameters. Asano and Sato (1980) calculated asymmetry parameters for randomly oriented spheroids which were greater than those for equivalent spheres; in contrast, Pollack and Cuzzi (1980) calculated asymmetry parameters for various nonspherical particles which were less. Therefore, it is not possible to state categorically that nonspherical particles are either more or less backscattering than equivalent spheres.

13.4.3   Polarization

Calculated and measured values of $P = -S_{12}/S_{11}$, the degree of linear polarization, for several nonspherical particles are shown in Fig. 13.9. The prolate and oblate spheroids, cubes, and irregular quartz particles have made their appearance already (Fig. 13.8); a new addition is NaCl cubes. Also shown are calculations for equivalent spheres.

There are marked differences between the polarization diagrams for spherical and nonspherical particles; moreover, a few generalizations appear to be possible. First, polarization diagrams for nonspherical particles do not have features near the rainbow angles (or their precursors for smaller particles) for equivalent spheres. Second, polarization tends to be positive over a wide range of angles for nonspherical particles. On the basis of extensive calculations for randomly oriented prolate spheroids with axial ratio 2, refractive index 1.44, and size parameters up to about 15, Asano and Sato (1980) concluded that the linear polarization of light scattered by nonspherical particles, in contrast with spheres, tends to be positive at middle scattering angles. Perry et al. (1978) measured scattering by nearly cubical NaCl particles of various sizes and found good agreement with calculations for small spheres ($x < 4$) but poor agreement for larger spheres (Fig. 13.9). They also found good agreement for all sizes up to $x = 12.4$ for rounded, but not spherical, ammonium sulfate particles. Similar positive polarization was observed in scattering by plate-like ice crystals (Sassen and Liou, 1979). Thus, for a variety of nonspherical particles, there appears to be a trend toward polarization opposite to that for spheres. Geometrical optics predicts that light externally and internally reflected by a sphere will contribute positively to linear polarization; note, for example, that the polarization is positive at the primary and secondary rainbow angles (Fig. 13.4), which are associated with rays internally reflected once and twice, respectively. Shape irregularities apparently distribute the positive polarization attributed to internal reflections over a larger range of angles (see, e.g., Coffeen, 1969).

Also shown in all but one of the examples of Fig. 13.9 is the normalized matrix element S_{22}/S_{11}, which is unity for all collections of spheres. Thus, the

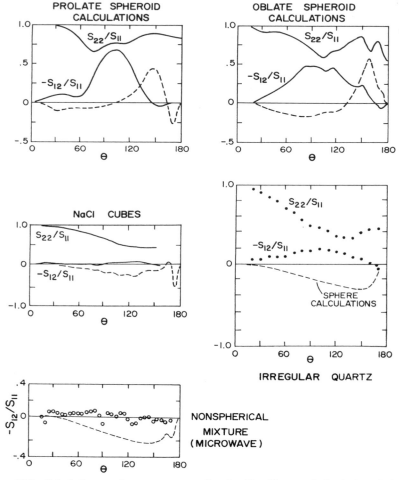

Figure 13.9 Calculations and measurements of $-S_{12}/S_{11}$ (linear polarization) and S_{22}/S_{11}: spheroid calculations from Asano and Sato (1980); NaCl measurements from Perry et al. (1978); quartz measurements from Holland and Gagne (1970) (calculated using their measurements of unnormalized matrix elements); and microwave measurements from Zerull et al. (1980). Dashed curves are calculations for equivalent spheres.

departure of S_{22}/S_{11} from unity is a measure of nonsphericity; the difference

$$\Delta = 1 - \frac{S_{22}}{S_{11}}$$

is sometimes called the *depolarization ratio*. In Section 13.1, however, $\frac{1}{2}(S_{11} - S_{22})$, which is the same as Δ except for the normalization factor S_{11}, was called the *cross polarization*. If Δ is zero, then under the assumption that the scattering matrix has the form (13.21) there is no depolarization of incident light polarized either parallel or perpendicular to the scattering plane (i.e., the scattered light is 100% polarized). However, Δ is measured by inserting linear polarizers with orthogonal transmission axes—crossed polarizers—fore and aft of the scattering medium. Thus, the term "depolarization ratio" for Δ reflects its physical significance, whereas the term "cross polarization" reflects the method by which it is measured.

All the examples in Fig. 13.9 show S_{22} approaching S_{11} in the forward direction, but S_{22}/S_{11} deviates appreciably from unity at large angles. For example, S_{22}/S_{11} for the NaCl cubes is less than about 0.5 at scattering angles greater than 90°. Calculations for prolate spheroids yield an interesting result: the depolarization ratio for spheroids with axial ratio 2 is greater than that for more elongated spheroids with axial ratio 5.

13.4.4 Implications for the Inverse Scattering Problem

In attempts to invert scattering data to infer properties of particles in the atmosphere, in interstellar space, and in the laboratory, differences between scattering by spherical and nonspherical particles can lead to conclusions that are quantitatively as well as qualitatively incorrect. For example, a flat response for S_{11} in back directions could be mistaken as an effect of large absorption in spheres rather than nonsphericity in weakly absorbing particles. Large positive polarizations could suggest very small particles rather than larger nonspherical particles. On the other hand, deviations of S_{22}/S_{11} from unity could be valuable indicators of nonsphericity, which would then signal caution in analyzing and interpreting experimental results with Mie theory.

13.5 PARTICLE SIZING

Particle sizing by elastic light scattering has been used widely because it is nondestructive and may be done rapidly. Kerker (1969, Chap. 7) has discussed in detail several particle sizing methods, including those for collections of particles distributed in size. The difficulty of inverting measurements to obtain a size distribution increases with its width. This drawback has led to greater use of methods in which light scattered by particles flowing singly through the scattering volume is analyzed to infer sizes. Pulses of light scattered into a set of directions are detected and sorted electronically into bins according to height; the size distribution is determined from the pulse height histogram by means of a calibration curve relating pulse height to particle size. Commercial instruments are now available for this purpose; they differ according to the directions for which scattered light is collected and the source of illumination, which can be a laser or focused light from an incandescent lamp.

The response R of a single-particle light scattering instrument is

$$R = \iint G \frac{dC_{sca}}{d\Omega} f \, d\lambda \, d\Omega;$$

G describes the illumination and collection geometry, f includes the source spectrum and spectral sensitivity of the photodetector, and the differential scattering cross section $dC_{sca}/d\Omega$ can be calculated from Mie theory if the particles are spheres. Response as a function of particle size for several commercial instruments has been published by various authors, including Cooke and Kerker (1975) and Pinnick and Auvermann (1979). Calculations for spheres by the latter authors are shown in Fig. 13.10; ripple and interference structure (see Chapters 4 and 11) are evident in the curves for the nonabsorbing spheres. These curves demonstrate some of the problems that must be faced in the design and use of particle sizing instruments. Because of the interference structure, response is a multivalued function of size: spheres of three different sizes can give the same response. This can be mitigated to some extent for certain kinds of aerosols by choosing the electronic discriminator setting to avoid regions of multivaluedness. If spheres of unknown or different refractive indices are to be sized, the variation of response with refractive index can be a problem.

Two instruments considered by Cooke and Kerker (1975) had single-valued response functions, presumably because of broad-band (white) light sources and large angular apertures for both incident and scattered light.

Calibration is invariably based on spheres, which means that an instrument can be used with confidence only for such particles. Of course, a response will duly be recorded if a nonspherical particle passes through the scattering volume. But what is the meaning of the "equivalent" radius corresponding to that response? Is it the radius of a sphere of equal cross-sectional area? Or equal surface area? Or equal volume? Or perhaps equal mean chord length? Answers to these questions depend on the particular instrument and nonspherical particle; comprehensive answers do not come easily because it is difficult to do calculations for nonspherical particles, even those of regular shape.

According to scalar diffraction theory (Section 4.4) the scattering amplitude in the forward direction is proportional to the cross-sectional area of the particle, regardless of its shape, and is independent of refractive index. To the extent that diffraction theory is a good approximation, therefore, the radius corresponding to the response of an instrument that collects light scattered near the forward direction by a nonspherical particle is that of a sphere with equal cross-sectional area. The larger the particle, however, the more the

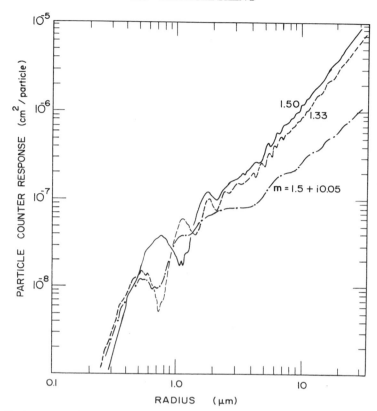

Figure 13.10 Response calculations for a particle counter that collects He–Ne laser light scattered between 4 and 22°. Reprinted with permission from R. G. Pinnick and J. J. Auvermann, *J. Aerosol Sci.*, **10** (1979), 55–74; copyright 1979, Pergamon Press, Ltd.

scattered light is peaked in the forward direction and the more difficult it is to discriminate between scattered and unscattered (incident) light. Nevertheless, the possible advantages of refractive index independence and insensitivity to shape make low-angle scattering an attractive method of particle sizing. Based on calculations of response functions for various instruments Heyder and Gebhart (1979) concluded that size distributions of particles with unknown refractive indices can be determined accurately by low-angle scattering of polychromatic light.

Experimental evaluation of various widely used commercial instruments has been undertaken by several investigators, including Liu et al. (1974), who used the vibrating orifice technique to generate spherical aerosols with narrow size distributions. Similarly, Pinnick and Auvermann (1979) generated droplets of a solution which were dried to make solid particles for evaluating the response of certain instruments to nonspherical aerosols. It appears from these evaluations that, depending on the instrument, spherical particles in the size range between about 0.2 and 15 μm can be sized accurately by light scattering if their refractive index is known. If, however, the particles are nonspherical, or of unknown refractive index, or if the aerosol is composed of particles with different refractive indices, accurate sizing may be less attainable.

13.6   SCATTERING MATRIX SYMMETRY

All the information about angular scattering by a medium (a single particle or a collection of particles) is contained in the 16 elements of its scattering matrix. In many instances—media with rotational symmetry, for example—all the matrix elements are not independent, and higher symmetry often further reduces the number of independent, nonzero elements. Even without considering its explicit angular dependence, therefore, the *form* of the scattering matrix reflects general properties of the scattering medium. The two major treatments of scattering matrix symmetry, those of Perrin (1942) and van de Hulst (1957, Chap. 5), arrive at similar results but by somewhat different paths: van de Hulst begins with 2×2 amplitude scattering matrices for single particles and derives 4×4 scattering matrix symmetries under various assumptions about the distribution of particle orientation and shape; Perrin, however, considers the scattering matrix directly without going through the intermediate step of appealing to symmetries of the 2×2 matrices. Because 2×2 amplitude scattering matrices for single particles are conceptually simpler than 4×4 scattering matrices for collections of particles, our approach to scattering matrix symmetry follows that of van de Hulst.

The most general amplitude scattering matrix for a *single* particle

$$\begin{pmatrix} S_2(\theta,\phi) & S_3(\theta,\phi) \\ S_4(\theta,\phi) & S_1(\theta,\phi) \end{pmatrix} \tag{13.4}$$

contains eight parameters, the real and imaginary parts of each element, or equivalently, the amplitude and phase, which depend on the scattering angle θ and the azimuthal angle ϕ. Because only *relative* phases are physically significant, there are in fact only seven *independent* parameters in (13.4). The 4×4 scattering matrix corresponding to (13.4) follows from (3.16); although all of its 16 elements may be nonzero, only seven of them are independent. Thus, there are nine independent relations among the matrix elements; these were given by Abhyankar and Fymat (1969). However, the scattering matrix for a collection of *different* particles, different by virtue of orientation or shape or composition, is not so constrained: all of its 16 elements may be independent. Nevertheless, there is still a set of independent *inequalities* relating the matrix elements (Fry and Kattawar, 1981). For example, if the particles are spherical, we have

$$S_{11}^2 \geqslant S_{12}^2 + S_{33}^2 + S_{34}^2.$$

Equality holds for a single sphere or a collection of identical spheres; inequality holds if they are distributed in size or composition. This inequality was used by Hunt and Huffman (1973), for example, as an indicator of dispersion in suspensions of spherical particles. It was pointed out by Fry and Kattawar (1981) that the inequalities they derived are useful consistency checks on measurements of all 16 scattering matrix elements.

In the following paragraphs we shall first discuss scattering matrices corresponding to specific particles for which analytical solutions to the scattering problem exist. Then, by appealing to very general symmetry relations, we shall obtain the form of scattering matrices without regard to the detailed nature of the particles.

13.6.1 Single-Particle Symmetries

We first point out a few scattering matrix symmetries for single particles based on results in previous chapters.

For spheres sufficiently small that Rayleigh theory (Chapter 5) is applicable, or for arbitrarily shaped particles that satisfy the requirements of the Rayleigh–Gans approximation (Chapter 6), incident light with electric field components parallel and perpendicular to the scattering plane may be scattered with different amplitudes; however, there is no phase shift between the two components. Hence, the amplitude scattering matrix has the form

$$\begin{pmatrix} FA_2 & 0 \\ 0 & FA_1 \end{pmatrix},$$

where F, the common factor, is complex, and A_1, A_2 are real. The corresponding scattering matrix is, from (3.16),

$$\begin{pmatrix} S_{11} & S_{12} & 0 & 0 \\ S_{12} & S_{11} & 0 & 0 \\ 0 & 0 & S_{33} & 0 \\ 0 & 0 & 0 & S_{33} \end{pmatrix}. \qquad (13.5)$$

Note that the cross polarization is zero and incident unpolarized light does not acquire a degree of circular polarization upon scattering.

Both the amplitudes and relative phase of the field components parallel and perpendicular to the scattering plane can be changed upon scattering by a sphere of arbitrary size and composition. Symmetry, however, precludes any mechanism for transforming parallel to perpendicularly polarized light, and vice versa. Thus, the cross polarization is zero, and the amplitude scattering matrix must have the form

$$\begin{pmatrix} S_2 & 0 \\ 0 & S_1 \end{pmatrix},$$

which leads to the 4×4 matrix

$$\begin{pmatrix} S_{11} & S_{12} & 0 & 0 \\ S_{12} & S_{11} & 0 & 0 \\ 0 & 0 & S_{33} & S_{34} \\ 0 & 0 & -S_{34} & S_{33} \end{pmatrix}. \qquad (13.6)$$

The only difference between (13.6) and (13.5) is that for an arbitrary sphere the matrix element S_{34} does not, in general, vanish. The effect of this matrix element is the same as that of a retarder such as a quarter-wave plate: the phase of one orthogonal component of the incident light is retarded relative to the other, which gives a degree of circular polarization to the scattered light if the incident light is polarized obliquely to the scattering plane. A consequence of the zero off-diagonal elements of the amplitude scattering matrix is that S_{11} and S_{22} are equal as are S_{33} and S_{44}. The scattering matrix for a normally illuminated cylinder (Section 8.4) has the same symmetry as (13.6).

The off-diagonal elements of the amplitude scattering matrix for an optically active sphere (Section 8.3)

$$\begin{pmatrix} S_2 & S_3 \\ -S_3 & S_1 \end{pmatrix} \qquad (13.7)$$

do not, in general, vanish; neither do those for an infinite cylinder illuminated at oblique incidence (Section 8.4). The 10-parameter scattering matrix corresponding to (13.7) is

$$\begin{pmatrix} S_{11} & S_{12} & S_{13} & S_{14} \\ S_{12} & S_{22} & S_{23} & S_{24} \\ -S_{13} & -S_{23} & S_{33} & S_{34} \\ S_{14} & S_{24} & -S_{34} & S_{44} \end{pmatrix}. \qquad (13.8)$$

It is the off-diagonal elements of (13.7) that give rise to cross polarization ($S_{11} \neq S_{22}$) as well as the nonzero elements S_{13}, S_{14}, S_{23}, and S_{24} in (13.8). That S_3 and S_4 should be nonzero for optically active particles follows from elementary physical reasoning: optical rotatory power in a homogeneous medium causes the direction of vibration to be rotated upon transmission of linearly polarized light by the medium. However, optical activity of the bulk parent material is merely sufficient, not necessary, for a particle to have nonzero cross polarization; for example, it is nonzero for isotropic, nonactive, elongated particles, such as prolate spheroids or circular cylinders, illuminated obliquely to their symmetry axes. The scattering matrices for an obliquely illuminated infinite circular cylinder are identical in form with those for an optically active sphere; it should be recalled, however, that the orthogonal basis vectors for specifying the incident electric field are parallel and perpendicular to the plane determined by the incident beam and the cylinder axis rather than by the incident beam and the scattering direction.

We have about exhausted possible scattering matrix symmetries obtained by appealing to approximate or exact solutions to specific scattering problems. But more can be said about particles, regardless of their shape, size, and composition, without explicit solutions in hand. The scattering matrix for a given particle implies those for particles obtained from this particle by the symmetry operations of rotation and reflection. We shall consider each of these symmetry operations in turn.

Consider an arbitrary particle illuminated by a plane wave *with unit amplitude* and linearly polarized along the direction $\hat{\mathbf{q}}$. The direction of propagation of this wave is specified by the unit vector $\hat{\mathbf{e}}_i$ and that of the scattered wave—that is, the direction of observation—by $\hat{\mathbf{e}}_s$. We denote by $\mathbf{A}(\hat{\mathbf{q}}; \hat{\mathbf{e}}_i, \hat{\mathbf{e}}_s)$ the amplitude of the scattered wave, where $\hat{\mathbf{e}}_i \cdot \hat{\mathbf{q}} = 0$ and $\hat{\mathbf{e}}_s \cdot \mathbf{A}(\hat{\mathbf{q}}; \hat{\mathbf{e}}_i, \hat{\mathbf{e}}_s) = 0$. The Maxwell equations (2.1)–(2.4) are invariant under time reversal: they have the same form if t is replaced by $-t$, $\mathbf{H}$ by $-\mathbf{H}$, and $\mathbf{J}_F$ by $-\mathbf{J}_F$. Reversal in time implies that an incident photon becomes a scattered photon and conversely. Or, stated another way, if a plane wave incident on a particle gives rise to a scattered outgoing spherical wave, then an incoming spherical wave will give rise to a scattered outgoing plane wave. This is not quite sufficient for our purposes because we are interested in scattering of plane waves, not spherical waves. However, if only the *directions* of incident and scattered waves are interchanged, then the amplitudes are connected by a *reciprocity relation* proved by Saxon (1955ab).

$$\hat{\mathbf{q}}' \cdot \mathbf{A}(\hat{\mathbf{q}}; \hat{\mathbf{e}}_i, \hat{\mathbf{e}}_s) = \hat{\mathbf{q}} \cdot \mathbf{A}(\hat{\mathbf{q}}'; -\hat{\mathbf{e}}_s, -\hat{\mathbf{e}}_i), \tag{13.9}$$

where $\hat{\mathbf{q}}' \cdot \hat{\mathbf{e}}_s = 0$ and $\hat{\mathbf{e}}_i \cdot \mathbf{A}(\hat{\mathbf{q}}'; -\hat{\mathbf{e}}_s, -\hat{\mathbf{e}}_i) = 0$. This reciprocity relation is the key to obtaining the form of the amplitude scattering matrices for two particles identical except for orientation.

It is convenient to introduce a coordinate system where the positive z direction coincides with $\hat{\mathbf{e}}_i$ (Fig. 13.11). Without loss of generality we may take the scattering plane to be the yz plane, in which instance the scattering amplitudes for two orthogonal polarization states of the incident light are (see Chapter 3, particularly Fig. 3.3 and Section 3.4)

$$\mathbf{A}(\hat{\mathbf{e}}_x; \hat{\mathbf{e}}_i, \hat{\mathbf{e}}_s) = S_3 \hat{\mathbf{e}}_{\|s} + S_1 \hat{\mathbf{e}}_{\perp s},$$

$$\mathbf{A}(\hat{\mathbf{e}}_y; \hat{\mathbf{e}}_i, \hat{\mathbf{e}}_s) = S_2 \hat{\mathbf{e}}_{\|s} + S_4 \hat{\mathbf{e}}_{\perp s}. \tag{13.10}$$

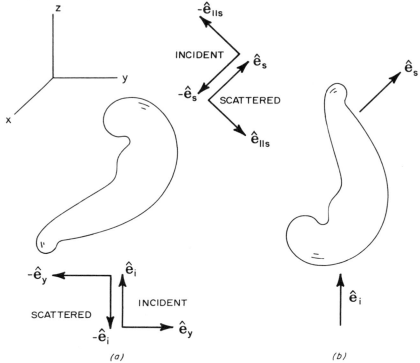

Figure 13.11 If the particle on the left is rotated through $180° − \theta$ about the x axis and then rotated through $180°$ about the z axis, the result is as shown on the right.

If the direction of the incident wave is now taken to be $-\hat{\mathbf{e}}_s$, the amplitudes of the waves scattered in the direction $-\hat{\mathbf{e}}_i$ for two orthogonal polarization states of the incident wave are

$$\mathbf{A}(-\hat{\mathbf{e}}_{\|s}; -\hat{\mathbf{e}}_s, -\hat{\mathbf{e}}_i) = S_2'(-\hat{\mathbf{e}}_y) + S_4'\hat{\mathbf{e}}_x,$$

$$\mathbf{A}(\hat{\mathbf{e}}_{\perp s}; -\hat{\mathbf{e}}_s, -\hat{\mathbf{e}}_i) = S_1'\hat{\mathbf{e}}_x + S_3'(-\hat{\mathbf{e}}_y). \tag{13.11}$$

It now follows trivially from (13.10), (13.11) and the reciprocity relation (13.9) that $S_1' = S_1$, $S_2' = S_2$, $S_3' = -S_4$, and $S_4' = -S_3$. The physical significance of the S_j' is that if the amplitude scattering matrix for an arbitrary particle is

$$\begin{pmatrix} S_2 & S_3 \\ S_4 & S_1 \end{pmatrix}, \tag{13.12}$$

then that for an identical particle rotated to bring $-\hat{\mathbf{e}}_s$ into congruence with $\hat{\mathbf{e}}_i$, and $-\hat{\mathbf{e}}_i$ with $\hat{\mathbf{e}}_s$, is

$$\begin{pmatrix} S_2 & -S_4 \\ -S_3 & S_1 \end{pmatrix}. \tag{13.13}$$

The relative orientation of the two particles is shown in Fig. 13.11; note that the pairwise correspondence between the matrix (13.12) for the particle in position (a) and the matrix (13.13) for the same particle rotated into position (b) is specific to a given direction of observation $\hat{\mathbf{e}}_s$. If the direction of observation changes, then so, in general, does the orientation of the rotated particle; so the matrix (13.13) does not apply to a *single* rotated particle in a fixed orientation but rather to a *set* of identical particles in different orientations, one for every direction of observation.

To check (13.13) we can invoke the solution to the scattering problem for an optically active sphere. Such a sphere is symmetric under all rotations, but the off-diagonal elements of its scattering matrix (13.7) do not vanish identically. As required by symmetry and the matrices (13.12) and (13.13), (13.7) is invariant with respect to interchange and sign reversal of its off-diagonal elements.

We now direct our attention to the *mirror image* particle, obtained by reflection in the scattering plane, corresponding to an arbitrary particle. We showed in Section 8.3 that the circularly polarized components of the incident and scattered fields are related by

$$\begin{pmatrix} E_{Ls} \\ E_{Rs} \end{pmatrix} = \begin{pmatrix} S_{2c} & S_{3c} \\ S_{4c} & S_{1c} \end{pmatrix} \begin{pmatrix} E_{Li} \\ E_{Ri} \end{pmatrix}, \tag{13.14}$$

where we omit a multiplicative factor. The matrix elements in the circular polarization representation are connected to those in the linear polarization representation by a unitary transformation:

$$\begin{pmatrix} S_{1c} \\ S_{2c} \\ S_{3c} \\ S_{4c} \end{pmatrix} = \frac{1}{2} \begin{pmatrix} 1 & 1 & i & -i \\ 1 & 1 & -i & i \\ -1 & 1 & i & i \\ -1 & 1 & -i & -i \end{pmatrix} \begin{pmatrix} S_1 \\ S_2 \\ S_3 \\ S_4 \end{pmatrix}. \tag{13.15}$$

Reflection transforms right-circularly polarized light to left-circularly polarized light and conversely. Thus, under reflection $E_{Ls} \to E_{Rs}$, $E_{Rs} \to E_{Ls}$, and (13.14) becomes

$$\begin{pmatrix} E_{Rs} \\ E_{Ls} \end{pmatrix} = \begin{pmatrix} S'_{2c} & S'_{3c} \\ S'_{4c} & S'_{1c} \end{pmatrix} \begin{pmatrix} E_{Ri} \\ E_{Li} \end{pmatrix}. \tag{13.16}$$

It therefore follows that the matrix elements for the particle and its image are related by

$$\begin{pmatrix} S_{1c} \\ S_{2c} \\ S_{3c} \\ S_{4c} \end{pmatrix} = \begin{pmatrix} 0 & 1 & 0 & 0 \\ 1 & 0 & 0 & 0 \\ 0 & 0 & 0 & 1 \\ 0 & 0 & 1 & 0 \end{pmatrix} \begin{pmatrix} S'_{1c} \\ S'_{2c} \\ S'_{3c} \\ S'_{4c} \end{pmatrix}. \qquad (13.17)$$

It is now not difficult to show from (13.15) and (13.17) that if (13.12) is the scattering matrix for an arbitrary particle, then that for its mirror image is

$$\begin{pmatrix} S_2 & -S_3 \\ -S_4 & S_1 \end{pmatrix}. \qquad (13.18)$$

Again, the exact solution to the optically active sphere problem provides us with an independent check of our analysis. Under reflection, the refractive indices m_L and m_R exchange roles and, as a consequence, the matrix element S_3 in (13.7) becomes $-S_3$; this is in accord with what is predicted by (13.18).

We shall need one more matrix, that for a particle obtained by the symmetry operations of reflection *and* rotation; this follows readily from (13.13) and (13.18):

$$\begin{pmatrix} S_2 & S_4 \\ S_3 & S_1 \end{pmatrix}. \qquad (13.19)$$

The amplitude scattering matrices (13.12), (13.13), (13.18), and (13.19) for *single* particles related by the symmetry operations of reflection and rotation will find their greatest use in the following paragraphs on *collections* of particles.

13.6.2 Symmetries for Collections of Particles

The 4×4 scattering matrix for a collection of particles is the sum of all the scattering matrices for the individual particles in the collection provided that there is no systematic relation among the phases of the individual scattered waves (i.e., scattering is completely incoherent). In general, therefore, all 16 elements are independent. Symmetry, however, reduces the number of independent, nonzero matrix elements. In what follows we derive the form of the scattering matrix for collections of particles under very general assumptions, proceeding from less to more symmetry.

Although the term "collection" usually implies a system of many particles, our symmetry considerations apply equally well to a single particle randomized by tumbling or some other means. That is, our averages can equally well be taken as either time averages or ensemble averages. Suppose that in a collection of identical particles all orientations are equally probable. This implies that, for every scattering direction, if there is a particle with amplitude scattering matrix (13.12), it is certain that another particle is oriented so that its scattering matrix has the form (13.13). Thus, to determine 4×4 scattering matrix symmetries, we need consider only pairs of particles. One example will suffice. It follows from (3.16), (13.12), and (13.13) that both the S_{12} and S_{21} matrix elements for a pair of particles are equal to $|S_2|^2 - |S_1|^2$. In a like manner we can find all the other symmetry relations. The result is a 10-parameter scattering matrix:

$$\begin{pmatrix} S_{11} & S_{12} & S_{13} & S_{14} \\ S_{12} & S_{22} & S_{23} & S_{24} \\ -S_{13} & -S_{23} & S_{33} & S_{34} \\ S_{14} & S_{24} & -S_{34} & S_{44} \end{pmatrix}. \tag{13.20}$$

This is the form of the scattering matrix for any medium with rotational symmetry even if all the particles are not identical in shape and composition. A collection of optically active spheres is perhaps the simplest example of a particulate medium which is symmetric under all rotations but not under reflection. Mirror asymmetry in a collection of randomly oriented particles can arise either from the shape of the particles (corkscrews, for example) or from optical activity (circular birefringence and circular dichroism).

Let us now increase the degree of symmetry. As in the previous paragraph, the particles are identical and randomly oriented, but with the additional property that each particle can be brought into congruence with its reflection in any scattering plane by a suitable rotation. This excludes particles with intrinsic optical activity, although linear birefringence and dichroism are permissible. With these assumptions, the collection can be grouped into quartets each particle of which is associated with one of the amplitude scattering matrices (13.12), (13.13), (13.18), or (13.19). Thus, the sum of the scattering matrices for these four particles gives us the form of the scattering matrix for the collection. Again, one example should be sufficient: both S_{34} and $-S_{43}$ for a quartet are equal to $2\,\mathrm{Im}\{S_2 S_1^*\}$. All the other matrix symmetries follow from a similar set of calculations obtained with (3.16) and the four amplitude scattering matrices. The final result is a matrix with eight nonzero elements, six of which are independent:

$$\begin{pmatrix} S_{11} & S_{12} & 0 & 0 \\ S_{12} & S_{22} & 0 & 0 \\ 0 & 0 & S_{33} & S_{34} \\ 0 & 0 & -S_{34} & S_{44} \end{pmatrix}. \tag{13.21}$$

Scattering media to which this matrix applies include randomly oriented anisotropic spheres of substances such as calcite or crystalline quartz (uniaxial) or olivine (biaxial). Also included are isotropic cylinders and ellipsoids of substances such as glass and cubic crystals. An example of an exactly soluble system to which (13.21) applies is scattering by randomly oriented isotropic spheroids (Asano and Sato, 1980). Elements of (13.21) off the block diagonal vanish. Some degree of alignment is implied, therefore, if these matrix elements do not vanish for a collection of isotropic elongated particles with mirror symmetry.

The matrix (13.21) is not restricted, however, to collections of particles each of which is congruent with its mirror image; it applies equally well to any medium that is invariant under rotation and reflection, which includes the possibility of mirror asymmetric particles each of which is paired with its image in the appropriate orientation.

Scattering matrices for collections of isotropic spherical particles have the greatest degree of symmetry. Regardless of size and composition, the scattering matrix for such particles—this includes inhomogeneous particles—has the form (13.6). There are eight nonzero elements; four of them, at most, are independent. Examples are spherical liquid droplets and solid spheres composed either of amorphous solids such as glass and silica or crystalline solids with cubic symmetry such as MgO and NaCl. Note that S_{11} and S_{22} are equal (as are S_{33} and S_{44}) in (13.6). If, therefore, cross polarization is detected in a measurement of angular scattering, this is sufficient to rule out isotropic spheres.

It is sometimes asserted that randomly oriented nonspherical particles are somehow equivalent to spherical particles—it is just a matter of choosing the correct size distribution. This reasoning is based, perhaps, on the realization that both systems are spherically symmetrical. But an inescapable conclusion to be drawn from the preceding paragraphs is that this assertion cannot be strictly true: regardless of the size and composition of the "equivalent" spheres, symmetry precludes full equivalence.

13.7   MEASUREMENT TECHNIQUES FOR THE SCATTERING MATRIX

The simplest, and probably most obvious, way to measure scattering matrix elements is with a conventional nephelometer (Fig. 13.5) and various optical elements fore and aft of the scattering medium. Recall that we introduced Stokes parameters in Section 2.11 by way of a series of conceptual measurements of differences between irradiances with different polarizers in the beam. Although we did not specify the origin of the beam, it could be light scattered in any direction. Combinations of scattering matrix elements can therefore be extracted from these kinds of measurements. There are now, however, two beams—incident and scattered—and many possible pairs of optical elements; these are discussed below.

We list in Table 13.1 all the possible measured irradiances (for unit incident irradiance) with a polarizer before the scattering medium and an analyzer before the detector. The light transmitted by an ideal polarizer P_s is polarized in state s: R and L denote right-circular and left-circular polarization; $\parallel$ and $\perp$ denote light polarized parallel and perpendicular to the scattering plane; $+$ and $-$ denote light polarized obliquely to the scattering plane at $+45°$ and $-45°$. U denotes the absence of a polarizer or analyzer; if U is indicated as

Table 13.1 Combinations of Scattering Matrix Elements That Result from Measurements with a Polarizer P_s Forward of the Scattering Medium and an Analyzer A_s aft[a]

U	U	S_{11}	$P_\perp$	U	$\frac{1}{2}(S_{11} - S_{12})$
U	$A_\parallel$	$\frac{1}{2}(S_{11} + S_{21})$	$P_\perp$	$A_\parallel$	$\frac{1}{4}(S_{11} - S_{12} + S_{21} - S_{22})$
U	$A_\perp$	$\frac{1}{2}(S_{11} - S_{21})$	$P_\perp$	$A_\perp$	$\frac{1}{4}(S_{11} - S_{12} - S_{21} + S_{22})$
U	A_+	$\frac{1}{2}(S_{11} + S_{31})$	$P_\perp$	A_+	$\frac{1}{4}(S_{11} - S_{12} + S_{31} - S_{32})$
U	A_-	$\frac{1}{2}(S_{11} - S_{31})$	$P_\perp$	A_-	$\frac{1}{4}(S_{11} - S_{12} - S_{31} + S_{32})$
U	A_R	$\frac{1}{2}(S_{11} - S_{41})$	$P_\perp$	A_R	$\frac{1}{4}(S_{11} - S_{12} - S_{41} + S_{42})$
U	A_L	$\frac{1}{2}(S_{11} + S_{41})$	$P_\perp$	A_L	$\frac{1}{4}(S_{11} - S_{12} + S_{41} - S_{42})$
$P_\parallel$	U	$\frac{1}{2}(S_{11} + S_{12})$	P_+	U	$\frac{1}{2}(S_{11} + S_{13})$
$P_\parallel$	$A_\parallel$	$\frac{1}{4}(S_{11} + S_{12} + S_{21} + S_{22})$	P_+	$A_\parallel$	$\frac{1}{4}(S_{11} + S_{13} + S_{21} + S_{23})$
$P_\parallel$	$A_\perp$	$\frac{1}{4}(S_{11} + S_{12} - S_{21} - S_{22})$	P_+	$A_\perp$	$\frac{1}{4}(S_{11} + S_{13} - S_{21} - S_{23})$
$P_\parallel$	A_+	$\frac{1}{4}(S_{11} + S_{12} + S_{31} + S_{32})$	P_+	A_+	$\frac{1}{4}(S_{11} + S_{13} + S_{31} + S_{33})$
$P_\parallel$	A_-	$\frac{1}{4}(S_{11} + S_{12} - S_{31} - S_{32})$	P_+	A_-	$\frac{1}{4}(S_{11} + S_{13} - S_{31} - S_{33})$
$P_\parallel$	A_R	$\frac{1}{4}(S_{11} + S_{12} - S_{41} - S_{42})$	P_+	A_R	$\frac{1}{4}(S_{11} + S_{13} - S_{41} - S_{43})$
$P_\parallel$	A_L	$\frac{1}{4}(S_{11} + S_{12} + S_{41} + S_{42})$	P_+	A_L	$\frac{1}{4}(S_{11} + S_{13} + S_{41} + S_{43})$
P_-	U	$\frac{1}{2}(S_{11} - S_{13})$	P_L	U	$\frac{1}{2}(S_{11} - S_{14})$
P_-	$A_\parallel$	$\frac{1}{4}(S_{11} - S_{13} + S_{21} - S_{23})$	P_L	$A_\parallel$	$\frac{1}{4}(S_{11} - S_{14} + S_{21} - S_{24})$
P_-	$A_\perp$	$\frac{1}{4}(S_{11} - S_{13} - S_{21} + S_{23})$	P_L	$A_\perp$	$\frac{1}{4}(S_{11} - S_{14} - S_{21} + S_{24})$
P_-	A_+	$\frac{1}{4}(S_{11} - S_{13} + S_{31} - S_{33})$	P_L	A_+	$\frac{1}{4}(S_{11} - S_{14} + S_{31} - S_{34})$
P_-	A_-	$\frac{1}{4}(S_{11} - S_{13} - S_{31} + S_{33})$	P_L	A_-	$\frac{1}{4}(S_{11} - S_{14} - S_{31} + S_{34})$
P_-	A_R	$\frac{1}{4}(S_{11} - S_{13} - S_{41} + S_{43})$	P_L	A_R	$\frac{1}{4}(S_{11} - S_{14} - S_{41} + S_{44})$
P_-	A_L	$\frac{1}{4}(S_{11} - S_{13} + S_{41} - S_{43})$	P_L	A_L	$\frac{1}{4}(S_{11} - S_{14} + S_{41} - S_{44})$
P_R	U	$\frac{1}{2}(S_{11} + S_{14})$			
P_R	$A_\parallel$	$\frac{1}{4}(S_{11} + S_{14} + S_{21} + S_{24})$			
P_R	$A_\perp$	$\frac{1}{4}(S_{11} + S_{14} - S_{21} - S_{24})$			
P_R	A_+	$\frac{1}{4}(S_{11} + S_{14} + S_{31} + S_{34})$			
P_R	A_-	$\frac{1}{4}(S_{11} + S_{14} - S_{31} - S_{34})$			
P_R	A_R	$\frac{1}{4}(S_{11} + S_{14} - S_{41} - S_{44})$			
P_R	A_L	$\frac{1}{4}(S_{11} + S_{14} + S_{41} + S_{44})$			

[a] U indicates the absence of a polarizer or analyzer.

being before the scattering medium, we assume that the incident light is completely unpolarized, which is an idealization. The linear analyzers $A_\parallel, \ldots, A_-$ are identical with the linear polarizers $P_\parallel, \ldots, P_-$. The circular analyzers A_R and A_L, however, are the circular polarizers P_R and P_L taken in reverse order. A circular polarizer or analyzer is a composite of two optical elements—a linear polarizer and a $\pm 90°$ retarder—the order of which in the optical train is important: if the beam first encounters the linear polarizer, then the composite is a circular polarizer; the reverse order gives a circular analyzer. We have not made a distinction between polarizers and analyzers in previous sections of this book. In this instance, however, such a distinction seems desirable to avoid confusion. Mueller matrices for polarizers and retarders are given in Section 2.11.

The possible outcomes of measurements—combinations of scattering matrix elements—listed in Table 13.1 follow from multiplication of three matrices: those representing the polarizer, the scattering medium, and the analyzer. If U is an element in the optical train, then the measured irradiance depends on only two matrix elements. In general, however, there are four elements in a combination, so that four measurements are required to obtain one matrix element.

Scattering matrix elements were measured in this manner by Pritchard and Elliot (1960) and by Holland and Gagne (1970); the former authors gave results similar to those in Table 13.1 for a medium with scattering matrix (13.21). Although this technique is straightforward, albeit possibly laborious, relative errors can be appreciable when two large signals are subtracted to obtain small matrix elements—S_{13} and S_{14}, for example. Subtraction can be avoided by amplifying the detector signal modulated by rotating a polarizer or retarder in either the incident or scattered beam. Sekera (1957) described such a method for measuring skylight polarization. It can be an improvement over subtractive techniques. But it has disadvantages: the element cannot always be rotated as fast as desired; imperfections, such as scratches or dust on the rotating element, spuriously modulate the signal, which limits sensitivity. There is, however, a polarization modulation technique, without rotating elements, for measuring angular scattering; this is described in the following paragraphs.

The heart of the polarization-modulated nephelometer is a photoelastic modulator, developed by Kemp (1969) and by Jasperson and Schnatterly (1969). The latter used their instrument for ellipsometry of light reflected by solid surfaces (the application described here could be considered as ellipsometry of scattered light). Kemp first used the modulation technique in laboratory studies but soon found a fertile field of application in astrophysics: the modulator, coupled with a telescope, allowed circular polarization from astronomical objects to be detected at much lower levels than previously possible.

The photoelastic modulator is composed of a piezoelectric transducer coupled to a block of amorphous quartz. The transducer, a quartz crystal, is driven by an electric field oscillating at a characteristic frequency set by the crystal dimensions (typically 50 kHz). Periodic stress birefringence is therefore induced in the amorphous quartz block. If light incident on the modulator is linearly polarized at 45° to its axis, the polarization state of the transmitted light alternates between left circular and right circular provided that the amplitude of the induced stress is sufficient to give 90° retardance. This rapid polarization-modulation technique permits phase-sensitive detection of discrete Fourier components of the scattered light and avoids problems commonly encountered with rotating elements.

A photoelastic modulator mated to a nephelometer is shown schematically in Fig. 13.12 (Hunt and Huffman, 1973, 1975; Perry et al., 1978). The transmission axis of the linear polarizer P is fixed at 45° relative to the stress axis of the modulator M, which may be at either $0°$ or $45°$ to the scattering plane. Light transmitted by the modulator is scattered by the sample S and then detected after (possibly) passing through the analyzer A mounted on the scanning arm. We shall illustrate the operation of this system in the configuration for determining the matrix elements S_{12} and S_{14}: the modulator is at $45°$ to the scattering plane and there is no analyzer. The sequence of matrix multiplications representing the transformations of the light as it traverses the system is given below.

$$\underbrace{\begin{pmatrix} S_{11} & S_{12} & S_{13} & S_{14} \\ S_{21} & S_{22} & S_{23} & S_{24} \\ S_{31} & S_{32} & S_{33} & S_{34} \\ S_{41} & S_{42} & S_{43} & S_{44} \end{pmatrix}}_{\text{Scattering Sample}} \underbrace{\begin{pmatrix} 1 & 0 & 0 & 0 \\ 0 & \cos\delta & 0 & -\sin\delta \\ 0 & 0 & 1 & 0 \\ 0 & \sin\delta & 0 & \cos\delta \end{pmatrix}}_{\text{Retarder}} \underbrace{\frac{1}{2}\begin{pmatrix} 1 & 1 & 0 & 0 \\ 1 & 1 & 0 & 0 \\ 0 & 0 & 0 & 0 \\ 0 & 0 & 0 & 0 \end{pmatrix}}_{\text{Polarizer}}$$

The incident beam, which we may take to be unpolarized, encounters three optical elements, each of which is represented by a matrix; we recall from Section 2.11 that matrices for polarizers and retarders depend on their orientation. The Stokes parameters of the light emerging successively from the polarizer, the modulator, and the scattering sample are (from right to left)

$$\underbrace{\frac{1}{2}\begin{pmatrix} S_{11} + S_{12}\cos\delta + S_{14}\sin\delta \\ S_{21} + S_{22}\cos\delta + S_{24}\sin\delta \\ S_{31} + S_{32}\cos\delta + S_{34}\sin\delta \\ S_{41} + S_{42}\cos\delta + S_{44}\sin\delta \end{pmatrix}}_{\text{After }S} \quad \underbrace{\frac{1}{2}\begin{pmatrix} 1 \\ \cos\delta \\ 0 \\ \sin\delta \end{pmatrix}}_{\text{After }M} \quad \underbrace{\frac{1}{2}\begin{pmatrix} 1 \\ 1 \\ 0 \\ 0 \end{pmatrix}}_{\text{After }P}$$

The detector response is proportional to the first Stokes parameter

$$\tfrac{1}{2}C(S_{11} + S_{12}\cos\delta + S_{14}\sin\delta),$$

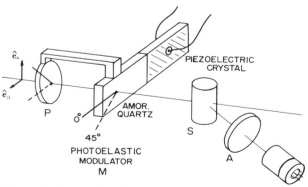

Figure 13.12   Schematic diagram of a photoelastic modulator mated to a polar nephelometer.

where C is a constant that incorporates such factors as light-collecting efficiency and detector sensitivity. The signal is time dependent because of the variable retardance

$$\delta = \frac{2\pi d}{\lambda} s \sin \omega t = A \sin \omega t,$$

where d is the thickness of the modulator, λ is the wavelength, ω is the frequency of the electric field driving the modulator, and s is proportional to the stress-optical constant and the amplitude of the induced stress (Born and Wolf, 1965, p. 705). The trigonometric functions of retardance can be expanded in series of Bessel functions

$$\sin(A \sin \omega t) = 2J_1(A)\sin \omega t + 2J_3(A)\sin 3\omega t + \cdots$$

$$\cos(A \sin \omega t) = J_0(A) + 2J_2(A)\cos 2\omega t + \cdots$$

If the modulator amplitude is adjusted so that $J_0(A) = 0$, the detector signal is

$$\tfrac{1}{2}C\{S_{11} + 2S_{12}J_2(A)\cos 2\omega t + 2S_{14}J_1(A)\sin \omega t + \cdots\}$$

$$= \frac{1}{2}CS_{11}\left\{1 + \frac{S_{12}}{S_{11}}2J_2(A)\cos 2\omega t + \frac{S_{14}}{S_{11}}2J_1(A)\sin \omega t + \cdots\right\}.$$

Factoring out of S_{11} is accomplished in practice by electronically servoing the dc signal to a constant value which, in effect, normalizes the terms by S_{11}. The signal therefore has components proportional to S_{14}/S_{11} and S_{12}/S_{11}, which oscillate at frequencies ω and 2ω, respectively. These two components are separated by phase-sensitive detection, that is, by "locking in" on either ω or 2ω. The instrument is calibrated by replacing the sample with a linear polarizer or a quarter-wave plate and adjusting the servo gain for 100% signal in the forward direction.

A great advantage of this type of system is that it makes possible measurements of relatively small matrix elements such as S_{14}: the signal can be amplified as necessary because only one component is proportional to S_{14}. Normalization by S_{11}, which is done electronically, eliminates fluctuations in particle number density and in the light source. If S_{11} is desired, the electronic servo can be turned off and the instrument used in the conventional way.

An alternative configuration is the polarizer–modulator placed before the detector rather than the sample. This might be useful for measurements in the atmosphere; for example, a searchlight as the light source with the modulator at or near the focus of a portable telescope followed by a detector. For laboratory investigations it is also sometimes convenient to place the modulator before the detector.

A disadvantage of the system described in the preceding paragraphs is that, in general, the time-harmonic components of the signal contain mixtures of matrix elements. With the modulator either forward or aft of the sample, only matrix elements in the first row and column can be measured individually; mixing occurs for the other nine elements. For example, an attempt to measure S_{34}/S_{11} yields $(S_{34} + S_{14})/(S_{11} + S_{31})$. This is not always a problem because S_{14} and S_{31} are zero for many collections of particles (see Section 13.6). If S_{14} and S_{31} are not zero they can, of course, be measured individually and used to extract S_{34}/S_{11} from measurements; but to a certain extent this nullifies the advantages of the modulation system.

Mixing of matrix elements can be avoided with more complicated systems having one or more modulators, driven at different frequencies, in both the entrance and exit beams. Azzam (1978) considered the general planning of such nephelometers for measuring all 16 elements simultaneously, which is possible only if both the entrance and exit beams are modulated. Thompson et al. (1980) constructed a system for measuring all 16 elements simultaneously. Four Pockels-cell modulators of different frequencies are used, two each in the entrance and exit beams. The resulting signal is a sum of time-harmonic terms where the amplitude of each term is proportional to a single matrix element.

13.8 SOME RESULTS FOR THE SCATTERING MATRIX

Few measurements or calculations of all 16 scattering matrix elements have been reported. There are only four nonzero independent elements for spherical particles and six for a collection of randomly oriented particles with mirror symmetry (Section 13.6). It is sometimes worth the effort, however, to determine if the expected equalities and zeros really occur. If they do not, this may signal interesting properties such as deviations from sphericity, unexpected asymmetry, or partial alignment; some examples are given in this section. But we begin with spherical particles.

13.8.1 Polystyrene Spheres and Water Droplets

Measurements of four matrix elements made with the polarization modulation instrument described in the preceding section are shown in Fig. 13.13. The curves on the left are for an aqueous suspension of polystyrene spheres narrowly distributed in size; on the right are the same matrix elements for a broad size distribution of water droplets made with an ultrasonic nebulizer (Hunt and Huffman, 1975). Many peaks and valleys appear in all the matrix elements for the polystyrene spheres; the curves in Figs. 13.1 and 13.2 demonstrated that such structure is very sensitive to size and refractive index and thus can be used to determine these quantities either for single spheres or narrow size distributions. Structure is largely absent from the water droplet curves, however, and only general features remain. Around 150°, for example, there is broad structure in all four matrix elements; this is the onset of the fog bow, or cloud bow, a feature that develops into the rainbow for larger droplets (Fig. 13.4). Note that the S_{11} signal is noisy because the number of particles in the scattering volume fluctuates; this noise is removed electronically from the other matrix elements by normalization. Because of normalization the number of droplets in the beam is not important provided that multiple scattering is negligible.

Quiney and Carswell (1972) made measurements on artificial fogs and reported that S_{33}/S_{11} and S_{34}/S_{11} showed more pronounced differences from one fog to another than did the more commonly measured matrix elements such as S_{12}/S_{11}. Hunt and Huffman (1975) suggested the possibility of using S_{34}/S_{11} at a single angle near 95° to monitor the mean size of nebulized water droplets. Because little use has been made of all matrix elements, however, a systematic study of their relative merits in determining size distributions has not been made.

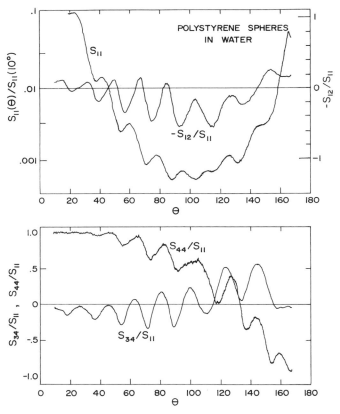

Figure 13.13 Measured matrix elements (left) for polystyrene spheres in water (mean radius 0.40 μm, wavelength 0.3250 μm). Measured matrix elements (right) for water droplets (mean radius 1.5 μm, wavelength 0.6328 μm). From Hunt and Huffman (1975).

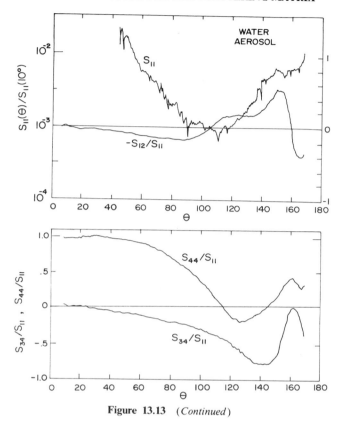

Figure 13.13   (*Continued*)

Sekera (1957) and Rozenberg (1960) emphasized the importance of measuring all matrix elements for atmospheric aerosols, and a few such measurements have been reported (Pritchard and Elliot, 1960; Beardsley, 1968; Golovanev et al., 1971). With sensitive modulation techniques it should indeed be possible to probe atmospheric particles remotely using the complete scattering matrix to infer not only size distributions but also refractive indices. Care must be exercised, however, because nonsphericity can lead to false inferences about absorption: analysis based on Mie theory cannot disentangle the two effects.

13.8.2   Nonspherical Particles

One of the few sets of measurements of all scattering matrix elements for nonspherical particles was made by Holland and Gagne (1970), who used various combinations of polarizers and retarders (see Section 13.7). They studied quartz (sand) particles with a fairly broad range of sizes. To investigate further the effects of nonsphericity on all matrix elements Perry et al. (1978),

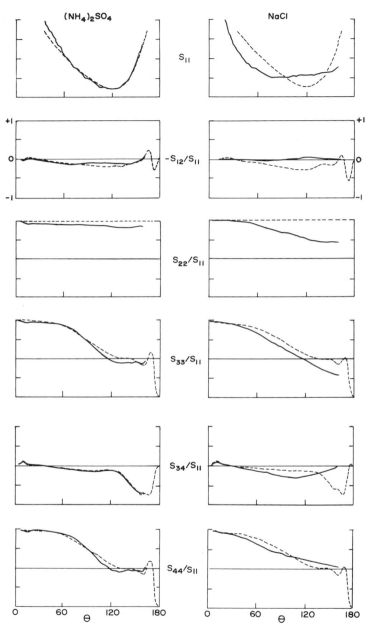

Figure 13.14 The six independent matrix elements for two aerosols. Solid lines are measurements from Perry et al. (1978); dashed lines are theoretical (we redid their calculations).

using the polarization modulation technique, studied two aerosols, both with a mean radius of about 0.63 μm and a fractional standard deviation of about 0.3, but differing in shape: rounded (but not perfectly spherical) ammonium sulfate particles and nearly cubic sodium chloride particles; they were made by nebulizing solutions of these salts and drying the droplets in a nitrogen stream. Eight matrix elements were determined to be zero within experimental accuracy, as expected; the remaining elements, for incident light of wavelength 0.3250 μm, are shown in Fig. 13.14. Because the size distributions and refractive indices of the two aerosols are similar, comparison of their matrix elements clearly shows the effect of shape.

Although S_{11} calculated from Mie theory has been adjusted arbitrarily relative to the measurements, it is obvious that cubic particles are poorly described by theory for spheres; this is in contrast with the data for the rounded particles, which agree well with calculations. S_{11} for the cubes is nearly constant at scattering angles greater than about 90° and does not rise sharply in the backscattering direction as it does for the rounded particles; this was pointed out in connection with Fig. 13.8. Positive polarization ($-S_{12}/S_{11}$) and appreciable deviations of S_{22}/S_{11} from unity for cubes were also discussed in Section 13.4. Note the small deviations of S_{22}/S_{11} from unity for the rounded particles. S_{22}/S_{11}, which is essentially the cross polarization, is one for spheres at all scattering angles; therefore, a single measurement of this quantity at some large angle could provide a simple and sensitive test of nonsphericity.

In the three matrix elements shown in the bottom half of Fig. 13.14, there appear to be less pronounced differences between spheres and nonspherical particles. Perry et al. pointed out that near the forward direction S_{34}/S_{11} is rather sensitive to the parameters of the size distribution, while measurements for both kinds of particles—rounded and cubic—agree quite well with calculations. Similar agreement between calculations for spheres and spheroids was noted by Asano and Sato (1980). This combination of sensitivity to size distribution and insensitivity to shape might be put to good use in particle sizing.

13.8.3 Clusters of Spheres

Shape effects caused by aggregation played an important part in our discussion of surface modes in Chapter 12. When two or more identical spheres aggregate into a cluster the resulting composite particle is nonspherical. As a consequence it must be expected to scatter light differently from its constituent spheres. In an experimental effort that reveals much about light scattering by nonspherical particles, Bottiger et al. (1980) measured all scattering matrix elements for single polystyrene spheres and clusters of two, three, and four similar spheres. A sphere or cluster of spheres was suspended in an electrostatic levitation chamber and all matrix elements were measured simultaneously using the instrument mentioned at the end of Section 13.7. Measured zeros for the eight elements off the block diagonal $(S_{13}, \ldots, S_{42})$ confirmed that the particles assumed all orientations during a measurement.

A series of measurements for single spheres and clusters is shown in Fig. 13.15. Large oscillations in S_{12}/S_{11} for a single sphere diminish in amplitude as the cluster size increases, although remnants of the single-particle signal are still evident. Other normalized matrix elements (S_{33}, S_{44}, S_{43}) show a similar trend toward subdued structure as the cluster size increases. The normalized S_{22} matrix element for the clusters decreases with increasing scattering angle, and the rate of decrease increases with increasing cluster size; this further confirms that deviation of S_{22}/S_{11} from unity is a clear indication of nonspher-

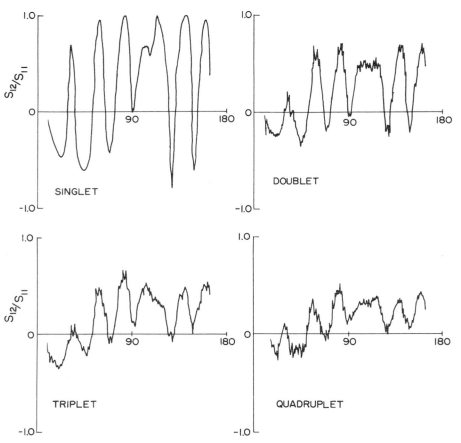

Figure 13.15 Angular dependence of S_{12}/S_{11} measured for a single polystyrene sphere of radius 1091 nm and for clusters of two, three, and four similar spheres; the wavelength of the incident light is 441.6 nm. From Bottiger et al. (1980).

icity. S_{33} and S_{44} are identical for spheres but show noticeable differences for the larger clusters.

13.8.4   Quartz Fibers

Bell (1981) (see also Bell and Bickel, 1981) measured all matrix elements for fused quartz fibers of a few micrometers in diameter with a photoelastic polarization modulator similar to that of Hunt and Huffman (1973); the HeCd (441.6 nm) laser beam was normal to the fiber axes. Advantages of fibers as single-particle scattering samples are their orientation is readily fixed and they can easily be manipulated and stored. Two of the four elements for a 0.96-μm-radius fiber are shown in Fig. 13.16; dots represent measurements and solid lines were calculated using an earlier version of the computer program in Appendix C. Bell was able to determine the fiber radius to within a few tenths of a percent by varying the radius in calculations, assuming a refractive index of $1.446 + i0.0$, until an overall best fit to the measured matrix elements was obtained.

13.8.5   Biological Particles

Angular light scattering is a widely used nondestructive technique for studying particles of biological origin such as viruses, bacteria, and eucaryotic cells. Measurements of all scattering matrix elements for such particles, however, are rare. An exception is the work of Bickel et al. (1976) and Bickel and Stafford (1980), who measured all matrix elements for a variety of biological particles. A striking result of this work is that S_{34}/S_{11} proved to be uniquely characteristic of each biological scatterer. Reproducible differences in S_{34}/S_{11} were found for particles that could not readily be distinguished by other common techniques.

Differences in the S_{34}/S_{11} signals for two mutant varieties of bacterial spores differing in a specific structural mutation are obvious in Fig. 13.17. Other matrix elements, however, are less obviously different for the two similar scatterers. To first approximation scattering by biological particles tends to be described well by Rayleigh–Gans theory (Chapter 6) for which $S_{34} = 0$. Within the framework of this theory elements off the block diagonal are also zero, as they are for larger and more refractive arbitrary particles provided that their scattering matrix has the symmetry (13.21). Thus, S_{34} is the matrix element that is likely to undergo the greatest relative change as a particle's characteristics deviate more and more from those for which the Rayleigh–Gans theory is valid. This may be the reason S_{34}/S_{11} is so sensitive to characteristics of biological scatterers.

13.8.6   Ocean Waters

All 16 matrix elements for natural ocean waters have been measured using combinations of polarizing and retarding elements (Kadyshevich et al., 1971,

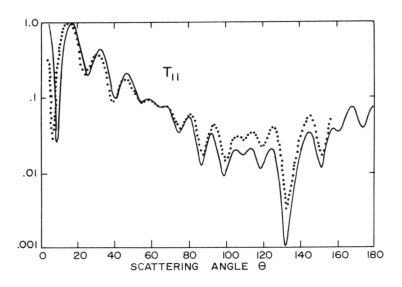

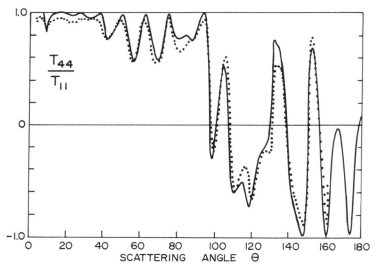

Figure 13.16 Two of the four matrix elements for a fused quartz fiber of radius 0.96 μm illuminated by light of wavelength 0.4416 μm. The solid curves are calculations, the dots are measurements. From Bell (1981).

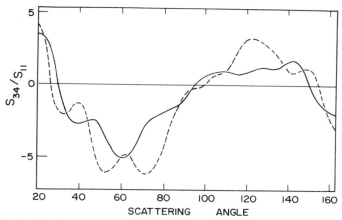

Figure 13.17 Angular dependence of S_{34}/S_{11} (percent) for two similar bacterial spores. From Bickel et al. (1976).

1976). A surprising result of this work is that the scattering matrix, particularly for Pacific Ocean waters, has none of the symmetries discussed in Section 13.6. Only the normalized elements S_{13}/S_{11} and S_{43}/S_{11} were determined to be zero within experimental accuracy; all others had appreciable values. Further, there were substantial differences among matrix elements and in matrix symmetry for different ocean waters (Atlantic and Pacific) and for Baltic Sea waters. The unusual symmetry of the scattering matrix seems to demand partial alignment of asymmetric particles by, for example, gravitational or magnetic fields.

Measurements by Thompson et al. (1978) on cultured populations of phytoplankton, such as might be expected to scatter light in ocean waters, revealed only a scattering matrix of the form (13.5), characteristic of particles for which either the Rayleigh (Chapter 5) or Rayleigh–Gans (Chapter 6) approximations are valid.

So we end this section with unanswered questions: Do the unusual scattering matrices measured by the Russian investigators really exist in nature, or are they merely experimental artifacts? If the former, how is this reconciled with the measurements of Thompson et al.?

13.9 SUMMARY: APPLICABILITY OF MIE THEORY

At the present time the electromagnetic scattering theory for a sphere, which we have called Mie theory, provides the only practical method for calculating light-scattering properties of finite particles of arbitrary size and refractive index. Clearly, however, many particles of interest are not spheres. It is therefore of considerable importance to know the extent to which Mie theory is applicable to nonspherical particles. To determine this requires generalizing from a large amount of experimental data and calculations. We summarize below the similarities and differences between scattering by spherical and nonspherical particles, based on the work cited in this chapter, in the hope that it will provide guidelines to the judicious use of Mie theory.

1. Scattering by single, or collections of oriented, nonspherical particles may, unlike scattering by spheres, be azimuthally dependent.

2. Large nonspherical particles scatter similarly to area-equivalent spheres near the forward direction.

3. Rainbow angles are not evident in matrix elements for nonspherical particles.

4. The scattering diagram (incident unpolarized light) for nonspherical particles tends to be flatter than that for spheres at angles greater then about 90°.

5. Scattering matrix elements off the block diagonal are zero for mirror-symmetric collections of randomly oriented particles, as they are for spheres.

6. S_{21}/S_{11} for nonspherical particles tends to be opposite in sign to that for spherical particles.

7. The inequalities $S_{22} \leqslant S_{11}$ and $S_{33} \neq S_{44}$ hold for nonspherical particles. The first inequality implies that the cross polarization does not necessarily vanish.

8. There is some evidence that S_{34}/S_{11} for spherical and nonspherical particles are in better agreement than other normalized matrix elements.

If there is one succinct conclusion to be made it is that nonspherical particles and area-equivalent spheres scatter similarly near the forward direction, but differences between the two tend to increase with increasing scattering angle.

NOTES AND COMMENTS

A good discussion of measuring angular scattering by particles is given by Stacey (1956). Techniques of particle production, as well as a wealth of information about the physical properties of particles, are discussed by Green and Lane (1964) in their superb book. Another good source of information about particles is *The Particle Atlas* (McCrone and Delly, 1973abcd; McCrone et al., 1979; McCrone et al., 1980).

For various reasons the microwave analog technique has seen only intermittent use in the United States since its inception over 20 years ago. But recently there has been an upsurge of activity—a hopeful sign. Some of the most recent measurements have been reported by Schuerman et al. (1981).

Additional measurements of the S_{34} matrix element for biological particles (red blood cells) have been reported by Kilkson et al. (1979).

Chapter 14

A Miscellany
of Applications

Applications are not totally absent from the preceding chapters; they appear now and then either to provide some relief from the mathematical development or to illustrate particular points. Some of them, and the sections where they are discussed, are listed below.

Red sunsets and blue moons (4.4).

Polarization of skylight (5.7).

Rainbow (7.2, 13.1).

Haloes (7.3).

Scattering by fibers (8.4).

Christiansen effect (11.2).

Sphere in a waveguide (11.4).

Levitation by radiation pressure (11.4).

Asymmetry of interstellar absorption bands (11.5).

Colors of colloidal gold (12.4).

Electron–hole droplets in germanium (12.4).

Particle sizing (13.5).

Biological particles (13.8).

Ocean waters (13.8).

This chapter merely contains a greater concentration of applications.

A separate volume could be devoted solely to applications. Indeed, it would be a volume almost without end because the rate at which we can evaluate and write about them is less than the rate at which new ones are being devised or old ones are being refined. This chapter is of necessity limited in scope. We have been selective, concentrating our efforts on those topics most familiar or of greatest interest to us. No attempt has been made to be comprehensive or to achieve a balanced treatment. We have, however, favored topics that illustrate points stressed particularly heavily in preceding chapters.

We begin with a discussion of optical constants, which are needed in almost all applications. A treatment of atmospheric aerosols is followed by two other topics from atmospheric science: inferences about noctilucent clouds—clouds at the edge of space—made on the basis of polarization measurements; and measuring rainfall with radar. A discussion of interstellar dust—particles in the "cosmic laboratory"—is followed by an earthbound laboratory investigation in which the advantages of small particles are exploited in high-pressure studies of optical spectra. The final topics have a biological flavor: the Giaever immunological slide, an application of surface modes in small metallic particles; and the effects of microwave radiation on macromolecules.

14.1 THE PROBLEM OF OPTICAL CONSTANTS

We have emphasized that the wavelength-dependent optical constants are the fundamental quantities that determine the macroscopic optical properties of matter. Because of this, optical constants are required in many applications of absorption and scattering of light by small particles: they are needed for determining optical properties of smog particles and mineral grains in the atmosphere, dust grains in interstellar space, phytoplankton in the ocean, and biological cells. It is little wonder then that scientists in many disciplines spend so much time searching through the physics, chemistry, and mineralogy journals (to name a few sources) for the optical constants they need.

14.1.1 Homogeneous Sample

Determination of optical constants is not necessarily an easy task, even for homogeneous solids and liquids at room temperature, particularly in spectral regions of very high or very low absorption. Optical constants are not directly measurable but must be inferred, sometimes rather circuitously, by analyzing primary measurements (e.g., transmission and reflection) with theoretical expressions (e.g., the Fresnel formulas). A brief survey of methods for determining optical constants of bulk matter has been given in Section 2.9. These techniques require homogeneous samples of a size and shape such that the conditions underlying the validity of the theory used for analyzing measurements are satisfied; examples are large, single crystals, slabs of glassy or amorphous solids, and liquid-filled cuvettes. Unfortunately, many solids are commonly obtainable only as small particles (i.e., powders). It is much more difficult to determine optical constants for these materials even if the powder sample contains only a single component. Lack of accurate optical constants for powdered materials arises not so much from want of effort as from the complexity of data analysis.

14.1.2 Powder Samples

To determine optical constants of powder samples requires one or more measurements of transmission, or diffuse reflection, or scattering, and an appropriate theoretical model, such as a theory of diffuse reflection or Mie theory. In principle these measurements are not difficult to make. But the appropriateness of the theoretical models used to analyze these measurements is sometimes doubtful. For example, if one desired infrared ($\sim 10 \ \mu$m) optical constants of quartz, one could measure transmission by a dilute suspension of fine quartz powder in a transparent KBr matrix and reflection from a pressed sample (e.g., Volz, 1972). Analysis of the transmission measurements might be based on Mie theory or one of its approximations, and the reflectance data might be analyzed with the Fresnel formulas. Neither of these two theories, however, is applicable. Quartz particles are highly irregular and their extinction spectrum is quite different from that of spheres (see Fig. 12.14). And the Fresnel formulas are not applicable because the quartz powder cannot be compressed into a homogeneous solid with a smooth surface. Even with accurate measurements, therefore, these two theories would quite likely yield highly erroneous optical constants. Pressed samples of softer materials, such as potassium bromide, may, however, approach single-crystal density and have rather smooth surfaces. For such materials these methods have been used with some success (Tomaselli et al., 1981). But for harder materials (and anisotropic solids) serious problems remain.

Optical constants can be inferred from analysis of measurements on single spheres or highly monodisperse collections of spheres. For example, angular scattering by colloidal suspensions of polystyrene spheres has been used frequently to determine both their size and refractive index at wavelengths where polystyrene is weakly absorbing. Measurements on single, homogeneous, perfectly spherical particles can be analyzed with Mie theory as confidently as measurements of reflectances at smooth plane interfaces can be analyzed with the Fresnel formulas. Pluchino et al. (1980), for example, determined the complex refractive index of carbon at a single wavelength in the visible by levitating a carbon sphere a few micrometers in diameter, measuring the angular light scattering, and fitting their data with Mie theory. A similar technique has been used by Wyatt (1980), although for inhomogeneous particles. As techniques for levitating single particles and measuring scattering patterns come into wider use, obtaining optical constants from measurements on single particles will surely become more common. In general, however, this is practically limited to particles of simple shape such as spheres or long, circular cylinders.

14.1.3 Optical Constants from Absorption Measurements

An approach widely used by atmospheric scientists is to infer the imaginary part of the refractive index k from measurements of the absorption coefficient α of particulate samples. Diffuse reflection, the photoacoustic effect, and integrating plates have been used for determining absorption even in the presence of considerable scattering; these methods are discussed briefly in the following section. The relation (2.52) between α and k, $\alpha = 4\pi k/\lambda$, is, of course, strictly valid only for homogeneous media. But under some circumstances its use for inhomogeneous media is justified. For example, it follows from (3.47), (3.48), and (12.9) that transmission by a dilute suspension of spheres small compared with the wavelength is approximately

$$\frac{I}{I_0} = \exp\left[-\frac{4\pi k}{\lambda} \frac{9n}{(n^2 + 2)^2} d \right] \qquad (14.1)$$

provided that extinction is dominated by absorption, k is small, and n is not too large; $d = fh$, where f is the volume fraction of particles and h is the sample thickness. The expression containing n is about 1 at visible wavelengths for many common substances. For example, it is about 0.75 for $n = 1.5$, and can be made even closer to 1 by suitably choosing the medium in which the particles are suspended: n is the *relative* refractive index. Measurement of transmission and calculation of k from (14.1) is therefore approximately correct for dilute suspensions of Rayleigh spheres of many common materials with small k—hence the popularity of this procedure. But for several reasons, some of which have been discussed by Toon et al. (1976), (14.1) may be inapplicable:

1. Scattering may not be negligible, giving rise to apparent additional absorption.

2. The particles may not be small enough for the Rayleigh theory to be valid. For larger particles, the relation between absorption and size is more complicated.

3. The particles may not be spherical or, if spherical, they may agglomerate into nonspherical clumps. Contrary to much common opinion, absorption by Rayleigh particles can be quite shape dependent, as we have shown in Chapter 12.

4. The optical constants n and k are not independent: if k varies strongly, so must n. Either n must be measured by some other method or a theory of optical constants that couples them together properly must be used, such as the oscillator model (9.25) or the Kramers–Kronig relations [(2.49), (2.50)].

While inferring k from measurements of absorption by particulate samples may be valid for many kinds of solids at visible wavelengths, it may not be valid in spectral regions where the optical constants rapidly vary, such as the infrared or far ultraviolet.

14.1.4 Powders of Anisotropic Solids

More problems must be faced when trying to extract optical constants from measurements on particles of anisotropic solids. Random orientation of the particles averages somehow the two or three sets of optical constants. We showed in Section 5.6 that in the Rayleigh approximation the average extinction cross section of a randomly oriented anisotropic sphere is just the equally weighted sum of three cross sections, one for an isotropic sphere with each set of optical constants. But this prescription has not been shown to be correct for larger particles (see Section 8.2). Even if it were correct in general, however, it is still not clear how to extract from measurements absorption associated with each of the principal axes of the crystal. For example, the infrared extinction spectrum of randomly oriented anisotropic particles such as calcite will show various bands. But without knowing which band is associated with which principal axis, it does not appear possible to extract the different sets of optical constants; this problem was recognized and discussed by Aronson and Emslie (1980). To ignore anisotropy and simply determine "average" optical constants as though the solid were isotropic is only sweeping the problem under the rug. These optical constants could not necessarily be used to analyze experiments other than those from which they were derived. For example, they could not be used to calculate the reflectance of an oriented, single crystal of the material. Even the use of average optical constants to calculate extinction by randomly oriented particles of the same kind would give errors if the size distribution were appreciably different from that of the sample from which the optical constants were derived.

14.1.5 Mixtures of Particles

Up to this point we have had in mind only single-component particulate samples. But multicomponent mixtures are common, and because of the difficulty of properly treating each component the tendency has been to infer "average" or "effective" optical constants from measurements on such mixtures. Yet this is fraught with difficulties. The measurements from which these optical constants are obtained, such as reflection or scattering or absorption, are not additive. Nor are optical constants additive. Hence, optical constants of mixtures determined by, for example, measuring absorption by particles in a matrix and reflection by a pressed surface cannot necessarily be used to predict correctly angular scattering by the particles. Despite these difficulties and the warnings issued by some, the deceptively simple procedure of determining a single set of optical constants for complex multicomponent mixtures of possibly anisotropic particles and then using such optical constants in all kinds of predictive calculations has been widespread. Optical constants of moon rocks consisting of many individual mineral components have often been used by astronomers in calculations relating to interplanetary or interstellar dust. Optical constants obtained for atmospheric dust have been used to predict possible climatic change; this is discussed further in the following section.

To conclude, we venture to state that the problem of how to determine accurate optical constants from measurements on particulate samples, in contrast with homogeneous solids and liquids, has not been solved in general, even for single-component powders. This does not mean that there have not been successful determinations of optical constants for some particulate materials. It is simply that extracting optical constants from measurements on such materials is not generally possible in all spectral regions and for all kinds of samples, which may be multicomponent mixtures of nonspherical and anisotropic particles.

14.2   ATMOSPHERIC AEROSOLS

Atmospheric aerosols usually means the solid and liquid particles in the earth's atmosphere, excluding the solid and liquid water particles in clouds, fog, and rain. Although very tenuous and highly variable, they act as condensation nuclei for cloud droplets, alter the optical properties of clouds, and possibly play a role in the formation of smog and acid rain. And an understanding of their optical properties is needed for many applications:

Climate may be modified by both natural and human-made aerosols. Optical techniques are being used increasingly for remote sensing of the atmosphere as well as for communication. The military importance of aerosols has increased greatly because of surveillance, using both visible and infrared radiation, from satellites and because of missile guidance with light beams. This in turn has stimulated development of countermeasures—shielding against enemy surveillance and laser-guided missiles—and has renewed interest in an old trick—throwing out smoke screens.

From among all the possible applications of light scattering by atmospheric aerosols we have selected only a few. We begin with the intriguing question of what effect these particles may have on the earth's climate. Then we discuss some of what is known about their chemical composition and optical properties and how these are obtained by various methods: *in situ* measurements; collecting particles; and remotely sensing them. This is followed by a discussion of the problem of characterizing the complex mixture of particles that is the atmospheric aerosol. Finally, we briefly discuss the possibility of monitoring the global wind by measuring the Doppler shift of infrared radiation backscattered by atmospheric particles.

14.2.1   Effect of Aerosols on Climate

A controversy with possibly far-reaching consequences concerns the impact of atmospheric particles on the earth's climate. Temperature is one of the most easily monitored indicators of climatic change. Among the many discussions of the effects of aerosols on the global mean temperature, we direct the reader to the monograph by Twomey (1977) and the paper by Toon and Pollack (1980), from which some of the following is taken.

Typical changes in the global mean temperature over the past 1000 years seem to have been of order 1°C, although climate has changed considerably as in the "Little Ice Age" from about A.D. 1500 to 1900. The difference between average Ice Age (1.75 million to 10,000 years ago) temperatures and the present-day average is thought to be about 5°C. More recently, during the late Stone Age, temperatures were perhaps 2.5°C higher than at present. Since human beings can adapt readily to temperature changes of tens of degrees such small excursions might at first glance appear inconsequential. Ocean levels, however, may have lowered about 100 m during the Ice Age because of storage of vast quantities of water in continental ice sheets. Coastal residents can surely appreciate that sea level changes of this magnitude could be catastrophic. Global mean temperature changes of even a few tenths of a degree may thus be of consequence. Particulate pollution is increasing, and it is therefore prudent to understand how such particles may affect the heat balance of our planet. Given the present state of knowledge, however, it is not certain if man's impact will be to increase or to decrease the global mean temperature.

The simplest view is that absorption by particles tends to heat the atmosphere while scattering into the backward hemisphere tends to cool it; thus, the two most important optical properties of aerosols as far as climatic change is concerned are absorption and backscattering. This simple view is complicated somewhat by the fact that aerosols exist primarily in two distinct atmospheric layers, the troposphere (from the ground up to about 10–20 km) and the stratosphere (between about 10–20 and 50 km). The troposphere is thermally coupled to the earth's surface rather strongly, whereas the stratosphere is not. Because of their thermal isolation, therefore, stratospheric particles tend to cool the earth's surface both by backscattering and absorbing solar radiation. Of less importance, but still appreciable, are thermal infrared effects of particles, particularly in the 8- to 12-μm "window" region, where the major atmospheric gases are highly transparent. It is in this wavelength region that the Planck function peaks for normal terrestrial temperatures. Upwelling infrared radiation from the earth's surface is therefore partly hindered from escaping into space because some particles have strong absorption bands in this spectral region. Thus, aerosols contribute to the so-called "greenhouse effect." Infrared radiation emitted toward the earth's surface by stratospheric particles may cause some warming in the lower atmosphere, thereby countering to some extent the cooling they cause by backscattering and absorbing solar radiation.

Various models have been proposed to predict the effect of atmospheric aerosols on the global mean temperature. The single-scattering albedo (i.e., the ratio of scattering to extinction) of such particles is a crucial parameter in these models because a high value will result in cooling, whereas a low value will result in warming; critical values—the boundary between high and low—from about 0.7 to 0.95 can be found in the literature. A value of 0.85 derived by Hansen et al. (1979) is being widely used. Charlock and Sellers (1980) arrived at a critical single-scattering albedo of 0.81.

Because the single-scattering albedo depends sensitively on the imaginary part of the refractive index there has been keen interest in determining optical constants of atmospheric particles. These are used to calculate the important parameters in the heat balance problem for present and predicted aerosol loadings. A path commonly followed is to measure optical properties of atmospheric aerosols, either remotely (e.g., solar extinction and bistatic lidar scattering) or on collected samples (e.g., absorption and diffuse reflection), and infer optical constants from theoretical models. The optical constants so obtained are then used as input to Mie theory incorporated into heat balance calculations. But this path is strewn with many potential hazards, which may invalidate the final results. We shall discuss some of these hazards in following paragraphs. Directly measuring absorption and scattering by aerosols *in situ* is less fraught with pitfalls, but we delay our discussion of this until we have surveyed the chemical composition and absorptive properties of aerosol particles; this survey, taken from laboratory measurements on homogeneous samples, serves to give a feeling for expected values of k.

14.2.2 Absorption by Constituents of the Atmospheric Aerosol

Among the reasons it is difficult to predict the effect of particles on climate is that their composition and distribution are poorly known. The main types of known atmospheric particles are listed in Table 14.1.

The dominant aerosol in the lower atmosphere is windblown dust composed mostly of mineral particles together with some organic matter. Although they

Table 14.1   Constituents of the Atmospheric Aerosol

Constituent	Size	Comments

Windblown Surface Dust
 Quartz
 Calcite
 Oxides of iron
 Clay minerals
   Montmorillonite
   Illite
	$1-10 \, \mu m$	~ 30% globally

Sea Spray Salt Organic particles	$1-10 \, \mu m$	10–15% globally
Sulfur Compounds Ammonium sulfate Sulfuric acid	~ 0.1 μm	Mostly stratospheric; 50% globally
Volcanic Ash		Stratospheric; following volcanic activity
Others Anthropogenic and natural carbon Organic materials from vegetation Photochemical smog		

are fairly large, with consequently high settling rates, under some conditions these dust particles can spread far from their source and remain in the atmosphere for long periods. A notable example is dust from the Sahara, which spreads across much of the Atlantic Ocean between Africa and the Caribbean during summer months (Carlson and Benjamin, 1980). Dominating the marine aerosol are salt particles, together with some organic particles of marine origin; Blanchard (1967) has given a fascinating account of how these salt particles enter the atmosphere. The globally dominant mass of particles, found distributed over the entire earth, is composed of sulfur compounds, including ammonium sulfate and sulfuric acid. These particles are formed mostly in the stratosphere by complex chemical reactions. In periods of strong volcanic activity, volcanic ash may be appreciable, remaining in the atmosphere for several years after its introduction. Minor constituents of the atmospheric aerosol include terrestrial organic substances such as terpenes, and pollution such as carbon, metal oxides, and photochemical smog.

The many difficulties inherent in determining optical constants in regions of high absorption, especially for particulate samples, and the practical impossibility of obtaining accurate optical constants for complex mixtures of particles have been discussed in Section 14.1. These difficulties point to why it is worth considering measured optical constants of homogeneous materials that are known constituents of the atmospheric aerosol. Such measurements are not available for all constituents, in some instances because they are not obtainable in homogeneous bulk form. In Fig. 14.1 we show the imaginary part of the refractive index of several solids and liquids that are found as atmospheric particles. Results are given for water, ammonium sulfate, crystalline quartz, sulfuric acid, carbon, sodium chloride, and hematite (α-Fe_2O_3); to avoid clutter, only k values over limited spectral regions are given for most of these materials.

Except for carbon, there are two distinct spectral regions in which k is high (~ 1) for these materials: one in the infrared and the other in the ultraviolet, with a region of high transparency between. The reasons for this have been discussed in Section 10.4. To emphasize just how transparent these materials are in the intermediate region, we show with a dashed line the value of k for which a 1-cm-thick homogeneous sample would give 1% transmission, neglecting reflection losses at the boundaries. Only carbon, which has metal-like overlapping electronic energy bands (see Section 9.4), has consistently high k of order unity throughout the visible and infrared regions. The mineral hematite, although a very minor constituent of the atmospheric aerosol, is included because it is one of the very few highly absorbing (at visible wavelengths) substances known to be in the atmosphere.

The hatched region in Fig. 14.1 shows the approximate values of k at visible wavelengths inferred by various remote-sensing techniques (Grams et al., 1974; Reagan et al., 1980; and references cited therein). If we compare these values of k with those for individual constituents of the atmospheric aerosol, it seems

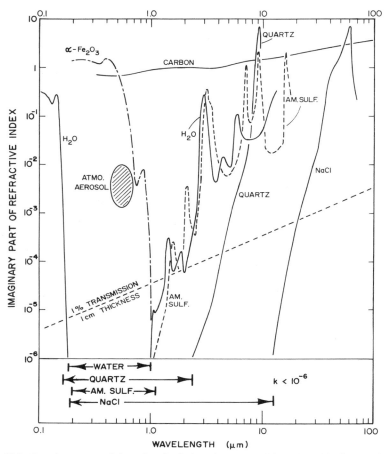

Figure 14.1 Imaginary part of the refractive index of several solids and liquids that are found as atmospheric particles.

clear that k determined remotely is some kind of average for a mixture containing a small amount of a strongly absorbing component such as carbon and much larger amounts of weakly absorbing components. Lindberg (1975) pointed this out after successfully matching measured spectral extinction by collected aerosol particles with calculations for mixtures of weakly absorbing minerals and about 0.5% carbon to give the required absorption. Thus the value $k = 0.001$ (for example) inferred from remote sensing should perhaps be looked upon as an average or effective k, but the averaging process and the subsequent uses to which this effective value is put must be examined critically. We shall return to this point later.

The possibility that carbon in small quantities is the dominant absorber in the atmospheric aerosol suggests looking for spectral features in carbon, which would provide a diagnostic test for this solid. Unfortunately, the absorption spectrum of bulk carbon is rather featureless in the infrared and visible. The one spectral region where carbon, in small-particle form at least, has a prominent absorption feature is between 2200 and 2500 Å. Although this feature is not apparent in Fig. 14.1, absorption by small spheres of both graphite and glassy carbon rises to a pronounced peak where the real parts of their dielectric function are closest to -2. This feature in graphite particles, the surface plasmon absorption discussed at length in Section 12.1, is thought to be responsible for the dominant peak in the interstellar extinction spectrum (Fig. 14.4). Although it would be difficult to observe by remote sensing, a feature between 2200 and 2500 Å might be very useful in laboratory analysis for carbon particles in aerosol samples collected on filters, for example.

Because of the importance of the 8- to 12-μm atmospheric "window" region to the earth's heat balance, and in view of the many potential applications of CO_2 laser beams (9–11 μm) propagating in the atmosphere, we call attention to k in this region of the infrared. Note that several constituents of the atmospheric aerosol, including quartz and ammonium sulfate, have intense absorption bands associated with lattice vibrations. Other silicate minerals, including the clay minerals listed in Table 14.1, also have strong bands near 10 μm. These vibrational absorption bands are strong enough to give rise to the kind of highly shape-dependent absorption and scattering discussed in Chapter 12. Indeed, Fig. 12.14 displayed measurements on crystalline quartz dust to illustrate the extreme sensitivity of these bands to particle shape. The implications of these absorption bands in quartz and ammonium sulfate particles for remote sensing are discussed later.

14.2.3 In Situ Measurements of Absorption and Backscattering

Predicting optical properties of atmospheric aerosols from calculations for homogeneous, spherical particles leaves much to be desired. Mie theory may be a gross oversimplification. In addition, there may not be accurate optical constants for the constituents, even those that are known; and they may not all be known. Yet even minor constituents can be major contributors to absorption.

The most direct way of obtaining the two parameters most relevant to climatic change, absorption and backscattering, is to measure them for actual atmospheric aerosols. Backscattering can be derived from angular scattering measured in situ with a polar nephelometer (e.g., Grams, 1981) or with an integrating nephelometer. Measured extinction could, in principle, yield absorption. But if the aerosol particles are weakly absorbing, as is common, this requires taking the difference of two large quantities—scattering and extinction— of similar magnitude.

Absorption can be measured directly by the photoacoustic method. Although used most commonly for dense, highly absorbing aerosols (Bruce and Pinnick, 1977; Japar and Killinger, 1979; Roessler and Faxvog, 1979b), it has been proven to be feasible for measuring weak absorption of visible light by ambient aerosols (Foot, 1979). A beam of direct sunlight was used as the source of modulated heating; and photoacoustic signals were measured alternately for ambient aerosols and for filtered air. This appears to be the only direct absorption measurement yet reported for atmospheric aerosols.

Direct measurement of absorption and backscattering at a sufficient number of sites to obtain a representative global average would be a step in the right direction toward assessing the impact of atmospheric particles on climate.

14.2.4   Sampling and Measurement

A less direct method of determining absorption by atmospheric particles is to collect them, on filters for example, and measure their absorption in the laboratory. In contrast with interstellar dust, atmospheric dust is sufficiently accessible to allow direct sampling and optical measurements. Yet the tenuousness of atmospheric aerosols leads to problems. Because of selective collecting lines and surfaces there may be differences between the sampled particles and those in the atmosphere. Evaporation of volatile compounds may occur, and contamination is always possible. And the state of aggregation will certainly change. Despite all this the advantages of a bird in the hand are substantial: time-consuming measurements can be done leisurely; particles can often be examined by electron or light microscopy to obtain their sizes and shapes; and chemical identification procedures can be used.

A major obstacle to determining absorption by collected particles is that scattering is often much greater than absorption. Some of the techniques devised to overcome this obstacle are shown schematically in Fig. 14.2; these are discussed very briefly in the following paragraphs.

Figure 14.2a shows the essential elements of a diffuse transmission technique (Fischer, 1970, 1973) that incorporates an integrating sphere—a sphere coated with a nonabsorbing, diffusely reflecting material (see Wendlandt and Hecht, 1966, Chap. 10). Measurements are made with and without the particulate sample (on a glass slide) in the light beam. From these measurements absorption can be obtained. A simpler arrangement (Lin et al., 1973) for accomplishing the same result, often called the integrating plate method, is shown in Fig. 14.2b. The particles are on an opal glass plate—a simple diffuser—which directs the light scattered into the forward hemisphere and the transmitted light to the detector. In this way opal glass takes the place of a much more expensive integrating sphere. Again, the signals both with and without the particles on the diffuser are compared. The simplicity of this technique has made it attractive to other investigators. In a modified version adapted to laser illumination (Fig. 14.2b), the diffusing plate method, the particle filter is both the substrate and the light diffuser (Rosen et al., 1978); it has been used to examine highly absorbing aerosols, identified as soot, from an urban environment. Rosen and Novakov (1982), in an analysis of this system, showed that backscattering by the particles does not, subject to restrictions on absorption by the filter and the particles, give rise to serious errors.

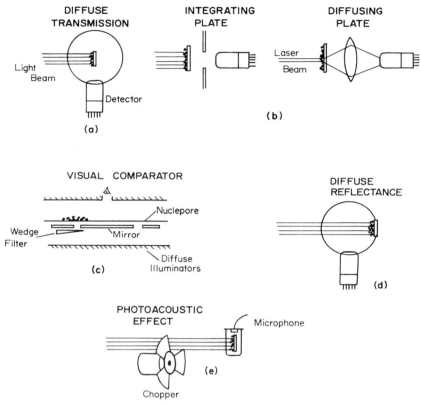

Figure 14.2 Techniques for measuring absorption by particles.

A simple scheme devised by Twomey (1980) does not even require a photoelectric detector—only the eye. A particle-laden portion of a filter and a clean portion of the same filter are each placed over holes in a thin, flat mirror, which is diffusely illuminated both from above and from below (Fig. 14.2c). Without the filter in place (assuming a perfect mirror) the brightness of a hole and its surroundings will be equal if the two illuminances are equal. But with the filter in place the illuminance from below will have to be reduced to obtain a brightness match, viewed from above, between a hole and its surroundings if there is any absorption. First a brightness match is obtained for the reference hole, the one over which there is only clean filter. Then a calibrated wedge filter is moved under the other hole, above which are the collected particles, until the brightness of the hole and its surroundings are matched. Knowing the wedge filter transmission, one can infer absorption by the particles.

In the diffuse reflectance technique (Fig. 14.2d), light scattered by a thick layer of particles is directed by the integrating sphere to a detector. Absorption by the sample particles decreases the detector signal; the resulting diffuse reflectance spectrum is therefore qualitatively similar to the transmission spectrum of the bulk solid. This assumes, of course, that there are no shape-dependent spectral effects such as those discussed in Chaper 12; for particles of insulating solids, these are usually restricted to infrared wavelengths. To extract quantitative absorption properties of the solids, however, requires a theory of diffuse reflection; this is often the Kubelka–Munk theory (see, e.g., Kortüm, 1969). Weakly absorbing powders can be analyzed directly by this technique, but highly absorbing powders may have to be diluted in a nonabsorbing powder (Lindberg and Laude, 1974).

For completeness, we include in this brief survey a schematic diagram of the photoacoustic method (Fig. 14.2e), which has been discussed in Section 11.7. Absorption of chopped light by the particles gives rise to periodic heating which is detected acoustically. Scattered light does not contribute to the signal.

Good summaries of these techniques for determining absorption by collected aerosol samples are given in the proceedings of a workshop held in August 1980 (Gerber and Hindman, 1982). The participants applied their separate techniques to measuring absorption of light by particles of solids with known optical constants. Results obtained by different techniques were generally in agreement within about a factor of 5. The accuracy of k measured for carbon, which is highly absorbing, was good. But for ammonium sulfate, which is weakly absorbing, measurements consistently overestimated k by factors of 100 or more.

In the study of atmospheric aerosols several techniques have been used to determine optical constants from measurements on particulate samples. And there have been many such measurements. Yet under pressure from funding agencies and from those waiting at computer terminals for optical constants of the complicated mixture that is the atmospheric aerosol, comparatively little effort has been expended on evaluating these techniques by applying them to particles of solids with known optical constants.

14.2.5 Remote Sensing

For determining optical properties of natural aerosols remote-sensing techniques avoid some of the drawbacks of collection and measurement. Because the particles are not disturbed by collection, possible changes such as evaporation of volatile components, agglomeration, and selective losses are avoided. Remote sensing, however, has its own drawbacks. Although vertical and horizontal probing can be done to a limited extent with some of the techniques, most of them require horizontal homogeneity of the atmosphere. And the measured optical properties may not be those of most direct interest. For example, in assessing the impact of aerosols on the earth's heat balance, what is wanted is how much they absorb solar radiation and scatter it into the backward hemisphere. Of less direct relevance are the optical depth, single-scattering albedo, and asymmetry parameter. Whereas optical depth may be determined directly, absorption must be inferred from other optical measurements, such as the extinction-to-backscatter ratio, together with a possibly oversimplified model (e.g., homogeneous, spherical particles of a single kind) the inapplicability of which may invalidate the results. Nevertheless, within limits remote sensing has excellent potential for global monitoring of some optical properties of atmospheric particles. Several remote-sensing techniques are listed below with brief comments.

Multiwavelength Solar Radiometry The sun is a light source of known spectral output. Measurement of the irradiance transmitted through the atmosphere therefore gives the total extinction from which extinction by the molecules can be subtracted to give that by the particles. Inversion techniques can then be used to infer particle size distributions from the extinction measurements (e.g., King et al., 1978). As shown in Fig. 13.8, Mie theory is a good approximation in the forward direction for sufficiently large nonspherical particles. If extinction is dominated by scattering about the forward direction, it is not very sensitive to particle shape.

Monostatic Lidar Remotely sensing atmospheric aerosols by measuring backscattering ($\theta \simeq 180°$) has become common since the ready availability of high-power pulsed lasers. Selection of various signal return times allows distance profiling. The extinction-to-backscatter ratio has been related to absorption properties of atmospheric particles using inversion techniques based on Mie theory (Spinhirne et al., 1980). It is evident from Fig. 13.8, however, that backscattering is quite sensitive to particle shape. So what is inferred as absorption perhaps should be attributed to nonsphericity. Inhomogeneity is also expected to have pronounced effects on backscattering because it is more sensitive to a particle's structure than forward scattering.

Bistatic Lidar Angular scattering by atmospheric particles can be measured by separating the collimated light source and the detector. The arrangement is much like that shown in Fig. 13.5; the volume sampled is determined by the intersection of the illuminating beam and the field of view of the detector. If a pulsed laser is used with time gating for the return signal the arrangement is called bistatic lidar. A similar system without time gating has been used for many years in searchlight probing of the atmosphere. For a survey of bistatic lidar, see Reagan et al. (1980) and references cited therein.

Linear polarization ratios have also been measured remotely. For example, results are given by Ward et al. (1973) and by Reagan et al. (1980). The sensitivity of scattering diagrams (Fig. 13.8), especially polarization (Fig. 13.9), to particle shape signals caution in inferring aerosol properties such as k from bistatic remote sensing.

14.2.6 Effective Refractive Index of Aerosols with Absorbers

The techniques just surveyed necessarily determine average refractive indices. And k values for atmospheric aerosols measured by these techniques suggest mixtures of vastly different kinds of particles. For we have noted that no common substances have k between 0.001 and 0.01 throughout the visible; yet this is what has been inferred. There is evidence that the smaller particles are the more highly absorbing (e.g., Lindberg and Gillespie, 1977). Also, carbon in the form of soot has been identified as a component of absorbing aerosols (Rosen et al., 1978). Since k of carbon is of order 1, the implication is that the observed absorption may be attributed to a small amount of carbon mixed with much greater amounts of weakly absorbing (at visible wavelengths) substances such as ammonium sulfate, sodium chloride, and crystalline quartz. Moreover, several investigators have pointed out that the way in which the absorber is mixed with the nonabsorbers can markedly affect optical properties such as single-scattering albedo, phase, function, and backscattering cross section. Bergstrom (1973) first called attention to some of the consequences of averaging optical properties, followed by Lindberg (1975), Toon et al. (1976), Gillespie et al. (1978), and Ackerman and Toon (1981). All of the last four papers present calculations based on Mie theory to show that quantities such as the single-scattering albedo can be quite different for a given amount of absorber, depending on whether it is distributed homogeneously throughout all the particles or whether it exists as discrete particles, separate from, inside, or on the surface of, the nonabsorbers. Because different size distributions and refractive indices were used it is difficult to compare the various results. Indeed, some of them conflict, as Ackerman and Toon (1981) pointed out in mentioning their failure to reproduce a result reported previously. To gain insight, therefore, we present our own simple back-of-the-envelope calculations for the single-scattering albedo of two different models of an absorbing aerosol.

One is a mixture of 1% by particle volume of spherical absorbers with radius 0.05 μm and refractive index $m_{bl} = 1.7 + i0.7$, which is appropriate to a form of carbon, and 99% by volume of larger particles with $m_{wh} = 1.55 + i10^{-6}$, which is roughly appropriate to several possible aerosol components. The subscripts bl (black) and wh (white) indicate how separate collections of these two kinds of particles would appear. Among the many optical constants reported for carbon in its various forms we have chosen those obtained from single-sphere measurements by Pluchino et al. (1980).

The second model aerosol is composed of "gray" spheres; that is, the absorber is incorporated in the nonabsorbing particles rather than separate from them. We might visualize this as small carbon spheres uniformly embedded in plum pudding fashion throughout much larger nonabsorbing spheres.

Our first problem is how to properly calculate the average optical constants when 1% of carbon by volume is uniformly distributed in a nonabsorbing medium. The usual procedure, as in the papers cited above, has been to simply volume-average n and k separately. But optical constants are not, in general, additive, so we have used the Maxwell Garnett expression (8.50). The result is $m_{gr} = 1.55 + i0.007$, which in this instance is identical, to the number of figures shown, with the result obtained by volume-averaging the refractive indices $1.55 + i0.0$ and $1.7 + i0.7$.

We now use several approximations discussed in previous chapters to estimate for the two model aerosols their single-scattering albedo $\bar{\omega}_0$, where

$$1 - \bar{\omega}_0 = \frac{C_{\text{abs}}}{C_{\text{ext}}}.$$

For a mixture of particles, C_{abs} and C_{ext} should be interpreted as number-weighted averages.

Extinction by carbon particles of radius 0.05 μm is dominated by absorption; moreover, the absorption cross section per unit particle volume in the Rayleigh limit (12.10) is independent of radius. So taking the carbon particles to be a single size is of no consequence provided that they are sufficiently small. Q_{ext} is approximately 2 for the large nonabsorbing particles. This approximation has the advantage that it excludes the complicated ripple and interference structures of single-sphere Mie theory, which are not very realistic for broad size distributions of irregularly shaped particles (see Fig. 11.20). With these approximations, therefore, we have for the black–white mixture,

$$1 - \bar{\omega}_0 \simeq \frac{f\alpha_{\text{bl}}}{f\alpha_{\text{bl}} + (1 - f)\alpha_{\text{wh}}}, \tag{14.2}$$

where the volume-normalized cross sections are

$$\alpha_{\text{bl}} \simeq 9.78 \ \mu\text{m}^{-1} \quad (\lambda = 0.55 \ \mu\text{m}); \qquad \alpha_{\text{wh}} \simeq \frac{1.5}{a_{\text{wh}}}.$$

To estimate $1 - \bar{\omega}_0$ for the gray particles we take $C_{\text{ext}} = 2\pi a_{\text{gr}}^2$ and use (7.2) for C_{abs}:

$$1 - \bar{\omega}_0 \simeq 0.142 a_{\text{gr}} \quad (\lambda = 0.55 \ \mu\text{m}), \tag{14.3}$$

where the factor multiplying the particle radius follows from the average refractive index $1.55 + i0.007$. The total particle volume is the same for both aerosols if $(1 + f)a_{\text{wh}}^3 = (1 - f)a_{\text{gr}}^3$, but a_{gr} and a_{wh} are nearly equal for $f = 0.01$.

In Fig. 14.3 we plot (14.2) and (14.3) as functions of large-particle radius. There are of course several restrictions to be kept in mind, including $2a\alpha \ll 1$ underlying the derivation of (7.2), which is only approximately satisfied for radii less than about 3 μm. To convince ardent Mie calculators that these simple expressions are approximately correct, we include single-size Mie calculations at 0.1-μm intervals. Except for the interference maxima and minima in the Mie calculations, which are unlikely to be observed in natural aerosols, the simple treatment is quite good.

The gray particles are more absorbing than the black–white mixture, and the difference is appreciable: the ratio of $1 - \bar{\omega}_0$ for the two model aerosols is about 3. We may roughly interpret this difference as arising from two causes. First, the absorption cross section of a small carbon sphere in a medium with $n = 1.55$ is about 1.6 times that of the same sphere in air. Second, there is a focusing effect; that is, more light is geometrically incident on a sphere when it is in a much larger transparent sphere than when it is in air. For example, ignoring external and internal reflections, it follows from geometrical optics that n^2 more light is incident on a small sphere when it is at the center of a much larger sphere then when it is in air. Based on this crude reasoning, therefore, we would expect $1 - \bar{\omega}_0$ for the gray particles to be at least 1.6 times, but not more than 3.8 times, greater than that for the black–white mixture. And this is indeed consistent with more detailed calculations.

Consider now a particle radius of 1.5 μm; from Fig. 14.3 it follows that $\bar{\omega}_0$ for the gray particles is about 0.75, whereas it is about 0.9 for the black–white mixture. And recall that a single-scattering albedo of 0.85 is used widely as the critical value separating a global cooling trend ($\bar{\omega}_0 > 0.85$) from a global warming trend ($\bar{\omega}_0 < 0.85$). For particles of this size as well as those somewhat smaller and larger, therefore, even the *direction* of the temperature change depends on just how the absorber is dispersed in the aerosol. We must conclude from this, therefore, that merely knowing the amount of absorbing material in the atmospheric aerosol is not sufficient to assess its potential impact on the global climate.

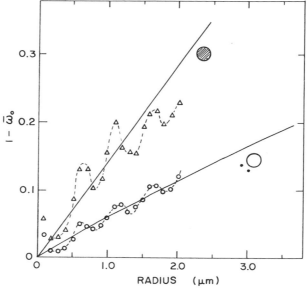

Figure 14.3 Absorption by gray spheres in which the absorber is uniformly distributed (upper) and by small absorbing spheres mixed with much larger nonabsorbing spheres (lower); the total amount of absorbing material is the same for both. Solid lines were calculated approximately and dashed lines connect points calculated with Mie theory.

14.2.7 Monitoring the Global Wind with Lidar Backscattering

It seems that we have cast atmospheric particles most often in the role of villains: they may cause cataclysmic climatic changes; they reduce visibility; they may aid in the formation of photochemical smog and acid rain. But there is a scheme afoot to make them do some useful work by acting as ubiquitous retro-reflectors for laser beams.

The Doppler shift of laser light backscattered by atmospheric particles carries information about the velocity of the air in which they are being swept along. Huffaker (1970) used a continuous CO_2 laser to demonstrate the practicability of the laser Doppler velocimeter (see Bilbro, 1980, for an overview). By using a pulsed laser and selecting signal returns at different times, particles at various distances from the source can be sampled; this is *lidar*—light detection and ranging. Doppler velocity measurements with lidar could therefore enable one to map out in direction and distance wind velocity components along the line of sight. Indeed, this concept is being considered for a future satellite-borne wind monitoring system (Abreu, 1980). As the satellite orbits the earth the infrared lidar system would scan conically looking downward into the atmosphere; depth profiling is accomplished by return-signal gating. Every chosen volume of air in the three-dimensional grid would be sampled from two different directions—forward and backward—within a short time as the satellite passes overhead. From these two line-of-sight velocity measurements the horizontal wind velocity can be extracted. Such a global wind monitoring system is quite ambitious, but would provide extremely useful meteorological data.

Small particles of some insulating solids, including those in the atmosphere, have Fröhlich modes (see Chapter 12) at certain wavelengths near 10 μm. Backscattering is also large at these wavelengths; this is evident from the efficiencies for absorption, scattering, and backscattering by a small sphere (Section 5.1), all of which contain the quantity $(m^2 - 1)/(m^2 + 2)$. Thus, the spectrum of backscattered light will peak at wavelengths near the extinction peak. Both quartz and ammonium sulfate, which are common aerosol constituents, have Fröhlich modes near 9 μm. This implies that choosing a wavelength near 9 μm rather than the more common 10.6-μm CO_2 laser radiation might give larger backscattering signals from spheres of these solids. But recall that surface modes are highly shape dependent. And backscattering should behave similarly to extinction. Figure 12.14 therefore indicates the effect of a distribution of shapes on the backscattering spectrum for quartz particles. Although CO_2 lasers are not quite tunable to the peak for quartz spheres near 9.0 μm, shape effects might actually improve the backscattering signal by spreading the spectrum of backscattered light into the region between about 9.1 and 9.2 μm where CO_2 lasers are able to operate. A Doppler lidar system operating at two different wavelengths might be able to discriminate between certain kinds of particles in the atmosphere and thus map out their distributions. But whatever use to which infrared Doppler lidar is put, it is clear from what we have said here and elsewhere that shape-dependent surface modes in small particles will have to be reckoned with.

14.3   NOCTILUCENT CLOUDS

As their name implies, noctilucent clouds are visible at night; the sun, after it has descended well below the horizon, is their source of illumination. They are a high-latitude (50–60°) summer phenomenon and differ from clouds observed during the day by their great height, about 82 km above the earth's surface, near the height (mesopause) where the atmosphere's temperature is lowest. Tenuousness also distinguishes them from the clouds of everyday experience: stars often can be seen through them. The term "noctilucent" predates satellite observations and by now may be somewhat of a misnomer: noctilucent clouds have been observed (from a satellite) during daytime (Donahue et al., 1972). Indeed, these observations indicate that what has been seen from the ground is merely the thin and ragged edge of a much thicker high-level cloud layer extending poleward. Those of us bound to earth, however, may still use the term "noctilucent" without contradiction.

Fogle and Haurwitz (1966) have given an excellent review of noctilucent clouds, to which we refer the reader for further details about their appearance as well as when, where, and for how long they may be observed.

Probing the nature and origin of noctilucent clouds is made difficult by their inaccessibility: they are too high to be visited by balloons and too low to be visited by satellites. Rockets, therefore, are the only means for directly sampling these clouds, but rocket flights are expensive and their visits to the upper atmosphere are necessarily fleeting and restricted to small regions. Although direct sampling is, of course, highly desirable, the most practical method for investigating large regions of noctilucent clouds over relatively long time intervals is by analyzing the sunlight scattered by them or the starlight transmitted through them. Indeed, before the advent of rockets, analysis of light from noctilucent clouds was the only means of inferring their properties.

Two characteristics of the light from noctilucent clouds may be observed with no more than one's eyes and a polarizing filter: its color and whether or not it is strongly polarized. This enabled Ludlum (1957) to estimate the size range of noctilucent cloud particles. Because of the observed strong polarization he set 0.16 μm as their upper size limit; on the basis of the observed color—white, silvery, sometimes bluish, but not sufficiently so as to indicate very small particles—he set 0.008 μm as their lower size limit. From other than optical evidence he also concluded that the particles were not ice, but were more likely to be volcanic, meteoric, or interplanetary dust.

Light from noctilucent clouds carries with it more than just information about how it was scattered by cloud particles, however. On its journey from the sun to the clouds, and thence to an observer, it must travel long atmospheric paths along which it suffers selective absorption and scattering by various gases and particles of uncertain kind and amount. This selective extinction must be subtracted from the observed spectrum to obtain the true spectrum of scattered light. Deirmendjian and Vestine (1959) corrected the spectral data of Grishin (1956) for extinction by a model atmosphere and concluded that the data were consistent with scattering by spherical particles with refractive index 1.33 and radius 0.4 μm.

14.3.1 Linear Polarization

Inferences made on the basis of the spectrum of light from noctilucent clouds are always going to be plagued by uncertainties about the corrections to be applied to ground-based observations. But the degree of polarization of light scattered by noctilucent cloud particles is insensitive to selective atmospheric extinction provided that the incident and scattered beams encounter no oriented particles (other than the noctilucent cloud particles themselves, of course). This was pointed out by Witt (1960), who measured linear polarization at scattering angles between about 20 and 60°; the degree of polarization increased monotonically with scattering angle to maximum values of about 0.4 for blue light ($\lambda = 4900$ Å) and 0.5 for red light ($\lambda = 6100$ Å). By using rockets the range of scattering angles has been extended. The high degree of linear polarization near 90° measured by Witt (1969), Tozer and Beeson (1974), and Witt et al. (1976) suggested to these authors that the upper limit of particle size is less than about 0.13 μm. Hummel and Olivero (1976) analyzed the satellite radiance measurements of Donahue et al. (1972) and concurred with this upper size limit.

The strongest evidence supporting small (< 0.1 μm) particles in noctilucent clouds is the high degree of measured linear polarization: it increases monotonically with scattering angle to almost unity near 90°. It is difficult to reconcile these observations with any conclusion other than that the particles are small.

Noctilucent cloud particles are now generally believed to be ice, although more by default—no serious competitor is still in the running—than because of direct evidence. The degree of linear polarization of visible light scattered by Rayleigh ellipsoids of ice is nearly independent of shape. This follows from (5.52) and (5.54): if the refractive index is 1.305, then $P(90°)$ is 1.0 for spheres, 0.97 for prolate spheroids, and 0.94 for oblate spheroids.

The applicability of Rayleigh-Gans theory to ice particles in the atmosphere is uncertain because $|m - 1|$ is close to values for which this theory begins to give large errors, at least for spheres (Kerker, 1969, p. 428). But if the Rayleigh–Gans theory is valid for ice particles, which requires them to be not larger than about 0.1 μm, they are necessarily highly polarizing: the degree of linear polarization at 90° is unity for all scatterers described by this theory regardless of their shape.

Some insight into how departures from Rayleigh theory affect linear polarization can be obtained from calculations of Asano and Sato (1980) for randomly oriented oblate spheroids with refractive index 1.33, which is near enough to that of ice, axial ratio $a/c = 5$, and size parameter $2\pi a/\lambda = 5$; for a wavelength of 0.5 μm this corresponds to an equal-volume sphere with radius 0.23 μm. Polarization increases monotonically with increasing scattering angle to a maximum near 90° and then monotonically decreases. That is, polarization is similar to that for Rayleigh spheres or spheroids except that the maximum polarization (0.89) is slightly less; it is also less than that measured in the light from noctilucent clouds, which provides additional evidence that the cloud particles are not larger than about 0.1 μm.

14.3.2   Circular Polarization

Circular polarization has also been observed in light from noctilucent clouds. In a set of 10 observations Gadsden (1975) measured positive values of V/I in the range 0.02–0.07. Subsequently, Gadsden (1977) measured both positive and negative V/I; most of the values were clustered between 0 and -0.05 but a few were appreciably outside this range. The most recent measurements, however, indicate a much smaller degree of circular polarization, about 0.005 (Gadsden et al., 1979), which may be atypical or it may reflect improved measurement techniques; from this vantage point it is difficult to say which.

One possible explanation for the observed circular polarization is that noctilucent cloud particles are partially oriented, which requires that they be nonspherical. Indeed, Gadsden (1978) has criticized previous size estimates because they were made under the assumption that noctilucent cloud particles are spherical, whereas there is evidence suggesting they are not. We have shown repeatedly in previous chapters that nonspherical particles often absorb and scatter light quite differently from "equivalent" spheres. So there are certainly good reasons for carefully considering Gadsden's criticisms. Before doing so, however, it will be helpful to consider the conditions under which unpolarized light acquires a degree of circular polarization upon scattering.

For incident unpolarized light to be (partially) circularly polarized upon scattering by a collection of particles, the scattering matrix element S_{41} must not be zero. It was shown in Section 13.6 that the scattering matrix for a collection (with mirror symmetry) of randomly oriented particles has the form

$$\begin{pmatrix} S_{11} & S_{12} & 0 & 0 \\ S_{12} & S_{22} & 0 & 0 \\ 0 & 0 & S_{33} & S_{34} \\ 0 & 0 & -S_{34} & S_{44} \end{pmatrix}. \tag{14.4}$$

If each particle is spherically symmetric, then $S_{33} = S_{44}$ and $S_{11} = S_{22}$. Such a collection of particles cannot circularly polarize unpolarized light or light polarized perpendicular (or parallel) to the scattering plane. But it can circularly polarize *obliquely* polarized incident light ($U_i \neq 0$) provided that S_{43} ($-S_{34}$) is not zero.

According to the ground rules laid down at the beginning of this book, multiple scattering is excluded from consideration. But it is not always prudent to pretend that multiple scattering does not exist. Fortunately, it is almost trivial—the mathematical apparatus of radiative transfer theory is unnecessary—to extend our treatment of scattering and circular polarization to multiple scattering media, and in this instance it is worth the small amount of effort required to do so.

Consider a particulate medium described by (14.4); for ease of visualization it may be taken to be a single particle. Unpolarized light will, upon scattering, become partially polarized either parallel or perpendicular to the scattering plane depending on the sign of S_{12}. This scattered light is now incident on another particle, but it is polarized obliquely, in general, to the various scattering planes determined by the directions of single-scattered (i.e., incident) and twice-scattered light. So unpolarized light can acquire a degree of circular polarization upon multiple scattering by randomly oriented particles.

Particles to which the Rayleigh–Gans theory is applicable, regardless of their shape, orientation, and composition, do not circularly polarize either unpolarized or linearly polarized light upon scattering, single or multiple, because $S_{41} = S_{42} = S_{43} = 0$. This is also true of *nonabsorbing* particles in the Rayleigh limit (Chapter 5), which follows from (3.16) and (5.47): the polarizability tensor is real. Ice is weakly absorbing at visible wavelengths (Grenfell and Perovich, 1981). Within the framework of both the Rayleigh and the Rayleigh–Gans theories, therefore, ice particles, regardless of their shape and orientation, cannot circularly polarize unpolarized or linearly polarized visible light.

We are still left with the question of the origin of the observed circular polarization of light from noctilucent clouds. Circular polarization of *unpolarized* incident light is possible if the cloud particles are appreciably larger than allowed by Rayleigh theory—provided that they are aligned. But no plausible alignment mechanism has yet been proposed. Gadsden (1975) suggested aerodynamic alignment, but this mechanism does not stand close examination: aerodynamic forces are very weak in the thin atmosphere at 82 km above the earth's surface. Moreover, whether or not elongated particles are aligned by falling in air is not independent of their size: alignment is counteracted by the disorienting tendency of Brownian rotation. Fraser (1979) has shown that, at sea level, the critical size below which particles are randomly oriented is about 10 μm. The critical size is not very sensitive to temperature and pressure, but it does increase slightly with altitude. Extrapolation from sea level to the mesopause where the mean free path of molecules is about 0.5 cm is, in general, invalid. But in this instance the critical size increases even faster than predicted by extrapolating Fraser's results. Indeed, Reid (1975) has calculated the force on an arbitrarily shaped particle, small compared with the mean free path (molecular flow regime), from which it follows that the net couple acting on such a particle is zero. Until an alignment mechanism is discovered, therefore, we can only assume that noctilucent cloud particles are randomly oriented and thus incapable of circularly polarizing unpolarized light.

A possible explanation for the observed circular polarization is that the light illuminating noctilucent clouds is partially polarized because of multiple scattering in the long paths it travels through the atmosphere. The incident light is therefore composed of unscattered, unpolarized (direct) light and multiply scattered, partially polarized (indirect) light:

$$
\begin{pmatrix} I_i \\ Q_i \\ U_i \\ V_i \end{pmatrix} = \begin{pmatrix} I_0 \\ 0 \\ 0 \\ 0 \end{pmatrix} + \begin{pmatrix} I_{ms} \\ Q_{ms} \\ U_{ms} \\ V_{ms} \end{pmatrix}
$$

incident   unscattered   multiply scattered
sunlight   sunlight
(direct)   (indirect)

If such light is incident on a medium described by the matrix (14.4), the degree of circular polarization of the scattered light is

$$
\frac{V_s}{I_s} = \frac{-\dfrac{S_{34}}{S_{11}}\dfrac{U_{ms}}{I_{ms}} + \dfrac{S_{44}}{S_{11}}\dfrac{V_{ms}}{I_{ms}}}{\dfrac{I_0}{I_{ms}} + 1 + \dfrac{S_{12}}{S_{11}}\dfrac{Q_{ms}}{I_{ms}}}. \tag{14.5}
$$

Note that the direct light is scattered through a single angle to an observer, whereas the indirect light, which is incident from many directions, is scattered through many angles; so S_{34}/S_{11}, U_{ms}/I_{ms}, and so on, should be interpreted as averages over the various directions of incidence.

If V_s/I_s is to be nonzero, either S_{34}/S_{11} and U_{ms}/I_{ms}, or S_{44}/S_{11} and V_{ms}/I_{ms}, or both, must not be zero. S_{34} is zero for nonabsorbing particles in the Rayleigh limit and for arbitrary particles in the Rayleigh–Gans approximation. Even for particles—spherical and nonspherical—of size comparable with the wavelength, however, S_{34} tends to be small, particularly in the forward direction (see Figs. 13.13 and 13.14).

If we neglect the product $(S_{34}/S_{11})(U_{ms}/I_{ms})$, then the degree of circular polarization of the scattered light is approximately

$$
\frac{V_s}{I_s} \simeq \frac{I_{ms}}{I_0}\frac{S_{44}}{S_{11}}\frac{V_{ms}}{I_{ms}},
$$

where we have also assumed that I_0/I_{ms} is the dominant term in the denominator of (14.5).

A necessary condition for the correctness of the multiple-scattering explanation of the observed circular polarization is that scattering by noctilucent cloud particles does not appreciably reduce the degree of circular polarization of the incident light. That this is so for randomly oriented Rayleigh ellipsoids is readily shown. M in (5.52) is nearly unity for ice ellipsoids, so to good approximation

$$\frac{S_{44}}{S_{11}} \simeq \frac{2\cos\theta}{1 + \cos^2\theta},$$

which is not less than 0.5 except for a small range of scattering angles centered about 90°.

It is required further that V_{ms}/I_{ms} be not smaller than the observed values of V_s/I_s; indeed, it should be a good bit larger because of the dilution factor I_{ms}/I_0. Gambling and Billard (1967) measured degrees of circular polarization as high as 0.67 in light from cloudless skies, although more typically they ranged between 0 and 0.17. These values may be taken to represent fairly the degree of circular polarization expected in the multiply scattered light illuminating noctilucent clouds.

Both V_{ms}/I_{ms} and I_{ms}/I_0 must be about 0.2 or higher if the highest values of circular polarization are the result of multiple scattering; if they both are about 0.07, the multiple scattering explanation is consistent with the lowest observed values. To put it another way, V_{ms}/I_0 must be not less than the observed degree of circular polarization; this is at least plausible—there are no strong reasons for rejecting it—but it has yet to be demonstrated unequivocally.

14.3.3 Summary

The high degree of linear polarization observed in light from noctilucent clouds strongly suggests that the particles cannot be much larger than about 0.1 μm. And if they are composed of ice, they need not be spherical. Regardless of their shape and orientation, small ice particles do not circularly polarize unpolarized light.

Aligned, hence nonspherical, particles have been suggested as the cause of the observed circular polarization in light from noctilucent clouds. But no plausible alignment mechanism has yet been proposed. Even if a mechanism were to be discovered, however, it is likely that the size required for alignment would not be consistent with the linear polarization data: particles are aligned by a couple—a force times a distance. Whatever the mechanism, therefore, alignment favors larger particles.

There are several possible explanations for the circular polarization data: they are in error; the particles are small, aligned, and highly absorbing; the light illuminating the clouds has acquired a degree of circular polarization by multiple scattering.

There is no reason for suspecting the data; we must take them at face value.

If the particles are highly absorbing, it may be difficult to reconcile this with both the observed color and linear polarization of the light from noctilucent clouds.

The third explanation is the most likely one, although further measurements and calculations are necessary to establish this with reasonable certainty. In the context of the multiple scattering explanation a statement by Witt et al. (1976) is apposite: "Although multiple scattering effects within the mesosphere itself need not be considered, the additional source of light from the lower atmosphere must be fully taken into account to allow the proper interpretation of upper-atmospheric polarization measurements."

14.4   RAINFALL MEASUREMENTS WITH RADAR

The ever-increasing number of papers presented at the radar meteorology conferences sponsored by the American Meteorological Society attest to the rapidly growing use of radar in meteorology; indeed, the collection of preprints will soon be beyond the ability of one person to lift, let alone read. To cover the various applications of radar meteorology adequately therefore requires more space than we can devote here. A more complete survey of the field is given in the standard work on radar meteorology by Battan (1973). More recently, Browning (1978) reviewed several applications of radar to meteorology, from which we have chosen one for discussion: measurement of rainfall by radar. A review of this topic is given by Wilson and Brandes (1979).

Let us assume for the moment that all raindrops have the same diameter D. The rainfall rate R, the rate at which the depth of water in a rain gauge of constant cross section increases with time, is the product of the total volume of water in a unit volume of air and the terminal velocity $V_t(D)$ of a raindrop:

$$R = N \frac{\pi}{6} D^3 V_t(D),$$

where N is the number of raindrops per unit volume of air. It is sometimes overlooked that V_t is the velocity *relative to the ground*, whereas various theoretical and empirical expressions are for terminal velocities *relative to the air* through which the droplet falls (Battan, 1976). If the raindrops were to fall in an updraft with velocity V_t, for example, the rainfall rate would be zero.

In principle the rainfall rate is determined by two quantities, the amount of water in the air and the rate at which it is falling to the ground. But these two quantities are correlated to some extent: V_t increases with increasing D as does, for a given N, the amount of water. Yet this correlation has not been established precisely. So it is perhaps best to assume that determining rainfall rates requires two independent measurements.

Raindrops, which have diameters in the range 1–5 mm, are small compared with the wavelengths of weather radars, usually 3, 5, or 10 cm. We showed in Section 5.1 that the backscattering cross section of a sphere is proportional to the sixth power of its radius in the Rayleigh approximation. So it is customary to define the *radar reflectivity factor*

$$Z = ND^6,$$

which determines the amount of power scattered by raindrops in a unit volume at a given distance back to a receiving antenna.

When a raindrop ceases to accelerate, its weight is balanced by the drag force (buoyancy is negligible):

$$\rho_w \frac{\pi}{6} D^3 g = \frac{1}{2} \rho_a V_t^2 C_d \frac{\pi}{4} D^2,$$

where ρ_a and ρ_w are the densities of air and water, g is the acceleration of gravity, and C_d is the drag coefficient. The drag coefficient of a raindrop is nearly independent of its diameter (Gunn and Kinzer, 1949), in which instance its terminal velocity is approximately proportional to the square root of its diameter.

If all these results are combined we obtain the following Z–R relation:

$$Z = KN^{-0.71}R^{1.71},$$

where K is a constant. Thus, there is a unique relation between the radar reflectivity factor and the rainfall rate provided that N is constant, or nearly so, from storm to storm, and all droplets are the same size. A Z–R relation puts the cart before the horse, however, for it is the rainfall rate that is to be determined by measuring the reflectivity factor, not the converse. Nevertheless, this is the convention, hallowed more by tradition than by ratiocination.

Up to this point we have assumed for simplicity that all raindrops are the same size. A more realistic assumption is that raindrop diameters are distributed according to a continuous function $N(D)$, in which instance the radar reflectivity factor and the rainfall rate are

$$Z = \int N(D)D^6\,dD, \qquad R = \frac{\pi}{6} \int N(D)V_t(D)D^3\,dD.$$

It has become the custom to express Z–R relations in the form

$$Z = aR^b,$$

where a and b are constants. They may be constant as far as individual investigators are concerned but among them there is no consensus: reported pairs of values of a and b differ in varying degree. Twomey (1953) lists eight different Z–R relations and adds one of his own to the pile; Battan (1973, pp. 90–92) lists nearly 70 different Z–R relations. Each year sees a new crop—and the end is not yet in sight.

With $Z–R$ relations having been steadily churned out for nearly 40 years, it is only natural to ask: How good are they for determining rainfall rates? Twomey (1953) concluded that "radar methods can give only an approximate measure of precipitation rate; the value deduced from the radar echo may be in error by a factor of 2 : 1 either way, and this randomly distributed error is independent of instrumentation or procedure adopted." With these words in mind it is indeed sobering to contemplate the latest word on $Z–R$ relations in the review article by Wilson and Brandes (1979): "With reasonable efforts, radar measurements... should be within a factor of two of the true rainfall about 75% of the time." In other words, not much progress has been made in over a quarter of a century. One cannot help feeling that this results from a failure to come to grips with a fundamental constraint imposed by nature: it really requires *two* independent measurements to determine rainfall with radar and no amount of statistical manipulation will ever change this.

We have assumed in previous paragraphs that raindrops are spherical, although this is not generally their shape. Nor are they shaped like teardrops despite the nearly universal habit of artists to so depict them. Their actual shape may be quite complicated; moreover, they oscillate, so the notion of a single shape for a raindrop is somewhat of an idealization. Sufficiently large raindrops are approximately spheroidal (oblate): they are flattened in the direction of fall. The forces shaping raindrops have been discussed by McDonald (1954). More recent treatments of raindrop shape are those by Pruppacher and Pitter (1971) and Green (1975). In the former there is a curve (Fig. 3) showing that the amount of deformation of a drop increases with its size. But so does its terminal velocity, which in turn implies that the deformation depends on the droplet velocity.

Radar backscattering by spheres is independent of the polarization state of the beam. But backscattering by spheroids (excluding illumination along their symmetry axes) is not. This is the physical basis for a method proposed by Seliga and Bringi (1976) for improving rainfall measurements by measuring two radar reflectivities, denoted by Z_H (horizontal polarization) and Z_V (vertical polarization), for orthogonally polarized beams. Ratios of the amount of radiation (singly) scattered for various polarization states of the incident beam are independent of the number of particles. As a consequence, the *differential reflectivity* Z_{DR}, defined by

$$Z_{DR} = 10 \log \frac{Z_H}{Z_V},$$

depends only on a mean size of the drops; Z_{DR} is zero, of course, if they are spheres. With this mean size and either Z_H or Z_V, the number of particles can be determined. So the two quantities necessary to determine rainfall rates, number and size, can be obtained in principle from the two reflectivity measurements.

The first tests of this proposed method have been encouraging. On the basis of comparisons between rainfall rates measured with the differential reflectivity technique and with a network of rain gauges, Seliga et al. (1981) concluded that "these first measurements of rainfall using the Z_{DR} technique support the theoretical expectations... that rainfall rate measurements with radar can be made with good accuracy." So it may yet be possible to accurately measure rainfall with radar—provided that measurements are made with two orthogonally polarized beams. This exemplifies one of the principal themes of this book: scattered polarized radiation contains information that may be put to good use.

14.5 INTERSTELLAR DUST

Interstellar dust, the small particles sparsely populating the regions between stars, has probably been studied as much as or more than any system of particles on earth, either laboratory produced or naturally occurring. Serious scientific study of interstellar dust has been going on since the early part of this century, and according to one of the latest review articles (Savage and Mathis, 1979), about 300 research papers that are in some way related to interstellar dust are being published yearly. Other review articles that survey this field and provide guides to its extensive literature are those of Wickramasinghe and Nandy (1972), Aannestad and Purcell (1973), and Huffman (1977); the monograph by Martin (1978) also covers the field well.

Advances in several areas of physics have followed from astronomical observations. Examples may be found in atomic spectroscopy, high-energy particle physics, and relativity. Even for optical studies of small particles the "cosmic laboratory" may have some advantages over laboratories bound to earth. We noted in Section 12.2 how difficult it is to produce submicrometer particles and to prevent them from aggregating; and it may also be necessary to maintain them in a high-vacuum environment while their absorption and scattering properties are being determined. But these requirements are admirably met by interstellar dust, which astronomers have studied carefully at wavelengths from far infrared to far ultraviolet and even x-rays.

The consequences of interstellar dust may be seen with the unaided eye. Under good seeing conditions dark patches can be observed in the Milky Way. These dark areas, we now know, do not result from the irregular distribution of stars in our galaxy but rather from the very effective obscuration of starlight by irregular clouds of small particles, the interstellar dust.

Several different types of this dust are distinguished by astronomers. On average, interstellar dust resides in widely separated *diffuse clouds*. But there are also dense regions of gas and dust into which little ultraviolet radiation can penetrate, thereby providing an environment for the formation of complex molecules; these are referred to as *molecular clouds*. Clouds of particles expelled by cooler stars into the regions around them are called *circumstellar shells.* Other dusty interstellar regions are found around novae, planetary nebulae, and in diffuse nebulae.

If all the matter in our galaxy, which is about evenly divided between the stars and the regions between them, were uniformly distributed, its average density would be about 6×10^{-24} g/cm^3. The solid component of the interstellar medium composes about 5% of its mass. But despite its low average density—about 1.5×10^{-26} g/cm^3—interstellar dust has an important effect on the distribution of electromagnetic radiation in our galaxy.

14.5.1 Extinction

The effectiveness of interstellar dust at extinguishing starlight is illustrated by the following example. If a kiloparsec (3.3×10^3 light years) column of dust were compressed into a homogeneous solid with a density of 2 g/cm^3, its thickness would be about 0.2 μm. Yet this small amount of dust transmits only about 6% of the visible light incident on it.

A great amount of effort has been expended on measuring the spectral dependence of interstellar extinction. Unlike laboratory samples, interstellar dust cannot be placed in and then taken out of a light beam, so astronomers must rely on finding two separate but similar light sources. These are two stars of similar spectral type, as judged by their emission spectra, but selected so that one has a large amount and the other a small amount of dust between it and the observer. Comparison of the light from the reddened star (I)—so named because its light is selectively depleted of longer wavelengths by extinction—with that from the unreddened star (I_0) enables the optical density $\log_{10}(I_0/I)$ to be determined.

The average interstellar extinction spectrum, which may not be representative of localized galactic regions, for a path length of 1 kiloparsec is shown in Fig. 14.4.

Extinction of visible light increases almost linearly with increasing photon energy (decreasing wavelength); interstellar dust, like the molecular and particulate constituents of our atmosphere, reddens starlight by extinction. Although this does not shed much light on the composition of the dust, it does imply that most of the particles responsible for interstellar extinction are small compared with visible wavelengths. Other features of the extinction spectrum are a prominent broad peak at about 5.7 eV (2170 Å)—referred to as the 2200-Å band—beyond which extinction continues to rise, and a less conspicuous knee in the curve near 2.8 eV (4500 Å).

Most explanations of the interstellar extinction curve, which are based on calculations for spherical particles (e.g., Mathis et al., 1977), require several different dust components: a special substance to give the 2200-Å band; very small particles for the far-ultraviolet upturn; and a larger size component to redden visible light. That very small spheres are needed for rising ultraviolet extinction is evident from Figs. 4.6 and 11.2. Particles of the proper size to match interstellar reddening in the visible would generally give rather neutral

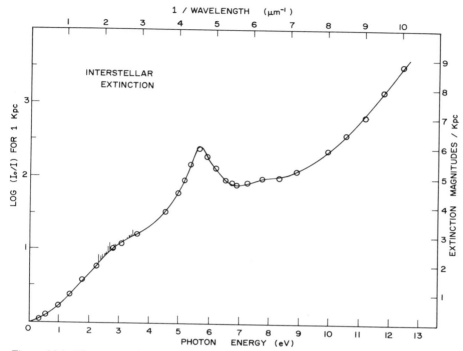

Figure 14.4 The average interstellar extinction spectrum. Data points are from Savage and Mathis (1979).

extinction at much shorter wavelengths: hence the apparent need for very small particles so that extinction continues to rise at these wavelengths. But there are laboratory measurements for graphite particles which suggest that the small-particle component may not be necessary (Day and Huffman, 1973).

Superimposed on the smoothly varying interstellar extinction curve is a series of some 39 or more narrow extinction bands ranging in width from a few angstroms to about 30 Å (Herbig, 1975; Snow et al., 1977). Although much is known about these bands—their positions, shapes, and relative strengths—not one of them has been satisfactorily explained. Because of their large widths compared with atomic and molecular absorption bands, astronomers call them the *diffuse* bands, although they are quite narrow compared with absorption bands in solids. The strongest band, near 4430 Å, was discovered between 1910 and 1920 and determined in the 1930s to be of interstellar origin. Since then the list of bands has continued to grow, with the spectroscopic information becoming ever better, yet with little definite progress in explaining their origins. At present the diffuse band mystery must surely be the most outstanding unsolved spectroscopic problem in astronomy, ranking with the longest-standing such problems of the last two centuries. The solution to this mystery is likely to elucidate greatly the nature of the interstellar medium and may lead to a major new investigative probe for astronomers.

Of more recent discovery are wide and shallow extinction bands with characteristic widths of about 500–1000 Å and extending from about 3400 to 11,000 Å (for a brief survey, see Huffman, 1977). This very broad structure (VBS) is too broad and weak to be seen in Fig. 14.4. Lack of correlation between the diffuse bands and the VBS suggests a different origin for the two.

14.5.2 The 2200-Å Band

Clearly, the dominant feature in the ultraviolet is the prominent peak near 2200 Å. This strong and generally occurring feature has increased the number of specific suggestions for the composition of interstellar dust beyond that possible if only reddening of visible light were known. The most widely held interpretation of this feature is that it is the result of extinction by small graphite particles, which, as pointed out by Gilra (1972ab), have a surface plasmon oscillation in the ultraviolet. Both Mie calculations using measured bulk optical constants of graphite and laboratory measurements (Day and Huffman, 1973; Stephens, 1980) have demonstrated the surface plasmon peak in small graphite particles, although neither agrees completely with the exact shape and position of the interstellar feature. In common with other such collective oscillations, discussed extensively in Chapter 12, the graphite feature might be expected to be quite sensitive to particle shape. But extinction by small graphite particles is not nearly as dependent on shape as that by small metallic (free electron) particles. We showed in Section 12.4 that the surface plasmon peak for aluminum particles can lie anywhere below the plasma frequency, which is in the ultraviolet, because this is the spectral region over which ϵ' for aluminum is negative. But for graphite there is only a relatively narrow range (4–7 eV) over which one of the principal components of its dielectric tensor is negative. Particle shape can thus be important but only within a restricted spectral region. Departures from sphericity may shift and broaden the sphere feature somewhat but they do not completely obliterate it. In this sense the extinction feature in small graphite particles, although the result of a collective electronic oscillation (surface plasmon), more nearly resembles collective lattice oscillations (surface phonons) in insulating particles such as quartz, silicon carbide, and MgO, for which the negative ϵ' region is relatively narrow (see Chapter 12).

There are several possible reasons why neither calculations nor measurements of graphite extinction agree in detail with the observed interstellar band. Calculations require accurate values of both sets of optical constants for graphite, which is highly anisotropic. But there are very large discrepancies among different measured optical constants for the experimentally difficult case when the electric field is parallel to the optic axis. Moreover, the optical constants of graphite depend on its degree of crystallinity. These factors have been discussed in the review by Huffman (1977). Even with accurate optical constants, however, the scheme for calculating extinction by anisotropic spherical particles (see Section 8.2) is of uncertain validity outside the Rayleigh limit. Finally, because graphite is anisotropic, particles of it are likely to be nonspherical, and nonsphericity is difficult to treat outside the Rayleigh limit. For these reasons failure to match the observed 2200-Å band with either calculations or measurements is perhaps not too surprising. Other explanations, of course, have been proffered; these are discussed in the articles cited in the first paragraph of this section.

14.5.3   Infrared Absorption Bands

Interstellar extinction features become increasingly difficult to observe as the wavelength is extended farther into the infrared because the cross section decreases, necessitating extremely long path lengths through the dust, while emission from heated dust around stars opposes absorption, filling in absorption bands and complicating interpretation. One of the best places to look for infrared absorption bands—absorption is nearly synonymous with extinction at infrared wavelengths if the particles are sufficiently small, say < 0.1 μm—is toward the center of the galaxy: the large concentration of stars near the galactic center and the long path lengths are favorable to detecting weaker absorption bands. The intensity spectrum in the direction of the galactic center shows a prominent absorption band at 9.7 μm (see Fig. 3 in Woolf, 1975, which is reproduced as Fig. 12 in Huffman, 1977). This band, although generally difficult to observe in interstellar extinction, has been observed as an excess emission hump in a variety of astronomical sources, including circumstellar shells, diffuse nebulae, and comet tails.

Soon after its discovery the broad and featureless 9.7-μm band was associated with Si—O vibrational stretching modes in silicates such as the mineral olivine (Mg, Fe)$_2$SiO$_4$. But when the optical constants of olivine were measured and extinction calculations performed it was evident that, in contrast with the observed structureless band, sharp structure would persist even for a distribution of particle sizes and shapes: there are several regions where ϵ' is negative, but sufficiently isolated from one another that shape does not cause the corresponding bands to overlap appreciably. Suggestions that disordering the crystal lattice would broaden such bands were confirmed by measurements of optical constants of highly disordered silicates together with small-particle extinction calculations and measurements (Day, 1976; Krätschmer and Huffman, 1979). This work shows that highly disordered silicates fit the featureless 9.7-μm interstellar band quite well. Absorption bands in disordered silicates are sufficiently weak compared with those in ionic solids that shape effects, such as those dominating the spectra discussed in Section 12.3, are not very important.

Although the 9.7-μm band is difficult to observe in typical interstellar regions, it has been found in the spectra of astronomical objects embedded in dense molecular clouds. Also commonly found in such clouds is an absorption band at 3.07 μm, usually attributed to ice, both water and ammonia. In water ice the band corresponds to the O—H stretching mode (see Section 10.3). Water was once thought to be a major component of the interstellar dust, but failure to detect a strong 3.1-μm ice band has somewhat weakened this view. The ratio of strengths of the 3.07- and 9.7-μm bands in molecular clouds varies widely, strongly suggesting that different substances are responsible for the two features. It may be that silicate grains become coated with a mantle of ice in the more protected regions of molecular clouds. To calculate extinction by such coated particles—assuming they are spherical—requires the theory of Section 8.1 and Appendix B; examples of such calculations are given by Aannestad (1975).

14.5.4   Emission Spectra of Circumstellar Shells

Stars further along in their life cycle are often cooler and redder than younger stars. Shells of dust that has been condensed from material ejected from these cool stars often surround them. Such circumstellar dust shells, heated by the stars, emit strongly in the infrared with a spectrum characteristic of absorption bands in the dust: the emissivity of a small particle is equal to its absorption efficiency (see Section 4.7). An excellent review of circumstellar dust has been given by Ney (1977).

Critical to the composition of the condensate is the carbon-to-oxygen ratio in the star. If carbon dominates (C/O > 1), the oxygen is tied up in gaseous carbon compounds, leaving excess carbon to combine with other elements to form solids. Probable condensation products of such carbon stars are SiC and solid carbon, perhaps in the form of graphite, amorphous carbon, or some intermediate structure. If the star is oxygen rich (C/O < 1), condensation is dominated by excess oxygen left over when the carbon is depleted, leading to probable solids such as oxides, including the silicates.

Two emission spectra of circumstellar shells are shown in Fig. 14.5, for an oxygen star in the bottom part of the figure and for a carbon star in the top.

Rising above the background emission, assumed to be that of a blackbody at 3000°K, is a broad 9.7-μm feature for the oxygen star. Between about 15 and 20 μm a much weaker excess emission feature is also evident. A measured absorption spectrum of amorphous olivine smoke particles (Krätschmer and Huffman, 1979) is shown for comparison.

Dust around the carbon star shows an excess emission feature between about 10.2 and 11.6 μm, clearly distinguishable in both shape and position from the 9.7-μm feature of the oxygen star, which has been attributed to small SiC particles. These particles cannot be spherical, however. According to the discussion in Section 12.2, shape effects spread an absorption band in small particles of materials like SiC between the transverse (ω_t) and longitudinal (ω_l) optical mode frequencies; these frequencies for SiC are indicated on the figure. This point was made by Treffers and Cohen (1974) using Gilra's unpublished calculations. To illustrate this further, calculations for a random distribution of

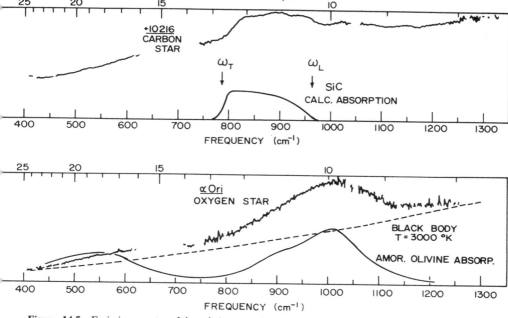

Figure 14.5 Emission spectra of dust shells around a carbon star (top) and around an oxygen star (bottom). From Treffers and Cohen (1974).

Rayleigh ellipsoids, arbitrarily normalized, are reproduced from Fig. 12.15. Spherical SiC particles match the circumstellar feature very poorly, but a wide distribution of shapes matches it quite well. This indicates that even shape information may sometimes be extracted from emission spectra of inaccessible particles.

14.5.5 Linear Polarization

Interstellar dust partially polarizes, as well as attenuates, the light it transmits, which places further constraints on particle size and composition. It is almost certain that this polarization is caused by asymmetric particles aligned in the galactic magnetic field, although the exact alignment mechanism is still uncertain (Aannestad and Purcell, 1973). The degree of linear polarization is as much as 10% and peaks between 0.4 and 0.8 μm. Although both the maximum polarization P_{max} and the corresponding wavelength λ_{max} vary from star to star, the normalized polarization P/P_{max} falls on a single curve (Fig. 14.6) when plotted against normalized wavelength λ_{max}/λ (Coyne et al., 1974). According to Martin (1974) this scaling relation implies that the optical constants of the particles at and near visible wavelengths do not vary apprecia-

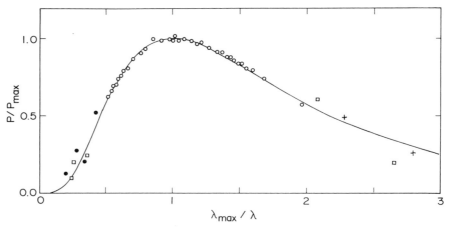

Figure 14.6   Observed linear polarization of light from several stars; λ_{max} is the wavelength at which the maximum polarization P_{max} occurs. From Coyne et al. (1974).

bly. This is evidence against graphite and in favor of insulating solids such as silicates, ices, SiC, etc. There does not appear to be a polarization feature associated with the 2200-Å extinction band generally attributed to graphite (Gehrels, 1974b); this indicates that the particles responsible for this band are not aligned.

Unpolarized incident light is partially polarized upon extinction by particles if the cross section depends on polarization; the degree of linear polarization is proportional to the difference in the cross sections for light polarized in two orthogonal directions. Aligned cylinders (Fig. 8.7) and aligned spheroids (Fig. 11.16) are examples of such particles. Note in Fig. 11.16 that the difference between the two cross sections for the oblate spheroid ($\zeta = 30°$) is greatest on the steeply rising part of the extinction curve. For a given particle size the degree of polarization is least for the smaller size parameters (longer wavelengths), increases with decreasing wavelength to a maximum, then decreases, and eventually reverses sign. If the particle size is increased, the polarization maximum shifts to a longer wavelength; if it is decreased, the maximum shifts to a shorter wavelength. Although the wavelength of maximum polarization depends on size, the degree of polarization depends on only the *ratio* of size to wavelength—provided that the optical constants vary weakly over the wavelength range of interest. This helps to explain Fig. 14.6.

14.5.6   Circular Polarization

Small degrees of circular polarization (V/I) in starlight, usually less than 1%, have been observed in recent years; the polarization modulation technique for observing such polarization was discussed in Section 13.7. There are at least two mechanisms for circularly polarizing starlight.

An interstellar dust cloud containing aligned particles may be looked upon as a linearly birefringent (and possibly linearly dichroic) medium (van de Hulst, 1957, p. 58): the cloud acts like a retarder. We showed at the end of Section 2.11 that linearly polarized light becomes circularly polarized upon transmission by a retarder. The first clear evidence for this kind of polarization mechanism was reported by Martin (1972), where light from the Crab Nebula was the source of linearly polarized incident light.

Interstellar grains may also circularly polarize unpolarized light. Such light is partially linearly polarized upon traversing a cloud of aligned grains. If this linearly polarized light is then incident on another cloud of aligned grains, where the alignment axis of the second set of grains is rotated relative to that of the first, the transmitted light will be partially circularly polarized. This is the simplest example of grain alignment changing along the line of sight, a mechanism for circular polarization discussed in detail by Martin (1974). A certain amount of this type of changing alignment is expected to be fairly common and provides the accepted explanation for the small degrees of circular polarization observed in light from a number of stars.

14.5.7 Scattering by Interstellar Dust

Light scattered by interstellar dust carries with it information about the grains. Such scattered light has been observed as diffuse galactic light (DGL), the faint but general glow of scattered starlight, and as reflection nebulae, the scattered light from particularly dense clouds of dust near bright stars or groups of stars. Diffuse galactic light is very weak compared with other sources of sky brightness, such as direct starlight, the zodiacal light from interplanetary dust in our solar system, and atmospheric airglow. But in spite of this, measurements of DGL together with extinction give the best values of the ratio of scattering to extinction—the single-scattering albedo—by interstellar grains. Reflection nebulae, like the one around the star Merope in the Pleiades and the large bright area in the constellation Orion, are much brighter than diffuse galactic light. Unfortunately, the configuration of source and dust is often poorly known. Even with the source of illumination in front of the dust cloud—perhaps the best configuration from the point of view of interpreting observations—it has been difficult to fit both angular intensity and polarization measurements with Mie calculations. Zellner (1973) concluded from this that Mie theory may be inapplicable to particles in reflection nebulae. In this connection refer to the discussion of scattering by nonspherical particles in Section 13.4, particularly Figs. 13.8 and 13.9. Recall that polarization by nonspherical particles can be opposite in sign to that by equivalent spheres of the same material (Fig. 13.9). And disagreement between the measured angular distribution of scattered light and that calculated by Mie theory (Fig. 13.8) is greatest in the backward directions; but these are the directions of light from reflection nebulae with the simplest configuration.

Extracting information about interstellar dust from analysis of scattered light, as opposed to transmitted light, is fraught with many difficulties.

14.5.8 Far-Infrared Emission

It is not difficult to show that the emissivity of small spherical particles, composed of both insulating and metallic crystalline solids, is expected to vary as $1/\lambda^2$ in the far infrared. For example, if the low-frequency limit of the dielectric function for a single Lorentz oscillator (9.16) is combined with (5.11), the resulting emissivity is

$$e = Q_{abs} \simeq \frac{48\pi^2 ac}{(\epsilon_0 + 2)^2} \frac{\gamma\omega_p^2}{\omega_0^4} \frac{1}{\lambda^2} \qquad (\omega \ll \omega_0)$$

where a is the sphere radius, c is the speed of light *in vacuo*, and $\epsilon_0 = \epsilon'(0)$. In this instance, therefore, the emissivity varies as $1/\lambda^2$ at frequencies well below the resonant frequency ω_0. Similarly, for a small metallic sphere with the Drude dielectric function (9.27), we have

$$e \simeq \frac{48\pi^2 ac\gamma}{\omega_p^2} \frac{1}{\lambda^2} \qquad (\omega \ll \gamma).$$

Again, the emissivity varies as $1/\lambda^2$ in the far infrared. But various observations of emission from interstellar dust suggest that the wavelength dependence of emissivity is closer to $1/\lambda$ than to $1/\lambda^2$ (Seki and Yamamoto, 1980). This may be a consequence of the failure of the conditions underlying the $1/\lambda^2$ dependence—the particles are *crystalline* and *spherical*—to be satisfied.

For the insulating solid it was assumed that far infrared frequencies were well below any absorption band. But the allowed optical transitions in an amorphous, insulating solid are not necessarily confined to bands; that is, there may be continuum absorption extending into the low-frequency region. The characteristics of amorphous grains as far-infrared emitters have been discussed by Seki and Yamamoto (1980), who also pointed out the $1/\lambda^2$ dependence discussed in the preceding paragraph. Measurements of far-infrared absorption by small particles of amorphous silicates (Day, 1979) and amorphous graphite (Koike et al., 1980) have in fact shown that the absorption efficiency varies approximately as $1/\lambda$.

Particle shape can strongly affect the far-infrared emission spectrum of metallic particles, for which the region of negative ϵ' is large; graphite also falls into this category. This is evident from measured absorption by small aluminum particles (Fig. 12.20), which we have interpreted using a distribution of ellipsoidal shapes. Note that the wavelength dependence of absorption by spheres is vastly different from that by a collection of particles distributed in shape.

From these considerations we conclude that the failure of the emission spectrum of interstellar dust to vary as $1/\lambda^2$ in the far infrared, which is predicted for small crystalline spheres, may be the result of either noncrystallinity or nonsphericity (or both). Therefore, the infrared emission spectrum may not prove to be as uniquely diagnostic of interstellar grain characteristics as it once was thought to be.

14.5.9   Summary of Observations and Their Interpretations

We summarize in Table 14.2 the observed characteristics of interstellar dust together with their most common interpretations. What is meant by a "common interpretation" is to some extent a matter of opinion, and this table naturally reflects ours. Other interpretations have been published in the many papers on interstellar dust; these can be found in the review articles cited at the beginning of this section.

Table 14.2   Summary of Interstellar Dust Observations

Observation	Interpretation
Reddening of visible light	Particles smaller than λ
Prominent extinction peak at 2170 Å (5.7 eV)	Surface plasmon in graphite
Continued rise in extinction to at least 12 eV	Extinction by the small size component
Extinction knee near 3 eV	Scattering by the large size component
Many narrow bands (diffuse bands) in the visible	Unidentified
Absorption and emission bands at 9.7 and 18 μm	Vibrational modes in disordered silicate grains
Emission band near 11 μm associated with carbon stars	Shape broadened vibrational modes in SiC
Linear polarization peaking in the visible	Asymmetric particles aligned in galactic magnetic fields
Circular polarization	Changing grain alignment; aligned grains acting like retarder
Absorption band at 3.07 μm	H_2O and NH_3 ice coatings on particles in protected regions
Far infrared emission decreasing as $\sim 1/\lambda$	Amorphous silicates or carbon; possible shape effects

14.6 PRESSURE DEPENDENCE OF INTRINSIC OPTICAL SPECTRA USING SMALL PARTICLES

Lattice parameters of crystalline solids can be varied over appreciable ranges by subjecting them to very high pressures. The development of diamond-anvil pressure cells (see, e.g., Block and Piermarini, 1976) and a simple technique for monitoring pressure using the fluorescence spectrum of a tiny chip of ruby inside a cell (Forman et al., 1972) has made it possible to study the effects of pressures up to more than 400 kbar on the optical properties of solids (and liquids). But the class of optical effects involving electronic excitations has been difficult to study in pressure cells. Absorption is so strong that very thin films must be used for transmission studies; but the structure of such films is often different from that of bulk crystalline solids because of the way films are grown on substrates. And specular reflection techniques are impractical for samples under pressure because of small apertures, many reflecting surfaces, and relatively long optical paths in pressure cells. It is possible, however, to exploit the advantages of small particles—it is easier to make small crystals than large crystals—as samples and study optical effects by measuring the dependence of spectral transmission on pressure.

An example of a spectral feature of possible interest is the exciton peak in MgO near 7.6 eV (see Figs. 9.5 and 10.1). This peak has been studied primarily by reflection from cleaved crystals of MgO. But careful inspection of Fig. 11.2 reveals that the exciton peak is also expected to appear clearly in extinction by MgO particles provided that they are sufficiently small.

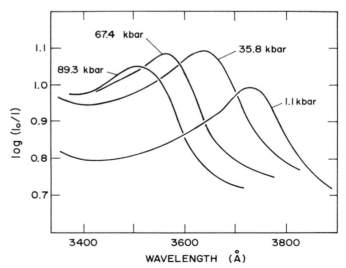

Figure 14.7 Optical density spectrum of ZnO particles for several pressures. From Huffman et al. (1982).

This idea of using small particles to obtain information about solids under pressure has been explored by Huffman et al. (1982), who studied the band gap exciton in ZnO at pressures up to 107 kbar. A smoke consisting of small crystalline particles (~ 0.1 μm) of ZnO was formed by arc-vaporizing metallic zinc in air and collecting the particles on the inner surface of one of the diamonds that form both the windows and the anvils of the pressure cell. Spectral transmission measurements taken by simply placing the sample-laden cell in a double-beam spectrophotometer clearly revealed the intrinsic band gap exciton (near 3730 Å at ambient pressure), the characteristics of which were then followed as pressure was increased to 107 kbar. A progressive shift of the band to about 3490 Å was observed (Fig. 14.7), and the exciton energy (position of the absorption edge) increased linearly with pressure. The primary aim of this experiment was not, however, to study the ZnO exciton but rather to demonstrate a new technique for studying optical effects in solids under pressure by using small particles.

14.7   GIAEVER IMMUNOLOGICAL SLIDE

A simple visual technique for investigating immunological reactions has been devised by Giaever (1973). A discontinuous film composed of small metallic particles, or islands, is evaporated onto a glass slide, and a monolayer of antigen adsorbed onto this surface; indium was first used, later an indium–gold alloy (Giaever and Laffin, 1974). The occurrence of an antigen–antibody reaction is indicated by darkening of the slide. The mechanism for the observed darkening was not discussed in detail; it was merely mentioned in passing that it was the result of increased scattering. But small metallic particles absorb light more strongly than they scatter it, and it seems clear that what is observed may be attributed to a shift of the surface mode frequency because of the coating on the metallic particles. The Giaever immunological slide is therefore a biomedical application of surface modes in small metallic particles, which were discussed at length in Chapter 12.

Treu (1976) measured transmission spectra (optical density) for indium films of the type used for immunological slides. His calculations based on Mie theory led him to the conclusion that this theory was not compatible with observations. In particular, calculations for spheres of diameter 1390 Å showed a feature that was not observed in experiments. Although calculations for spheres half this size did not show this feature, the wavelength of maximum optical density was much shorter than that observed.

It is not surprising, however, that Mie theory is inadequate in this instance: the indium particles are not spheres, they are more nearly oblate spheroids with (average) major and minor diameters of about 1390 and 368 Å.

Although the indium particles are not small compared with the wavelength in all directions, they are sufficiently small (368 Å) along the direction of propagation of the incident light that Rayleigh theory is a good approximation. The unobserved feature calculated for 1390-Å spheres is therefore easy to explain: such spheres are, at visible and ultraviolet frequencies, sufficiently large that higher-order modes are excited; this was shown in Fig. 12.1 by a set of calculations for progressively larger SiC particles. The disparity between the measured frequency of peak extinction and that calculated for 700-Å spheres is also readily explicable: this frequency is strongly dependent on shape and can extend all the way from the plasma frequency ω_p (typically in the ultraviolet) down to radio frequencies.

Interpretation of the observed transmission spectra of immunological slides is complicated by the close separation of the indium particles; Treu (1976), for example, reports areal coverages of 68%. Also, the particles, even if they are uncoated, are not embedded in a single homogeneous medium: there is air on one side of them and glass on the other. And while the particles are most definitely not spheres, they are not identical oblate spheroids: there is a distribution of shapes. But we have taken the view throughout this book that understanding small-particle effects begins with isolated, uncoated spheres. Each of these restrictions can then be successively removed to assess their relative importance in determining spectral features.

Indium is nearly a free-electron metal with a plasma frequency of about 11.5 eV, which corresponds to a wavelength λ_p of 1080 Å (Koyama et al., 1973). It follows from (12.28) that the wavelength λ_s of maximum absorption by a small metallic oblate spheroid illuminated by light incident along its symmetry axis is approximately

$$\lambda_s = \lambda_p \sqrt{\frac{\epsilon_m - L_1(\epsilon_m - 1)}{L_1}} , \qquad (14.6)$$

where ϵ_m is the dielectric function of the surrounding medium; the geometrical factor L_1 takes on all values between 0 (disk) and $\frac{1}{3}$ (sphere). The particles are on a glass slide, so let us take ϵ_m to be 1.625, the average of that of glass (2.25) and air. For a sphere, therefore, $\lambda_s = \lambda_F$ (the wavelength corresponding to the Fröhlich frequency) is about 2230 Å, which is close to the value 2500 Å calculated for a 700 Å sphere by Treu (1976).

As the sphere is flattened into a disk the position of maximum absorption shifts to longer wavelengths. For example, if $c/a = 368/1390$, it follows from Fig. 5.6 that L_1 is about 0.19 and from (14.6) that λ_s is 3040 Å. This is an appreciable shift—over 800 Å—but still short of the measured value 4100 Å. Our analysis, however, implicitly assumed isolated spheroids, a condition that was not satisfied in the experiments.

We can estimate the magnitude of the shift attributable to interaction between particles by appealing to the Maxwell Garnett theory (Section 8.5). This theory is strictly applicable only to a medium consisting of small particles distributed throughout a volume, whereas the slides consist of a single layer of particles on a surface. Nevertheless, for our limited purposes here the Maxwell Garnett theory is adequate.

It follows from (8.47) that the average dielectric function of a suspension of identical oblate spheroids is

$$\epsilon_{av} = \frac{(1-f)\epsilon_m + f\epsilon\lambda_1}{1-f+f\lambda_1}; \qquad \lambda_1 = \frac{\epsilon_m}{\epsilon_m + L_1(\epsilon - \epsilon_m)}, \qquad (14.7)$$

where the electric field is parallel to the major axes. If (14.7) is expanded to terms *linear* in the particle volume fraction f, the wavelength where ϵ_{av}'' is a maximum is given by (14.6), as expected. But if ϵ_{av}'' is truncated after the *quadratic* term in f, the wavelength of maximum absorption $\lambda_s(f)$ is given by

$$\lambda_s(f) = \lambda_s\left[1 + \frac{1}{2}\frac{f\epsilon_m}{L_1 + \epsilon_m(1 - L_1)}\right].$$

Thus, the effect of decreasing the separation between particles is to shift the position of maximum absorption to longer wavelengths, which is consistent with the experimental observations.

The absorption spectrum of isolated indium spheres differs from that of closely packed oblate spheroids in that the peak shifts from 2230 Å to 4100 Å; about half of this shift is attributable to particle shape and half to particle interaction. Indium particles on immunological slides are not identical, however, but are distributed in size and shape about some mean; this tends to broaden the spectrum.

The observed darkening of the indium slides results from a shift of the absorption peak because of the coating on the particles. Because of the cumbersomeness of the expressions for coated ellipsoids (Section 5.4) this shift can be understood most easily by appealing to (12.15), the condition for surface mode excitation in a coated sphere. For a small metallic sphere with dielectric function given by the Drude formula (9.26) and coated with a nonabsorbing material with dielectric function ϵ_2, the wavelength of maximum absorption is approximately

$$\lambda_s = \lambda_F\left[1 + \delta(\epsilon_2 - \epsilon_m)\frac{\epsilon_2 + 2\epsilon_m}{\epsilon_2 + 2\epsilon_m\epsilon_2}\right],$$

where terms of higher order in δ, the coating thickness relative to the core radius, are neglected. If ϵ_2 is greater than ϵ_m, therefore, the effect of the coating is to shift the position of peak absorption to longer wavelengths, which is also in accord with observations. Moreover, to first order at least, this shift is proportional to the thickness of the coating and increases with increasing ϵ_2.

Our assertion that the darkening of immunological slides, which are observed by transmitted light, is primarily an absorption rather than a scattering phenomenon can be supported by a simple calculation. The cross sections for absorption and scattering by a single, uncoated oblate spheroid illuminated along its symmetry axis are (see Section 5.5)

$$C_{abs} = k \, \text{Im}\{\alpha_1\}; \qquad C_{sca} = \frac{k^4}{6\pi}|\alpha_1|^2$$

$$\alpha_1 = 4\pi a^2 c \frac{\epsilon - \epsilon_m}{3\epsilon_m + 3L_1(\epsilon - \epsilon_m)}.$$

If we use the values of ϵ' and ϵ'' measured by Koyama et al. (1973), the ratio of maximum absorption to maximum scattering is approximately

$$\frac{C_{abs}}{C_{sca}} \simeq 6 \qquad (\lambda \simeq 3000 \text{ Å})$$

for $L_1 = 0.19$, $a = 695$ Å, and $c = 184$ Å. Absorption therefore dominates extinction in this instance.

Biology is an unlikely field in which to find an application of surface modes in small metallic particles. After all, the optical properties of particles of biological origin are very much unlike those of metals. Yet the Giaever immunological slide is just such an application. Moreover, it exemplifies many of the salient characteristics of surface modes which were discussed in Chapter 12: their strong dependence on particle shape, coating, and the surrounding medium.

14.8 MICROWAVE ABSORPTION BY MACROMOLECULES

A rather grisly example of biological effects of microwave radiation is provided by the story, possibly apocryphal, of the woman who put her cat in a microwave oven to dry after it had come in out of the rain. The cat exploded! And one hears tales of crows and seagulls being shot out of the sky with intense microwave beams.

Regardless of the truth of these stories, it is undeniable that microwave radiation can be hazardous to life. One damage mechanism is merely intense heating of the water bound up in all living organisms. It is clear, therefore, that the potential hazard of radiation of a given frequency depends on the optical constants, particularly ϵ'', of water at that frequency. For example, at room temperature the maximum value of ϵ'' occurs at about 20 GHz (see Fig. 9.15), that is, at the relaxation frequency $1/2\pi\tau$, where τ is the relaxation time in (9.41).

Although liquid water is usually looked upon as a single, unvarying substance, its structure near surfaces may be quite different from that of "free" water. Interaction between water molecules and an immediately adjacent surface orients these molecules, and this orientation can extend for several molecular diameters into the bulk liquid. A very clear and succinct description of this mechanism was given by Henniker (1949): "Although the powerful forces involved are very short range, they are transmitted by successive polarization of neighboring molecules to an impressive depth. The analogy is with a magnet of very limited range of appreciable direct interaction which nevertheless can lift or hold a piece of iron at a considerable distance if there is an intermediate chain of iron fillings or pieces of iron in between." Many papers have been written about the properties of liquids at interfaces, the earliest of which were reviewed by Henniker (1949); more recently, Drost-Hansen (1969) thoroughly examined the evidence for the existence of ordered water near solid interfaces.

If the structure of water depends on distance from a surface, so must its physical properties, including its dielectric function. We noted in Section 9.5 that at microwave frequencies the dielectric function of water changes markedly when the molecules are immobilized upon freezing; as a consequence, the relaxation frequency of ice is much less than that of liquid water. Water irrotationally bound to surfaces is therefore expected to have a relaxation frequency between that of water and ice.

Water molecules are oriented at the surfaces of macromolecules as well as at solid surfaces. For example, Bernal (1965) refers to a "regular formation of ice" surrounding most protein molecules, although by "ice" he does not mean free water ice. Bound water in hydration shells surrounding macromolecules in aqueous solutions is sometimes denoted as lattice-ordered or ice-like and has been taken into account in interpreting the dielectric functions of such solutions (Buchanan et al., 1952; Jacobson, 1955; Pennock and Schwan, 1969).

The possible consequences of irrotationally bound water for microwave absorption by biological materials are twofold. Calculations of the rate of energy deposition using the dielectric function of free water may be appreciably in error at frequencies well below its relaxation frequency, which is greater than that of bound water. Moreover, the bound water is confined to thin hydration shells around macromolecules, so the enhanced energy absorption in these shells may cause localized disorganization. Dawkins et al. (1979) put forward these ideas and supported them with calculations of power deposition for a simple model of a hydrated molecule: a sphere of radius 5 nm surrounded by a shell of water two molecules thick; these calculations were based on the theory of Section 8.1—only the first term in the series is needed because of the extremely small size parameter—and the dielectric function (9.41) with various relaxation frequencies. Although the validity of using macroscopic theory for a single macromolecule coated by only two layers of water molecules is uncertain, their results are likely to be qualitatively correct and, consequently, are worth contemplating.

The volumetric power deposition calculated for bound water was appreciably greater—up to about five times—than that for free water; the maximum difference occurs near the bound water relaxation frequency. This enhanced energy deposition is localized in the bound water shell and therefore may cause more damage than if it were distributed uniformly throughout the medium. But Dawkins et al. consider the enhancement of biological damage by localization still to be speculative because of uncertainties about the rate of energy transfer from the shell to its macromolecule host and to its surroundings. Nevertheless, ignoring the differences between the dielectric functions of bound and free water may have important consequences. Dawkins et al. illustrate this with a particular microwave injury the mechanism for which is well understood. Cataracts are induced by microwave radiation because it is absorbed by lens water and the resulting temperature increase denatures lens proteins. About 65% of the weight of the lens is water, at least 40% of which is bound water. If all the water were taken as free, as is usual in estimating microwave energy absorption in biological materials, the rate of heating by microwave radiation in the frequency range 0.3–3 GHz could be underestimated.

Appendixes

Computer Programs

The following appendixes contain subroutines for calculating scattering by a homogeneous sphere (BHMIE), a coated sphere (BHCOAT), and a normally illuminated infinite cylinder (BHCYL), together with their calling programs and sample calculations. A brief description accompanies each program. The final versions of the programs were tested on the CDC 7600 computer at the Los Alamos Scientific Laboratory. Although we tried to write the programs in standard Fortran so that they can be run on most computers, it is likely that they will have to be modified slightly. For example, the first statement in each calling program is not executable by many Fortran compilers; also, input, output, and format statements may have to be changed. Major changes, although unlikely, may be necessary.

We aimed at simplicity and lucidity rather than programming elegance so that users will be able to easily adapt these programs to their own needs. To this end Fortran variables were chosen to resemble or suggest the corresponding variables in the sections where the underlying theory is developed and discussed. The logic of the programs should be evident without prolonged study: there are no clever, but obscure, shortcuts. Little effort was made to optimize the programs; they are neither the best nor the brightest. A lifetime could be devoted to refining them, making them faster and more efficient, and extending the range of their applicability.

Any program will, if extended beyond its proper limits, give unreliable or nonsensical results. The programs described below are most likely to give trouble either for very small or very large size parameters. Computations based on exact theories, however, are often unnecessary in such extremes: Rayleigh theory is a good approximation for very small particles, and geometrical optics combined with diffraction theory is a good approximation for very large particles. To call for the exact theory in these instances is to call for an elephant gun to shoot a mouse.

The programs should not be used without an attitude of healthy skepticism. We tested them as much as possible, but they undoubtedly contain hidden flaws. In the following appendixes we discuss criteria that the programs were required to satisfy. Some of them are obvious: the extinction efficiency must not be less than the scattering efficiency, and both must be nonnegative; others are more subtle.

Each program is composed of two parts: a calling program, which requires as input refractive indices, particle size, and wavelength; and a subroutine—the workhorse—which computes scattering coefficients, scattering matrix elements, and efficiencies. The possible combinations of input and output variables are almost countless. Therefore, we have included calling programs merely to point the way to how the subroutines might be grafted onto calling programs more suited to the users' needs.

Appendix A

Homogeneous Sphere

The theory underlying this appendix is given in Chapter 4; some of the computational aspects of Mie theory are discussed in Section 4.8.

Perhaps the best known program for computing Mie scattering coefficients is that by Dave (1968)—it is certainly one of the earliest to have a wide distribution. We have profited greatly from this program, and we would be remiss if we did not acknowledge our indebtedness to Dave. The subroutine BHMIE described in this appendix is, however, sufficiently different that it should not be considered as merely a minor variant form. We have borrowed tricks from here and there as well as added a few of our own, all with the aim of writing a simple, efficient program, easy to understand and hence easy to modify.

One of the major departures from the Dave program is that in BHMIE convergence of series is not determined by iteration. With the wisdom of hindsight, iteration seems inefficient because there is little disagreement about the approximate number of terms required for convergence: slightly more than x terms are sufficient, where x is the size parameter. We have tried various criteria, based more or less on guessing. After BHMIE was written, however, an extensive study was published by Wiscombe (1979, 1980), and we have modified our programs in the light of his work. Thus, series in BHMIE are terminated after NSTOP terms, where NSTOP is the integer closest to

$$x + 4x^{1/3} + 2.$$

A similar criterion was used by Wiscombe, who was guided by a suggestion by Khare and extensive computations. This criterion can, of course, be changed. But lest the reader with a large computer budget be seduced by the idea that if a certain number of terms is good then even more are better, we must issue a warning. Computation of ψ_n by forward recurrence is unstable, and roundoff error will eventually become unacceptable. Provided that one does not generate more orders of ψ_n than are needed for reasonable convergence, and that ψ_n is a double-precision variable, problems are not likely to be encountered with a computer of moderate size. But an attempt to squeeze out a few more decimal places might lead to disaster: scattering coefficients of order appreciably greater than NSTOP might be computed inaccurately, and greatly so, even though they are not really needed.

$D_n(mx)$ in the coefficients (4.88) is computed by the downward recurrence relation (4.89) beginning with D_{NMX}. Provided that NMX is sufficiently greater than NSTOP and $|mx|$, logarithmic derivatives of order less than NSTOP are remarkably insensitive to the choice of D_{NMX}; this is a consequence of the stability of the downward recurrence scheme for ψ_n. For vastly different choices of D_{NMX}, and a range of arguments mx, computed values of D_{NMX-5} were independent of D_{NMX}. Thus, NMX is taken to be Max(NSTOP,$|mx|$) + 15 in BHMIE, and recurrence is begun with $D_{NMX} = 0.0 + i0.0$.

Both ψ_n and ξ_n, where $\xi_n = \psi_n - i\chi_n$, satisfy

$$\psi_{n+1}(x) = \frac{2n+1}{x}\psi_n(x) - \psi_{n-1}(x),$$

and are computed by this upward recurrence relation in BHMIE beginning with

$$\psi_{-1}(x) = \cos x, \qquad \psi_0(x) = \sin x,$$
$$\chi_{-1}(x) = -\sin x, \qquad \chi_0(x) = \cos x.$$

ψ_n is a double-precision and χ_n a single-precision variable.

The angle-dependent functions π_n and τ_n are computed by the upward recurrence relations (4.47). They need be computed only for scattering angles between 0 and 90° because of the relations (4.48).

Tests of BHMIE   We have tested BHMIE thoroughly, which gives credence to its impeccability but does not guarantee it. In particular, we have never encountered any appreciable differences between results from BHMIE and those from Dave's program DBMIE. Aside from comparing results from BHMIE with those tabulated elsewhere or computed by other subroutines, there are several independent checks on any scattering program:

Q_{ext} and Q_{sca} must not be negative, and Q_{ext} must be greater than Q_{sca} except for a nonabsorbing sphere, in which instance they are equal. For very large size parameters the extinction efficiency approaches the limit 2. This might seem to be a good test of a program for large x. Q_{ext} oscillates about 2, however, and one is never sure if a deviation from 2 is a natural oscillation or an indicator of incipient error. We have found that a much more sensitive test of a program for large x is the asymptotic expression (4.83) for the backscattering efficiency. This seems not to be widely recognized, and it is worth mentioning here because it can be used to test other programs.

As a check on the amplitude scattering matrix elements, we compute Q_{ext} in BHMIE from the optical theorem (4.76), whereas Q_{sca} is computed from the series (4.61). POL, the degree of polarization, must vanish for scattering angles of 0 and 180°, as must S_{34}. Also, the 4×4 scattering matrix elements must satisfy

$$\left(\frac{S_{12}}{S_{11}}\right)^2 + \left(\frac{S_{33}}{S_{11}}\right)^2 + \left(\frac{S_{34}}{S_{11}}\right)^2 = 1$$

for all scattering angles.

```
 1        PROGRAM CALLBH (INPUT=TTY,OUTPUT=TTY,TAPE5=TTY)
 2    C   :::::::::::::::::::::::::::::::::::::::::::::::::::::::::::::::::::::::::
 3    C      CALLBH CALCULATES THE SIZE PARAMETER (X) AND RELATIVE
 4    C      REFRACTIVE INDEX (REFREL) FOR A GIVEN SPHERE REFRACTIVE
 5    C      INDEX, MEDIUM REFRACTIVE INDEX, RADIUS, AND FREE SPACE
 6    C      WAVELENGTH. IT THEN CALLS BHMIE, THE SUBROUTINE THAT COMPUTES
 7    C      AMPLITUDE SCATTERING MATRIX ELEMENTS AND EFFICIENCIES
 8    C   :::::::::::::::::::::::::::::::::::::::::::::::::::::::::::::::::::::::::
 9        COMPLEX REFREL,S1(200),S2(200)
10        WRITE (5,11)
11    C   :::::::::::::::::::::::::::::::::::::::::::::::::::::::::::::::::::::::::
12    C      REFMED = (REAL) REFRACTIVE INDEX OF SURROUNDING MEDIUM
13    C   :::::::::::::::::::::::::::::::::::::::::::::::::::::::::::::::::::::::::
14        REFMED=1.0
15    C   :::::::::::::::::::::::::::::::::::::::::::::::::::::::::::::::
16    C      REFRACTIVE INDEX OF SPHERE = REFRE + I:REFIM
17    C   :::::::::::::::::::::::::::::::::::::::::::::::::::::::::::::::
18        REFRE=1.55
19        REFIM=0.0
20        REFREL=CMPLX(REFRE,REFIM)/REFMED
21        WRITE (5,12) REFMED,REFRE,REFIM
22    C   :::::::::::::::::::::::::::::::::::::::::::::::::::::
23    C      RADIUS (RAD) AND WAVELENGTH (WAVEL) SAME UNITS
24    C   :::::::::::::::::::::::::::::::::::::::::::::::::::::
25        RAD=.525
26        WAVEL=.6328
27        X=2.:3.14159265:RAD:REFMED/WAVEL
28        WRITE (5,13) RAD,WAVEL
29        WRITE (5,14) X
30    C   :::::::::::::::::::::::::::::::::::::::::::::::::::::::::::::::::::::
31    C      NANG = NUMBER OF ANGLES BETWEEN 0 AND 90 DEGREES
32    C      MATRIX ELEMENTS CALCULATED AT 2:NANG - 1 ANGLES
33    C      INCLUDING 0, 90, AND 180 DEGREES
34    C   :::::::::::::::::::::::::::::::::::::::::::::::::::::::::::::::::::::
35        NANG=11
36        DANG=1.570796327/FLOAT(NANG-1)
37        CALL BHMIE(X,REFREL,NANG,S1,S2,QEXT,QSCA,QBACK)
38        WRITE (5,65) QSCA,QEXT,QBACK
39        WRITE (5,17)
40    C   :::::::::::::::::::::::::::::::::::::::::::::::::::::::::::::::::::::
41    C      S33 AND S34 MATRIX ELEMENTS NORMALIZED BY S11.
42    C      S11 IS NORMALIZED TO 1.0 IN THE FORWARD DIRECTION
43    C      POL=DEGREE OF POLARIZATION (INCIDENT UNPOLARIZED LIGHT)
44    C   :::::::::::::::::::::::::::::::::::::::::::::::::::::::::::::::::::::
45        S11NOR=0.5:(CABS(S2(1)):2+CABS(S1(1)):2)
46        NAN=2:NANG-1
47        DO 355 J=1,NAN
48        AJ=J
49        S11=0.5:CABS(S2(J)):CABS(S2(J))
50        S11=S11+0.5:CABS(S1(J)):CABS(S1(J))
51        S12=0.5:CABS(S2(J)):CABS(S2(J))
52        S12=S12-0.5:CABS(S1(J)):CABS(S1(J))
53        POL=-S12/S11
54        S33=REAL(S2(J):CONJG(S1(J)))
55        S33=S33/S11
56        S34=AIMAG(S2(J):CONJG(S1(J)))
57        S34=S34/S11
58        S11=S11/S11NOR
59        ANG=DANG:(AJ-1.):57.2958
60    355 WRITE (5,75) ANG,S11,POL,S33,S34
```

```
 61      65 FORMAT (//,1X,"QSCA= ",E13.6,3X,"QEXT = ",E13.6,3X,
 62         2"QBACK = ",E13.6)
 63      75 FORMAT (1X,F6.2,2X,E13.6,2X,E13.6,2X,E13.6,2X,E13.6)
 64      11 FORMAT (/"SPHERE SCATTERING PROGRAM"//)
 65      12 FORMAT(5X,"REFMED = ",F8.4,3X,"REFRE =",E14.6,3X,
 66         3"REFIM = ",E14.6)
 67      13 FORMAT (5X,"SPHERE RADIUS = ",F7.3,3X,"WAVELENGTH = ", F7.4)
 68      14 FORMAT (5X,"SIZE PARAMETER =",F8.3/)
 69      17 FORMAT(//,2X,"ANGLE",7X,"S11",13X,"POL",13X,"S33",13X,"S34"//)
 70         STOP
 71         END
 72 C          ::::::::::::::::::::::::::::::::::::::::::::::::::::::::::::::::
 73 C          SUBROUTINE BHMIE CALCULATES AMPLITUDE SCATTERING MATRIX
 74 C          ELEMENTS AND EFFICIENCIES FOR EXTINCTION, TOTAL SCATTERING
 75 C          AND BACKSCATTERING FOR A GIVEN SIZE PARAMETER AND
 76 C          RELATIVE REFRACTIVE INDEX
 77 C          ::::::::::::::::::::::::::::::::::::::::::::::::::::::::::::::::
 78         SUBROUTINE BHMIE (X,REFREL,NANG,S1,S2,QEXT,QSCA,QBACK)
 79         DIMENSION AMU(100),THETA(100),PI(100),TAU(100),PI0(100),PI1(100)
 80         COMPLEX D(3000),Y,REFREL,XI,XI0,XI1,AN,BN,S1(200),S2(200)
 81         DOUBLE PRECISION PSI0,PSI1,PSI,DN,DX
 82         DX=X
 83         Y=X*REFREL
 84 C          ::::::::::::::::::::::::::::::::::::::::::::::::::::::::
 85 C          SERIES TERMINATED AFTER NSTOP TERMS
 86 C          ::::::::::::::::::::::::::::::::::::::::::::::::::::::::
 87         XSTOP=X+4.*X**.3333+2.0
 88         NSTOP=XSTOP
 89         YMOD=CABS(Y)
 90         NMX=AMAX1(XSTOP,YMOD)+15
 91         DANG=1.570796327/FLOAT(NANG-1)
 92         DO 555 J=1,NANG
 93         THETA(J)=(FLOAT(J)-1.)*DANG
 94     555 AMU(J)=COS(THETA(J))
 95 C          :::::::::::::::::::::::::::::::::::::::::::::::::::::::::::::::::
 96 C          LOGARITHMIC DERIVATIVE D(J) CALCULATED BY DOWNWARD
 97 C          RECURRENCE BEGINNING WITH INITIAL VALUE 0.0 + I*0.0
 98 C          AT J = NMX
 99 C          :::::::::::::::::::::::::::::::::::::::::::::::::::::::::::::::::
100         D(NMX)=CMPLX(0.0,0.0)
101         NN=NMX-1
102         DO 120 N=1,NN
103         RN=NMX-N+1
104     120 D(NMX-N)=(RN/Y)-(1./(D(NMX-N+1)+RN/Y))
105         DO 666 J=1,NANG
106         PI0(J)=0.0
107     666 PI1(J)=1.0
108         NN=2*NANG-1
109         DO 777 J=1,NN
110         S1(J)=CMPLX(0.0,0.0)
111     777 S2(J)=CMPLX(0.0,0.0)
112 C          ::::::::::::::::::::::::::::::::::::::::::::::::::::::::::::::::
113 C          RICCATI-BESSEL FUNCTIONS WITH REAL ARGUMENT X
114 C          CALCULATED BY UPWARD RECURRENCE
115 C          ::::::::::::::::::::::::::::::::::::::::::::::::::::::::::::::::
116         PSI0=DCOS(DX)
117         PSI1=DSIN(DX)
118         CHI0=-SIN(X)
119         CHI1=COS(X)
120         APSI0=PSI0
```

```
121          APSI1=PSI1
122          XI0=CMPLX(APSI0,-CHI0)
123          XI1=CMPLX(APSI1,-CHI1)
124          QSCA=0.0
125          N=1
126     200  DN=N
127          RN=N
128          FN=(2.*RN+1.)/(RN*(RN+1.))
129          PSI=(2.*DN-1.)*PSI1/DX-PSI0
130          APSI=PSI
131          CHI=(2.*RN-1.)*CHI1/X - CHI0
132          XI=CMPLX(APSI,-CHI)
133          AN=(D(N)/REFREL+RN/X)*APSI - APSI1
134          AN=AN/((D(N)/REFREL+RN/X)*XI-XI1)
135          BN=(REFREL*D(N)+RN/X)*APSI - APSI1
136          BN=BN/((REFREL*D(N)+RN/X)*XI - XI1)
137          QSCA=QSCA+(2.*RN+1.)*(CABS(AN)*CABS(AN)+CABS(BN)*CABS(BN))
138          DO 789 J=1,NANG
139          JJ=2*NANG-J
140          PI(J)=PI1(J)
141          TAU(J)=RN*AMU(J)*PI(J) - (RN+1.)*PI0(J)
142          P=(-1.)**(N-1)
143          S1(J)=S1(J)+FN*(AN*PI(J)+BN*TAU(J))
144          T=(-1.)**N
145          S2(J)=S2(J)+FN*(AN*TAU(J)+BN*PI(J))
146          IF(J.EQ.JJ) GO TO 789
147          S1(JJ)=S1(JJ) + FN*(AN*PI(J)*P+BN*TAU(J)*T)
148          S2(JJ)=S2(JJ)+FN*(AN*TAU(J)*T+BN*PI(J)*P)
149     789  CONTINUE
150          PSI0=PSI1
151          PSI1=PSI
152          APSI1=PSI1
153          CHI0=CHI1
154          CHI1=CHI
155          XI1=CMPLX(APSI1,-CHI1)
156          N=N+1
157          RN=N
158          DO 999 J=1,NANG
159          PI1(J)=((2.*RN-1.)/(RN-1.))*AMU(J)*PI(J)
160          PI1(J)=PI1(J)-RN*PI0(J)/(RN-1.)
161     999  PI0(J)=PI(J)
162          IF (N-1-NSTOP) 200,300,300
163     300  QSCA=(2./(X*X))*QSCA
164          QEXT=(4./(X*X))*REAL(S1(1))
165          QBACK=(4./(X*X))*CABS(S1(2*NANG-1))*CABS(S1(2*NANG-1))
166          RETURN
167          END
```

SPHERE SCATTERING PROGRAM

```
REFMED =   1.0000   REFRE =  .155000E+01   REFIM =   0.
SPHERE RADIUS =   .525   WAVELENGTH =   .6328
SIZE PARAMETER =   5.213
```

QSCA= .310543E+01   QEXT = .310543E+01   QBACK = .292534E+01

ANGLE	S11	POL	S33	S34
0.00	.100000E+01	0.	.100000E+01	0.
9.00	.785390E+00	-.459811E-02	.999400E+00	.343261E-01
18.00	.356897E+00	-.458541E-01	.986022E+00	.160184E+00
27.00	.766119E-01	-.364744E+00	.843603E+00	.394076E+00
36.00	.355355E-01	-.534997E+00	.686967E+00	-.491787E+00
45.00	.701845E-01	.959953E-02	.959825E+00	-.280434E+00
54.00	.574313E-01	.477927E-01	.985371E+00	.163584E+00
63.00	.219660E-01	-.440604E+00	.648043E+00	.621216E+00
72.00	.125959E-01	-.831996E+00	.203255E+00	-.516208E+00
81.00	.173750E-01	.341670E-01	.795354E+00	-.605182E+00
90.00	.124601E-01	.230462E+00	.937497E+00	.260742E+00
99.00	.679093E-02	-.713472E+00	-.717397E-02	.700647E+00
108.00	.954239E-02	-.756255E+00	-.394748E-01	-.653085E+00
117.00	.863419E-02	-.281215E+00	.536251E+00	-.795835E+00
126.00	.227421E-02	-.239612E+00	.967602E+00	.795798E-01
135.00	.543998E-02	-.850804E+00	.187531E+00	-.490882E+00
144.00	.160243E-01	-.706334E+00	.495254E+00	-.505781E+00
153.00	.188852E-01	-.891081E+00	.453277E+00	-.226817E-01
162.00	.195254E-01	-.783319E+00	-.391613E+00	.482752E+00
171.00	.301676E-01	-.196194E+00	-.962069E+00	.189556E+00
180.00	.383189E-01	0.	-.100000E+01	0.

Appendix B

Coated Sphere

As might be expected, adding a coating to a homogeneous sphere leads to several new computational problems, not all of which we were able to completely solve. As a consequence, BHCOAT, though similar in many ways to BHMIE, does not have as wide a range of applicability, and it should be used with more caution. The reasons for this will be discussed in the following paragraphs.

The mathematical form of all the scattering functions for a coated sphere—efficiencies and matrix elements—have the same form as those for a homogeneous sphere. Only the scattering coefficients (8.2) are different; these may be written in a form more suitable for computations:

$$a_n = \frac{\{\tilde{D}_n/m_2 + n/y\}\psi_n(y) - \psi_{n-1}(y)}{\{\tilde{D}_n/m_2 + n/y\}\xi_n(y) - \xi_{n-1}(y)},$$

$$b_n = \frac{\{m_2\tilde{G}_n + n/y\}\psi_n(y) - \psi_{n-1}(y)}{\{m_2\tilde{G}_n + n/y\}\xi_n(y) - \xi_{n-1}(y)},$$

$$\tilde{D}_n = \frac{D_n(m_2 y) - A_n\chi'_n(m_2 y)/\psi_n(m_2 y)}{1 - A_n\chi_n(m_2 y)/\psi_n(m_2 y)},$$

$$\tilde{G}_n = \frac{D_n(m_2 y) - B_n\chi'_n(m_2 y)/\psi_n(m_2 y)}{1 - B_n\chi_n(m_2 y)/\psi_n(m_2 y)},$$

$$A_n = \psi_n(m_2 x)\frac{mD_n(m_1 x) - D_n(m_2 x)}{mD_n(m_1 x)\chi_n(m_2 x) - \chi'_n(m_2 x)},$$

$$B_n = \psi_n(m_2 x)\frac{mD_n(m_2 x) - D_n(m_1 x)}{m\chi'_n(m_2 x) - D_n(m_1 x)\chi_n(m_2 x)}.$$

D_n is the logarithmic derivative ψ'_n/ψ_n and m is m_2/m_1. Because of the relation

$$\frac{1}{\psi_n} = \chi_n D_n - \chi'_n,$$

which follows from the Wronskian (4.60), only three of the four functions $\psi_n, \chi_n, \chi'_n, D_n$ are independent. We also have

$$\chi'_n(z) = \chi_{n-1}(z) - \frac{n\chi_n(z)}{z}.$$

A_n and B_n depend on the size parameter x of the *inner*, or core, sphere only; they are independent of the thickness of the coating. Moreover, they are similar in form to the scattering coefficients for a homogeneous sphere with size parameter x. Therefore, it is reasonable to expect that we need not compute A_n and B_n for orders n appreciably larger than x. Convergence of the series for efficiencies and matrix elements, however, is determined by the size parameter y of the *outer* sphere. Therefore, if y is much greater than x, A_n and B_n will be computed well beyond the range where they are needed unless special care is taken. If this were merely inefficient, it would be tolerable. But, as we pointed out in Appendix A, one cannot expect to reliably compute by upward recurrence Bessel functions of order much larger than their argument. We have included, therefore, four tests in BHCOAT: if all the inequalities

$$\text{DEL}|D_n(m_2 y)| > |A_n\chi'_n(m_2 y)/\psi_n(m_2 y)| \quad \text{and} \quad |B_n\chi'_n(m_2 y)/\psi_n(m_2 y)|$$

$$\text{DEL} > |A_n\chi_n(m_2 y)/\psi_n(m_2 y)| \quad \text{and} \quad |B_n\chi_n(m_2 y)/\psi_n(m_2 y)|$$

are satisfied for some index n, then $\tilde{D}_n$ and $\tilde{G}_n$ are set equal to $D_n(m_2 y)$ for all successive indices; note that when this occurs, the scattering coefficients are identical with those for a homogeneous sphere with size parameter y and relative refractive index m_2. The inner sphere convergence criterion DEL is 10^{-8} in BHCOAT; it can, of course, be changed.

The fact that the contributions from the inner sphere to the scattering coefficients converge more rapidly than those associated with the outer sphere is merely a nuisance which is easily brushed aside. A more serious obstacle—one which we failed to hurdle—to writing an "explosion-proof" program for an *arbitrary* coated sphere is that the scattering coefficients cannot be put in a form that avoids computing excessively large numbers. The scattering coefficients for a homogeneous sphere can be written in such a way that functions of complex arguments are ratios—the logarithmic derivative—and the Riccati–Bessel functions have *real* arguments. This cannot be done with the coated sphere: we must, in general, compute functions $\psi_n(z)$ and $\chi_n(z)$ of the *complex* variable $z = z_r + iz_i$. That this can lead to difficulties is easily demonstrated. Consider the zero-order function with which upward recurrence is begun:

$$\chi_0(z) = \cos z = \frac{e^{z_i} + e^{-z_i}}{2}\cos z_r - i\frac{e^{z_i} - e^{-z_i}}{2}\sin z_r.$$

Because of the factor $\exp(z_i)$ it is always possible to generate numbers that exceed the limits of any computer if the particle is sufficiently large and absorbing. For a given computer it is difficult to set precise upper bounds on the imaginary parts of the arguments of the functions computed in BHCOAT: although the zero-order functions might not exceed the limits of the computer, successive higher-order functions computed by upward recurrence might. We recommend that BHCOAT not be used if any of the imaginary parts of m_1x, m_2x, or m_2y exceeds 30. This is only a rough guide, however; we strongly urge the user to do a bit of experimenting before accepting as correct any numbers produced by BHCOAT.

The convergence criterion in BHCOAT is the same as that in BHMIE: series are terminated after $y + 4y^{1/3} + 2$ terms. Unlike BHMIE, however, all functions, including logarithmic derivatives, are computed by upward recurrence: it seemed pointless to compute these derivatives by downward recurrence when they are not the major obstacle to writing a program valid for an arbitrary coated sphere.

Tests of BHCOAT We subjected BHCOAT to the same tests as BHMIE. In addition, BHCOAT agrees with BHMIE when $m_1 = m_2$. We found that the asymptotic limit (4.83) for the backscattering efficiency was also a sensitive indicator of the health of the coated sphere program. The composition and size of the core is irrelevant if almost all the incident light is absorbed in the coating. On *physical* grounds, therefore, we expect Q_b for large y to be approximately equal to the reflectance of a slab with refractive index m_2 if the coating is sufficiently thick and absorbing; this was verified by computations.

As a further test we compared efficiencies computed by BHCOAT with those given by Fenn and Oser (1965); there was good agreement in all instances.

```
 1          PROGRAM COAT (INPUT=TTY,OUTPUT=TTY,TAPE5=TTY)
 2 C        ::::::::::::::::::::::::::::::::::::::::::::::::::::::::::::::::::::
 3 C           COAT IS THE CALLING PROGRAM FOR BHCOAT, THE SUBROUTINE
 4 C           THAT CALCULATES EFFICIENCIES FOR A COATED SPHERE.
 5 C           FOR GIVEN RADII AND REFRACTIVE INDICES OF INNER AND
 6 C           OUTER SPHERES, REFRACTIVE INDEX OF SURROUNDING
 7 C           MEDIUM, AND FREE SPACE WAVELENGTH, COAT CALCULATES SIZE
 8 C           PARAMETERS AND RELATIVE REFRACTIVE INDICES
 9 C        ::::::::::::::::::::::::::::::::::::::::::::::::::::::::::::::::::::
10 C
11 C                     ::::::::::::CAUTION::::::::::::::
12 C
13 C           BHCOAT SHOULD NOT BE USED FOR LARGE, HIGHLY ABSORBING
14 C           COATED SPHERES
15 C           X::REFIM1, X::REFIM2, AND Y::REFIM2 SHOULD BE LESS THAN ABOUT 3
16 C
17 C                     ::::::::::::CAUTION::::::::::::::
18 C
19 C
20          COMPLEX RFREL1,RFREL2
21          WRITE (5,11)
22 C        ::::::::::::::::::::::::::::::::::::::::::::::::::::::::::::::::::::
23 C           REFMED = (REAL) REFRACTIVE INDEX OF SURROUNDING MEDIUM
24 C        ::::::::::::::::::::::::::::::::::::::::::::::::::::::::::::::::::::
25          REFMED=1.0
26 C        :::::::::::::::::::::::::::::::::::::::::::::::::::::::::::::
27 C           REFRACTIVE INDEX OF CORE = REFRE1 + I::REFIM1
28 C           REFRACTIVE INDEX OF COAT = REFRE2 + I::REFIM2
29 C        ::::::::::::::::::::::::::::::::::::::::::::::::::::::::::::::
30          REFRE1=1.59
31          REFIM1=.66
32          REFRE2=1.409
33          REFIM2=.1747
34 C        ::::::::::::::::::::::::::::::::::::::::::::::::
35 C           RADCOR = RADIUS OF CORE
36 C           RADCOT = RADIUS OF COAT
37 C           RADCOR, RADCOT, WAVEL SAME UNITS
38 C        ::::::::::::::::::::::::::::::::::::::::::::::::
39          RADCOR=.171
40          RADCOT=6.265
41          WAVEL=3.
42          WRITE (5,12) REFMED,REFRE1,REFIM1,REFRE2,REFIM2
43          WRITE (5,13) RADCOR,RADCOT,WAVEL
44          RFREL1=CMPLX(REFRE1,REFIM1)/REFMED
45          RFREL2=CMPLX(REFRE2,REFIM2)/REFMED
46          PI=3.14159265
47          X=2.::PI::RADCOR::REFMED/WAVEL
48          Y=2.::PI::RADCOT::REFMED/WAVEL
49          WRITE (5,14) X,Y
50          CALL BHCOAT (X,Y,RFREL1,RFREL2,QEXT,QSCA,QBACK)
51          WRITE (5,67) QSCA,QEXT,QBACK
52       11 FORMAT (//"COATED SPHERE SCATTERING PROGRAM"//)
53       12 FORMAT (//5X,"REFMED = ",F8.4/5X,"REFRE1 =",E14.6,
54          13X,"REFIM1 =",E14.6/5X,"REFRE2 =",E14.6,3X,"REFIM2 =",E14.6)
55       13 FORMAT (5X,"CORE RADIUS =",F7.3,3X,"COAT RADIUS =",F7.3/
56          15X,"WAVELENGTH =",F7.4)
57       14 FORMAT (5X,"CORE SIZE PARAMETER = ",F8.3,3X,"COAT SIZE"
58          1" PARAMETER =",F8.3)
59       67 FORMAT (/,1X,"QSCA =",E13.6,3X,"OEXT =",E13.6,3X,
60          1"QBACK =",E13.6//)
```

```
 61          STOP
 62          END
 63          SUBROUTINE BHCOAT (X,Y,RFREL1,RFREL2,QEXT,QSCA,QBACK)
 64   C      ※※※※※※※※※※※※※※※※※※※※※※※※※※※※※※※※※※※※※※※※※※
 65   C        SUBROUTINE BHCOAT CALCULATES EFFICIENCIES FOR
 66   C        EXTINCTION, TOTAL SCATTERING, AND BACKSCATTERING
 67   C        FOR GIVEN SIZE PARAMETERS OF CORE AND COAT AND
 68   C        RELATIVE REFRACTIVE INDICES
 69   C        ALL BESSEL FUNCTIONS COMPUTED BY UPWARD RECURRENCE
 70   C      ※※※※※※※※※※※※※※※※※※※※※※※※※※※※※※※※※※※※※※※※※※
 71          COMPLEX RFREL1,RFREL2,X1,X2,Y2,REFREL
 72          COMPLEX D1X1,D0X1,D1X2,D0X2,D1Y2,D0Y2
 73          COMPLEX XI0Y,XI1Y,XIY,CHI0Y2,CHI1Y2,CHIY2,CHI0X2,CHI1X2,CHIX2
 74          COMPLEX CHIPX2,CHIPY2,ANCAP,BNCAP,DNBAR,GNBAR,AN,BN,CRACK,BRACK
 75          COMPLEX XBACK,AMESS1,AMESS2,AMESS3,AMESS4
 76          DEL=1.0E-8
 77   C      ※※※※※※※※※※※※※※※※※※※※※※※※※※※※※※※※※※※※※※※※※※
 78   C        DEL IS THE INNER SPHERE CONVERGENCE CRITERION
 79   C      ※※※※※※※※※※※※※※※※※※※※※※※※※※※※※※※※※※※※※※※※※※
 80          X1=RFREL1※X
 81          X2=RFREL2※X
 82          Y2=RFREL2※Y
 83          YSTOP = Y + 4.※Y※※.3333 + 2.
 84          REFREL=RFREL2/RFREL1
 85          NSTOP=YSTOP
 86   C      ※※※※※※※※※※※※※※※※※※※※※※※※※※※※※※※※※※※※※※※※※※
 87   C        SERIES TERMINATED AFTER NSTOP TERMS
 88   C      ※※※※※※※※※※※※※※※※※※※※※※※※※※※※※※※※※※※※※※※※※※
 89          D0X1=CCOS(X1)/CSIN(X1)
 90          D0X2=CCOS(X2)/CSIN(X2)
 91          D0Y2=CCOS(Y2)/CSIN(Y2)
 92          PSI0Y=COS(Y)
 93          PSI1Y=SIN(Y)
 94          CHI0Y=-SIN(Y)
 95          CHI1Y=COS(Y)
 96          XI0Y=CMPLX(PSI0Y,-CHI0Y)
 97          XI1Y=CMPLX(PSI1Y,-CHI1Y)
 98          CHI0Y2=-CSIN(Y2)
 99          CHI1Y2=CCOS(Y2)
100          CHI0X2=-CSIN(X2)
101          CHI1X2=CCOS(X2)
102          QSCA=0.0
103          QEXT=0.0
104          XBACK=CMPLX(0.0,0.0)
105          N=1
106          IFLAG=0
107    200   RN=N
108          PSIY=(2.※RN-1.)※PSI1Y/Y-PSI0Y
109          CHIY=(2.※RN-1.)※CHI1Y/Y-CHI0Y
110          XIY=CMPLX(PSIY,-CHIY)
111          D1Y2=1./(RN/Y2-D0Y2)-RN/Y2
112          IF (IFLAG.EQ.1) GO TO 999
113          D1X1=1./(RN/X1-D0X1)-RN/X1
114          D1X2=1./(RN/X2-D0X2)-RN/X2
115          CHIX2=(2.※RN-1.)※CHI1X2/X2-CHI0X2
116          CHIY2=(2.※RN-1.)※CHI1Y2/Y2-CHI0Y2
117          CHIPX2=CHI1X2-RN※CHIX2/X2
118          CHIPY2=CHI1Y2-RN※CHIY2/Y2
119          ANCAP=REFREL※D1X1-D1X2
120          ANCAP=ANCAP/(REFREL※D1X1※CHIX2-CHIPX2)
```

```
121        ANCAP=ANCAP/(CHIX2*D1X2-CHIPX2)
122        BRACK=ANCAP*(CHIY2*D1Y2-CHIPY2)
123        BNCAP=REFREL*D1X2-D1X1
124        BNCAP=BNCAP/(REFREL*CHIPX2-D1X1*CHIX2)
125        BNCAP=BNCAP/(CHIX2*D1X2-CHIPX2)
126        CRACK=BNCAP*(CHIY2*D1Y2-CHIPY2)
127        AMESS1=BRACK*CHIPY2
128        AMESS2=BRACK*CHIY2
129        AMESS3=CRACK*CHIPY2
130        AMESS4=CRACK*CHIY2
131        IF(CABS(AMESS1).GT.DEL*CABS(D1Y2)) GO TO 999
132        IF(CABS(AMESS2).GT.DEL) GO TO 999
133        IF(CABS(AMESS3).GT.DEL*CABS(D1Y2)) GO TO 999
134        IF(CABS(AMESS4).GT.DEL) GO TO 999
135        BRACK=CMPLX(0.0,0.0)
136        CRACK=CMPLX(0.0,0.0)
137        IFLAG=1
138    999 DNBAR=D1Y2-BRACK*CHIPY2
139        DNBAR=DNBAR/(1.-BRACK*CHIY2)
140        GNBAR=D1Y2-CRACK*CHIPY2
141        GNBAR=GNBAR/(1.-CRACK*CHIY2)
142        AN=(DNBAR/RFREL2+RN/Y)*PSIY-PSI1Y
143        AN=AN/((DNBAR/RFREL2+RN/Y)*XIY-XI1Y)
144        BN=(RFREL2*GNBAR+RN/Y)*PSIY-PSI1Y
145        BN=BN/((RFREL2*GNBAR+RN/Y)*XIY-XI1Y)
146        QSCA=QSCA+(2.*RN+1.)*(CABS(AN)*CABS(AN)+CABS(BN)*CABS(BN))
147        XBACK=XBACK+(2.*RN+1.)*(-1.)**N*(AN-BN)
148        QEXT=QEXT+(2.*RN+1.)*(REAL(AN)+REAL(BN))
149        PSI0Y=PSI1Y
150        PSI1Y=PSIY
151        CHI0Y=CHI1Y
152        CHI1Y=CHIY
153        XI1Y=CMPLX(PSI1Y,-CHI1Y)
154        CHI0X2=CHI1X2
155        CHI1X2=CHIX2
156        CHI0Y2=CHI1Y2
157        CHI1Y2=CHIY2
158        D0X1=D1X1
159        D0X2=D1X2
160        D0Y2=D1Y2
161        N=N+1
162        IF(N-1-NSTOP) 200,300,300
163    300 QSCA=(2./(Y*Y))*QSCA
164        QEXT=(2./(Y*Y))*QEXT
165        QBACK=XBACK*CONJG(XBACK)
166        QBACK=(1./(Y*Y))*QBACK
167        RETURN
168        END
```

SHELL

COATED SPHERE SCATTERING PROGRAM


```
     REFMED =    1.0000
     REFRE1 =    .159000E+01    REFIM1 =    .660000E+00
     REFRE2 =    .140900E+01    REFIM2 =    .174700E+00
     CORE RADIUS =    .171    COAT RADIUS =  6.265
     WAVELENGTH = 3.0000
     CORE SIZE PARAMETER =      .358    COAT SIZE PARAMETER =   13.121

 QSCA =  .114341E+01    QEXT =  .232803E+01    QBACK =  .285099E-01
```

Appendix C

Normally Illuminated Infinite Cylinder

The program described in this appendix is an improved version of one contained in a report by Bohren and Timbrell (1979); the underlying theory and definitions of symbols are given in Section 8.4.

The coefficients (8.38) are still not in a form suitable for computations: indices are required to be positive integers in most versions of Fortran. We can get around this easily enough by defining $G_n = D_{n-1}$ ($n = 1, 2, \ldots$), which satisfies the recurrence relation

$$G_{n-1}(z) = \frac{n-2}{z} - \frac{1}{\dfrac{n-1}{z} + G_n(z)}. \qquad (C.1)$$

If we use the relations $J_{-n} = (-1)^n J_n$ and $Y_{-n} = (-1)^n Y_n$, the coefficients a_0 and b_0 may be written

$$a_0 = \frac{G_1(mx)BJ_1(x)/m + BJ_2(x)}{G_1(mx)BH_1(x)/m + BH_2(x)}, \qquad b_0 = \frac{mG_1(mx)BJ_1(x) + BJ_2(x)}{mG_1(mx)BH_1(x) + BH_2(x)},$$

where we define $BJ_n = J_{n-1}$ and $BH_n = H^{(1)}_{n-1}$ ($n = 1, 2, \ldots$). For $n \neq 0$ we have

$$a_n = \frac{\{G_{n+1}(mx)/m + n/x\}BJ_{n+1}(x) - BJ_n(x)}{\{G_{n+1}(mx)/m + n/x\}BH_{n+1}(x) - BH_n(x)},$$

$$b_n = \frac{\{mG_{n+1}(mx) + n/x\}BJ_{n+1}(x) - BJ_n(x)}{\{mG_{n+1}(mx) + n/x\}BH_{n+1}(x) - BH_n(x)}.$$

As in the previous programs, series for scattering matrix elements and efficiencies are truncated after NSTOP terms, where NSTOP $= x + 4x^{1/3} + 2$. $G_n(mx)$ is computed by (C.1): beginning with G_{NMX}, successive lower-order logarithmic derivatives $G_{\text{NMX}-1}, \ldots, G_1$ are computed by downward recurrence. Provided that NMX is sufficiently greater than NSTOP and $|mx|$, G_p for

$p \leqslant$ NSTOP is insensitive to the choice of G_{NMX}. We varied G_{NMX} over five orders of magnitude for a range of arguments mx; in each instance $G_{\text{NMX}-5}$ was independent of G_{NMX}. Therefore, we take NMX = Max(NSTOP, $|mx|$) + 15, and recurrence begins with $G_{\text{NMX}} = 0.0 + i0.0$.

Computation of Bessel Functions The Bessel functions J_n and Y_n pose more computational problems than the logarithmic derivative. In BHCYL these functions are computed by an algorithm credited to Miller (British Association, 1952, p. xvii), further details of which are given by Stegun and Abramowitz (1957) and by Goldstein and Thaler (1959); we outline this scheme in the following paragraph.

To calculate $J_n(x)$ we generate a sequence of functions $F_p(x)$—in BHCYL $F(P)$ is F_{p-1}—by *downward* recurrence:

$$F_{p-1}(x) = \frac{2pF_p(x)}{x} - F_{p+1}(x) \qquad (p = M - 1, \dots, 1) \qquad \text{(C.2)}$$

beginning with $F_M = 0.0$ and $F_{M-1} = 10^{-32}$, where M is greater than both n and x. Although F_{M-1} is arbitrary, it should be small to avoid generating excessively large numbers when $x < 1$. J_p and Y_p also satisfy (C.2), and for all p sufficiently smaller than M we have

$$J_p(x) \simeq \frac{F_p(x)}{\alpha},$$

where α is independent of p; this constant is evaluated by using

$$J_0 + 2 \sum_{m=1}^{\infty} J_{2m} = 1,$$

from which it follows that

$$\alpha \simeq F_0(x) + 2 \sum_{m=1}^{\frac{1}{2}M-1} F_{2m}(x).$$

$Y_n(x)$ is computed by *upward* recurrence

$$Y_{p+1}(x) = \frac{2pY_p(x)}{x} - Y_{p-1}(x),$$

beginning with Y_0 and Y_1; the first of these is obtained from

$$Y_0(x) = \frac{2}{\pi} \left[\left(\log\frac{x}{2} + \gamma \right) J_0(x) - 2 \sum_{m=1}^{\infty} \frac{(-1)^m J_{2m}(x)}{m} \right],$$

where Euler's constant γ is $0.57721566\ldots$; the second is calculated from the Wronskian

$$J_1(x)Y_0(x) - J_0(x)Y_1(x) = \frac{2}{\pi x}.$$

The choice of $M = \text{MST}$ in the recurrence relation (C.2) is sometimes determined by iteration. This may be appropriate in a program to compute $J_n(x)$ for arbitrary n and x. But our needs are more limited, and we can avoid iterating—thereby designing a more efficient program—by properly choosing MST. NSTOP, which is greater than x, is the largest order of the Bessel functions required in scattering calculations. Therefore, we take MST to be the even integer closest to NSTOP + NDELTA. To determine NDELTA we did a series of calculations: NDELTA was incremented until successive values of J_{NSTOP} agreed to nine decimal places; the size parameter was varied from 0.001 to 5000. Our results are summarized by the empirical relation

$$\text{NDELTA} = (101 + x)^{0.499}.$$

Tests of BHCYL   Computed values of $J_n(x)$ and $Y_n(x)$ for various n and x were compared with values tabulated by Olver (1964). Computed logarithmic derivatives were also compared with values calculated from tabulated Bessel functions. In all instances there was agreement to as many decimal places as were given in the tables. This gives us some confidence that BHCYL does what it was designed to do.

Scattering coefficients for absorbing cylinders over a limited range of m and x have been tabulated by Libelo (1962). These were compared with coefficients computed by BHCYL; in all instances there was agreement to as many decimal places as were given in the tables.

There are a few obvious checks on the scattering and extinction efficiencies: (1) they are nonnegative for all x and m; and (2) Q_{sca} and Q_{ext} are equal when the cylinder is nonabsorbing. BHCYL has never failed to pass these tests. Scattering efficiencies are calculated in BHCYL as the sum of squares of the scattering coefficients, whereas extinction efficiencies are calculated from the real part of the forward amplitude scattering matrix elements T_1 and T_2. This was done purposely to provide a check on the amplitude scattering matrix elements.

As a further check, we compared efficiencies computed by BHCYL with values given by Larkin and Churchill (1959); in all instances, there was good agreement.

```
1          PROGRAM CALCYL (INPUT=TTY,OUTPUT=TTY,TAPE5=TTY)
2    C     ::::::::::::::::::::::::::::::::::::::::::::::::::::::::::::::::
3    C        CALCYL CALCULATES THE SIZE PARAMETER (X) AND RELATIVE
4    C        REFRACTIVE INDEX (REFREL) FOR A GIVEN CYLINDER REFRACTIVE
5    C        INDEX, MEDIUM REFRACTIVE INDEX, RADIUS, AND FREE SPACE
6    C        WAVELENGTH. IT THEN CALLS BHCYL, THE SUBROUTINE THAT COMPUTES
7    C        AMPLITUDE SCATTERING MATRIX ELEMENTS AND EFFICIENCIES
8    C     ::::::::::::::::::::::::::::::::::::::::::::::::::::::::::::::::
9          COMPLEX REFREL,T1(200),T2(200)
10         DIMENSION ANG(200)
11         WRITE (5,11)
12   C     ::::::::::::::::::::::::::::::::::::::::::::::::::::::::::::::::
13   C        REFMED = (REAL) REFRACTIVE INDEX OF SURROUNDING MEDIUM
14   C     ::::::::::::::::::::::::::::::::::::::::::::::::::::::::::::::::
15         REFMED=1.0
16   C     ::::::::::::::::::::::::::::::::::::::::::::::::::::::::::::::
17   C        REFRACTIVE INDEX OF CYLINDER = REFRE + I::REFIM
18   C     ::::::::::::::::::::::::::::::::::::::::::::::::::::::::::::::
19         REFRE=1.55
20         REFIM=0.0
21         REFREL=CMPLX(REFRE,REFIM)/REFMED
22         WRITE (5,12) REFMED,REFRE,REFIM
23         PI=3.14159265
24   C     ::::::::::::::::::::::::::::::::::::::::::::::::
25   C        RADIUS (RAD) AND WAVELENGTH (WAVEL) SAME UNITS
26   C     ::::::::::::::::::::::::::::::::::::::::::::::::
27         RAD=.525
28         WAVEL=.6328
29         X=2.::PI::RAD::REFMED/WAVEL
30         WRITE (5,13) RAD,WAVEL
31         WRITE (5,14) X
32   C     ::::::::::::::::::::::::::::::::::::::::::::::::::::::::::::
33   C        FIN = FINAL ANGLE (DEGREES)
34   C        INTANG = NUMBER OF INTERVALS BETWEEN 0 AND FIN
35   C     ::::::::::::::::::::::::::::::::::::::::::::::::::::::::::::
36         FIN=180.
37         INTANG=20
38         WRITE (5,15)
39         CALL BHCYL (X,REFREL,T1,T2,QSCPAR,QSCPER,QEXPAR,QEXPER,
40        1FIN,INTANG,ANG)
41         NPTS=INTANG+1
42         T11NOR=0.5::(CABS(T1(1))::CABS(T1(1)))
43         T11NOR=T11NOR+0.5::(CABS(T2(1))::CABS(T2(1)))
44   C     ::::::::::::::::::::::::::::::::::::::::::::::::::::::::::::::::
45   C        T33 AND T34 MATRIX ELEMENTS NORMALZIED BY T11
46   C        T11 IS NORMALIZED TO 1.0 IN THE FORWARD DIRECTION
47   C        POL = DEGREE OF POLARIZATION (INCIDENT UNPOLARIZED LIGHT)
48   C     ::::::::::::::::::::::::::::::::::::::::::::::::::::::::::::::::
49         DO 107 J=1,NPTS
50         TPAR=CABS(T1(J))
51         TPAR=TPAR::TPAR
52         TPER=CABS(T2(J))
53         TPER=TPER::TPER
54         T11=0.5::(TPAR+TPER)
55         T12=0.5::(TPAR-TPER)
56         POL=T12/T11
57         T33=REAL(T1(J)::CONJG(T2(J)))
58         T34=AIMAG(T1(J)::CONJG(T2(J)))
59         T33=T33/T11
60         T34=T34/T11
```

```
61          T11=T11/T11NOR
62      107 WRITE (5,68) ANG(J),T11,POL,T33,T34
63          WRITE   (5,67)   QSCPAR,QEXPAR,QSCPER,QEXPER
64       67 FORMAT (//,"QSCPAR =",E14.6,3X,"QEXPAR =",E14.6/
65          1"QSCPER =",E14.6,3X,"QEXPER =",E14.6//)
66       68 FORMAT (1X,F8.2,2X,E13.6,2X,E13.6,2X,E13.6,2X,E13.6)
67       11 FORMAT (/"CYLINDER PROGRAM: NORMALLY INCIDENT LIGHT"//)
68       12 FORMAT (5X,"REFMED =",F8.4,3X,"REFRE =",E14.6,3X
69          1"REFIM =",E14.6)
70       13 FORMAT (5X,"CYLINDER RADIUS =",F7.3,3X,"WAVELENGTH =",F7.4)
71       14 FORMAT (5X,"SIZE PARAMETER =",F8.3/)
72       15 FORMAT (//2X,"ANGLE",7X,"T11",13X,"POL",13X,"T33",13X,"T34"//)
73          STOP
74          END
75  C       :::::::::::::::::::::::::::::::::::::::::::::::::::::::::::::::::::
76  C       SUBROUTINE BHCYL CALCULATES AMPLITUDE SCATTERING MATRIX
77  C       ELEMENTS AND EFFICIENCIES FOR EXTINCTION AND SCATTERING
78  C       FOR A GIVEN SIZE PARAMETER AND RELATIVE REFRACTIVE INDEX
79  C       THE INCIDENT LIGHT IS NORMAL TO THE CYLINDER AXIS
80  C       PAR:ELECTRIC FIELD PARALLEL TO CYLINDER AXIS
81  C       PER:ELECTRIC FIELD PERPENDICULAR TO CYLINDER AXIS
82  C       :::::::::::::::::::::::::::::::::::::::::::::::::::::::::::::::::::
83          SUBROUTINE BHCYL (X,REFREL,T1,T2,QSCPAR,QSCPER,QEXPAR,QEXPER,
84          2FIN,INTANG,ANG)
85          COMPLEX REFREL,Y,AN,BN,A0,B0
86          COMPLEX G(1000),BH(1000),T1(200),T2(200)
87          DIMENSION THETA(200),ANG(200),BJ(1000),BY(1000),F(1000)
88          Y=X*REFREL
89          XSTOP=X+4.*X**.3333+2.
90  C       :::::::::::::::::::::::::::::::::::::::::::::::::::::::
91  C       SERIES TERMINATED AFTER NSTOP TERMS
92  C       :::::::::::::::::::::::::::::::::::::::::::::::::::::::
93          NSTOP=XSTOP
94          YMOD=CABS(Y)
95          NMX=AMAX1(XSTOP,YMOD)+15
96          NPTS=INTANG+1
97          DANG=FIN/FLOAT(INTANG)
98          DO 555 J=1,NPTS
99          ANG(J)=(FLOAT(J)-1.)*DANG
100     555 THETA(J)=ANG(J)*0.017453292
101 C       ::::::::::::::::::::::::::::::::::::::::::::::::::::::::::::::::::::
102 C       LOGARITHMIC DERIVATIVE G(J) CALCULATED BY DOWNWARD
103 C       RECURRENCE BEGINNING WITH INITIAL VALUE 0.0 + I*0.0
104 C       AT J = NMX
105 C       ::::::::::::::::::::::::::::::::::::::::::::::::::::::::::::::::::::
106         G(NMX)=CMPLX(0.0,0.0)
107         NN=NMX-1
108         DO 120 N=1,NN
109         RN=NMX-N+1
110         K=NMX-N
111     120 G(K)=((RN-2.)/Y)-(1./(G(K+1)+(RN-1.)/Y))
112 C       ::::::::::::::::::::::::::::::::::::::::::::::::::::::::::::::::::::
113 C       BESSEL FUNCTIONS J(N) COMPUTED BY DOWNWARD RECURRENCE
114 C       BEGINNING AT N = NDELTA
115 C       BESSEL FUNCTIONS Y(N) COMPUTED BY UPWARD RECURRENCE
116 C       BJ(N+1) = J(N),   BY(N+1) = Y(N)
117 C       ::::::::::::::::::::::::::::::::::::::::::::::::::::::::::::::::::::
118         NDELTA=(101.+X)**.499
119         MST=NSTOP+NDELTA
120         MST=(MST/2)*2
```

```
121        F(MST+1)=0.0
122        F(MST)=1.0E-32
123        M1=MST-1
124        DO 201 L=1,M1
125        ML=MST-L
126   201  F(ML)=2.*FLOAT(ML)*F(ML+1)/X-F(ML+2)
127        ALPHA=F(1)
128        M2=MST-2
129        DO 202 LL=2,M2,2
130   202  ALPHA=ALPHA+2.*F(LL+1)
131        M3=M2+1
132        DO 203 N=1,M3
133   203  BJ(N)=F(N)/ALPHA
134        BY(1)=BJ(1)*(ALOG(X/2.)+.577215664)
135        M4=MST/2-1
136        DO 204 L=1,M4
137   204  BY(1)=BY(1)-2.*((-1.)**L)*BJ(2*L+1)/FLOAT(L)
138        BY(1)=.636619772*BY(1)
139        BY(2)=BJ(2)*BY(1)-.636619772/X
140        BY(2)=BY(2)/BJ(1)
141        NS=NSTOP-1
142        DO 205 KK=1,NS
143   205  BY(KK+2)=2*FLOAT(KK)*BY(KK+1)/X-BY(KK)
144        NN=NSTOP+1
145        DO 715 N=1,NN
146   715  BH(N)=CMPLX(BJ(N),BY(N))
147        A0=G(1)*BJ(1)/REFREL+BJ(2)
148        A0=A0/(G(1)*BH(1)/REFREL+BH(2))
149        B0=REFREL*G(1)*BJ(1)+BJ(2)
150        B0=B0/(REFREL*G(1)*BH(1)+BH(2))
151        QSCPAR=CABS(B0)*CABS(B0)
152        QSCPER=CABS(A0)*CABS(A0)
153        DO 101 K=1,NPTS
154        T1(K)=B0
155   101  T2(K)=A0
156        DO 123 N=1,NSTOP
157        RN=N
158        AN=(G(N+1)/REFREL+RN/X)*BJ(N+1)-BJ(N)
159        AN=AN/((G(N+1)/REFREL+RN/X)*BH(N+1)-BH(N))
160        BN=(REFREL*G(N+1)+RN/X)*BJ(N+1)-BJ(N)
161        BN=BN/((REFREL*G(N+1)+RN/X)*BH(N+1)-BH(N))
162        DO 102 J=1,NPTS
163        C=COS(RN*THETA(J))
164        T1(J)=2.*BN*C+T1(J)
165   102  T2(J)=2.*AN*C+T2(J)
166        QSCPAR=QSCPAR+2.*CABS(BN)*CABS(BN)
167   123  QSCPER=QSCPER+2.*CABS(AN)*CABS(AN)
168        QSCPAR=(2./X)*QSCPAR
169        QSCPER=(2./X)*QSCPER
170        QEXPER=(2./X)*REAL(T2(1))
171        QEXPAR=(2./X)*REAL(T1(1))
172        RETURN
173        END
```

```
CYL

CYLINDER PROGRAM: NORMALLY INCIDENT LIGHT


     REFMED =  1.0000   REFRE =   .155000E+01   REFIM =  0.
     CYLINDER RADIUS =   .525   WAVELENGTH =  .6328
     SIZE PARAMETER =   5.213


     ANGLE        T11              POL              T33              T34

      0.00    .100000E+01     .734486E-01     .997149E+00    -.172894E-01
      9.00    .686631E+00     .291477E-01     .999432E+00    -.169025E-01
     18.00    .217683E+00    -.135736E+00     .986700E+00     .894351E-01
     27.00    .144205E+00     .103749E+00     .931604E+00     .348352E+00
     36.00    .259646E+00     .162651E+00     .977440E+00     .134744E+00
     45.00    .231162E+00    -.894687E-02     .997329E+00     .724853E-01
     54.00    .132150E+00    -.179789E+00     .953175E+00     .243175E+00
     63.00    .839900E-01    -.349048E-01     .900228E+00     .434018E+00
     72.00    .669177E-01     .504876E-01     .937414E+00     .344536E+00
     81.00    .622477E-01    -.823535E-02     .942424E+00     .334320E+00
     90.00    .482920E-01    -.510106E-01     .967653E+00     .247076E+00
     99.00    .199993E-01    -.606254E+00     .782214E+00     .143519E+00
    108.00    .244164E-01    -.141679E+00     .173427E+00     .974602E+00
    117.00    .416869E-01     .476291E+00     .534335E+00     .698307E+00
    126.00    .200601E-01     .488882E+00     .839228E+00    -.238100E+00
    135.00    .186030E-01    -.671603E+00    -.708250E+00    -.217558E+00
    144.00    .655546E-01    -.676521E-01    -.325732E+00     .943039E+00
    153.00    .632725E-01     .262420E-01    -.223743E+00     .974295E+00
    162.00    .168029E-01    -.282769E-01    -.771987E+00     .635010E+00
    171.00    .333764E-01     .956354E+00    -.135136E+00     .259084E+00
    180.00    .673014E-01     .899741E+00     .641930E-01     .431676E+00


QSCPAR =  .209716E+01   QEXPAR =   .209716E+01
QSCPER =  .192782E+01   QEXPER =   .192782E+01
```

References

Aannestad, P. A., and E. M. Purcell, 1973. Interstellar grains, *Annu. Rev. Astron. Astrophys.*, **11**, 309–362.

Aannestad, P. A., 1975. Absorptive properties of silicate core–mantle grains, *Astrophys. J.*, **200**, 30–41.

Abelès, F. (Ed.), 1972. *Optical Properties of Solids*, Elsevier, New York.

Abeles, B., and J. I. Gittleman, 1976. Composite material films: optical properties and applications, *Appl. Opt.*, **15**, 2328–2332.

Abhyankar, K. D., and A. L. Fymat, 1969. Relations between the elements of the phase matrix for scattering, *J. Math. Phys.*, **10**, 1935–1938.

Abramowitz, M., and I. A. Stegun (Eds.), 1964. *Handbook of Mathematical Functions*, National Bureau of Standards, Washington, D.C.

Abreu, V. J., 1980. Lidar from orbit, *Opt. Eng.*, **19**, 489–493.

Ackerman, T. P., and O. B. Toon, 1981. Absorption of visible radiation in atmosphere containing mixtures of absorbing and nonabsorbing particles, *Appl. Opt.*, **20**, 3661–3667.

Acquista, C., 1976. Light scattering by tenuous particles: a generalization of the Rayleigh–Gans–Rocard approach, *Appl. Opt.*, **15**, 2932–2936.

Acquista, C., A. Cohen, J. A. Cooney, and J. Wimp, 1980. Asymptotic behavior of the efficiencies in Mie scattering, *J. Opt. Soc. Am.*, **70**, 1023–1025.

Aden, A. L., 1951. Electromagnetic scattering from spheres with sizes comparable to the wavelength, *J. Appl. Phys.*, **22**, 601–605.

Aden, A. L., and M. Kerker, 1951. Scattering of electromagnetic waves from two concentric spheres, *J. Appl. Phys.*, **22**, 1242–1246.

Affolter, P., and B. Eliasson, 1973. Electromagnetic resonances and Q-factors of lossy dielectric spheres, *IEEE Trans. Microwave Theory Tech.*, **MTT-21**, 573–578.

Agranovich, V. M., and V. L. Ginzburg, 1966. *Spatial Dispersion and the Theory of Excitons*, Wiley-Interscience, New York.

Antosiewicz, H. A., 1964. Bessel functions of fractional order, in *Handbook of Mathematical Functions*, M. Abramowitz and I. A. Stegun (Eds.), National Bureau of Standards, Washington, D.C., pp. 435–478.

Aronson, J. R., and A. G. Emslie, 1975. Composition of the Martian dust as derived by infrared spectroscopy from Mariner 9, *J. Geophys. Res.*, **80**, 4925–4931.

Aronson, J. R., and A. G. Emslie, 1980. Effective optical constants of anisotropic materials, *Appl. Opt.*, **24**, 4128–4129.

Asano, S., 1979. Light scattering properties of spheroidal particles, *Appl. Opt.*, **18**, 712–723.

Asano, S., and M. Sato, 1980. Light scattering by randomly oriented spheroidal particles, *Appl. Opt.*, **19**, 962–974.

Asano, S., and G. Yamamoto, 1975. Light scattering by a spheroidal particle, *Appl. Opt.*, **14**, 29–49.

Ashkin, A., 1970. Acceleration and trapping of particles by radiation pressure, *Phys. Rev. Lett.*, **24**, 156–159.

Ashkin, A., and J. M. Dziedzic, 1971. Optical levitation by radiation pressure, *Appl. Phys. Lett.*, **19**, 283–285.

Ashkin, A., and J. M. Dziedzic, 1976. Optical levitation in high vacuum, *Appl. Phys. Lett.*, **28**, 333–335.

Ashkin, A., and J. M. Dziedzic, 1977. Observation of resonances in the radiation pressure on dielectric spheres, *Phys. Rev. Lett.*, **38**, 1351–1354.

Axe, J. D., and G. D. Pettit, 1966. Infrared dielectric dispersion and lattice dynamics of uranium dioxide and thorium dioxide, *Phys. Rev.*, **151**, 676–680.

Azzam, R. M. A., 1978. Photopolarimetric measurement of the Mueller matrix by Fourier analysis of a single detected signal, *Opt. Lett.*, **2**, 148–150.

Azzam, R. M. A., and N. M. Bashara, 1977. *Ellipsometry and Polarized Light*, North-Holland, Amsterdam.

Baltes, H. P., 1976. On the validity of Kirchhoff's law of heat radiation for a body in a nonequilibrium environment, *Prog. Opt.*, **13**, 1–25.

Baltes, H. P. (Ed.), 1978. *Inverse Source Problems in Optics*, Springer-Verlag, Berlin.

Baltes, H. P. (Ed.), 1980. *Inverse Scattering Problems in Optics*, Springer-Verlag, Berlin.

Bancroft, W. D., 1919. The color of water, *J. Franklin Inst.*, **187**, 249–271, 459–485.

Banderman, L. W., and J. C. Kemp, 1973. Circular polarization by single scattering of unpolarized light from loss-less non-spherical particles, *Mon. Not. R. Astron. Soc.*, **162**, 367–377.

Barber, P., and C. Yeh, 1975. Scattering of electromagnetic waves by arbitrarily shaped dielectric bodies, *Appl. Opt.*, **14**, 2864–2872.

Barker, A. S., 1973. Infrared absorption of localized longitudinal-optical phonons, *Phys. Rev.*, **B7**, 2507–2520.

Bartholdi, M., G. C. Salzman, R. D. Hiebert, and M. Kerker, 1980. Differential light scattering photometer for rapid analysis of single particles in flow, *Appl. Opt.*, **19**, 1573–1581.

Bates, R. H. T., 1975. Analytic constraints on electromagnetic field computations, *IEEE Trans. Microwave Theory Tech.*, **MTT-23**, 605–623.

Battan, L. J., 1973. *Radar Observations of the Atmosphere*, rev. ed., University of Chicago Press, Chicago.

Battan, L. J., 1976. Vertical air motions and the $Z-R$ relation, *J. Appl. Meteorol.*, **15**, 1120–1121.

Beardsley, G. F., 1968. Mueller scattering matrix of sea water, *J. Opt. Soc. Am.*, **58**, 52–57.

Beckmann, P., and A. Spizzichino, 1963. *The Scattering of Electromagnetic Waves from Rough Surfaces*, Macmillan, New York.

Bell, B., 1981. The entire scattering matrix for a single fiber, M.S. thesis, The University of Arizona.

Bell, B. W., and W. S. Bickel, 1981. Single fiber light scattering matrix: an experimental determination, *Appl. Opt.*, **20**, 3874–3879.

Bell, E. E., 1967. Optical constants and their measurement, *Handbuch der Physik*, Vol. 25/2a, S. Flügge (Ed.), Springer-Verlag, Berlin, pp. 1–58.

Bendow, B., 1978. Multiphonon infrared absorption in the highly transparent frequency regime of solids, *Solid State Phys.*, **33**, 249–316.

Bennett, H. S., and G. J. Rosasco, 1978. Resonances in the efficiency factors for absorption: Mie scattering theory, *Appl. Opt.*, **17**, 491–493.

Bergland, R. N., and B. Y. H. Liu, 1973. Generation of monodisperse aerosol standards, *Environ. Sci. Technol.*, **7**, 147–153.

Bergstrom, R. W., 1973. Bemerkung zur Bestimmung des Absorptionskoeffizienten atmosphärischen Aerosols, *Beitr. Phys. Atmos.*, **46**, 198–202.

Bernal, J. D., 1965. The structure of water and its biological implications, in *The State and Movement of Water in Living Organisms*, Symposium 19 of the Society for Experimental Biology, Academic, New York, pp. 17–32.

Bhagavantam, S., 1942. *Scattering of Light and the Raman Effect*, Chemical Publishing Co., New York.

Bickel, W. S., and M. E. Stafford, 1980. Biological particles as irregularly shaped particles, in *Light Scattering by Irregularly Shaped Particles*, D. Schuerman (Ed.), Plenum, New York, pp. 299–305.

Bickel, W. S., J. F. Davidson, D. R. Huffman, and R. Kilkson, 1976. Application of polarization effects in light scattering: a new biophysical tool, *Proc. Natl. Acad. Sci. USA*, **73**, 486–490.

Bilbro, J. W., 1980. Atmospheric laser Doppler velocimetry: an overview, *Opt. Eng.*, **19**, 533–542.

Blanchard, D. C., 1967. *From Raindrops to Volcanoes: Adventure with Sea Surface Meteorology*, Doubleday, New York.

Block, S., and G. Piermarini, 1976. Diamond cell aids high-pressure research, *Phys. Today*, **29**(9), 44–55.

Bohm, D., and E. P. Gross, 1949a. Theory of plasma oscillations: A. Origin of medium-like behavior, *Phys. Rev.*, **75**, 1851–1864.

Bohm, D., and E. P. Gross, 1949b. Theory of plasma oscillations: B. Excitation and damping of oscillations, *Phys. Rev.*, **75**, 1864–1876.

Bohm, D., and D. Pines, 1951. A collective description of electron interactions: I. Magnetic interactions, *Phys. Rev.*, **82**, 625–634.

Bohm, D., and D. Pines, 1953. A collective description of electron interactions: III. Coulomb interactions in a degenerate electron gas, *Phys. Rev.*, **92**, 609–625.

Bohren, C. F., 1974. Light scattering by an optically active sphere, *Chem. Phys. Lett.*, **29**, 458–462.

Bohren, C. F., 1975. Scattering of electromagnetic waves by an optically active spherical shell, *J. Chem. Phys.*, **62**, 1566–1571.

Bohren, C. F., 1977. Circular dichroism and optical rotatory dispersion spectra of arbitrarily shaped optically active particles, *J. Theor. Biol.*, **65**, 755–767.

Bohren, C. F., 1978. Scattering of electromagnetic waves by an optically active cylinder, *J. Colloid Interface Sci.*, **66**, 105–109.

Bohren, C. F., and B. R. Barkstrom, 1974. Theory of the optical properties of snow, *J. Geophys. Res.*, **79**, 4527–4535.

Bohren, C. F., and L. J. Battan, 1980. Radar backscattering by inhomogeneous precipitation particles, *J. Atmos. Sci.*, **37**, 1821–1827.

Bohren, C. F., and L. J. Battan, 1981. Backscattering of microwaves by spongy ice spheres, *Preprints, 20th Conference on Radar Meteorology (Boston)*, American Meteorological Society, Boston, pp. 385–388.

Bohren, C. F., and G. M. Brown, 1981. Once in a blue moon, *Weatherwise*, **34**, 129–130.

Bohren, C. F., and D. P. Gilra, 1979. Extinction by a spherical particle in an absorbing medium, *J. Colloid Interface Sci.*, **72**, 215–221.

Bohren, C. F., and B. M. Herman, 1979. On the asymptotic scattering efficiency of a large sphere, *J. Opt. Soc. Am.*, **69**, 1615–1616.

Bohren, C. F., and D. R. Huffman, 1981. Absorption cross-section maxima and minima in IR absorption bands of small ionic ellipsoidal particles, *Appl. Opt.*, **20**, 959–962.

Bohren, C. F., and A. J. Hunt, 1977. Scattering of electromagnetic waves by a charged sphere, *Can. J. Phys.*, **55**, 1930–1935.

Bohren, C. F., and V. Timbrell, 1979. Computer programs for calculating scattering and absorption by normally illuminated infinite cylinders, Project No. 38, Institute of Occupational and Environmental Health, Montreal.

Bohren, C. F., and N. C. Wickramasinghe, 1977. On the computation of optical properties of heterogeneous grains, *Astrophys. Space Sci.*, **50**, 461–472.

Borghese, E., P. Denti, G. Toscano, and O. I. Sindoni, 1979. Electromagnetic scattering by a cluster of spheres, *Appl. Opt.*, **18**, 116–120.

Born, M., and E. Wolf, 1965. *Principles of Optics*, 3rd ed., Pergamon, Oxford.

Bottiger, J. R., E. S. Fry, and R. C. Thompson, 1980. Phase matrix measurements for electromagnetic scattering by sphere aggregates, in *Light Scattering by Irregularly Shaped Particles*, D. Schuerman (Ed.), Plenum, New York, pp. 283–290.

Box, M. A., and B. H. J. McKellar, 1978. Direct evaluation of aerosol mass loadings from multispectral extinction data, *Q. J. R. Meteorol. Soc.*, **104**, 775–781.

Brain, J. P., 1972. Halo phenomena—an investigation, *Weather*, **27**, 409–410.

Brillouin, L., 1949. The scattering cross section of spheres for electromagnetic waves, *J. Appl. Phys.*, **20**, 1110–1125.

Brillouin, L., 1960. *Wave Propagation and Group Velocity*, Academic, New York.

British Association for the Advancement of Science, 1952. *Mathematical Tables*, Vol. X; *Bessel Functions*, Part II: *Functions of Positive Integer Order*, Cambridge University Press, Cambridge.

Brown, F. C., 1967. *The Physics of Solids*, W. A. Benjamin, New York.

Brown, F. C., 1974. Ultraviolet spectroscopy of solids with the use of synchrotron radiation, *Solid State Phys.*, **29**, 1–73.

Browning, K. A., 1978. Meteorological applications of radar, *Rep. Prog. Phys.*, **41**, 761–806.

Bruce, C. W., and R. G. Pinnick, 1977. *In situ* measurements of aerosol absorption with a resonant CW laser spectrophone, *Appl. Opt.*, **16**, 1762–1765.

Bruggeman, D. A. G., 1935. Berechnung verschiedener physikalischer Konstanten von heterogenen Substanzen. I. Dielektrizitätskonstanten und Leitfähigkeiten der Mischkörper aus isotropen Substanzen, *Ann. Phys. (Leipzig)*, **24**, 636–679.

Bryant, H. C., and A. J. Cox, 1966. Mie theory and the glory, *J. Opt. Soc. Am.*, **56**, 1529–1532.

Bryksin, V. V., Yu. M. Gerbshtein, and D. N. Mirlin, 1971. Optical effects in alkali-halide crystals due to surface vibrations, *Sov. Phys.-Solid State*, **13**, 1342–1347.

Buchanan, T. J., G. H. Haggis, J. B. Hasted, and B. G. Robinson, 1952. The dielectric estimation of protein hydration, *Proc. R. Soc. Lond.*, **A213**, 379–391.

Burns, R. G., 1970. *Mineralogical Applications of Crystal Field Theory*, Cambridge University Press, Cambridge.

Carlon, H. R., D. H. Anderson, M. E. Milham, T. L. Tarnove, R. H. Frickel, and I. Sindoni, 1977. Infrared extinction spectra of some common liquid aerosols, *Appl. Opt.*, **16**, 1598–1605.

Carlson, T. N., and S. G. Benjamin, 1980. Radiative heating rates for Saharan dust, *J. Atmos. Sci.*, **37**, 193–213.

Champeney, D. C., 1973. *Fourier Transforms and Their Physical Applications*, Academic, New York.

Chandrasekhar, S., 1950. *Radiative Transfer*, Oxford University Press, London (reprinted by Dover, New York, 1960).

Charlock, T. P., and W. D. Sellers, 1980. Aerosol effects on climate calculations with time-dependent and steady-state radiative–convective models, *J. Atmos. Sci.*, **37**, 1327–1341.

Chen, T. S., F. W. de Wette, and L. Kleinman, 1978. Infrared absorption in MgO microcrystals, *Phys. Rev.*, **B18**, 958–962.

Chen, Y., and W. A. Sibley, 1969. A study of zero phonon lines in electron-irradiated, neutron-irradiated, and additively colored MgO, *Philos. Mag.*, **20**, 217–223.

Christy, R. W., 1972. Classical theory of optical dispersion, *Am. J. Phys.*, **40**, 1403–1419.

Chu, B., 1974. *Laser Light Scattering*, Academic, New York.

Churchill, R. V., 1960. *Complex Variables and Applications*, McGraw-Hill, New York.

Chýlek, P., 1975. Asymptotic limits of the Mie-scattering characteristics, *J. Opt. Soc. Am.*, **65**, 1316–1318.

Chýlek, P., 1976. Partial-wave resonances and the ripple structure in the Mie normalized extinction cross section, *J. Opt. Soc. Am.*, **66**, 285–287.

Chýlek, P., J. T. Kiehl, and M. K. W. Ko, 1978a. Narrow resonance structure in the Mie scattering characteristics, *Appl. Opt.*, **17**, 3019–3021.

Chýlek, P., J. T. Kiehl, and M. K. W. Ko, 1978b. Optical levitation and partial-wave resonances, *Phys. Rev.*, **A18**, 2229–2233.

Clarke, D., 1974. Polarimetric definitions, in *Planets, Stars and Nebulae Studied with Photopolarimetry*, T. Gehrels, (Ed.), University of Arizona Press, Tucson, Ariz, pp. 45–53.

Clarke, D., and J. F. Grainger, 1971. *Polarized Light and Optical Measurements*, Pergamon, Oxford.

Coffeen, D. L., 1969. Wavelength dependence of polarization: XVI. Atmosphere of Venus, *Astron. J.*, **74**, 446–460.

Cole, K. S., and R. H. Cole, 1941. Dispersion and absorption in dielectrics: I. Alternating current characteristics, *J. Chem. Phys.*, **9**, 341–351.

Collett, E., 1968. The description of polarization in classical physics, *Am. J. Phys.*, **22**, 351–362.

Cook, H. F., 1952. A comparison of the dielectric behaviour of pure water and human blood at microwave frequencies, *Br. J. Appl. Phys.*, **3**, 249–255.

Cooke, D. D., and M. Kerker, 1975. Response calculations for light-scattering aerosol particle counters, *Appl. Opt.*, **14**, 734–739.

Corn, M., and N. A. Esmen, 1976. Aerosol generation, in *Handbook on Aerosols*, R. Dennis (Ed.), ERDA Technical Information Center, Oak Ridge, Tenn., pp. 9–39.

Courant, R., and D. Hilbert, 1953. *Methods of Mathematical Physics*, Vol. I, Wiley-Interscience, New York.

Coyne, G. V., T. Gehrels, and K. Serkowski, 1974. Wavelength dependence of polarization: XXVI. The wavelength of maximum polarization as a characteristic parameter of interstellar grains, *Astron. J.*, **79**, 581–589.

Crosswhite, H. M. and H. W. Moos (Eds.), 1967. *Optical Properties of Ions in Crystals*, Interscience, New York.

Dave, J. V., 1968. *Subroutines for Computing the Parameters of the Electromagnetic Radiation Scattered by a Sphere*, IBM Order Number 360D-17.4.002.

Davies, Kenneth, 1969. *Ionospheric Radio Waves*, Blaisdell, Waltham, Mass.

Davis, E. J., and A. K. Ray, 1977. Determination of diffusion coefficients by submicron droplet evaporation, *J. Chem. Phys.*, **67**, 414–419.

Dawkins, A. W. J., N. R. V. Nightingale, G. P. South, R. J. Sheppard, and E. H. Grant, 1979. The role of water in microwave absorption by biological material with particular reference to microwave hazards, *Phys. Med. Biol.*, **24**, 1168–1176.

Day, K. L., 1976. Further measurements of amorphous silicates, *Astrophys. J.*, **210**, 614–617.

Day, K. L., 1979. Mid-infrared optical properties of vapor-condensed magnesium silicates, *Astrophys. J.*, **234**, 158–161.

Day, K. L., and D. R. Huffman, 1973. Measured extinction efficiency of graphite smoke in the region 1200–6000 Å, *Nature Phys. Sci.*, **243**, 50–51.

Dayawansa, I. J., and C. F. Bohren, 1978. The effect of substrate and aggregation on infrared extinction spectra of MgO particles, *Phys. Status Solidi (b)*, **86**, K27–K30.

Debye, P., 1909. Der Lichtdruck auf Kugeln von beliebigem Material, *Ann. Phys.*, **30**, 57–136.

Debye, P., 1929. *Polar Molecules*, Chemical Catalog Co., New York.

de Groot, S. R., 1969. *The Maxwell Equations*, North-Holland, Amsterdam.

Deirmendjian, D., 1969. *Electromagnetic Scattering on Spherical Polydispersions*, Elsevier, New York.

Deirmendjian, D., and E. H. Vestine, 1959. Some remarks on the nature and origin of noctilucent cloud particles, *Planet. Space Sci.*, **1**, 146–153.

Devore, J. R., and A. H. Pfund, 1947. Optical scattering by dielectric powders of uniform particle size, *J. Opt. Soc. Am.*, **37**, 826–832.

DiBartolo, B. (Ed.), 1974. *Optical Properties of Ions in Solids*, Plenum, New York.

Diehl, S. R., D. T. Smith, and M. Sydor, 1979. Analysis of suspended solids by single-particle scattering, *Appl. Opt.*, **18**, 1653–1657.

Dixon, J. R., 1969. Electric-susceptibility mass of free carriers in semiconductors, in *Optical Properties of Solids*, S. Nudelman and S. S. Mitra (Eds.), Plenum, New York, pp. 61–83.

Djerassi, C., 1960. *Optical Rotatory Dispersion*, McGraw-Hill, New York.

Dobbins, R. A., and T. I. Eklund, 1977. Ripple structure of the extinction coefficient, *Appl. Opt.*, **16**, 281–282.

Donahue, T. M., B. Guenther, and J. E. Blamont, 1972. Noctilucent clouds in daytime: circumpolar particulate layers near the summer mesopause, *J. Atmos. Sci.*, **29**, 1205–1209.

Doremus, R. H., 1964. Optical properties of small gold particles, *J. Chem. Phys.*, **40**, 2389–2396.

Doyle, W. T., 1958. Absorption of light by colloids in alkali halide crystals, *Phys. Rev.*, **111**, 1067–1077.

Drost-Hansen, W., 1969. Structure of water near solid interfaces, *Ind. Eng. Chem.*, **61**, 10–47.

Eastwood, E., 1967. *Radar Ornithology*, Methuen, London.

Ehrenreich, H., and H. R. Philipp, 1962. Optical properties of Ag and Cu, *Phys. Rev.*, **128**, 1622–1629.

Ehrenreich, H., H. R. Philipp, and B. Segall, 1963. Optical properties of aluminum, *Phys. Rev.*, **132**, 1918–1928.

Emeis, C. A., L. J. Oosterhoff, and G. de Vries, 1967. Numerical evaluation of Kramers–Kronig relations, *Proc. R. Soc. Lond.*, **A297**, 54–65.

Erma, V. A., 1968a. An exact solution for the scattering of electromagnetic waves from conductors of arbitrary shape: I. Case of cylindrical symmetry, *Phys. Rev.*, **173**, 1243–1257.

Erma, V. A., 1968b. Exact solution for the scattering of electromagnetic waves from conductors of arbitrary shape: II. General case, *Phys. Rev.*, **176**, 1544–1553.

Erma, V. A., 1969. Exact solution for the scattering of electromagnetic waves from bodies of arbitrary shape: III. Obstacles with arbitrary electromagnetic properties, *Phys. Rev.*, **179**, 1238–1246.

Eversole, J. D., and H. P. Broida, 1977. Size and shape effects in light scattering from small silver, copper, and gold particles, *Phys. Rev.*, **B15**, 1644–1655.

Fabelinskii, I. L., 1968. *Molecular Scattering of Light*, Plenum, New York.

Faraday, M., 1857. Experimental relations of gold (and other metals) to light, *Philos. Trans.*, **147**, 145 (reprinted in M. Faraday, *Experimental Researches in Chemistry and Physics*, William Francis, London, 1859, pp. 391–443).

Faxvog, F. R., and D. M. Roessler, 1979. Optoacoustic measurements of diesel particulate emission, *J. Appl. Phys.*, **50**, 7880–7882.

Faxvog, F. R., and D. M. Roessler, 1981. Optical absorption in thin slabs and spherical particles, *Appl. Opt.*, **20**, 729–731.

Fenn, R. W., and H. Oser, 1965. Scattering properties of concentric soot-water spheres for visible and infrared light, *Appl. Opt.*, **4**, 1504–1509.

Fischer, K., 1970. Measurements of absorption of visible radiation by aerosol particles, *Beitr. Phys. Atmos.*, **43**, 244–254.

Fischer, K., 1973. Mass absorption coefficient of natural aerosol particles in the 0.4–2.4 μm wavelength interval, *Beitr. Phys. Atmos.*, **46**, 89–100.

Fogle, B., and B. Haurwitz, 1966. Noctilucent clouds, *Space Sci. Rev.*, **6**, 279–340.

Foley, J. T., and D. N. Pattanayak, 1974. Electromagnetic scattering from a spatially dispersive sphere, *Opt. Commun.*, **12**, 113–117.

Foot, J. S., 1979. Spectrophone measurements of the absorption of solar radiation by aerosol, *Q. J. R. Meteorol. Soc.*, **105**, 275–283.

Forman, R. A., G. J. Piermarini, J. D. Barnett, and S. Block, 1972. Pressure measurements made by the utilization of ruby sharp-line luminescence, *Science*, **176**, 284–285.

Fouquart, Y., W. M. Irvine, and J. Lenoble, 1980. Standard procedures to compute atmospheric radiative transfer in a scattering atmosphere, Vol. 2, Int. Assoc. Met. Atmos. Phys. (prepared for publication by S. Ruttenberg, National Center for Atmospheric Research, Boulder, Colo.).

Fraser, A. B., 1972. Inhomogeneities in the color and intensity of the rainbow, *J. Atmos. Sci.*, **29**, 211–212.

Fraser, A. B., 1979. What size of ice crystals causes the halos? *J. Opt. Soc. Am.*, **69**, 1112–1118.

Fröhlich, H., 1949. *Theory of Dielectrics*, Oxford University Press, London.

Fröhlich, H., and H. Pelzer, 1955. Plasma oscillations and energy loss of charged particles in solids, *Proc. Phys. Soc. Lond.*, **A68**, 525–529.

Fry, E. S., and G. W. Kattawar, 1981. Relationships between elements of the Stokes matrix, *Appl. Opt.*, **20**, 2811–2814.

Fuchs, R., 1975. Theory of the optical properties of ionic crystal cubes, *Phys. Rev.*, **B11**, 1732–1740.

Fuchs, R., 1978. Infrared absorption in MgO microcrystals, *Phys. Rev.*, **B18**, 7160–7162.

Fuchs, R., and K. L. Kliewer, 1968. Optical modes of vibration in an ionic crystal sphere, *J. Opt. Soc. Am.*, **58**, 319–330.

Gadsden, M., 1975. Observations of the colour and polarization of noctilucent clouds, *Ann. Geophys.*, **31**, 507–516.

Gadsden, M., 1977. The polarization of noctilucent clouds, *Ann. Geophys.*, **33**, 363–366.

Gadsden, M., 1978. The sizes of particles in noctilucent clouds: implications for mesospheric water vapor, *J. Geophys. Res.*, **83**, 1155–1156.

Gadsden, M., P. Rothwell, and M. J. Taylor, 1979. Detection of circularly polarised light from noctilucent clouds, *Nature*, **278**, 628–629.

Gambling, D. J., and B. Billard, 1967. A study of the polarization of skylight, *Aust. J. Phys.*, **20**, 675–681.

Garbuny, M., 1965. *Optical Physics*, Academic, New York.

Gehrels, T. (Ed.), 1974a. *Planets, Stars and Nebulae Studied with Photopolarimetry*, University of Arizona Press, Tucson, Ariz.

Gehrels, T., 1974b. Wavelength dependence of polarization: XXVII. Interstellar polarization from 0.22 to 2.2 μm, *Astron. J.*, **79**, 590–593.

Genzel, L., 1969. Impurity-induced lattice absorption, in *Optical Properties of Solids*, S. Nudelman and S. S. Mitra (Eds.), Plenum, New York, pp. 453–487.

Genzel, L., 1974. Aspects of the physics of microcrystals, in *Festkörper probleme (Advances in Solid State Physics)*, Vol. XIV, O. Madelung (Ed.), Pergamon-Vieweg, Braunschweig, pp. 183–203.

Genzel, L., and T. P. Martin, 1972. Infrared absorption by small ionic crystals, *Phys. Status Solidi (b)*, **51**, 91–99.

Genzel, L., and T. P. Martin, 1973. Infrared absorption by surface phonons and surface plasmons in small crystals, *Surf. Sci.*, **34**, 33–49.

Gerber, H., and E. Hindman, 1982. *Light Absorption by Aerosol Particles*, Spectrum Press, Hampton, Va.

Gevers, M., 1946. The relation between the power factor and the temperature coefficient of the dielectric constant of solid dielectrics, I, *Philips Res. Rep.*, **1**, 197–224.

Giaever, I., 1973. The antibody-antigen reaction: a visual observation, *J. Immunol.*, **110**, 1424–1426.

Giaever, I., and R. J. Laffin, 1974. Visual detection of hepatitis B Antigen, *Proc. Natl. Acad. Sci. USA*, **71**, 4533–4535.

Gillespie, J. B., S. G. Jennings, and J. D. Lindberg, 1978. Use of an average complex refractive index in atmospheric propagation calculations, *Appl. Opt.*, **17**, 989–991.

Gilra, D. P., 1972a. Collective excitations and dust particles in space, in *The Scientific Results from the Orbiting Astronomical Observatory OAO-2*, A. D. Code (Ed.), NASA SP-310, pp. 295–319.

Gilra, D. P., 1972b. Collective excitations in small solid particles and astronomical applications, Ph.D. thesis, University of Wisconsin.

Glick, A. J., and E. D. Yorke, 1978. Theory of far-infrared absorption by small metallic particles, *Phys. Rev.*, **B18**, 2490–2493.

Glicksman, M., 1971. Plasmas in solids, *Solid State Phys.*, **26**, 275–427.

Goldberger, M. L., 1960. Introduction to the theory and applications of dispersion relations, in *Dispersion Relations and Elementary Particles*, C. de Witt and R. Omnes (Eds.), Wiley, New York, pp. 15–157.

Goldstein, M., and R. M. Thaler, 1959. Recurrence techniques for the calculation of Bessel functions, *Math. Tables Other Aids Comput.*, **13**, 102–108.

Golovanev, V. A., G. I. Gorchakov, A. S. Yemilenko, and V. N. Sidorov, 1971. Climatic variability of light scattering matrices of the atmosphere, *Izv. Acad. Sci. USSR Atmos. Ocean. Phys.*, **7**, 1318–1322.

Goody, R. M., 1964. *Atmospheric Radiation*, Vol. 1: *Theoretical Basis*, Oxford University Press, London.

Gor'kov, L. P., and G. M. Eliashberg, 1965. Minute metallic particles in an electromagnetic field, *Sov. Phys.-JETP*, **21**, 940–947.

Grams, G. W., 1981. *In-situ* measurements of scattering phase functions of stratospheric aerosol particles in Alaska during July 1979, *Geophys. Res. Lett.*, **8**, 13–14.

Grams, G. W., I. H. Blifford, D. A. Gillette, and P. B. Russell, 1974. Complex index of refraction of airborne soil particles, *J. Appl. Meteorol.*, **13**, 459–471.

Granqvist, C. G., and O. Hunderi, 1977. Optical properties of ultrafine gold particles, *Phys. Rev.*, **B16**, 3513–3534.

Granqvist, C. G., R. A. Buhrman, J. Wyns, and A. J. Sievers, 1976. Far-infrared absorption in ultrafine Al particles, *Phys. Rev. Lett.*, **37**, 625–629.

Grant, E. H., T. J. Buchanan, and H. F. Cook, 1957. Dielectric behavior of water at microwave frequencies, *J. Chem. Phys.*, **26**, 156–161.

Green, A. W., 1975. An approximation for the shapes of large raindrops, *J. Appl. Meteorol.*, **14**, 1578–1583.

Green, H. L., and W. R. Lane, 1964. *Particulate Clouds: Dust, Smokes, and Mists*, 2nd ed., Van Nostrand, New York.

Greenaway, D. L., and G. Harbeke, 1968. *Optical Properties and Band Structure of Solids*, Pergamon, Oxford.

Greenberg, J. M., and S. S. Hong, 1974. Effects of particle shape on the shape of extinction and polarization bands in grains, in *Planets, Stars and Nebulae Studied with Photopolarimetry*, T. Gehrels (Ed.), University of Arizona Press, Tucson, Ariz., pp. 916–925.

Greenberg, J. M., and R. Stoeckly, 1971. Shape of the diffuse interstellar bands, *Nature Phys. Sci.*, **230**, 15–16.

Greenberg, J. M., N. E. Pedersen, and J. C. Pedersen, 1961. Microwave analog to the scattering of light by nonspherical particles, *J. Appl. Phys.*, **32**, 233–242.

Greenler, R., 1980. *Rainbows, Halos, and Glories*, Cambridge University Press, Cambridge.

Grenfell, T. C., and D. K. Perovich, 1981. Radiation absorption coefficients of polycrystalline ice from 400–1400 nm, *J. Geophys. Res.*, **86**, 7447–7450.

Grishin, N. I., 1956. Research on the continuous spectrum of noctilucent clouds, *All Union Ast. Geod. Soc. Bull. USSR*, **19**, 3–16.

Gucker, F. T., and J. J. Egan, 1961. Measurement of the angular variation of light scattered from single aerosol droplets, *J. Colloid Sci.*, **16**, 68–84.

Gunn, R., and G. D. Kinzer, 1949. The terminal velocity of fall for water droplets in stagnant air, *J. Meteorol.*, **6**, 243–248.

Hadni, A., 1970a. Far infrared electronic transitions in rare earth ions, in *Far-Infrared Properties of Solids*, S. S. Mitra and S. Nudelman (Eds.), Plenum, New York, pp. 535–560.

Hadni, A., 1970b. Lattice absorption in the very far infrared, in *Far-Infrared Properties of Solids*, S. S. Mitra and S. Nudelman (Eds.), Plenum, New York, pp. 561–587.

Häfele, H. G., 1963. Die optischen Konstanten von Magnesiumoxid im Infraroten, *Ann. Phys.*, **10**, 321–326.

Hagemann, H. J., W. Gudat, and C. Kunz, 1974. Optical constants from the far infrared to the x-ray region: Mg, Al, Cu, Ag, Au, Bi, C, and Al_2O_3, Deutsches Elektronen-Synchrotron DESY SR-74/7.

Hale, G. M., and M. R. Querry, 1973. Optical constants of water in the 200-nm to 200-μm wavelength region, *Appl. Opt.*, **12**, 555–563.

Hansen, J. E., and J. W. Hovenier, 1974. Interpretation of the polarization of Venus, *J. Atmos. Sci.*, **31**, 1137–1160.

Hansen, J. E., and L. D. Travis, 1974. Light scattering in planetary atmospheres, *Space Sci. Rev.*, **16**, 527–610.

Hansen, J. E., W. C. Wang, and A. A. Lacis, 1979. Climatic effects of atmospheric aerosols, Proc. Conf. on Aerosols: Urban and Rural Characteristics, Source and Transport Studies, New York Academy of Sciences, New York.

Harbeke, G., 1972. Optical properties of semiconductors, in *Optical Properties of Solids*, F. Abelès (Ed.), Elsevier, New York, pp. 21–92.

Harrick, N. J., 1967. *Internal Reflection Spectroscopy*, Wiley-Interscience, New York.

Hart, R. W., and E. P. Gray, 1964. Determination of particle structure from light scattering, *J. Appl. Phys.*, **35**, 1408–1415.

Hass, G., and R. Tousey, 1959. Reflecting coatings for the extreme ultraviolet, *J. Opt. Soc. Am.*, **49**, 593–602.

Henniker, J. C., 1949. The depth of the surface zone of a liquid, *Rev. Mod. Phys.*, **21**, 322–341.

Hensel, J. C., T. G. Philips, and G. A. Thomas, 1977. The electron-hole liquid in semiconductors: experimental aspects, *Solid State Phys.*, **32**, 88–314.

Herbig, G. H., 1975. The diffuse interstellar bands: IV. The region 4400–6850 Å, *Astrophys. J.*, **196**, 129–160.

Herman, B. M., 1962. Infra-red absorption, scattering, and total attenuation cross-sections for water spheres, *Q. J. R. Meteorol. Soc.*, **88**, 143–150.

Herman, B. M., and L. J. Battan, 1961. Calculations of Mie back-scattering of microwaves from ice spheres, *Q. J. R. Meteorol. Soc.*, **87**, 223–230.

Herzberg, G., 1945. *Infrared and Raman Spectra of Polyatomic Molecules*, Van Nostrand, New York.

Heyder, J., and J. Gebhart, 1979. Optimization of response functions of light scattering instruments for size evaluation of aerosol particles, *Appl. Opt.*, **18**, 705–711.

Hobbs, P. V., 1974. *Ice Physics*, Oxford University Press, London.

Hodgson, J. N., 1970. *Optical Absorption and Dispersion in Solids*, Chapman & Hall, London.

Hodkinson, J. R., 1963. Light scattering and extinction by irregular particles larger than the wavelength, in *Electromagnetic Scattering*, M. Kerker (Ed.), Macmillan, New York, pp. 87–100.

Hodkinson, J. R., and I. Greenleaves, 1963. Computations of light-scattering and extinction by spheres according to diffraction and geometrical optics and some comparisons with the Mie theory, *J. Opt. Soc. Am.*, **53**, 577–588.

Holland, A. C., and J. S. Draper, 1967. Analytical and experimental investigation of light scattering from polydispersions of Mie particles, *Appl. Opt.*, **6**, 511–518.

Holland, A. C., and G. Gagne, 1970. The scattering of polarized light by polydisperse systems of irregular particles, *Appl. Opt.*, **9**, 1113–1121.

Huebner, R. H., E. T. Arakawa, R. A. MacRae, and R. R. Hamm, 1964. Optical constants of vacuum-evaporated silver films, *J. Opt. Soc. Am.*, **54**, 1434–1437.

Huffaker, R. M., 1970. Laser Doppler detection system for gas velocity measurement, *Appl. Opt.*, **9**, 1026–1039.

Huffman, D. R., 1977. Interstellar grains: the interaction of light with a small-particle system, *Adv. Phys.*, **26**, 129–230.

Huffman, D. R., and C. F. Bohren, 1980. Infrared absorption spectra of non-spherical particles treated in the Rayleigh-ellipsoid approximation, in *Light Scattering by Irregularly Shaped Particles*, D. Schuerman (Ed.), Plenum, New York, pp. 103–111.

Huffman, D. R., L. A. Schwalbe, and D. Schiferl, 1982. Use of smoke samples in diamond-anvil cells to measure pressure dependence of optical spectra: application to the ZnO exciton, *Solid State Commun.* (in press).

Hummel, J. R., and J. J. Olivero, 1976. Satellite observation of the mesospheric scattering layer and implied climatic consequences, *J. Geophys. Res.*, **81**, 3177–3178.

Hunt, A. J., and D. R. Huffman, 1973. A new polarization-modulated light scattering instrument, *Rev. Sci. Instrum.*, **44**, 1753–1762.

Hunt, A. J., and D. R. Huffman, 1975. A polarization-modulated instrument for determining liquid aerosol properties, *Jpn. J. Appl. Phys.*, **14** (Suppl. 14-1), 435–440.

Hunt, A. J., T. R. Steyer, and D. R. Huffman, 1973. Infrared surface modes in small NiO particles, *Surf. Sci.*, **36**, 454–461.

Hurwitz, H., 1945. The statistical properties of unpolarized light, *J. Opt. Soc. Am.*, **35**, 525–531.

Inagaki, T., E. T. Arakawa, R. N. Hamm, and M. W. Williams, 1977. Optical properties of polystyrene from the near-infrared to the x-ray region and convergence of the optical sum rules, *Phys. Rev.*, **B15**, 3243–3253.

Irani, G. B., T. Huen, and F. Wooten, 1971. Optical properties of Ag and α-phase Ag-Al alloys, *Phys. Rev.*, **3B**, 2385–2390.

Irvine, W. M., and J. B. Pollack, 1968. Infrared optical properties of water and ice spheres, *Icarus*, **8**, 324–360.

Jackson, J. D., 1975. *Classical Electrodynamics*, 2nd ed., Wiley, New York.

Jacobson, B., 1955. On the interpretation of dielectric constants of aqueous macromolecular solutions. Hydration of macromolecules, *J. Am. Chem. Soc.*, **77**, 2919–2926.

Japar, S. M., and D. K. Killinger, 1979. Photoacoustic and absorption spectrum of airborne carbon particulate using a tunable dye laser, *Chem. Phys. Lett.*, **66**, 207–209.

Jasperse, J. R., A. Kahan, J. N. Plendl, and S. S. Mitra, 1966. Temperature dependence of infrared dispersion in ionic crystals LiF and MgO, *Phys. Rev.*, **146**, 526–542.

Jasperson, S. N., and S. E. Schnatterly, 1969. An improved method for high reflectivity ellipsometry based on a new polarization modulation technique, *Rev. Sci. Instrum.*, **40**, 761–767.

Jenkins, F. A., and H. E. White, 1957. *Fundamentals of Optics*, 3rd ed., McGraw-Hill, New York.

Jones, D. S., 1955. On the scattering cross section of an obstacle, *Philos. Mag.*, **46**, 957–962.

Jones, D. S., 1964. *The Theory of Electromagnetism*, Pergamon, Oxford.

Jones, D. S., and M. Kline, 1958. Asymptotic expansions of multiple integrals and the method of stationary phase, *J. Math. Phys.*, **37**, 1–28.

Jones, R. C., 1945. A generalization of the dielectric ellipsoid problem, *Phys. Rev.*, **68**, 93–96.

June, K. R., 1972. Error in the KBr dispersion equation, *Appl. Opt.*, **11**, 1655.

Kadyshevich, Ye. A., Yu. S. Lyubovtseva, and I. N. Plakhina, 1971. Measurements of matrices for light scattered by sea water, *Izv. Acad. Sci. USSR Atmos. Ocean. Phys.*, **7**, 367–371.

Kadyshevich, Ye. A., Yu. S. Lyubovtseva, and G. V. Rozenberg, 1976. Light-scattering matrices of Pacific and Atlantic Ocean waters, *Izv. Acad. Sci. USSR Atmos. Ocean. Phys.*, **12**, 106–111.

Kälin, R., and F. Kneubühl, 1976. Size effects on the spectral thermal emissivity of alkali halides, *Infrared Phys.*, **16**, 491–508.

Kattawar, G. W., and M. Eisner, 1970. Radiation from a homogeneous isothermal sphere, *Appl. Opt.*, **9**, 2685–2690.

Kattawar, G. W., and T. J. Humphreys, 1980. Electromagnetic scattering from two identical pseudospheres, in *Light Scattering by Irregularly Shaped Particles*, D. Schuerman (Ed.), Plenum, New York, pp. 177–190.

Kattawar, G. W., and G. N. Plass, 1967. Electromagnetic scattering from absorbing spheres, *Appl. Opt.*, **6**, 1377–1382.

Kemp, J. C., 1969. Piezo-optical birefringence modulators: new use for a long-known effect, *J. Opt. Soc. Am.*, **59**, 950–954.

Kerker, M., 1969. *The Scattering of Light and Other Electromagnetic Radiation*, Academic, New York.

Kerker, M., D. Cooke, W. A. Farone, and R. T. Jacobsen, 1966. Electromagnetic scattering from an infinite circular cylinder at oblique incidence: I. Radiance functions for $m = 1.46$, *J. Opt. Soc. Am.*, **56**, 487–491.

Kerker, M., P. Scheiner, and D. D. Cooke, 1978. The range of validity of the Rayleigh and Thomson limits for Lorenz–Mie scattering, *J. Opt. Soc. Am.*, **68**, 135–137.

Kerr, G. D., R. N. Hamm, M. W. Williams, R. D. Birkhoff, and L. R. Painter, 1972. Optical and dielectric properties of water in the vacuum ultraviolet, *Phys. Rev.*, **A5**, 2523–2527.

Khare, V., and H. M. Nussenzveig, 1977. Theory of the glory, *Phys. Rev. Lett.*, **38**, 1279–1282.

Kilkson, R., W. S. Bickel, W. S. Jetter, and M. E. Stafford, 1979. Influence of absorption on polarization effects in light scattering from human red blood cells, *Biochim. Biophys. Acta*, **584**, 175–179.

King, M. D., B. M. Herman, and J. A. Reagan, 1978. Aerosol size distributions obtained by inversion of spectral optical depth measurements, *J. Atmos. Sci.*, **35**, 2153–2167.

Kirchner, F., and R. Zsigmondy, 1904. Über die Ursachen der Farbernänderunden von Gold-Gelatinepräparaten, *Ann. Phys.*, **15**, 573–595.

Kittel, C., 1962. *Elementary Solid State Physics*, Wiley, New York.

Kittel, C., 1971. *Introduction to Solid State Physics*, 4th ed., Wiley, New York.

Koike, C., H. Hasegawa, and A. Manabe, 1980. Extinction coefficients of amorphous carbon grains from 2100 Å to 340 μm, *Astrophys. Space Sci.*, **67**, 495–502.

Kortüm, G. F., 1969. *Reflectance Spectroscopy*, Springer-Verlag, Berlin.

Koyama, R. Y., N. V. Smith, and W. E. Spicer, 1973. Optical properties of indium, *Phys. Rev.*, **B8**, 2426–2431.

Krätschmer, W., and D. R. Huffman, 1979. Infrared extinction of heavy ion irradiated and amorphous olivine, with applications to interstellar dust, *Astrophys. Space Sci.*, **61**, 195–203.

Kreibig, U., 1974. Electronic properties of small silver particles: the optical constants and their temperature dependence, *J. Phys. F.*, **4**, 999–1014.

Kreibig, U., and C. von Fragstein, 1969. The limitation of electron mean free path in small silver particles, *Z. Phys.*, **224**, 307–323.

Landau, L. D., and E. M. Lifshitz, 1960. *Electrodynamics of Continuous Media*, Pergamon, Oxford.

Landauer, R., 1952. The electrical resistance of binary metallic mixtures, *J. Appl. Phys.*, **23**, 779–784.

Lane, J. A., and J. A. Saxton, 1952. Dielectric dispersion in pure polar liquids at very high radio frequencies: I. Measurements on water, methyl and ethyl alcohols, *Proc. R. Soc. Lond.*, **A213**, 400–408.

Langbein, D., 1976. Normal modes of small cubes and rectangular particles, *J. Phys. A*, **9**, 627–644.

Larkin, B., and S. W. Churchill, 1959. Scattering and absorption of electromagnetic radiation by infinite cylinders, *J. Opt. Soc. Am.*, **49**, 188–190.

Latimer, P., A. Brunsting, B. E. Pyle, and C. Moore, 1978. Effects of asphericity on single particle scattering, *Appl. Opt.*, **17**, 3152–3158.

Lenoble, J. (Ed.), 1977. Standard procedures to compute atmospheric radiative transfer in a scattering atmosphere, Vol. I, Int. Assoc. Met. Atmos. Phys. (prepared for publication by S. Ruttenberg, National Center for Atmospheric Research, Boulder, Colo.).

Libelo, L. F., 1962. Light scattering by partially absorbing cylinders—coefficients, NOLTR 62-142. U.S. Naval Ordnance Lab., White Oak, Md.

Lin, C.-I., M. Baker, and R. J. Charlson, 1973. Absorption coefficient of atmospheric aerosol: a method for measurement, *Appl. Opt.*, **12**, 1356–1363.

Lind, A. C., and J. M. Greenberg, 1966. Electromagnetic scattering by obliquely oriented cylinders, *J. Appl. Phys.*, **37**, 3195–3203.

Lindberg, J. D., 1975. The composition and optical absorption coefficient of atmospheric particulate matter, *Opt. Quant. Electron.*, **7**, 131–139.

Lindberg, J. D., and J. B. Gillespie, 1977. Relationship between particle size and imaginary refractive index in atmospheric dust, *Appl. Opt.*, **16**, 2628–2630.

Lindberg, J. D., and L. Laude, 1974. Measurement of absorption coefficient of atmospheric dust, *Appl. Opt.*, **13**, 1923–1927.

Liu, B. Y. H, R. N. Bergland, and J. K. Agarwal, 1974. Experimental studies of particle counters, *Atmos. Environ.*, **8**, 717–732.

Lobdell, D. D., 1968. Particle size–amplitude relations for the ultrasonic atomizer, *J. Acoust. Soc. Am.*, **43**, 229–231.

Logan, N. A., 1965. Survey of some early studies of the scattering of plane waves by a sphere, *Proc. IEEE*, **53**, 773–785.

Ludlum, F. H., 1957. Noctilucent clouds, *Tellus*, **9**, 341–363.

Lushnikov, A. A., V. V. Maksimenko, and A. Ya. Simonov, 1978. Absorption of low-frequency electromagnetic radiation by small metallic particles, *Sov. Phys.-Solid State*, **20**, 292–295.

Luxon, J. T., D. J. Montgomery, and R. Summitt, 1969. Effect of particle size and shape on the infrared absorption of magnesium oxide powders, *Phys. Rev.*, **188**, 1345–1356.

Maksimenko, V. V., A. J. Simonov, and A. A. Lushnikov, 1977. Far-infrared absorption in small metallic particles, *Phys. Status Solidi (b)*, **83**, 377–382.

Marion, J. B., 1970. *Classical Dynamics of Particles and Systems*, Academic, New York.

Markham, J. M., 1966. F-centers in alkali halides, *Solid State Phys.*, Suppl. 8.

Marshall, T. R., C. S. Parmenter, and M. Seaver, 1976. Characterization of polymer latex aerosols by rapid measurement of 360° light scattering patterns from individual particles, *J. Colloid Interface Sci.*, **55**, 624–636.

Martin, P. G., 1972. Interstellar circular polarization, *Mon. Not. R. Astron. Soc.*, **159**, 179–190.

Martin, P. G., 1974. Interstellar polarization from a medium with changing grain alignment, *Astrophys. J.*, **187**, 461–472.

Martin, P. G., 1978. *Cosmic Dust*, Oxford University Press, London.

Martin, T. P., 1969. Infrared absorption in small KCl crystals, *Phys. Rev.*, **177**, 1349–1350.

Martin, T. P., 1970. Interaction of finite NaCl crystals with infrared radiation, *Phys. Rev.*, **B8**, 3480–3488.

Martin, T. P., 1971. Infrared absorption by surface modes in small KBr crystals, *Solid State Commun.*, **9**, 623–625.

Martin, T. P., 1973. Lattice dynamics of ionic microcrystals, *Phys. Rev.*, **B7**, 3906–3912.

Martin, T. P., and H. Schaber, 1977. Infrared absorption of aggregated and isolated microcrystals, *Phys. Status Solidi (b)*, **81**, K41–K45.

Mathis, J. S., W. Rumpl, and K. H. Nordsieck, 1977. The size distribution of interstellar grains, *Astrophys. J.*, **217**, 425–433.

Matumura, O., and M. Cho, 1981. Thermal-emission spectra of coagulated small MgO crystals, *J. Opt. Soc. Am.*, **71**, 393–396.

Maxwell Garnett, J. C., 1904. Colours in metal glasses and in metallic films, *Philos. Trans. R. Soc.*, **A203**, 385–420.

McCartney, E. J., 1976. *Optics of the Atmosphere*, Wiley, New York.

McClure, D. S., 1959a. Electronic spectra of molecules and ions in crystals: Part I. Molecular crystals, *Solid State Phys.*, **8**, 1–47.

McClure, D. S., 1959b. Electronic spectra of molecules and ions in crystals: Part II. Spectra of ions in crystals, *Solid State Phys.*, **9**, 399–525.

McClure, D. S., 1974. Spectroscopy of magnetic insulators, in *Optical Properties of Ions in Solids*, B. DiBartolo (Ed.), Plenum, New York, pp. 259–305.

McCombie, C. W., 1970. Modification of lattice vibrations by impurities, in *Far-Infrared Properties of Solids*, S. S. Mitra and S. Nudelman (Eds.), Plenum, New York, pp. 297–359.

McCrone, W. C., and J. G. Delly, 1973a. *The Particle Atlas*, 2nd ed., Vol. I: *Principles and Techniques*, Ann Arbor Science Publishers, Ann Arbor, Mich.

McCrone, W. C., and J. G. Delly, 1973b. *The Particle Atlas*, 2nd ed., Vol. II: *The Light Microscopy Atlas*, Ann Arbor Science Publishers, Ann Arbor, Mich.

McCrone, W. C., and J. G. Delly, 1973c. *The Particle Atlas*, 2nd ed., Vol. III: *The Electron Microscopy Atlas*, Ann Arbor Science Publishers, Ann Arbor, Mich.

McCrone, W. C., and J. G. Delly, 1973d. *The Particle Atlas*, 2nd ed., Vol. IV: *The Particle Analyst's Handbook*, Ann Arbor Science Publishers, Ann Arbor, Mich.

McCrone, W. C., J. G. Delly, and S. J. Palenik, 1979. *The Particle Atlas*, 2nd ed., Vol. V: *Light Microscopy Atlas and Techniques*, Ann Arbor Science Publishers, Ann Arbor, Mich.

McCrone, W. C., J. A. Brown, and I. M. Stewart, 1980. *The Particle Atlas*, 2nd ed., Vol VI: *Electron Optical Atlas and Techniques*, Ann Arbor Science Publishers, Ann Arbor, Mich.

McDonald, J. E., 1954. The shape and aerodynamics of large raindrops, *J. Meteorol.*, **11**, 478–494.

McDonald, J. E., 1962. Large-sphere limit of the radar back-scattering coefficient, *Q. J. R. Meteorol. Soc.*, **88**, 183–186.

McKellar, B. H. J., 1976. What property of a haze is determined by light scattering? 2. Nonuniform particles of arbitrary shape, *Appl. Opt.*, **15**, 2464–2467.

McMaster, W. H., 1954. Polarization and the Stokes parameters, *Am. J. Phys.*, **22**, 351–362.

Merzbacher, E., 1970. *Quantum Mechanics*, 2nd ed., Wiley, New York.

Mie, G., 1908. Beitrage zur Optik trüber Medien speziell kolloidaler Metallösungen, *Ann. Phys.*, **25**, 377–445.

Milne, E. A., 1930. Thermodynamics of the stars, in *Handbuch der Astrophysik*, Vol. 3/1, G. Eberhard, A. Kohlschütter, and H. Ludendorff (Eds.), Springer-Verlag, Berlin, pp. 65–255.

Minnaert. M., 1954. *The Nature of Light and Colour in the Open Air*, Dover, New York.

Mitra, S. S., 1969. Infrared and Raman spectra due to lattice vibrations, in *Optical Properties of Solids*, S. Nudelman and S. S. Mitra (Eds.), Plenum, New York, pp. 333–451.

Mitra, S. S., and B. Bendow (Eds.), 1975. *Optical Properties of Highly Transparent Solids*, Plenum, New York.

Mitra, S. S., and S. Nudelman (Eds.), 1970. *Far-Infrared Properties of Solids*, Plenum, New York.

Moravec, T. J., J. C. Rife, and R. N. Dexter, 1976. Optical constants of nickel, iron, and nickel–iron alloys in the vacuum ultraviolet, *Phys. Rev.*, **B13**, 3297–3306.

Morris, S. J., H. A. Shultens, M. A. Hellweg, G. Striker, and T. M. Jovin, 1979. Dynamics of structural changes in biological particles from rapid light scattering measurements, *Appl. Opt.*, **18**, 303–311.

Moss, T. S., 1959. *Optical Properties of Semiconductors*, Butterworth, Kent, England.

Mueller, H., 1948. The foundation of optics, *J. Opt. Soc. Am.*, **38**, 661.

Neuberger, H., 1951. *Introduction to Physical Meteorology*, Pennsylvania State University, State College, Pa.

Neuroth, N., 1956. Über die Bestimmung der optischen Konstanten n, κ aus Reflexionsmessungen, *Z. Phys.*, **144**, 85–90.

Newton, R. G., 1976. Optical theorem and beyond, *Am. J. Phys.*, **44**, 639–642.

Ney, E. P., 1977. Star dust, *Science*, **195**, 541–546.

Nielsen, M. L., P. M. Hamilton, and R. J. Walsh, 1963. Ultrafine metal oxides by decomposition of salts in a flame, in *Ultrafine Particles*, W. E. Kuhn, H. Lamprey, and C. Sheer (Eds.), Wiley, New York, pp. 181–195.

Nikitine, S., 1969. Excitons, in *Optical Properties of Solids*, S. Nudelman and S. S. Mitra (Eds.), Plenum, New York, pp. 197–237.

Niklasson, G. A., C. G. Granqvist, and O. Hunderi, 1981. Effective medium models for the optical properties of inhomogeneous media, *Appl. Opt.*, **20**, 26–30.

Nilsson, P. O., 1974. Optical properties of metals and alloys, *Solid State Phys.*, **29**, 139–234.

Nudelman, S., and S. S. Mitra (Eds.), 1969. *Optical Properties of Solids*, Plenum, New York.

Nussenzveig, H. M., 1972. *Causality and Dispersion Relations*, Academic, New York.

Nussenzveig, H. M., 1979. Complex angular momentum theory of the rainbow and the glory, *J. Opt. Soc. Am.*, **69**, 1068–1079.

Nussenzveig, H. M., and W. J. Wiscombe, 1980. Efficiency factors in Mie scattering, *Phys. Rev. Lett.*, **45**, 1490–1494.

Nye, J. F., 1957. *Physical Properties of Crystals, Their Representation by Tensors and Matrices*, Oxford University Press, London.

Oguchi, T., 1973. Attenuation and phase rotation of radio waves due to rain: calculations at 19.3 and 34.8 GHz, *Radio Sci.*, **8**, 31–38.

Olver, F. W. J., 1964. Bessel functions of integer order, in *Handbook of Mathematical Functions*, M. Abramowitz and I. A. Stegun (Eds.), National Bureau of Standards, Washington, D.C., pp. 355–436.

O'Neill, P., and A. Ignatiev, 1978. Influence of microstructure on the optical properties of particulate materials: gold black, *Phys. Rev.*, **B18**, 6540–6548.

Palmer, R. E., and S. E. Schnatterly, 1971. Observation of surface plasmons and measurement of optical constants for sodium and potassium, *Phys. Rev.*, **B4**, 2329–2339.

Penndorf, R., 1953. On the phenomenon of the colored sun, especially the "blue" sun of September 1950, Geophysical Research Paper No. 20 (AFCRC Tech. Rep. 53-7), Air Force Cambridge Research Center, Cambridge, Mass.

Penndorf, R. B., 1958. An approximation method to the Mie theory for colloidal spheres, *J. Phys. Chem.*, **62**, 1537–1542.

Pennock, B. E., and H. P. Schwan, 1969. Further observations on the electrical properties of hemoglobin-bound water, *J. Phys. Chem.*, **73**, 2600–2610.

Perrin, F., 1942. Polarization of light scattered by isotropic opalescent media, *J. Chem. Phys.*, **10**, 415–427.

Perry, R. J., A. J. Hunt, and D. R. Huffman, 1978. Experimental determinations of Mueller scattering matrices for nonspherical particles, *Appl. Opt.*, **17**, 2700–2710.

Philipp, H. R., and H. Ehrenreich, 1963. Optical properties of semiconductors, *Phys. Rev.*, **129**, 1550–1560.

Phillips, D. T., P. J. Wyatt, and R. M. Berkman, 1970. Measurement of the Lorenz–Mie scattering of a single particle: polystyrene latex, *J. Colloid Interface Sci.*, **34**, 159–162.

Phillips, J. C., 1966. The fundamental optical spectra of solids, *Solid State Phys.*, **18**, 55–164.

Pines, D., and D. Bohm, 1952. A collective description of electron interactions: II. Collective vs. individual particle aspects of the interaction, *Phys. Rev.*, **85**, 338–353.

Pinnick, R. G., and J. J. Auvermann, 1979. Response characteristics of Knollenberg light-scattering aerosol counters, *J. Aerosol Sci.*, **10**, 55–74.

Pinnick R. G., J. M. Rosen, and D. J. Hoffmann, 1973. Measured light-scattering properties of individual aerosol particles compared to Mie scattering theory, *Appl. Opt.*, **12**, 37–41.

Pinnick, R. G., D. E. Carroll, and D. J. Hoffmann, 1976. Polarized light scattered from monodisperse randomly oriented nonspherical aerosol particles: measurements, *Appl. Opt.*, **15**, 384–393.

Pippard, A. B., 1978. *The Physics of Vibration*, Vol. 1, Cambridge University Press, Cambridge.

Planck, M., 1913. *Theorie der Wärmestrahlung*, 2nd ed., Johann Ambrosius Barth, Leipzig (reprinted by Dover, New York, 1959).

Plendl, J. N., 1970. Characteristic energy absorption spectra of solids, in *Far-Infrared Properties of Solids*, S. S. Mitra and S. Nudelman (Eds.), Plenum, New York, pp. 387–450.

Pluchino, A. B., S. S. Goldberg, J. N. Dowling, and C. M. Randall, 1980. Refractive-index measurements of single micron-sized carbon particles, *Appl. Opt.*, **19**, 3370–3372.

Pollack, J. B., and J. N. Cuzzi, 1980. Scattering by nonspherical particles of size comparable to a wavelength: a new semi-empirical theory and its application to tropospheric aerosols, *J. Atmos. Sci.*, **37**, 868–881.

Porch, W. M., D. S. Ensor, R. J. Charlson, and J. Heintzenberg, 1973. Blue moon: Is this a property of background aerosol? *Appl. Opt.*, **12**, 34–36.

Pritchard, B. S., and W. G. Elliot, 1960. Two instruments for atmospheric optics measurements, *J. Opt. Soc. Am.*, **50**, 191–202.

Procter, T. D., and D. Barker, 1974. The turbidity of suspensions of irregularly shaped particles, *Aerosol Sci.*, **5**, 91–99.

Procter, T. D., and G. W. Harris, 1974. The turbidity of suspensions of irregular quartz particles, *Aerosol Sci.*, **5**, 81–90.

Pruppacher, H. R., and R. L. Pitter, 1971. A semi-empirical determination of the shape of cloud and rain drops, *J. Atmos. Sci.*, **28**, 86–94.

Pultz, W. W., and W. Hertl, 1966. Perturbations in the infrared spectra of submicroscopic size, single crystal, fibrous SiC, *Spectrochim. Acta*, **22**, 573–575.

Purcell, E. M., 1963. *Electricity and Magnetism*, McGraw-Hill, New York.

Purcell, E. M., 1969. On the absorption and emission of light by interstellar grains, *Astrophys. J.*, **158**, 433–440.

Purcell, E. M., and C. R. Pennypacker, 1973. Scattering and absorption of light by non-spherical dielectric grains, *Astrophys. J.*, **186**, 705–714.

Quiney, R. G., and A. I. Carswell, 1972. Laboratory measurements of light scattering by simulated atmospheric aerosols, *Appl. Opt.*, **11**, 1611–1618.

Ramachandran, G. N., and S. Ramaseshan, 1961. Crystal optics, *Handbuch der Physik*, Vol. 25/1, S. Flügge (Ed.), Springer-Verlag, Berlin, pp. 1–217.

Raman, C. V., 1922. On the molecular scattering of light, *Proc. R. Soc. Lond.*, **A101**, 63–80.

Rasigni, M., and G. Rasigni, 1977. Optical constants of lithium deposits as determined from Kramers–Kronig analysis, *J. Opt. Soc. Am.*, **67**, 54–59.

Rathmann, J., 1981. The extinction by small aluminum particles from the far infrared to the vacuum ultraviolet, Ph.D. thesis, University of Arizona.

Ray, P. S., 1972. Broadband complex refractive indices of ice and water, *Appl. Opt.*, **11**, 1836–1844.

Rayleigh, Lord, 1871. On the light from the sky, its polarization and colour, *Philos. Mag.*, **41**, 107–120, 274–279 (reprinted in *Scientific Papers by Lord Rayleigh*, Vol. I: 1869–1881, No. 8, Dover, New York, 1964).

Reagan, J. A., D. M. Byrne, M. D. King, J. D. Spinhirne, and B. M. Herman, 1980. Determination of the complex refractive index of atmospheric particulates from bistatic–monostatic lidar and solar radiometer measurements, *J. Geophys. Res.*, **85**, 1591–1599.

Rebane, K. K., 1974. Some problems of the vibrational structure of optical spectra of impurities in solids, in *Optical Properties of Ions in Solids*, B. DiBartolo (Ed.), Plenum, New York, pp. 247–258.

Reid, G. C., 1975. Ice clouds at the summer polar mesopause, *J. Atmos. Sci.*, **32**, 523–535.

Reif, F., 1965. *Statistical and Thermal Physics*, McGraw-Hill, New York.

Reynolds, D. C., 1969. Excitons in II–IV compounds, in *Optical Properties of Solids*, S. Nudelman and S. S. Mitra (Eds.), Plenum, New York, pp. 239–286.

Rice, T. M., 1977. The electron-hole liquid in semiconductors: theoretical aspects, *Solid State Phys.*, **32**, 1–86.

Richtmyer, R. D., 1939. Dielectric resonators, *J. Appl. Phys.*, **10**, 391–398.

Ritchie, R. H., 1957. Plasma losses by fast electrons in thin films, *Phys. Rev.*, **106**, 874–881.

Robin, G. deQ., S. Evans, and J. T. Bailey, 1969. Interpretation of radio echo sounding in polar ice sheets, *Philos. Trans. R. Soc. Lond.*, **A265**, 437–505.

Robinson, F. N. H., 1973. *Macroscopic Electromagnetism*, Pergamon, Oxford.

Roessler, D. M., and F. R. Faxvog, 1979a. Optoacoustic measurement of optical absorption in acetylene smoke, *J. Opt. Soc. Am.*, **69**, 1699–1704.

Roessler, D. M., and F. R. Faxvog, 1979b. Opacity of black smoke: calculated variation with particle size and refractive index, *Appl. Opt.*, **18**, 1399–1403.

Roessler, D. M., and W. C. Walker, 1967. Optical constants of magnesium oxide and lithium fluoride in the far ultraviolet, *J. Opt. Soc. Am.*, **57**, 835–836.

Rose, J. H., H. B. Shore, and T. M. Rice, 1978. Infrared absorption and scattering by electron-hole droplets in Ge, *Phys. Rev.*, **B17**, 752–757.

Rosen, H., and T. Novakov, 1982. Lawrence Berkeley Laboratory laser transmission method, in *Light Absorption by Aerosol Particles*, H. Gerber and E. Hindman (Eds.), Spectrum Press, Hampton, Va.

Rosen, H., A. D. A. Hansen, L. Gundel, and T. Novakov, 1978. Identification of the graphitic carbon component of source and ambient particulates by Raman spectroscopy and an optical attenuation technique, *Appl. Opt.*, **17**, 3859–3861.

Rosencwaig, A., 1980. *Photoacoustics and Photoacoustic Spectroscopy*, Wiley, New York.

Roth, J., and M. J. Dignam, 1973. Scattering and extinction cross sections for a spherical particle with an oriented molecular layer, *J. Opt. Soc. Am.*, **63**, 308–311.

Rozenberg, G. V., 1960. Light scattering in the earth's atmosphere, *Sov. Phys.-Usp.*, **3**, 346–371.

Rukhadze, A. A., and V. P. Silin, 1961. Electrodynamics of media with spatial dispersion, *Sov. Phys.-Usp.*, **4**, 459–484.

Ruppin, R., 1975. Optical properties of small metal spheres, *Phys. Rev.*, **B11**, 2871–2876.

Ruppin, R., 1979. Far-infrared absorption in small metallic particles, *Phys. Rev.*, **B19**, 1318–1321.

Ruppin, R., 1981. Optical properties of spatially dispersive dielectric spheres, *J. Opt. Soc. Am.*, **6**, 755–758.

Ruppin, R., and R. Englman, 1970. Optical phonons of small crystals, *Rep. Prog. Phys.*, **33**, 149–196.

Russakoff, G., 1970. A derivation of the macroscopic Maxwell equations, *Am. J. Phys.*, **38**, 1188–1195.

Sassen, K., and K. N. Liou, 1979. Scattering of polarized laser light by water droplet, mixed-phase and ice crystal clouds: Part I. Angular scattering patterns, *J. Atmos. Sci.*, **36**, 838–852.

Saunders, M. J., 1970. Near-field backscattering measurements from a microscopic water droplet, *J. Opt. Soc. Am.*, **60**, 1359–1365.

Saunders, M. J., 1980. The effect of an electric field on the backscattered radiance of a single water droplet, in *Light Scattering by Irregularly Shaped Particles*, D. Schuerman (Ed.), Plenum, New York, pp. 237–242.

Savage, B. D., and J. S. Mathis, 1979. Observed properties of interstellar dust, *Annu. Rev. Astron. Astrophys.*, **17**, 73–111.

Saxon, D. S., 1955a. Lectures on the Scattering of Light, University of California at Los Angeles, Department of Meteorology, Scientific Report No. 9.

Saxon, D. S., 1955b. Tensor scattering matrix for the electromagnetic field, *Phys. Rev.*, **100**, 1771–1775.

Scadron, M. D., 1979. *Advanced Quantum Theory*, Springer-Verlag, New York.

Schuerman, D. (Ed.), 1980. *Light Scattering by Irregularly Shaped Particles*, Plenum, New York.

Schuerman, D. W., R. T. Wang, B. A. S. Gustafson, and R. W. Schaefer, 1981. Systematic studies of light scattering: 1. Particle shape, *Appl. Opt.*, **20**, 4039–4050.

Sekera, Z., 1957. Polarization of skylight, in *Handbuch der Physik*, Vol. 48, S. Flügge (Ed.), Springer-Verlag, Berlin, pp. 288–328.

Seki, J., and T. Yamamoto, 1980. Amorphous interstellar grains: wavelength dependence of far-infrared emission efficiency, *Astrophys. Space Sci.*, **72**, 79–86.

Seliga, T. A., and V. N. Bringi, 1976. Potential use of radar differential reflectivity measurements at orthogonal polarizations for measuring precipitation, *J. Appl. Meteorol.*, **15**, 69–76.

Seliga, T. A., V. N. Bringi, and H. H. Al-Khatib, 1981. A preliminary study of comparative measurements of rainfall rate using the differential reflectivity technique and a raingauge network, *J. Appl. Meteorol.*, **20**, 1362–1368.

Sherwood, P. M. A., 1972. *Vibrational Spectroscopy of Solids*, Cambridge University Press, Cambridge.

Shifrin, K. S., 1951. *Scattering of Light in a Turbid Medium*, NASA Technical Translation, NASA TT F-477, National Aeronautics and Space Administration, Washington, D.C.

Shurcliff, W. A., 1962. *Polarized Light*, Harvard University Press, Cambridge, Mass.

Silsbee, R. H., 1969. Optical analogue of the Mossbauer effect, in *Optical Properties of Solids*, S. Nudelman and S. S. Mitra (Eds.), Plenum, New York, pp. 607–624.

Šimánek, E., 1977. Far-infrared absorption in ultrafine Al particles, *Phys. Rev. Lett.*, **38**, 1161–1163.

Skillman, D. C., and C. R. Berry, 1968. Effect of particle shape on the spectral absorption of colloidal silver in gelatin, *J. Chem. Phys.*, **48**, 3297–3304.

Smith, D. Y., 1976. Comments on the dispersion relations for the complex refractive index of circularly and elliptically polarized light, *J. Opt. Soc. Am.*, **66**, 454–460.

Smith, R. A., F. E. Jones, and R. P. Chasmar, 1968. *The Detection and Measurement of Infra-Red Radiation*, 2nd ed., Oxford University Press, London.

Snow, T. P., D. G. York, and D. E. Welty, 1977. Catalogue of diffuse interstellar band measurements, *Astron. J.*, **82**, 113–128.

Sokolnikoff, I. S., and R. M. Redheffer, 1958. *Mathematics of Physics and Modern Engineering*, McGraw-Hill, New York.

Spinhirne, J. D., J. A. Reagan, and B. M. Herman, 1980. Vertical distribution of aerosol extinction cross section and inference of aerosol imaginary index in the troposphere by lidar technique, *J. Appl. Meteorol.*, **19**, 426–438.

Spitzer, L., 1948. The temperature of interstellar matter, I, *Astrophys. J.*, **107**, 6–33.

Spitzer, W. G., and D. A. Kleinman, 1961. Infrared lattice bands of quartz, *Phys. Rev.*, **121**, 1324–1335.

Spitzer, W. G., D. Kleinman, and D. Walsh, 1959. Infrared properties of hexagonal silicon carbide, *Phys. Rev.*, **113**, 127–132.

Stacey, K., 1956. *Light Scattering in Physical Chemistry*, Academic, New York.

Stegun, I. A., and M. Abramowitz, 1957. Generation of Bessel functions on high speed computers, *Math. Tables Other Aids Comput.*, **11**, 255–257.

Steinmann, W., 1968. Optical plasma resonances in solids, *Phys. Status Solidi*, **28**, 437–462.

Stephens, J. R., 1980. Visible and ultraviolet (800–130 nm) extinction of vapor-condensed silicate, carbon, and silicon carbide smokes and the interstellar extinction curve, *Astrophys. J.*, **237**, 450–461.

Stephens, R. E., E. K. Plyler, W. S. Rodney, and R. J. Spindler, 1953. Refractive index of potassium bromide for infrared radiant energy, *J. Opt. Soc. Am.*, **43**, 110–112.

Stern, E. A., and R. A. Ferrell, 1960. Surface plasma oscillations of a degenerate electron gas, *Phys. Rev.*, **120**, 130–136.

Stern, F., 1963. Elementary theory of the optical properties of solids, *Solid State Phys.*, **15**, 299–408.

Steyer, T. R., K. L. Day, and D. R. Huffman, 1974. Infrared absorption by small amorphous quartz spheres, *Appl. Opt.*, **13**, 1586–1590.

Stokes, G. G., 1852. On the composition and resolution of streams of polarized light from different sources, *Trans. Camb. Philos. Soc.*, **9**, 399–416 (reprinted in *Mathematical and Physical Papers*, Vol. III, Cambridge University Press, Cambridge, 1901).

Stone, J. M., 1963. *Radiation and Optics*, McGraw-Hill, New York.

Stookey, S. D., and R. J. Araujo, 1968. Selective polarization of light due to absorption by small elongated silver particles in glass, *Appl. Opt.*, **7**, 777–780.

Stookey, S. D., G. H. Beall, and J. E. Pierson, 1978. Full-color photosensitive glass, *J. Appl. Phys.*, **49**, 5114–5123.

Stratton, J. A., 1941. *Electromagnetic Theory*, McGraw-Hill, New York.

Ström, S., 1975. On the integral equations for electromagnetic scattering, *Am. J. Phys.*, **43**, 1060–1069.

Stroud, D., 1975. Generalized effective-medium approach to the conductivity of an inhomogeneous medium, *Phys. Rev.*, **B12**, 3368–3373.

Swindell, W. (Ed.), 1975. *Polarized Light*, Dowden, Hutchinson & Ross, Stroudsburg, Pa.

Taft, E. A., and H. R. Phillip, 1965. Optical properties of graphite, *Phys. Rev.*, **138**, A197–A202.

Tang, I. N., and H. R. Munkelwitz, 1978. Optical size determination for single cubic particles suspended in a laser beam, *J. Colloid Interface Sci.*, **63**, 297–303.

Tanner, D. B., A. J. Sievers, and R. A. Buhrman, 1975. Far-infrared absorption in small metallic particles, *Phys. Rev.*, **B11**, 1330–1341.

Tauc, J. (Ed.), 1966. *The Optical Properties of Solids*, Academic, New York.

Tauc, J., 1972. Optical properties of noncrystalline solids, in *Optical Properties of Solids*, F. Abelès (Ed.), Elsevier, New York, pp. 277–313.

Thompson, R. C., J. R. Bottiger, and E. S. Fry, 1978. Scattering matrix measurements of oceanic hydrosols, *Proc. Soc. Photo-Opt. Instrum. Eng.*, **160**, 43–48.

Thompson, R. C., J. R. Bottiger, and E. S. Fry, 1980. Measurement of polarized light interactions via the Mueller matrix, *Appl. Opt.*, **19**, 1323–1332.

Timusk, T., and A. Silin, 1975. Far-infrared absorption of electron-hole drops in pure and doped germanium, *Phys. Status Solidi (b)*, **69**, 87–91.

Tinkham, M., 1970a. Far infrared absorption in ordered magnetic systems, in *Far-Infrared Properties of Solids*, S. S. Mitra, and S. Nudelman (Eds.), Plenum, New York, pp. 196–222.

Tinkham, M., 1970b. Far infrared absorption in superconductors, in *Far-Infrared Properties of Solids*, S. S. Mitra and S. Nudelman (Eds.), Plenum, New York, pp. 223–246.

Toll, J. S., 1956. Causality and the dispersion relation: logical foundations, *Phys. Rev.*, **104**, 1760–1770.

Tomaselli, V. P., R. Rivera, D. C. Edewaard, and K. D. Möller, 1981. Infrared optical constants of black powders determined from reflection measurements, *Appl. Opt.*, **20**, 3961–3967.

Tonks, L., and I. Langmuir, 1929. Oscillations in ionized gases, *Phys. Rev.*, **33**, 195–210.

Toon, O. B., and J. B. Pollack, 1980. Atmospheric aerosols and climate, *Am. Sci.*, **68**, 268–278.

Toon, O. B., J. B. Pollack, and B. N. Khare, 1976. The optical constants of several atmospheric aerosol species: ammonium sulfate, aluminum oxide and sodium chloride, *J. Geophys. Res.*, **81**, 5733–5748.

Tozer, W. F., and D. E. Beeson, 1974. Optical model of noctilucent cloud based on polarimetric measurements from two sounding rocket campaigns, *J. Geophys. Res.*, **79**, 5607–5612.

Treffers, R., and M. Cohen, 1974. High-resolution spectra of cool stars in the 10- and 20-micron regions, *Astrophys. J.*, **188**, 545–552.

Treu, J. I., 1976. Mie scattering, Maxwell Garnett theory, and the Giaever immunology slide, *Appl. Opt.*, **15**, 2746–2750.

Tricker, R. A. R., 1970. *Introduction to Meteorological Optics*, Elsevier, New York.

Tricker, R. A. R., 1979. *Ice Crystal Haloes*, Optical Society of America, Washington, D.C.

Tsai, W-C., and R. J. Pogorzelski, 1975. Eigenfunction solution of the scattering of beam radiation fields by spherical objects, *J. Opt. Soc. Am.*, **65**, 1457–1463.

Turkevich, J., G. Garton, and P. C. Stevenson, 1954. The color of colloidal gold, *J. Colloid Sci.*, **9** (Suppl. 1), 26–35.

Turner, L., 1973. Rayleigh–Gans–Born light scattering by ensembles of randomly oriented anisotropic particles, *Appl. Opt.*, **12**, 1085–1090.

Twersky, V., 1964. Rayleigh scattering, *Appl. Opt.*, **3**, 1150–1162.

Twomey, S., 1953. On the measurement of precipitation intensity by radar, *J. Meteorol.*, **10**, 66–67.

Twomey, S. A., 1977. *Atmospheric Aerosols*, Elsevier, New York.

Twomey, S., 1980. Direct visual photometric technique for estimating absorption in collected aerosol samples, *Appl. Opt.*, **19**, 1740–1741.

Twomey, S., and C. F. Bohren, 1980. Simple approximations for calculations of absorption in clouds, *J. Atmos. Sci.*, **37**, 2086–2094.

van de Hulst, H. C., 1947. A theory of the anti-corona, *J. Opt. Soc. Am.*, **37**, 16–22.

van de Hulst, H. C., 1949. On the attenuation of plane waves by obstacles of arbitrary size and form, *Physica*, **15**, 740–746.

van de Hulst, H. C., 1957. *Light Scattering by Small Particles*, Wiley, New York.

van de Hulst, H. C., 1980. *Multiple Light Scattering*, Vols. 1 and 2, Academic, New York.

Volz, F. E., 1972. Infrared refractive index of atmospheric aerosol substance, *Appl. Opt.*, **11**, 755–759.

Walker, J. D., 1976. Multiple rainbows from single drops of water and other liquids, *Am. J. Phys.*, **44**, 421–433.

Walker, M. J., 1954. Matrix calculus and the Stokes parameters of polarized radiation, *Am. J. Phys.*, **22**, 170–174.

Wang, D-S., and P. W. Barber, 1979. Scattering by inhomogeneous nonspherical objects, *Appl. Opt.*, **18**, 1190–1197.

Wang, D-S., H. C. H. Chen, P. W. Barber, and P. J. Wyatt, 1979. Light scattering by polydisperse suspensions of inhomogeneous nonspherical particles, *Appl. Opt.*, **18**, 2672–2678.

Wangsness, R. K., 1963. *Introduction to Theoretical Physics: Classical Mechanics and Electrodynamics*, Wiley, New York.

Ward, G., K. M. Cushing, R. D. McPeters, and A. E. S. Green, 1973. Atmospheric aerosol index of refraction and size–altitude distribution from bistatic laser scattering and solar aureole measurements, *Appl. Opt.*, **12**, 2582–2592.

Waterman, P. C., 1965. Matrix formulation of electromagnetic scattering, *Proc. IEEE*, **53**, 805–812.

Waterman, P. C., 1971. Symmetry, unitarity, and geometry in electromagnetic scattering, *Phys. Rev.*, **D3**, 825–839.

Watson, G. N., 1958. *A Treatise on the Theory of Bessel Functions*, 2nd ed., Cambridge University Press, Cambridge.

Weaver, J., C. Krafka, D. Lynch, and E. Koch, 1981a. *Physics Data: Optical Properties of Metals, Pt. I: The Transition Metals*, Fach-Informations-Zentrum, Karlsruhe.

Weaver, J., C. Krafka, D. Lynch, and E. Koch, 1981b. *Physics Data: Optical Properties of Metals, Pt. II: Noble Metals, Aluminum, Scandium, Yttrium, the Lanthanides and the Actinides*, Fach-Informations-Zentrum, Karlsruhe.

Welker, T., 1978. Optical absorption of matrix isolated silver aggregates and microcrystals, *Ber. Bunsenges. Phys. Chem.*, **82**, 40–41.

Wendlandt, W. W., and H. G. Hecht, 1966. *Reflectance Spectroscopy*, Wiley-Interscience, New York.

Weyl, W. A., 1951. *Colored Glasses*, The Society of Glass Technology, Sheffield, England.

Wickramasinghe, N. C., and K. Nandy, 1970. The shape of the interstellar absorption band, *Astrophys. Space Sci.*, **6**, 154–156.

Wickramasinghe, N. C., and K. Nandy, 1972. Recent work on interstellar grains, *Rep. Prog. Phys.*, **35**, 157–234.

Wiegel, E., 1954. Über die Farben des kolloiden Silbers und die Miesche Theorie, *Z. Phys.*, **136**, 642–653.

Wilson, E. B., J. C. Decius, and P. C. Cross, 1955. *Molecular Vibrations*, McGraw-Hill, New York (Dover edition published in 1980).

Wilson, J. W., and E. A. Brandes, 1979. Radar measurement of rainfall—a summary, *Bull. Am. Meteorol. Soc.*, **60**, 1048–1058.

Wilson, R., 1951. The blue sun of 1950 September, *Mon. Not. R. Astron. Soc.*, **111**, 478–489.

Wiscombe, W. J., 1979. Mie scattering calculations: advances in technique and fast, vector-speed computer codes, NCAR/TN-140 + STR, National Center for Atmospheric Research, Boulder, Colo.

Wiscombe, W. J., 1980. Improved Mie scattering algorithms, *Appl. Opt.*, **19**, 1505–1509.

Witt, G., 1960. Polarization of light from noctilucent clouds, *J. Geophys. Res.*, **65**, 925–933.

Witt, G., 1969. The nature of noctilucent clouds, *Space Res.*, **9**, 157–169.

Witt, G., J. E. Dye, and N. Wilhelm, 1976. Rocket-borne measurements of scattered sunlight in the mesosphere, *J. Atmos. Terr. Phys.*, **38**, 223–238.

Wood, E. A., 1977. *Crystals and Light*, 2nd rev. ed., Dover, New York.

Woolf, N. J., 1975. Circumstellar dust, in *The Dusty Universe*, G. B. Field and A. G. W. Cameron (Eds.), Neale Watson, New York, pp. 60–87.

Wooten, F., 1972. *Optical Properties of Solids*, Academic, New York.

Wyatt, P. J., 1968. Differential light scattering: a physical method for identifying living bacterial cells, *Appl. Opt.*, **7**, 1879–1896.

Wyatt, P. J., 1980. Some chemical, physical, and optical properties of fly ash particles, *Appl. Opt.*, **19**, 975–983.

Wyatt, P. J., and D. T. Phillips, 1972. Structure of single bacteria from light scattering, *J. Theor. Biol.*, **37**, 493–501.

Yeh, C., 1964. Perturbation approach to the diffraction of electromagnetic waves by arbitrarily shaped dielectric obstacles, *Phys. Rev.*, **135**, A1193–A1201.

Yeh, C., and K. K. Mei, 1980. On the scattering from arbitrarily shaped inhomogeneous particles—exact solutions, in *Light Scattering by Irregularly Shaped Particles*, D. Schuerman (Ed.), Plenum, New York, pp. 201–206.

Yildiz, A., 1963. Scattering of plane plasma waves from a plasma sphere, *Nuovo Cimento*, **30**, 1182–1207.

Young, A. T., 1982. Rayleigh scattering, *Phys. Today*, **35**(1), 42–48.

Zellner, B., 1973. Dust grains in reflection nebulae, in *Interstellar Dust and Related Topics*, IAU Symposium No. 52, J. M. Greenberg and H. C. van de Hulst (Eds.), D. Reidel, Dordrecht, Holland, pp. 109–113.

Zerull, R., and R. H. Giese, 1974. Microwave analogue studies, in *Planets, Stars and Nebulae Studied with Photopolarimetry*, T. Gehrels (Ed.), University of Arizona Press, Tucson, Ariz., pp. 901–915.

Zerull, R. H., R. H. Giese, and K. Weiss, 1977. Scattering functions of nonspherical dielectric and absorbing particles vs. Mie theory, *Appl. Opt.*, **16**, 777–778.

Zerull, R. H., R. H. Giese, S. Schwill, and K. Weiss, 1980. Scattering by particles of non-spherical shape, in *Light Scattering by Irregularly Shaped Particles*, D. W. Schuerman (Ed.), Plenum, New York, pp. 273–282.

Ziman, J. M., 1972. *Principles of the Theory of Solids*, Cambridge University Press, Cambridge.

Zolotarev, V. M., 1970. The optical constants of amorphous SiO_2 and GeO_2 in the valence band region, *Opt. Spectrosc.*, **29**, 34–37.

Index

Absorption:
  by atmospheric aerosols, 436–446
  dominance of, in Rayleigh limit, 309
  effect on ripple and interference structure,
    306, 318
  mechanisms in bulk matter, summary, 282
  temperature effects on, 281
Absorption coefficient, bulk matter:
  defined, 29
  measurement of, 29, 30
Absorption cross section:
  of cubes, 368
  defined, 71
  of distribution of ellipsoids, 353–356
  of ellipsoids, Rayleigh limit, 150–152, 345,
    348
  integrated, of ellipsoid, 347
  in Rayleigh-Gans approximation, 161, 162
  of sphere, Rayleigh limit, 140
  of spheroid, 313
  of weakly absorbing large sphere, 166–169
Absorption edge, *see* Band gap
Absorption efficiency:
  defined, 72
  and emissivity, 125
  greater than 1, 339–342
  of small sphere, 136
  of sphere, asymptotic limit, 172, 173
  of water droplets, 170, 171
Aggregation of particles:
  effect on extinction spectra:
    aluminum, 376
    amorphous quartz, 361
    gold, 372
    magnesium oxide, 366–368
  effect on optical properties, 81, 315, 342,
    258
  and sampling of atmospheric aerosols, 440
  scattering by, spherical clusters, 423–425
Albedo, single scattering:
  defined, 445
  and global climate, 435, 442, 443–446

Alexander's dark band, 177
Alignment of particles:
  in interstellar dust clouds, 463–465
  in noctilucent clouds, 451
  in ocean waters, 427
Aluminum:
  bulk optical properties of, 225, 271–273,
    353
  particles, extinction by:
    ellipsoids, 346
    measurements, 374–377
    spheres, 294, 295, 310, 338, 375
  surface plasmons in, 338, 339
Amorphous solids:
  and far-infrared emission, 446
  as oxide coating, 376
  particles of, in interstellar dust, 462–466
Amplitude scattering matrix:
  for anisotropic dipole, 154
  for cylinder, 202, 205, 408
    Rayleigh limit, 208
  defined, 63
  for optically active particles, 189, 408
  in Rayleigh-Gans approximation, 161, 407
  for sphere, 112
    Rayleigh limit, 132, 140, 407
  and sum rule, 154
  symmetry of, 406–412
Angle-dependent functions, in Mie theory:
  defined, 94
  polar plots of, 96
  recurrence relations for, 95, 119
Angular scattering:
  calculations:
    for spheres, 114, 115
      distributed in size, 387–389
      of increasing size, 384–387
      in Rayleigh limit, 133, 134
    for spheroids, 397–400
  measurements:
    absolute and relative, 391
    applicability of Mie theory to, 427, 428